3rd International Conference on Computational Mathematics and Engineering Sciences (CMES 2018)

ITM Web of Conferences Volume 22 (2018)

Girne, Turkey
4 – 6 May 2018

Editors:

C. Cattani
A. Atangana
H. Bulut
Z. Hammouch
H.M. Baskonus
T. Mekkaoui
S. Agoujil

ISBN: 978-1-5108-7315-5

Printed from e-media with permission by:

Curran Associates, Inc.
57 Morehouse Lane
Red Hook, NY 12571

Some format issues inherent in the e-media version may also appear in this print version.

Printed by Curran Associates, Inc. (2018)

For additional information, please contact EDP Sciences – Web of Conferences at the address below.

EDP Sciences – Web of Conferences
17, Avenue du Hoggar
Parc d'Activité de Courtabœuf
BP 112
F-91944 Les Ulis Cedex A
France

Phone: +33 (0) 1 69 18 75 75
Fax: +33 (0) 1 69 28 84 91

contact-edps@webofconferences.org

Additional copies of this publication are available from:

Curran Associates, Inc.
57 Morehouse Lane
Red Hook, NY 12571 USA
Phone: 845-758-0400
Fax: 845-758-2633
Email: curran@proceedings.com
Web: www.proceedings.com

TABLE OF CONTENTS

THE THIRD INTERNATIONAL CONFERENCE ON COMPUTATIONAL MATHEMATICS AND ENGINEERING SCIENCES (CMES-2018), GIRNE, 04-06 MAY 2018

The **Third International Conference on Computational Mathematics and Engineering Sciences (CMES-2018)** has been held in Final International University from May 04 to 06, 2018 in Girne/CYPRUS. It provides an ideal academic platform for researchers and professionals to discuss recent developments in both theoretical, applied mathematics and engineering sciences. This event aims also to stimulate interactions between researchers in the field of computational mathematics and their applications in science and engineering, to present the development reached in this areas, and to showcase the computational expertise of our invited speakers and participants.

The organizing Committee

EDITORS

Carlo Cattani, Tuscia University, Italy
Abdon Atangana, University of the Free State, South Africa
Hasan Bulut, Firat University, Elazig, Turkey
Zakia Hammouch, FST Errachidia, Moulay Ismail University Morocco
Haci Mehmet Baskonus, Munzur University, Tunceli, Turkey
Toufik Mekkaoui, E3MI FST Errachidia, Moulay Ismail University, Morocco
Said Agoujil, Moulay Ismail University, Morocco

Statement of Peer review

In submitting conference proceedings to *Web of Conferences*, the editors of the proceedings certify to the Publisher that

1. They adhere to its **Policy on Publishing Integrity** in order to safeguard good scientific practice in publishing.
2. All articles have been subjected to peer review administered by the proceedings editors.
3. Reviews have been conducted by expert referees, who have been requested to provide unbiased and constructive comments aimed, whenever possible, at improving the work.
4. Proceedings editors have taken all reasonable steps to ensure the quality of the materials they publish and their decision to accept or reject a paper for publication has been based only on the merits of the work and the relevance to the journal.

Title, date and place of the conference

The Third International Conference on Computational Mathematics and Engineering Sciences (CMES-2018), 04-06 May 2018, Girne/Cyprus (Turkish Republic of Northern of Cyprus)

Proceedings editor(s):

Carlo Cattani, Tuscia University, Italy

Abdon Atangana, Free State University, South africa

Hasan Bulut, Firat University, Turkey

Zakia Hammouch, Moulay Ismail University, Morocco

Haci Mehmet Baskonus, Munzur University, Turkey

Toufik Mekkaoui, Moulay Ismail University, Morocco

Said Agoujil, Moulay Ismail University, Morocco

Date and editor's signature

17.09.2018 Prof.Dr.Hasan BULUT

ITM Web of Conferences **22**, 01001 (2018) https://doi.org/10.1051/itmconf/20182201001
CMES-2018

On Majorization Problems Associated with the Subclass $Q(j,\lambda,\alpha,n)$ of Starlike Functions with Positive Coefficients

Hüseyin Baba[1,*] *and Ekrem Kadıoğlu*[2]

[1]Department of Mathematics, Çölemerik Vocational School, University of Hakkari, Hakkari, Turkey

[2]Department of Mathematics, University of Atatürk, Erzurum, Turkey

Abstract. We consider the subclass $Q(j,\lambda,\alpha,n)$ of starlike functions by using the differential D^n operator and functions of the form $f(z) = z + \sum_{k=j+1}^{\infty} a_k z^k$ which are analytic in the open unit disk. In this paper is to investigate an majorization problem for the subclass $Q(j,\lambda,\alpha,n)$. Relevant connections of the main result obtained in this paper with those given by earlier workers on the subject are also pointed out.

1 Introduction

Let the functions $f(z)$ and $g(z)$ be analytic in the open unit disk

$$U = \{z : z \in \mathbb{C} \; and \; |z| < 1\}. \tag{1}$$

It is called that $f(z)$ is majorized by $g(z)$ in U and write

$$f(z) \ll g(z)(z \epsilon U) \tag{2}$$

if there exists a function $\vartheta(z)$, analytic in U, such that

$$\vartheta(z) \leq 1 \text{ and } f(z) = \vartheta(z)g(z)(z \in U) \tag{3}$$

Let A denote the family of functions f of the form

$$f(z) = z + \sum_{n=2}^{\infty} a_n z^n \tag{4}$$

that are analytic in the open unit disk U. A function $f \in A$ is said to be starlike of order $\alpha(0 \leq \alpha < 1)$ if and only if

*Correspondingauthor: huseyinbaba@hakkari.edu.tr

$$Re\left\{\frac{zf'(z)}{f(z)}\right\} > \alpha, \quad (z\epsilon U). \tag{5}$$

We denote by $S^*(\alpha)$, the class of all such functions. On the other hand, a function $f \in A$ is said to be convex of order α $(0 \leq \alpha < 1)$ if and only if

$$Re\left[1+\frac{zf''(z)}{f'(z)}\right] > \alpha \ (z\epsilon U). \tag{6}$$

Let $C(\alpha)$ denote the class of all those functions which are convex of order α in U. Note that $S^*(0) = S^*$ and $C(0) = C$ are, respectively, the classes of starlike and convex functions in U.

Let $A(j)$ denote the class of functions of the form:

$$f(z) = z + \sum_{k=j+1}^{\infty} a_k z^k \ (j \in \mathbb{N} = \{1,2,3,\dots\}) \tag{7}$$

which are analytic in the open unit disk U.

For a function $f(z)$ in $A(j)$, we define

$$D^0 f(z) = f(z)$$

$$D^1 f(z) = Df(z) = zf'(z) = z + \sum_{k=j+1}^{\infty} k a_k z^k$$

$$\vdots$$

$$D^n f(z) = D\left(D^{n-1} f(z)\right) = z + \sum_{k=j+1}^{\infty} k^n a_k z^k \qquad (n \in \mathbb{N}).$$

The differential operator D^n was introduced by Sălăgean [1]. With the help of the differential operator D^n, we say that a function $f(z)$ belonging to $A(j)$ is in the class $Q(j,\lambda,\alpha,n)$ if and only if

$$Re\left\{\frac{(1-\lambda)z\left(D^n f(z)\right)' + \lambda z\left(D^{n+1} f(z)\right)'}{(1-\lambda)D^n f(z) + \lambda D^{n+1} f(z)}\right\} > \alpha \tag{8}$$

for some α $(0 \leq \alpha < 1)$ and λ $(0 \leq \lambda \leq 1)$, and for all $z\epsilon U$. The subclass $Q(j,\lambda,\alpha,n)$ is defined by [2].

2 Majorization Problems ForThe Class $Q(j,\lambda,\alpha,n)$

Theorem Let the function $f(z)$ be in the $A(j)$ and suppose that $g(z) \in Q(j,\lambda,\alpha,n)$. If

$D^n f(z)$ is majorized by $D^n g(z)$ in U then

$$|(D^n f(z))'| \le |(D^n g(z))'| \text{ or } |D^{n+1} f(z)| \le |D^{n+1} g(z)| \quad (|z| < r), \tag{9}$$

where

$$r = r(\lambda, \alpha) = \frac{2(1+\lambda\alpha) - \sqrt{4(1+\lambda\alpha)^2 - (1-\lambda)^2}}{1-\lambda} \tag{10}$$

$(j \in \mathbb{N}; n \in \mathbb{N}_0; 0 \le \alpha < 1; 0 \le \lambda < 1;\ z \in U)$.

The result is sharp.

Proof Since $g(z) \in Q(j, \lambda, \alpha, n)$, we readily find from (8), if

$$k(z) = \frac{(1-\lambda)z(D^n g(z))' + \lambda z(D^{n+1} g(z))'}{(1-\lambda)D^n g(z) + \lambda z D^{n+1} g(z)} \text{ and } p(z) = \frac{k(z)-\alpha}{1-\alpha} \tag{11}$$

Then

$$p(z) = 1 + \cdots \text{ and } \Re\{p(z)\} > 0 \tag{12}$$

and

$$p(z) = \frac{1-w(z)}{1+w(z)},\ (w \in \Omega) \tag{13}$$

where Ω denotes the well-known class of bounded analytic functions in U, which satisfy the conditions [[3], p.58]

$$w(0) = 0 \text{and } |w(z)| \le |z| (z \in U). \tag{14}$$

Making use of (11) and (13), as readily obtain

$$\begin{aligned}
\lambda(1+w(z))z^2(D^n g(z))'' + [(1-\lambda\alpha)(1+w(z)) - (1-\alpha)(1-w(z))\lambda]z(D^n g(z))' \\
= [(1-\alpha)(1-\lambda)(1-w(z)) + \alpha(1-\lambda)(1+w(z))]D^n g(z) \\
\Rightarrow (1+|w(z)|)(1+\lambda+2\lambda\alpha)|D^n g(z)| \\
\ge (1-|w(z)|)(1-\lambda)|D^n g(z)|
\end{aligned}$$

or

$$(1-|w(z)|)(1-\lambda)|D^n g(z)| \le (1+|w(z)|)(1+\lambda+2\lambda\alpha)|D^{n+1} g(z)| \tag{15}$$

which, in view of (14), immediately yields the inequality

$$|D^n g(z)| \le \frac{(1+|z|)(1+\lambda+2\lambda\alpha)}{(1-|z|)(1-\lambda)} |D^{n+1} g(z)| (z \in U). \tag{16}$$

Next, since $D^n f(z)$ is majorized by $D^n g(z)$ in U, from (3) we have

$$D(D^n f(z)) = D(\vartheta(z) D^n g(z))$$

$$D^{n+1} f(z) = \vartheta'(z) z D^n g(z) + \vartheta(z) D^{n+1} g(z)$$

$$|D^{n+1} f(z)| = |\vartheta'(z)||z||D^n g(z)| + |\vartheta(z)||D^{n+1} g(z)|. \qquad (17)$$

Thus, observing that $\vartheta \in \Omega$ satisfiees the inequality

$$|\vartheta'(z)| \leq \frac{1 - |\vartheta(z)|^2}{1 - |z|^2} \qquad (18)$$

and applying (16) and (18) in (15), we get

$$|D^{n+1} f(z)| \leq \left[\frac{1 - |\vartheta(z)|^2}{1 - |z|^2} \frac{(1 + |z|)(1 + \lambda + 2\lambda\alpha)}{(1 - |z|)(1 - \lambda)} |z| + |\vartheta(z)| \right] |D^{n+1} g(z)| \qquad (19)$$

which upon setting

$|z| = r$ and $|\vartheta(z)| = \rho (0 \leq \rho \leq 1)$

leads us to the inequality

$$|D^{n+1} f(z)| \leq \frac{\theta(\rho)}{(1 - r^2)(1 - \lambda)} |D^{n+1} g(z)| \qquad (20)$$

where the function $\theta(\rho)$ defined by

$$\theta(\rho) = (1 - \rho^2)(1 + \lambda + 2\lambda\alpha) + (1 - r)^2 (1 - \lambda)\rho \qquad (21)$$

takes on its maximum value at $\rho = 1$ with

$$r = r(\lambda, \alpha)$$

given by (10). Furthermore, if

$$0 \leq \sigma \leq r(\lambda, \alpha)$$

where $r(\lambda, \alpha)$ is given by (10) then the function $A(\rho)$ defined

$$A(\rho) = (1 - \rho^2)(1 + \lambda + 2\lambda\alpha) + (1 - \sigma)^2 (1 - \lambda)\rho \qquad (22)$$

is seen to be in increasing function on the interval $0 \leq \rho \leq 1$, so that

$$A(\rho) \leq A(1) = (1-\sigma)^2(1-\lambda)$$
$$0 \leq \rho \leq 1\,;\, 0 \leq \sigma \leq r(\lambda, \alpha)$$

Hence, by setting $\rho = 1$ in (21), we conclude that (9) of Theorem holds true for $|z| \leq r(\lambda, \alpha)$, where $r(\lambda, \alpha)$ is given by (10). This completes the proof of Theorem.

3 Application

From the expression of the Theorem means that the radius is independent of n. If we set $n = 0$ in Theorem, we find the same radius.The following application refers to this.

Application Let the function $f(z)$ be in the $A(j)$ and suppose that $g(z) \in Q(j, \lambda, \alpha, 0)$. If $f(z)$ is majorized by $g(z)$ in U then

$$|Df(z)| \leq |Dg(z)| \text{or} |f'(z)| \leq |g'(z)| \quad (|z| < r),$$

where

$$r = r(\lambda, \alpha) = \frac{2(1+\lambda\alpha) - \sqrt{4(1+\lambda\alpha)^2 - (1-\lambda)^2}}{1-\lambda}$$

$(j \in \mathbb{N}; n = 0; 0 \leq \alpha < 1; 0 \leq \lambda < 1;\ z \in U)$.

Corollary If we set $\lambda = 0$ in Theorem, we immediately obtain. (MacGregor [[4], p.296, Therom 1]). Let the functions $f(z)$ be analytic in U and suppose that $g \in S^* = S^*(0)$, where S denotes the class of (normalized) analytic and univalent functions in U . If $f(z)$is majorized $g(z)$ in, then

$$|f'(z)| \leq |g'(z)| \quad (|z| \leq 2-\sqrt{3}).$$

4 Conclusion

Let S denote the family of functions $f(z)$ which are analytic and univalent in the open unit disc U and which are normalized so that $f(0) = 0$, $f'(0) = 1$.Let $F(z)$ be an arbitrary function in S. If $f(z) << F(z)$in U, determine the largest r for which $f'(z) << F'(z)$ in $|z| < r$ [5].

In this paper, we obtain solution of majorization problems for $Q(j, \lambda, \alpha, n)$ class. So if $D^n f(z) << D^n g(z)$ in U, determine the largest r for which $D^{n+1} f(z) << D^{n+1} g(z)$ in $|z| < r$. Relevant connections of the main result obtained in this paper with those given by earlier workers on the subject are also pointed out.

References

1. G. Şt. Sălăgean, Subclasses of univalent functions, in "ComplexAnalysis: FifthRomanian-FinnishSeminar." Part I (Bucharest, 1981), pp. 362-372. Lecture Notes in Mathematics, Vol. **1013**, Springer - Verlag, Berlin/Newyork, 1983.
2. M. K.Aouf, and H. M. Srivastava, Somefamilies of starlikefunctionswithnegative coefficients, J. Math. Anal. Appl. **203**(1996), 762-790, Article No:0411.
3. A.W.Goodman, *Univalent Functions*, vol. **I**, Mariner PublishingCompany, Tapma, FL, 1983.
4. T. H.MacGregor, Majorization by univalent functions, Ann. Math. **37**(1936) 374-408.
5. T. H. MacGregor, Majorization by univalent functions, Duke Math. J. **34** (1967), 95-102. MR 34 #6062.

High Order Iterative Methods for Matrix Inversion and Regularized Solution of Fredholm Integral Equation of First Kind with Noisy Data

Suzan Cival Buranay[1,*] and *Ovgu Cidar* Iyikal[1]

[1]Department of Mathematics, Faculty of Arts and Sciences, Eastern Mediterranean University, Famagusta, North Cyprus, via Mersin 10, Turkey

Abstract. The motivation of the present work is to propose high order iterative methods with a recurrence formula for approximate matrix inversion and provide regularized solution of Fredholm integral equation of first kind with noisy data by an algorithm using the proposed methods. From the given family of methods of orders $p = 7,11,15,19$ are applied to solve problems of Fredholm integral equation of first kind. From the literature, iterative methods of same orders are used to solve the considered problems and numerical comparisons are shown through tables and figures.

1 Introduction

Consider the system of linear equations

$$Ax = b, \quad A \in R^{N_1 \times N_1} \tag{1}$$

when, A is nonsingular dense stored matrix. Such a system is commonly solved by Gaussian elimination with pivoting and the accuracy of the computed solution depends primarily on the condition number of the matrix A [1]. The discretization of the ill-posed problems such as the Fredholm integral equations of first kind with smooth kernel results algebraic linear systems possessing a severely ill-conditioned or singular coefficent matrix and these problems are known as linear discrete ill-posed problems. In these problems the solution is very sensitive to perturbations in the data such as measurement or approximation errors. In order to determine a meaningful approximation of the solution of (1) when the coefficient matrix A is severely ill-conditioned one typically replaces the linear system (1) by a nearby system

$$\widetilde{A}y = \widetilde{b}, \tag{2}$$

that is less sensitive to the perturbations of the right-hand side b and the coefficient matrix A (see [2]-[8]) . Another approach to find a smooth solution to the ill-conditioned system (1) when A is nonsingular is to use a preconditioner matrix V that approximates A^{-1}. The p-th order hyperpower methods $p \geq 2$ are a family of well known methods used to approximate the inverse of a nonsingular matrix [9]. Recent studies on iterative methods for computing approximate inverse are but not limited to [10]-[12]. The factorization of hyperpower method of order 7 for computing generalized inverse $A_{T,S}^{(2)}$

$$V_{m+1} = V_m(I + (T_m + T_m^{\ 4})(I + T_m + T_m^{\ 2})), \ m = 0,1,\dots . \tag{3}$$

* Corresponding author: suzan.buranay@emu.edu.tr

is proposed in [11]. In [12] several systematic algorithms for the factorizatios of the hyperpower methods of orders *11, 15*, and *19*

$$V_{m+1} = V_m(I + T_m(I + (T_m + T_m^2 + T_m^3)(I + T_m^3 + T_m^6))),\ m = 0,1,\dots, \quad (4)$$

$$V_{m+1} = V_m(I + (T_m + T_m^2)(I + (T_m^2 + T_m^4)(I + T_m^4 + T_m^8))),\ m = 0,1,\dots, \quad (5)$$

$$V_{m+1} = V_m(I + (T_m + T_m^2)(I + T_m^2 + T_m^4)(I + T_m^6 + T_m^{12})),\ m = 0,1,\dots, \quad (6)$$

are given, respectively to approximate the outer inverse.

In this study for a given integer $k \geq 1$, we propose high order iterative methods of orders $p = 4k + 3$, using a recurrence formula to compute approximate matrix inverse and use the methods of orders *p=7,11,15,19* from this family to obtain regularized solution of Fredholm Integral Equations of first kind such as the harmonic continuation problem.

2 A Family of Methods with Recurrence Formula

Let I denote the $N_1 \times N_1$ unit matrix and $A \in R^{N_1 \times N_1}$ be nonsingular matrix. We denote the approximate inverse of A at the m-th iteration by V_m, and the residual matrix by $T_m = I - AV_m$. Let

$$\begin{aligned} &\aleph(T_m) = T_m T_m,\ \Omega(T_m) = T_m + \aleph(T_m), \\ &\Gamma(T_m) = \aleph(T_m)\aleph(T_m),\ \Psi(T_m) = \aleph(T_m) + \Gamma(T_m). \end{aligned} \quad (7)$$

We call Family Generator Function to the matrix valued function $\Phi_p(T_m)$, defined as

$$\Phi_p(T_m) = \Omega(T_m)(I + \Psi(T_m)[I + \Gamma(T_m)(I + \Gamma(T_m)(I + \Gamma(T_m)(\dots)))]), \quad (8)$$

with $(I + \Gamma(T_m))$ is 1 nested loop, $(I + \Gamma(T_m))(I + \Gamma(T_m))$ is 2 nested loop, (8) can be given by the following recurrence formula

$$\begin{aligned} &P_0 = I,\ P_j = I + \Gamma(T_m)P_{j-1},\ j = 1,\dots,k-1, \\ &\hat{\Phi}_p(T_m) = \Phi_p(T_m) = \Omega(T_m)(I + \Psi(T_m)P_{k-1}). \end{aligned} \quad (9)$$

For an integer $k \geq 1$ and the corresponding Family Generator Function $\hat{\Phi}_p(T_m),\ p = 4k+1$, we proposed

$$V_{m+1} = V_m(I + \hat{\Phi}_p(T_m)),\ m = 0,1,\dots \quad (10)$$

the family of methods for matrix inversion.

Theorem : *Let* $A \in R^{N_1 \times N_1}$ *be a nonsingular matrix. For a given* $k \geq 1$ *integer if the method (10) is used, the necessary and sufficient condition for the convergence of (10) to* A^{-1} *is that* $\rho(T_0) < 1$ *holds, where ρ is spectral radius,* $T_0 = I - AV_0$ *and* V_0 *is the initial approximation.*

3 Computational Complexity and Algorithm for Regularized Solution

When the two factorizations of p-th order hyperpower method have same asymptotic convergence factor ACF [10] values then it is necessary to use an other quantity which measures the efficiency of the hyperpower method both with respect to matrix by matrix multiplications (MMs) and matrix by matrix additions (Mas). Let κ be the number of MMs and p be the order of the hyperpower method. Let ν be the number of Mas and γ be the number of matrix additions by identity. We define the Asymptotic Convergence Values by

$$ACV = \left(\frac{\kappa}{\ln p}, \frac{\upsilon}{\ln p}, \frac{\gamma}{\ln p} \right). \quad (11)$$

Table 1 shows the rounded ACV of the proposed methods for the orders $p = 7,11,15,19$ and of the methods (3), (4), (5), (6). We denote the methods (10) for $k=1$, $k=2$, $k=3$, $k=4$ by M_1, M_3, M_5, M_7, respectively and the methods (3), (4), (5), (6) by M_2, M_4, M_6, M_8, respectively.

Method	κ	υ	γ	ACV
M_1	5	2	3	(2.5695, 1.0278, 1.5417)
M_2	5	2	3	(2.5695, 1.0278, 1.5417)
$p=7$ in [10]	6	1	4	(3.083, 0.5139, 2.0556)
M_3	6	2	4	(2.5022, 0.8341, 1.6681)
M_4	7	3	4	(2.9102, 1.2511, 1.6681)
M_5	7	2	5	(2.5849, 0.7385, 1.8463)
M_6	7	3	4	(2.5849, 1.1078, 1.4771)
M_7	8	2	6	(2.7170, 0.6792, 2.0377)
M_8	8	3	4	(2.7170, 1.0189, 1.3584)
$p=19$ in [10]	10	2	5	(3.3962, 0.6792, 1.6981)

Table 1. ACV for the proposed family of methods (10) and for the methods from the literature [10], [11] and [12] of the orders *p=7,11,15,19*.

We consider the perturbed system (2) in the following form

$$(A+\Delta A)y = b+\Delta b, \quad (12)$$

where $\tilde{A} = A+\Delta A$, $\tilde{b} = b+\Delta b$ and ΔA is the perturbation in A, Δb is the perturbation in b. The applicability of the proposed methods for solving the perturbed system (12) is given by the following algorithm.

Algorithm 1 : When $A \in R^{N_1 \times N_1}$ is nonsingular matrix for a given integer $k \geq 1$ and a pre-described accuracy $\varepsilon > 0$, the approximate inverse $V_{\tilde{m}}$ of $\tilde{A}$ in (12) is found by using the methods (10) with $\tilde{m}$ iterations to reach the accuracy $\frac{\|r_{\tilde{m}}\|_\infty}{\|\tilde{b}\|_\infty} \leq \varepsilon$ where $r_{\tilde{m}} = \tilde{b} - \tilde{A}y_{\tilde{m}}$. Here $y_{\tilde{m}} = V_{\tilde{m}}b$ is the regularized solutions. This algorithm is given analogous to Algorithm 1 in [10].

4 Numerical Results

The numerical analysis of the proposed methods (10) of orders $p = 7,11,15,19$ are provided by the application on two examples of Fredholm integral equation of first kind. All the computations are performed using a personal computer with properties Intel(R) Core(TM)-2DUO CPUT5870@ 2.00GHz and Fortran programs in double precision.

Harmonic Continuation Problem: We consider the harmonic continuation problem [1]:

$$\frac{1}{2\pi} \int_0^{2\pi} \frac{1-r^2}{1-2r\cos(\theta-\varphi)+r^2} h(\varphi)d\varphi = f(\theta), \quad (13)$$

for the same data used by Franklin [8] where, for $|z| = r = 0.5$,

$$h(\varphi)=\cos 3\varphi-\cos\varphi+[\sin(\cos\varphi)][\cos(\sin\varphi)],\ f(\theta)=\tfrac{1}{8}\cos 3\theta-\tfrac{1}{2}\cos\theta+[\sin(\tfrac{1}{2}\cos\theta)][\cos(\tfrac{1}{2}\sin\theta)], \tag{14}$$

Using the quadrature nodes $\varphi_j=\frac{2\pi j}{N_1}$, $j=1,\dots,N_1$, evaluating $f(\theta)$ at the points $\theta_i=\frac{2\pi i}{N_1}$, $i=1,\dots,N_1$ and by discretizing (13) we obtain the algebraic linear system (1) as

$$A_{i,j}=\frac{3}{N_1(5-4\cos 2\pi(i-j)/N_1)},\ b_j=f\left(\frac{2\pi j}{N_1}\right). \tag{15}$$

Consider a noisy vector $\tilde{b}$ and a noisy matrix $\tilde{A}$ with available noise levels δ_b and δ_A, respectively such that

$$\left\|\tilde{b}-b\right\|_2=\left\|\Delta b\right\|_2\le\delta_b \text{ and } \left\|\tilde{A}-A\right\|_2=\left\|\Delta A\right\|_2\le\delta_A. \tag{16}$$

We take $N_1=50$, $\delta_b=10^{-11},\delta_A=0.5(\delta_b)^{1.5}$ and apply the Algorithm 1 for the methods $M_i, i=1,\dots,8$ when $V_0=\frac{\tilde{A}^T}{\|\tilde{A}\|_1\|\tilde{A}\|_\infty}$ and $\varepsilon=5\times10^{-11}$. Table 2 presents the CPU time in seconds, iteration numbers $\tilde{m}$ and relative norms of the errors obtained by Algorithm 1 using the methods $M_i, i=1,\dots,8$ for the harmonic continuation problem.

Methods	*CPU*	$\tilde{m}$	$\left\|x-y_{\tilde{m}}\right\|_\infty$	$\frac{\left\|x-y_{\tilde{m}}\right\|_2}{\left\|x\right\|_2}$
M_1	0.094	8	7.5780278E-009	7.6071543E-009
M_2	0.094	8	7.5778557E-009	7.6070067E-009
M_3	0.094	7	3.7195719E-010	3.2294976E-010
M_4	0.140	7	3.7184901E-010	3.2282504E-010
M_5	0.094	6	2.0519901E-009	1.9954705E-009
M_6	0.095	6	2.0521896E-009	1.9954422E-009
M_7	0.109	6	1.0521583E-010	9.7355378E-011
M_8	0.140	6	1.0616301E-010	9.7465166E-011

Table.2. CPU times in seconds, iteration numbers $\tilde{m}$ and relative norms of the errors obtained by Algorithm 1 using the methods $M_i, i=1,\dots,8$ for the harmonic continuation problem, when $N_1=50$ and $\delta_b=10^{-11},\delta_A=0.5(\delta_b)^{1.5}$.

Phillip's Problem: We consider the Fredholm integral equation of first kind $\int_{-6}^{6}k(s,t)h(t)dt=f(s)$ discussed in [13], where its solution, kernel and the right-hand side are

$$h(t)=\begin{cases}1+\cos(\frac{\pi}{3}t), & \text{if } |t|<3\\ 0, & \text{otherwise}\end{cases} \tag{17}$$

$$k(s,t)=h(s-t),\ f(s)=(6-|s|)\left(1+\tfrac{1}{2}\cos(\tfrac{\pi}{3}s)\right)+\tfrac{9}{2\pi}\sin(\tfrac{\pi}{3}|s|),$$

respectively. We discretize (17) by taking the quadrature nodes $t_j=-6+\frac{12j}{N_1}$, $j=1,\dots,N_1$ and evaluating $f(s)$ at the points $s_i=-6+\frac{12i}{N_1}$, $i=1,\dots,N_1$ for $N_1=64$. We apply Algorithm 1 using the methods $M_i, i=1,\dots,8$ for $\delta_b=10^{-7}$, $\delta_A=0.5(\delta_b)^{1.5}$ when

$V_0 = \frac{\tilde{A}^T}{\|\tilde{A}\|_1 \|\tilde{A}\|_\infty}$, $\varepsilon = 5\times 10^{-7}$. The CPU time in seconds, iteration numbers $\tilde{m}$ and relative norms of the errors for the Phillip's problem are given in Table 3.

Methods	*CPU*	$\tilde{m}$	$\|x-y_{\tilde{m}}\|_\infty$	$\frac{\|x-y_{\tilde{m}}\|_2}{\|x\|_2}$
M_1	0.078	8	5.0460857E-004	3.6367029E-004
M_2	0.078	8	5.0460858E-004	3.6367029E-004
M_3	0.078	7	2.8448691E-004	1.7447789E-004
M_4	0.093	7	2.8448691E-004	1.7447789E-004
M_5	0.077	6	3.4897422E-004	2.4563953E-004
M_6	0.078	6	3.4897422E-004	2.4563953E-004
M_7	0.077	6	1.5296893E-004	1.0051624E-004
M_8	0.078	6	1.5296893E-004	1.0051624E-004

Table.3. CPU times in seconds, iteration numbers $\tilde{m}$ and relative norms of the errors obtained by Algorithm 1 using the methods $M_i, i=1,\ldots,8$ for the Phillip's problem when $N_1 = 64$ and $\delta_b = 10^{-7}, \delta_A = 0.5(\delta_b)^{1.5}$.

5 Conclusion

The best error of numerical solution occurred in [1] for the Harmonic continuation problem by singular value decomposition when $A \in R^{50\times 50}$ was approximately 10^{-3} using single precision. The minimum relative error given in [13] for Phillip's problem when $A \in R^{64\times 64}$ is 1.84×10^{-3}. Comparing with the results in Table 2 and Table 3 proposed Algorithm 1 gives accurate results. Furthermore, the methods M_3, M_5, M_7 require less CPU times then the methods M_4, M_6, M_8, respectively.

References

1. J.M.Varah, SIAM, **10(2)** , 257-267 (1973).
2. P.C. Hansen, SIAM, **4** (1998).
3. D. Colton, R. Kress, Springer, New York (1992).
4. D.D. Ang, R. Gorenflo, V.K. Le, D.D. Trong, 1792 Springer Berlin (2002).
5. F.S. V. Bazán, J.Sci Comput **63**, 163-184 (2015).
6. C.W. Groetsch, Research notes in Mathematics 105, Pitman, Boston (1984).
7. A.N. Tikhonov, Soviet Math. Dokl. **4**, 1036-1038 (1963).
8. J.N. Franklin, Journal of Mathematical Analysis and Appl., **31**, 682-716 (1970).
9. A. Ben-Israel, T. N. E.Greville, Theory and Applications, Springer, (2003).
10. S.C. Buranay, D. Subasi, O.C. Iyikal, Numer. Linear Algebra Appl., **24 (6)**, (2017).
11. F. Soleymani, Numerical Algorithms, **69** 569-578 (2015).
12. F.Soleymani, P.S.Stanimirović, F.K.Haghani, Linear Algebra & its Appl. **484**, 477-495 (2015).
13. M. Rezghi, S. M. Hosseini, Journal of Computational and Applied Mathematics, **231**, 914-924 (2009) .

ITM Web of Conferences **22**, 01003 (2018) https://doi.org/10.1051/itmconf/20182201003
CMES-2018

Pentagonal Cone b-metric Spaces over Banach Algebras and Fixed Point Theorem of Generalized Lipschitz Mapping

Abba Auwalu[1,*]

[1]Department of Mathematics, Near East University, Nicosia, via Mersin 10, Turkey

Abstract. In this paper, we introduce the concept of pentagonal cone b-metric space over Banach algebras as a generalization of cone metric space over Banach algebras and many of its generalizations. Furthermore, we prove Banach fixed point theorem in such a space. Our result unify, complement and/or generalized some recent results in the papers [1–7], and many others. We provide some examples to elucidate the validity and superiority of our results.

1 Introduction

In 2013, Huang and Shao [1] instigated the idea of cone metric space with a Banach algebra by replacing the complete normed space in cone metric space with a Banach algebra. They established that cone metric spaces over Banach algebras are not equal to metric spaces in terms of existence of the fixed points of mappings. Further, they proved some fixed point results for various contractive conditions in such a space. Subsequently, many authors established interesting and significant results in a cone metric space over Banach algebras and its generalizations (see; [2], [3], [4], [5], [7] and references therein).

2 Preliminaries

In this part, we provide some definitions and results required in the consequence.

Definition 2.1 ([1]) *A subset $\mathcal{P}$ of Banach algebra $\mathcal{B}$ is called a cone if:*
(1) $\mathcal{P}$ is nonempty closed and $\{\theta, e\} \subset \mathcal{P}$; (2) $k_1\mathcal{P} + k_2\mathcal{P} \subset \mathcal{P}$ for all nonnegative real numbers k_1, k_2; (3) $\mathcal{P}^2 = \mathcal{P}\mathcal{P} \subset \mathcal{P}$; (4) $\mathcal{P} \cap (-\mathcal{P}) = \{\theta\}$, where θ and e denote the zero and unit elements of the Banach algebra $\mathcal{B}$, respectively. For a given cone $\mathcal{P} \subset \mathcal{B}$, we write $y \preccurlyeq z$ if and only if $z - y \in \mathcal{P}$, where $\preccurlyeq$ is a partial order relation defined on $\mathcal{P}$. We shall write $y \prec z$ to indicate that $y \preccurlyeq z$ but $y \neq z$, while $y \ll z$ will stand for $z - y \in int\mathcal{P}$, where $int\mathcal{P}$ denotes the interior of $\mathcal{P}$. If $int\mathcal{P} \neq \emptyset$ then $\mathcal{P}$ is called a solid cone.

Definition 2.2 ([2]) *Let $\mathcal{P}$ be a solid cone in a Banach algebra $\mathcal{B}$. A sequence $\{z_i\} \subset \mathcal{P}$ is said to be a c-sequence if for each $\theta \ll c$ there exists $n_0 \in \mathbb{N}$ such that $z_i \ll c$ for all $i > n_0$.*

Lemma 2.3 ([2]) *Let $\mathcal{P}$ be a solid cone in a Banach algebra $\mathcal{B}$ and $\{z_i\} \subset \mathcal{P}$ be a sequence with $\|z_i\| \to 0$ $(i \to \infty)$, then for each $\theta \ll c$, there exists $n_0 \in \mathbb{N}$ such that for all $i > n_0$, we have $z_i \ll c$.*

*e-mail: abba.auwalu@neu.edu.tr, abbaauwalu@yahoo.com

Lemma 2.4 ([8]) *Let $\mathcal{B}$ be a Banach algebra with a unit e and $\tau \in \mathcal{B}$. If the spectral radius $\delta(\tau)$ of τ is less than one, i.e. $\delta(\tau) = \lim_{n\to\infty} \|\tau^n\|^{\frac{1}{n}} = inf_{n\in\mathbb{N}}\|\tau^n\|^{\frac{1}{n}} < 1$, then $(e-\tau)$ is invertible in $\mathcal{B}$. Moreover, $(e-\tau)^{-1} = \sum_{j=0}^{\infty} \tau^j$.*

Remark 2.5 ([2]) *If the spectral radius $\delta(\tau) < 1$, then $\|\tau^i\| \to 0$ $(i \to \infty)$.*

Lemma 2.6 ([4]) *Let $\mathcal{B}$ be a real Banach algebra with a solid cone $\mathcal{P}$. For $a, b, c, \tau \in \mathcal{P}$, if (1) $a \preccurlyeq b \ll c$, then $a \ll c$. (2) $a \preccurlyeq \tau a$ and $\delta(\tau) < 1$, then $a = \theta$.*

3 Main Results

In this part, we firstly instigate the idea of a pentagonal cone b-metric space over a Banach algebra and provide an example showing that this concept is very tremendous than pentagonal cone metric space over a Banach algebra [7]. Secondly, we describe convergence and a Cauchy sequence in a pentagonal cone b-metric space over a Banach algebra and then prove Banach fixed point theorem in this space.

Definition 3.1 *Let $\mathcal{Z}$ be a nonempty set and $\mathcal{B}$ be a Banach algebra. Suppose that $\rho : \mathcal{Z}\times\mathcal{Z} \to \mathcal{B}$ is a mapping satisfying, for all $x, y, z, u, v \in \mathcal{Z}$, $x \neq z, z \neq u, u \neq v, v \neq y$, the following conditions:*
(Pb_1) $\theta \prec \rho(x,z)$ and $\rho(x,z) = \theta \Leftrightarrow x = z$;
(Pb_2) $\rho(x,z) = \rho(z,x)$;
(Pb_3) there exist $k \in \mathcal{P}$, $e \preccurlyeq k$ such that $\rho(x,y) \preccurlyeq k[\rho(x,z) + \rho(z,u) + \rho(u,v) + \rho(v,y)]$.
Then ρ is called a pentagonal cone b-metric on $\mathcal{Z}$, and $(\mathcal{Z},\rho)$ is called a pentagonal cone b-metric space over Banach algebra $\mathcal{B}$.

Remark 3.2 *A pentagonal cone b-metric space over a Banach algebra $\mathcal{B}$ is also pentagonal cone metric space over Banach algebra $\mathcal{B}$ where $k = 1$. For $k > 1$, the converse is not necessarily true, see the following example.*

Example 3.3 *Let $\mathcal{B} = \{a = (a_{mn})_{2\times 2} : a_{mn} \in \mathbb{R}, 1 \le m, n \le 2\}$, $\|a\| = \sum_{1\le m,n\le 2} |a_{mn}|$. Then $\mathcal{B}$ is a real Banach algebra with a unit element $e = \begin{pmatrix}1 & 0\\ 0 & 1\end{pmatrix}$ and zero element $\theta = \begin{pmatrix}0 & 0\\ 0 & 0\end{pmatrix}$. The set $\mathcal{P} = \{a \in \mathcal{B} : a_{mn} \ge 0, 1 \le m, n \le 2\}$ is a solid cone in $\mathcal{B}$. Let $\mathcal{Z} = \{1,2,3,4,5\}$, and mapping $\rho : \mathcal{Z}\times\mathcal{Z} \to \mathcal{B}$ is define as:*

$$\rho(z,z) = \begin{pmatrix}0 & 0\\ 0 & 0\end{pmatrix}, \text{ for all } z \in \mathcal{Z};$$

$$\rho(1,2) = \rho(2,1) = \begin{pmatrix}6 & 6\\ 10 & 10\end{pmatrix};$$

$$\rho(1,3) = \rho(3,1) = \rho(1,4) = \rho(4,1) = \rho(2,3) = (3,2) = \rho(2,4) = \rho(4,2) = \rho(3,4) = \rho(4,3) = \begin{pmatrix}\frac{1}{2} & \frac{1}{2}\\ 1 & 1\end{pmatrix};$$

$$\rho(1,5) = \rho(5,1) = \rho(2,5) = \rho(5,2) = \rho(3,5) = \rho(5,3) = \rho(4,5) = \rho(5,4) = \begin{pmatrix}2 & 2\\ 3 & 3\end{pmatrix}.$$

Then $(\mathcal{Z},\rho)$ is a pentagonal cone b-metric space over Banach algebra $\mathcal{B}$ with coefficient $k = \begin{pmatrix}2 & 0\\ 0 & 2\end{pmatrix}$. But $(\mathcal{Z},\rho)$ is not a pentagonal cone metric space over Banach algebra $\mathcal{B}$

because it lacks the pentagonal property (P_3) *in [7] as follows:*

$$\begin{aligned}\begin{pmatrix}6&6\\10&10\end{pmatrix} &= \rho(1,2) > \rho(1,3)+\rho(3,4)+\rho(4,5)+\rho(5,2)\\ &= \begin{pmatrix}\frac{1}{2}&\frac{1}{2}\\1&1\end{pmatrix}+\begin{pmatrix}\frac{1}{2}&\frac{1}{2}\\1&1\end{pmatrix}+\begin{pmatrix}2&2\\3&3\end{pmatrix}+\begin{pmatrix}2&2\\3&3\end{pmatrix}=\begin{pmatrix}5&5\\8&8\end{pmatrix},\\ \text{as } \begin{pmatrix}6&6\\10&10\end{pmatrix} &- \begin{pmatrix}5&5\\8&8\end{pmatrix} = \begin{pmatrix}1&1\\2&2\end{pmatrix}\in\mathcal{P}.\end{aligned}$$

Definition 3.4 *Let* $(\mathcal{Z},\rho)$ *be a pentagonal cone b-metric space with Banach algebra* $\mathcal{B}$, $z\in\mathcal{Z}$ *and* $\{z_i\}$ *be a sequence in* $(\mathcal{Z},\rho)$. *Then we say*
(1) $\{z_i\}$ *converges to* z *if, for each* $c\in\mathcal{B}$ *with* $\theta\ll c$, *there is a natural number* n_0 *such that* $\rho(z_i,z)\ll c$ *for all* $i\geq n_0$. *We denote this by* $z_i\to z$ $(i\to\infty)$.
(2) $\{z_i\}$ *is a Cauchy sequence if, for each* $c\in\mathcal{B}$ *with* $\theta\ll c$, *there is a natural number* n_0 *which is independent of n such that* $\rho(z_i,z_{i+n})\ll c$ *for all* $i\geq n_0$.
(3) $(\mathcal{Z},\rho)$ *is said to be complete if every Cauchy sequence in* $(\mathcal{Z},\rho)$ *is convergent.*

Theorem 3.5 *Let* $(\mathcal{Z},\rho)$ *be a complete pentagonal cone b-metric space over Banach algebra* $\mathcal{B}$ *with* $\theta\preccurlyeq k$ *and* $\mathcal{P}$ *be a solid cone in* $\mathcal{B}$. *Suppose that* $F:\mathcal{Z}\to\mathcal{Z}$ *is a mapping satisfying the generalized Lipschitz condition:*

$$\rho(Fz,Fy)\preccurlyeq\tau\rho(z,y),\ \text{for all } z,y\in\mathcal{Z}, \tag{1}$$

where $\tau\in\mathcal{P}$ *such that the spectral radius* $\delta(\tau)<1$. *Then F has a unique fixed point* z^* *in* $\mathcal{Z}$. *Moreover, for any* $z_0\in\mathcal{Z}$, *the iterative sequence* $\{F^iz_0\}$ $(i\in\mathbb{N})$ *converges to* z^*.

Proof. Let z_0 be arbitrary point in $\mathcal{Z}$. We define a sequence $\{z_i\}$ in $(\mathcal{Z},\rho)$ by

$$z_{i+1}=Fz_i=F^{i+1}z_0,\ \text{for all } i\in\mathbb{N}. \tag{2}$$

Suppose $z_j=z_{j+1}$ for some $j\in\mathbb{N}$, then $z^*=z_j=F^jz_0$ is a fixed point of F. Further, for any $z_0\in\mathcal{Z}$, the iterative sequence $\{F^jz_0\}$ $(j\in\mathbb{N})$ converges to z^* and the result is proved. Hence, we assume that $z_i\neq z_{i+1}$ for all $i\in\mathbb{N}$. We shall show that $\{z_i\}$ is a Cauchy sequence in $(\mathcal{Z},\rho)$. To achieve our goal, we split the proof via three cases as follows:
Case 1. Let $\delta(\tau)\in[0,\frac{1}{k})$ for $k>1$. Now, using (1) and (2), we have

$$\rho(z_i,z_{i+1})=\rho(Fz_{i-1},Fz_i)\preccurlyeq\tau\rho(z_{i-1},z_i)\prec\rho(z_{i-1},z_i), \tag{3}$$

that is, the sequence $\{\rho(z_i,z_{i+1})\}$ is strictly decreasing, and from this it follows that $\rho(z_i,z_{i+1})\neq\rho(z_j,z_{j+1})$ whenever $i\neq j$. Hence, from (1), (2) and (3), we obtain

$$\begin{aligned}\rho(z_i,z_{i+1})&\preccurlyeq\tau\rho(z_{i-1},z_i)=\tau\rho(Fz_{i-2},Fz_{i-1})\preccurlyeq\tau^2\rho(z_{i-2},z_{i-1})=\tau^2\rho(Fz_{i-3},Fz_{i-2})\\ &\preccurlyeq\tau^3\rho(z_{i-3},z_{i-2})\preccurlyeq\cdots\preccurlyeq\tau^i\rho(z_0,z_1),\ \text{for all } i\in\mathbb{N}.\end{aligned} \tag{4}$$

Similarly, $$\rho(z_i,z_{i+2})\preccurlyeq\tau^i\rho(z_0,z_2),\ \text{for all } i\in\mathbb{N}. \tag{5}$$

Since $\delta(\tau)<\frac{1}{k}$, then $\delta(k\tau^3)\leq k\delta(\tau^3)\leq k\delta(\tau)\delta(\tau)\delta(\tau)<\frac{1}{k^2}<1$ and by Lemma 2.4, it follows that $(e-k\tau^3)$ is invertible in $\mathcal{B}$. Moreover,

$$(e-k\tau^3)^{-1}=\sum_{j=0}^{\infty}(k\tau^3)^j,\ \text{and similarly } \left(\frac{e}{k}-\tau\right)^{-1}=\sum_{j=0}^{\infty}k^{j+1}\tau^j. \tag{6}$$

Also since $\delta(\tau) < \dfrac{1}{k} < 1$, by Remark 2.5, we obtain that $\quad \|\tau^i\| \to 0\ (i \to \infty). \qquad (7)$

Now, we consider $\rho(z_i, z_{i+n})$ in two cases:

Case 1.1. If n is odd, say $n = 2j + 1$ $(j \in \mathbb{N})$, then, by using (Pb_3), (4), (6) and the fact that $\rho(z_i, z_{i+1}) \neq \rho(z_j, z_{j+1})$ whenever $i \neq j$, we obtain that

$$\begin{aligned}
\rho(z_i, z_{i+2j+1}) &\preccurlyeq k[\rho(z_i, z_{i+1}) + \rho(z_{i+1}, z_{i+2}) + \rho(z_{i+2}, z_{i+3})] + k\rho(z_{i+3}, z_{i+2j+1}) \\
&\preccurlyeq k[\rho(z_i, z_{i+1}) + \rho(z_{i+1}, z_{i+2}) + \rho(z_{i+2}, z_{i+3})] \\
&\quad + k^2[\rho(z_{i+3}, z_{i+4}) + \rho(z_{i+4}, z_{i+5}) + \rho(z_{i+5}, z_{i+6})] + k^2\rho(z_{i+6}, z_{i+2j+1}) \\
&\preccurlyeq k[\rho(z_i, z_{i+1}) + \rho(z_{i+1}, z_{i+2}) + \rho(z_{i+2}, z_{i+3})] \\
&\quad + k^2[\rho(z_{i+3}, z_{i+4}) + \rho(z_{i+4}, z_{i+5}) + \rho(z_{i+5}, z_{i+6})] \\
&\quad + k^3[\rho(z_{i+6}, z_{i+7}) + \rho(z_{i+7}, z_{i+8}) + \rho(z_{i+8}, z_{i+9})] + \cdots \\
&\quad + k^{\frac{2j}{3}}[\rho(z_{i+2j-3}, z_{i+2j-2}) + \rho(z_{i+2j-2}, z_{i+2j-1}) + \rho(z_{i+2j-1}, z_{i+2j})] \\
&\quad + k^{\frac{2j}{3}}\rho(z_{i+2j}, z_{i+2j+1}) \\
&\preccurlyeq k[\tau^i\rho(z_0, z_1) + \tau^{i+1}\rho(z_0, z_1) + \tau^{i+2}\rho(z_0, z_1)] \\
&\quad + k^2[\tau^{i+3}\rho(z_0, z_1) + \tau^{i+4}\rho(z_0, z_1) + \tau^{i+5}\rho(z_0, z_1)] \\
&\quad + k^3[\tau^{i+6}\rho(z_0, z_1) + \tau^{i+7}\rho(z_0, z_1) + \tau^{i+8}\rho(z_0, z_1)] + \cdots \\
&\quad + k^j[\tau^{i+2j-3}\rho(z_0, z_1) + \tau^{i+2j-2}\rho(z_0, z_1) + \tau^{i+2j-1}\rho(z_0, z_1)] + k^j\tau^{i+2j}\rho(z_0, z_1) \\
&\preccurlyeq k\tau^i\Big(e + k\tau^3 + (k\tau^3)^2 + \cdots\Big)\rho(z_0, z_1) + k\tau^{i+1}\Big(e + k\tau^3 + (k\tau^3)^2 + \cdots\Big)\rho(z_0, z_1) \\
&\quad + k\tau^{i+2}\Big(e + k\tau^3 + (k\tau^3)^2 + \cdots\Big)\rho(z_0, z_1) \\
&= k\tau^i\Big(\sum_{j=0}^{\infty}(k\tau^3)^j\Big)\Big(e + \tau + \tau^2\Big)\rho(z_0, z_1)
\end{aligned}$$

$$\therefore\quad \rho(z_i, z_{i+2j+1}) \preccurlyeq k\tau^i(e - k\tau^3)^{-1}\Big(e + \tau + \tau^2\Big)\rho(z_0, z_1).$$

By using (7), we have $\|k\tau^i(e - k\tau^3)^{-1}\big(e + \tau + \tau^2\big)\rho(z_0, z_1)\| \to 0$ $(i \to \infty)$, and it follows, by Lemma 2.3, that for any $c \in \mathcal{B}$ with $\theta \ll c$, there exists $n_1 \in \mathbb{N}$ such that

$$\rho(z_i, z_{i+2j+1}) \preccurlyeq k\tau^i(e - k\tau^3)^{-1}\Big(e + \tau + \tau^2\Big)\rho(z_0, z_1) \ll c, \text{ for all } i > n_1.$$

Therefore, by Lemma 2.6(1), we get

$$\rho(z_i, z_{i+2j+1}) \ll c, \text{ for all } i > n_1, \tag{8}$$

Case 1.2. If n is even, say $n = 2j$ $(j \in \mathbb{N})$, then, by using (Pb_3), (4), (5), (6) and the fact that $\rho(z_i, z_{i+1}) \neq \rho(z_j, z_{j+1})$ whenever $i \neq j$, we obtain that

$$\begin{aligned}
\rho(z_i, z_{i+2j}) &\preccurlyeq k[\rho(z_i, z_{i+1}) + \rho(z_{i+1}, z_{i+2}) + \rho(z_{i+2}, z_{i+3})] + k\rho(z_{i+3}, z_{i+2j}) \\
&\preccurlyeq k[\rho(z_i, z_{i+1}) + \rho(z_{i+1}, z_{i+2}) + \rho(z_{i+2}, z_{i+3})] \\
&\quad + k^2[\rho(z_{i+3}, z_{i+4}) + \rho(z_{i+4}, z_{i+5}) + \rho(z_{i+5}, z_{i+6})] + k^2\rho(z_{i+6}, z_{i+2j}) \\
&\preccurlyeq k[\rho(z_i, z_{i+1}) + \rho(z_{i+1}, z_{i+2}) + \rho(z_{i+2}, z_{i+3})] \\
&\quad + k^2[\rho(z_{i+3}, z_{i+4}) + \rho(z_{i+4}, z_{i+5}) + \rho(z_{i+5}, z_{i+6})] \\
&\quad + k^3[\rho(z_{i+6}, z_{i+7}) + \rho(z_{i+7}, z_{i+8}) + \rho(z_{i+8}, z_{i+9})] + \cdots \\
&\quad + k^{\frac{2j-2}{3}}[\rho(z_{i+2j-5}, z_{i+2j-4}) + \rho(z_{i+2j-4}, z_{i+2j-3}) + \rho(z_{i+2j-3}, z_{i+2j-2})] \\
&\quad + k^{\frac{2j-2}{3}}\rho(z_{i+2j-2}, z_{i+2j})
\end{aligned}$$

$$
\begin{aligned}
\rho(z_i, z_{i+2j}) &\preccurlyeq k[\tau^i\rho(z_0,z_1) + \tau^{i+1}\rho(z_0,z_1) + \tau^{i+2}\rho(z_0,z_1)] \\
&\quad + k^2[\tau^{i+3}\rho(z_0,z_1) + \tau^{i+4}\rho(z_0,z_1) + \tau^{i+5}\rho(z_0,z_1)] \\
&\quad + k^3[\tau^{i+6}\rho(z_0,z_1) + \tau^{i+7}\rho(z_0,z_1) + \tau^{i+8}\rho(z_0,z_1)] + \cdots \\
&\quad + k^{j-1}[\tau^{i+2j-5}\rho(z_0,z_1) + \tau^{i+2j-4}\rho(z_0,z_1) + \tau^{i+2j-3}\rho(z_0,z_1)] + k^{j-1}\tau^{i+2j-2}\rho(z_0,z_2) \\
&\preccurlyeq k\tau^i\left(e + k\tau^3 + (k\tau^3)^2 + \cdots\right)\rho(z_0,z_1) + k\tau^{i+1}\left(e + k\tau^3 + (k\tau^3)^2 + \cdots\right)\rho(z_0,z_1) \\
&\quad + k\tau^{i+2}\left(e + k\tau^3 + (k\tau^3)^2 + \cdots\right)\rho(z_0,z_1) + k^{j-1}\tau^{i+2j-2}\rho(z_0,z_2) \\
&= k\tau^i\left(\sum_{j=0}^{\infty}(k\tau^3)^j\right)\left(e + \tau + \tau^2\right)\rho(z_0,z_1) + k^{j-1}\tau^{i+2j-2}\rho(z_0,z_2) \\
&\preccurlyeq k\tau^i(e - k\tau^3)^{-1}\left(e + \tau + \tau^2\right)\rho(z_0,z_1) + k^{2j-1}\tau^{2j-2}\tau^i\rho(z_0,z_2) \\
&\preccurlyeq k\tau^i(e - k\tau^3)^{-1}\left(e + \tau + \tau^2\right)\rho(z_0,z_1) + \tau^i\sum_{m=0}^{\infty}k^{m+1}\tau^m\rho(z_0,z_2)
\end{aligned}
$$

$$
\therefore\ \rho(z_i, z_{i+2j}) \preccurlyeq k\tau^i\left[(e - k\tau^3)^{-1}\left(e + \tau + \tau^2\right)\rho(z_0,z_1) + \left(\frac{e}{k} - \tau\right)^{-1}\rho(z_0,z_2)\right].
$$

By using (7), we have $\left\|k\tau^i\left[(e - k\tau^3)^{-1}\left(e + \tau + \tau^2\right)\rho(z_0,z_1) + \left(\frac{e}{k} - \tau\right)^{-1}\rho(z_0,z_2)\right]\right\| \to 0\ (i \to \infty)$, and it follows, by Lemma 2.3, that for any $c \in \mathcal{B}$ with $\theta \ll c$, there exists $n_2 \in \mathbb{N}$ such that, for all $i > n_2$

$$
\rho(z_i, z_{i+2j}) \preccurlyeq k\tau^i\left[(e - k\tau^3)^{-1}\left(e + \tau + \tau^2\right)\rho(z_0,z_1) + \left(\frac{e}{k} - \tau\right)^{-1}\rho(z_0,z_2)\right] \ll c.
$$

Therefore, by Lemma 2.6(1), we get

$$
\rho(z_i, z_{i+2j}) \ll c, \text{ for all } i > n_2, \tag{9}
$$

Let $n_0 := \max\{n_1, n_2\}$, hence, from (8) and (9), we have that

$$
\rho(z_i, z_{i+n}) \ll c, \text{ for all } i > n_0,
$$

which implies, by Definition 3.4, that $\{z_i\}$ is a Cauchy sequence in $(\mathcal{Z}, \rho)$ and since $(\mathcal{Z}, \rho)$ is complete, the sequence $\{z_i\}$ converges i.e. there exists z^* in $\mathcal{Z}$ such that $z_i \to z^*\ (i \to \infty)$. Next, we shall show that z^* is a fixed point of F. By using (Pb_3), (1), (2) and (4), we have

$$
\begin{aligned}
\rho(z^*, Fz^*) &\preccurlyeq k[\rho(z^*, z_i) + \rho(z_i, z_{i+1}) + \rho(z_{i+1}, z_{i+2}) + \rho(z_{i+2}, Fz^*)] \\
&\preccurlyeq k[\rho(z^*, z_i) + \tau^i\rho(z_0,z_1) + \tau^{i+1}\rho(z_0,z_1) + \tau\rho(z_{i+1}, z^*)] \\
\therefore\ \rho(z^*, Fz^*) &\preccurlyeq k[\rho(z^*, z_i) + \tau^i(e + \tau)\rho(z_0,z_1) + \tau\rho(z_{i+1}, z^*)] \to \theta\ (i \to \infty).
\end{aligned}
$$

Thus, $Fz^* = z^*$. Hence, z^* is a fixed point of F. Next we shall establish that this fixed point is unique. Assume that y^* is another fixed point of F, i.e. $Fy^* = y^*$, then using (1), we have that

$$
\rho(z^*, y^*) = \rho(Fz^*, Fy^*) \preccurlyeq \tau\rho(z^*, y^*).
$$

By lemma 2.6(2), we have that $\rho(z^*, y^*) = \theta$ and so $z^* = y^*$ i.e. the fixed point of F is unique.
Case 2. If $\delta(\tau) \in \left[\frac{1}{k}, 1\right)$ for $k > 1$. Since $\delta(\tau) < 1$ in this case, we have $[\delta(\tau)]^i \to \theta\ (i \to \infty)$, and so there exists i_0 such that $[\delta(\tau)]^{i_0} < \frac{1}{k}$. Note that $\delta(\tau^{i_0}) \leq [\delta(\tau)]^{i_0} < \frac{1}{k}$. Also by (1)

$$
\begin{aligned}
\rho(F^{i_0}z, F^{i_0}y) &= \rho\left(F(F^{i_0-1})z, F(F^{i_0-1})y\right) \preccurlyeq \tau\rho(F^{i_0-1}z, F^{i_0-1}y) = \tau\rho\left(F(F^{i_0-2})z, F(F^{i_0-2})y\right) \\
&\preccurlyeq \tau^2\rho(F^{i_0-2}z, F^{i_0-2}y) \preccurlyeq \cdots \preccurlyeq \tau^{i_0}\rho(z, y), \text{ for all } i_0 \in \mathbb{N}.
\end{aligned}
$$

Hence, by **Case 1**, the mapping F^{i_0} has a unique fixed point z^{**} in $\mathcal{Z}$. Now, observe that

$$F^{i_0}(Fz^{**}) = F^{i_0+1}(z^{**}) = F(F^{i_0}z^{**}) = Fz^{**},$$

that is, Fz^{**} is also a fixed point of F^{i_0}. Thus, by uniqueness of the fixed point of F^{i_0}, we get that $Fz^{**} = z^{**}$, which imply that z^{**} is also a fixed point of F. Now suppose that $z^{***} \in \mathcal{Z}$ is another fixed point of F i.e. $Fz^{***} = z^{***}$. Then

$$F^{i_0}z^{**} = F^{i_0-1}(Fz^{**}) = F^{i_0-1}z^{**} = \cdots = Fz^{**} = z^{**},$$
$$F^{i_0}z^{***} = F^{i_0-1}(Fz^{***}) = F^{i_0-1}z^{***} = \cdots = Fz^{***} = z^{***},$$

which means that z^{**} and z^{***} are fixed point of F^{i_0}. Thus, by uniqueness of the fixed point of F^{i_0}, we get that $z^{**} = z^{***}$ i.e. the fixed point of F is only one of its kind.
Case 3. If $k = 1$. Since $\delta(\tau) < 1$, the proof follows from **Case 1**. This completes the proof.

Example 3.6 *In Example 3.3, $(\mathcal{Z},\rho)$ is a pentagonal cone b-metric space over Banach algebra $\mathcal{B}$ with coefficient $k = \begin{pmatrix} 2 & 0 \\ 0 & 2 \end{pmatrix}$. Now we define a mapping $F : \mathcal{Z} \to \mathcal{Z}$ as follows:*

$$Fz = \begin{cases} 4, & \text{if } z \neq 5; \\ 2, & \text{if } z = 5. \end{cases}$$

Hence, $\rho(F1, F2) = \rho(F1, F3) = \rho(F1, F4) = \rho(F2, F3) = \rho(F2, F4) = (F3, F4) = \begin{pmatrix} 0 & 0 \\ 0 & 0 \end{pmatrix}$. And in all other cases we have that $\rho(Fz, Fy) = \begin{pmatrix} \frac{1}{2} & \frac{1}{2} \\ 1 & 1 \end{pmatrix}$, $\rho(z, y) = \begin{pmatrix} 2 & 2 \\ 3 & 3 \end{pmatrix}$. Let $\tau \in \mathcal{P}$ where $\tau := \begin{pmatrix} \frac{1}{3} & 0 \\ 0 & \frac{1}{3} \end{pmatrix}$. It is easy to prove that $\|\tau\| < 1$, which implies that $\delta(\tau) < 1$. Hence, the mapping F satisfies all the conditions of Theorem 3.5 and $z^* = 4 \in \mathcal{Z}$ is the only one of its kind fixed point of F.

4 Conclusion

In this piece of writing, we instigate the idea of pentagonal cone b-metric space over Banach algebras and prove Banach fixed point theorem in such a space. Our result, Theorem 3.5, generalizes Theorem 2.1 in [1], Theorem 3.1 in [2], Theorem 2.1(i) in [3], Theorem 5 in [4], Theorem 3.5 in [5], Theorem 3.10 in [6], and Theorem 3.1 in [7] to a more general pentagonal cone b-metric space over a Banach algebra. We hope that this result will be useful in fixed point theory and may be generalized in further spaces with efficient conditions.

References

[1] L. Huang, Xu. Shao, Fixed Point Theory Appl. **320**, 1–10 (2013)
[2] S. Xu, S. Radenović, Fixed Point Theory Appl. **102**, 1–12 (2014)
[3] H. Huang, S. Radenović, J. Comput. Anal. Appl. **20**, 566–583 (2016)
[4] S. Shukla, S. Balasubra, M. Pavlovic, Bull. Mal. Math. Sci. Soc. **39**, 1529–1539 (2016)
[5] R. George, H.A. Nabwey, R. Raj, S. Radenović, Fixed Point Appl. **14**, 1–15 (2017)
[6] M. Garg, S. Agarwal, J. Adv. Stud. Topol. **3**, 12–18 (2012)
[7] A. Auwalu, E. Hınçal, Int. Conf. Anal. Appl. Math. (ICAAM), (Accepted) (2018)
[8] W. Rudin, *Functional Analysis* (McGraw-Hill Inc., New York, 1991) 245–272

Numerical Investigation of Acoustics Performance of Low- Pressure Ducted Axial Fan by Using Different Turbulence Models

Aykut Bacak* and *Ali* Pinarbasi

Yıldız Technical University, Mechanical Engineering Department, 34349 Besiktas, Istanbul, Turkey

*Corresponding Author: bacakaykut@gmail.com

Abstract. In this article, capacity and acoustics parameters of low pressure ducted axial fan is numerically investigated with Realizable k-epsilon, k-w SST and DES turbulence models by using computational fluid dynamics software. One slice of six bladed axial fan operating at 3000 RPM is simulated periodically as low pressure ducted axial ventilation fan. Simulations are run for operating point on the performance curve for each turbulence models. Investigation of acoustics parameters are obtained Ffowcs-Williams Hawkings acoustic model to calculate sound pressure levels for related frequencies. Numerical results are compared with the experimental results provided from blade manufacturer company.

1 Introduction

Axial flow fans are the continuous flow machines that air or any other gas flows through in an approximately axial direction by using rotational blades inside them. Axial fans are generally used at buildings, industrial areas cars and public transport vehicles ventilating and/or air conditioning equipment [1]. The term of ducted fan is defined by the sum of air or other gas which it forces through the system against resistance representing skin friction, pressure loses and secondary flows. Ducted axial fans contain blades with aerodynamic cross-section, electric motor, construction support elements and other auxiliary equipment [2]. Axial fans generate high or low noise levels while operating at high or low rotational speed for different reasons. The increase in the use of high-rise buildings in big cities, and therefore the increased use of axial fan, will lead to noise pollution. Therefore, while designing the axial fan blades, it is necessary to examine the acoustic performance while examining parameters such as flow rate, pressure and efficiency.

Sound is always related with fast small-scale pressure fluctuations exceeding the normal atmospheric pressure. These fluctuations are radiated from sources and travel as waves through the medium at sound speed. Sound may be reflected, particularly absorbed or reduced before getting at the human eardrum where it generates a sensation of hearing rely on the amplitude of the sound wave [3]. Neise [4], defined aeroacoustics sound generation mechanism for fans like shown in Figure 1. According that, the most common noise mechanism occurs from the unsteady-rotating forces some of them can be regards as performance loses such as secondary flows and vortex shedding.

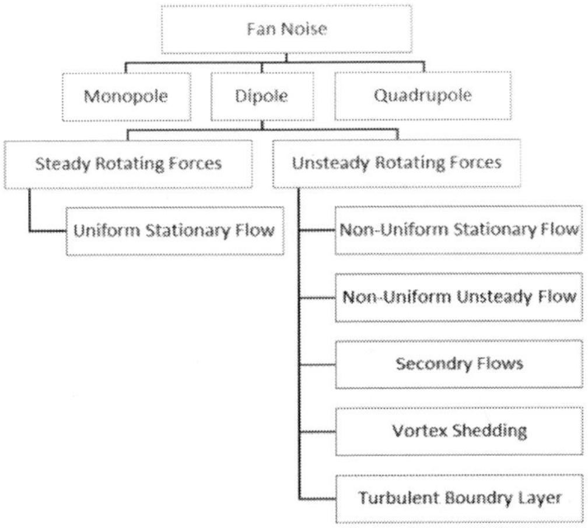

Fig. 1. Overview of the aeroacoustics sound generation mechanisms for fans.

Noise generated by axial fan can be reduced by optimizing electrical motor noise and aerodynamically generated noise. Any kind of blade design with different size and shape can be investigated by using numerical methods with different case studies to research noise generating parameters such as sound pressure and power levels to optimize aero-acoustically. There are many great efforts both in industrial area and scientific society to reduce ventilation fan noise levels by using different type of computational fluid dynamics software and numerical methods in literature. Sturm et al. [5], investigated tonal sound pressures of axial fan rotor by using Lattice Boltzmann approach for geometrically different test cases. They obtained relatively close SPL results for all cases and similar results to experimental ones. Reese at al. [6], investigated the acoustics performance of an axial fan by using different turbulence models such as URANS, DES, SAS and LES. They obtained sound pressure power spectral density results by using virtual sound receptor located at one meter far-away from the fan with 45° angle from the axis. They obtained the closest results to experimental studies at LES turbulence model. Perot et al. [7], investigated an axial fan aerodynamic and aeroacoustics performance numerically and experimentally with Lattice-Boltzmann method. They consider the influence of inlet flow on BPF noise in their studies. They stated importance of modeling the geometrical details of a hood to capture the BPF. They attained a great agreement for numerical and experimental results. Marsan et al. [8], investigated the tip leakage and its impact of low-speed fan numerically with Lattice-Boltzmann Method as three different voxel sizes and compared numerical sound pressure level of same fan with reference experimental results.

2 Numerical Method

In this paper, all simulations are made by using Ansys 18.1 Fluent, computational fluid dynamics software. Simulating process consists of design, meshing, fluid flow modeling and solution process. Basis steps of simulation process such as defining properties of simulation, applying boundary conditions and selecting appropriate turbulence model are applied at flow modeling section. In this paper, one slice of six bladed axial fan operating at 3000 RPM is simulated periodically with Realizable k-ε, k-w SST and DES turbulence models at operating point.

Dimensions of domain is taken from ANSI/ AMCA 320-08 [9]. Upstream and downstream length is taken 10D and the radius of expansion domain is taken 3D where D is the hydraulic diameter of axial fan. Fluid domain consists of velocity inlet, pressure outlet, periodically rotating hub and blade walls, stationary and periodic walls as boundary conditions. Rotating blade and hub walls with shroud surface defined as noise sources like shown in Figure 1 in a large scale. Around 800,000 triangular mesh elements and 10 inflation layers with 0,008 mm first layer thickness near the wall are used for all simulations. Static pressure measurement is done just after and before from rotating domain and velocity measurement is taken from 8.5D faraway from fan in the downstream flow. As it is defined at ANSI/AMCA 210-07 Standard [10], velocity measurements are taken at defined location with identified measurement points. Mass-weighted Average method at that location is used for velocity measurement instead of measuring identified measurement points. Static pressure drop is also measured with Mass-weighted average method since hub prevents the fluid flow around the axis.

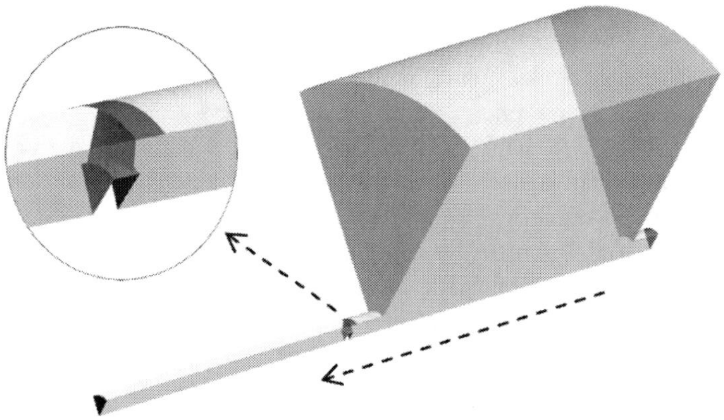

Fig. 2. Fluid domain of numerical simulation.

Ffows-William Hawkings acoustic module is used to obtain acoustic parameters with 2E-05 Pa reference acoustic pressure, 340 m/s sound speed and 1.225 kg/m^3 air density as boundary conditions. Virtual noise measurement sensors are placed 2m far-away with 45° angle from the axis of the fan upstream in the expansion domain not to be influenced from the fluid flow like shown in Figure 3 as experimental measurement.

Fig. 3. Experimental acoustic performance measurement provided by manufacturer.

Continuity and Navier-Stokes Equations coupled with k-ε, k-ω SST and DES turbulence models are used in fan simulations. Continuity, momentum and k-ε, k-ω SST and DES turbulence model equations are given below;

The Mass Conservation Equation;

$$\frac{\partial \rho}{\partial t} + \nabla(\rho\vec{v}) = 0 \tag{1}$$

where v is velocity and ρ is the density [11].

Momentum Conservation Equation;

$$\frac{\partial(\rho\vec{v})}{\partial t} + \nabla(\rho\vec{v}\vec{v}) = \nabla P + \nabla.(\overline{\tau}) + \rho\vec{g} + \vec{F} \tag{2}$$

$$\overline{\tau} = \mu\left[\left(\nabla\vec{v} + \nabla\vec{v}^T\right) - \frac{2}{3}\nabla\vec{v}I\right] \tag{3}$$

where P is the static pressure, $\overline{\tau}$ is the stress tensor, $\rho\vec{g}$ and $\vec{F}$ are the gravitational body force and external body forces respectively, μ is the molecular viscosity and I is the unit tensor. [11].

Transport equations for realizable k-ε turbulence model;

$$\frac{\partial(\rho k)}{\partial t} + \frac{\partial(\rho k u_j)}{\partial x_j} = \frac{\partial}{\partial x_j}\left[\left(\mu + \frac{\mu_t}{\sigma_k}\right)\frac{\partial k}{\partial x_j}\right] + G_k + G_b - \rho\varepsilon - Y_M + S_k \tag{4}$$

$$\frac{\partial(\rho\varepsilon)}{\partial t} + \frac{\partial(\rho\varepsilon u_j)}{\partial x_j} = \frac{\partial}{\partial x_j}\left[\left(\mu + \frac{\mu_t}{\sigma_\varepsilon}\right)\frac{\partial\varepsilon}{\partial x_j}\right] + \rho C_1 S\varepsilon - \rho C_2\frac{\varepsilon^2}{k + \sqrt{\upsilon\varepsilon}} + C_1\varepsilon\frac{\varepsilon}{k}C_{3\varepsilon}G_b + S_\varepsilon \tag{5}$$

$$C_1 = max\left[0.43, \frac{\eta}{\eta+5}\right] \tag{6}$$

$$\eta = S\frac{k}{\varepsilon} \tag{7}$$

$$S = \sqrt{2S_{ij}S_{ij}} \tag{8}$$

Where G_k represents the generation of turbulence kinetic energy due to mean velocity gradients, G_b, is the generation of turbulence kinetic energy due to buoyancy, Y_M, represents the contribution of the fluctuating dilatation in compressible turbulence to the overall dissipation rate, C_2 and $C_{1\varepsilon}$ are constants, σ_k and σ_ε are the turbulence Prandtl numbers for k and ε, respectively. S_k and S_ε are user-defined source terms [11].

Transport equations for k-ω SST turbulence model;

$$\frac{\partial(\rho k)}{\partial t} + \frac{\partial(\rho k u_i)}{\partial x_i} = \frac{\partial}{\partial x_j}\left(\Gamma_k \frac{\partial k}{\partial x_j}\right) + G_k - Y_k + S_k \tag{9}$$

$$\frac{\partial(\rho \omega)}{\partial t} + \frac{\partial(\rho \omega u_j)}{\partial x_j} = \frac{\partial}{\partial x_j}\left(\Gamma_\omega \frac{\partial \omega}{\partial x_j}\right) + G_\omega - Y_\omega + S_\omega \tag{10}$$

where G_k represents the generation of turbulence kinetic energy due to mean velocity gradients, G_ω, represents the generation of ω, Γ_k and Γ_ω represent the effective diffusivity of k and ω, respectively, Y_k and Y_ω represent the dissipation of k and ω due to turbulence, S_k and S_ω are user-defined source terms [11].

DES with the SST k-ω model equations;

$$Y_k = \rho\beta^* k\omega F_{DES} \tag{11}$$

$$F_{DES} = max\left(\frac{L_t}{C_{DES}\Delta_{max}}, 1\right) \tag{12}$$

$$L_t = \frac{\sqrt{k}}{\beta^* \omega} \tag{13}$$

$$F_{DES} = max\left(\frac{L_t}{C_{DES}\Delta_{max}}(1 - F_{SST}, 1\right) \tag{14}$$

where Δ_{max} is the maximum local grid spacing $(\Delta_x, \Delta_y, \Delta_z)$, $\boldsymbol{C_{DES}}$ is a calibration constant used in DES model has a value of 0.61 [11].

$$\frac{1}{a_0^2}\frac{\partial^2 p'}{\partial t^2} - \nabla^2 p' = \frac{\partial^2\left(T_{ij}H(f)\right)}{\partial x_i \partial x_j} - \frac{\partial}{\partial x_j}\left(\left[P_{ij}n_j + \rho u_i(u_n - u_v)\right]\delta(f)\right) + \frac{\partial}{\partial t}\left(\left[\rho_0 v_n + \rho(u_n - u_v)\right]\delta(f)\right) \tag{15}$$

where u_i is fluid velocity component in the x_i direction, u_n is fluid velocity component normal to the surface f=0, v_i is surface velocity components in the x_i direction, v_n is surface velocity component normal to the surface, δ(f) is the dirac delta function and H(f) is Heaviside function.

3 Result and Discussion

In this study an axial fan operating at around 6800 m^3/h and 285 Pa (Static) is numerically investigated to obtain sound pressure levels between 63 and 8000 Hz by using k-epsilon, k-w SST and DES turbulence models. It has been seen that pressure drop values measured just after and before from the fan rotor and axial velocity values calculated as mass average weighted from indicated location described previously are in acceptable levels for all cases. Sound pressure levels, shown in Figure 3, obtained by virtual receivers located 2 m far away from the fan rotor and numerical simulations run for around 100 seconds. Sound pressure levels at different octave band are obtained from Fast Fourier Transform (FFT) in same CFD Software. As it can be seen from Figure 3, total sound pressure level values are in acceptable values with provided values from manufacturers.

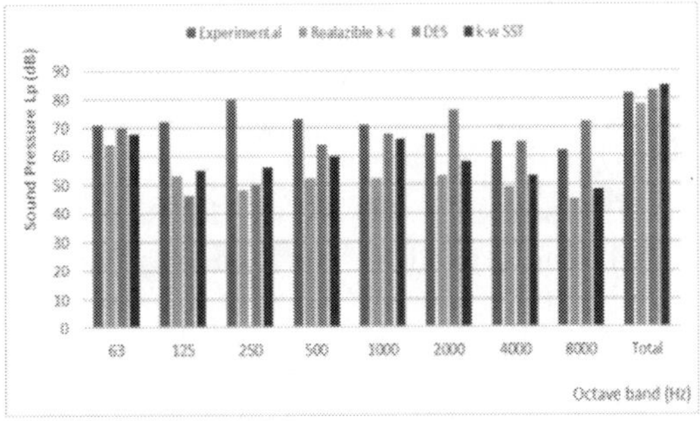

Fig. 4. Sound pressure levels results at octave band.

Dipole source strength, turbulence kinetic energy and turbulence eddy dissipation contours obtained by realizable k-epsilon turbulence model at z=0 location (center plane) are shown in Figure 4. It is useful to investigate dipole source strength contour at center plane, since most of noise generated by axial fan is dipole as it is described first chapter as table. At it is examined turbulence kinetic energy contour, it is seen that kinetic energy distribution is similar the dipole source strength distribution. So, reducing the kinetic energy around fan blade may reduce the noise levels. Turbulence eddy dissipation contour is also similar the dipole source strength distribution. Reducing eddy dissipation around fan blade may be another solution for reducing the noise levels of fans.

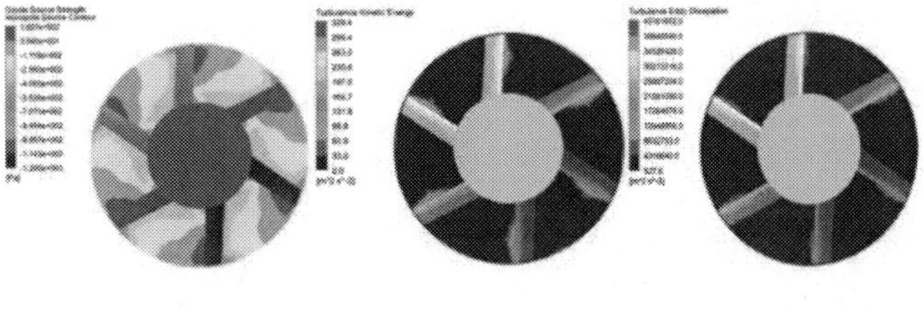

Fig. 5. Dipole source strength, turbulence kinetic energy and turbulence eddy dissipation contour showing at the same plane

 https://doi.org/10.1051/itmconf/20182201004

4 Conclusion

In this study, some numerical investigations on the airflow and acoustics performances of an axial ventilation fan are performed. The analysis is based on the aero-acoustic analogy where the CFD computations have been performed using Ansys Fluent 18.1. Numerical sound pressure level values obtained by CFD calculations are with values provided by axial fan blade manufacturer. Acceptable agreement has been observed between numerical and provided sound pressure levels. Good agreement has also been observed for pressure drop and flow rate provided values.

REFERENCES

[1] F. P. Bleier, Fan Handbook Selection, Application, and Design, The USA: McGraw-Hill, (1997).

[2] R. A. Wallis, Axial Flow Fans Design and Practice, New York and London: Academic Press, (1961).

[3] S. Wagner, R. Bareib, G. Guidati, Wind Turbin Noise, Germany: Springer, (1996).

[4] W. Neise, "Review of Fan Noise Generation Mechanicsm and Control Methods," in Fan Noise Symposium CETIM, France , (1992).

[5] Sturm M., Sanjose M., Moreau S., Carolus T., "Aeroacoustic Simulation of an Axial Fan Including the Full Test Rig by Using Lattice Boltzmann Method," in Fan 2015 Conference, (2015).

[6] Reese H., Carolus T., Kato C., "Numerical Prediction of the Aeroacoustic Sound Sources in a Low-Pressure Axial Fan with Inflow Distortion," in Fan Conference 2007, Lyon, France, (2007).

[7] Perot F., Kim M. S., Moreau S., Henner M., "Axial Fan Noise Aeroacoustics Predictions and Inflow Effect on Tonal Noise using LBM," in CFD Canada 2013 Conference, Canada, (2013).

[8] Marsan A., Daniels D. L., Sanose M., Moreaus S., "Tip Leakage Flow and its Implication on the Acoustic Signature of a Low-Speed Fan," in Fan 2018 Conference, Germany, (2018).

[9] ANSI/AMCA 320-08 Laboratory Methods of Sound Testing of Fans Using Sound Intensity, (2013).

[10] ANSI/AMCA 210-07 Laboratory Methdos of Testing Fans for Certified Aerodynamic Performance Rating, (2007).

[11] ANSYS Fluent Thery Guide, ANSYS, Inc., (2013).

ITM Web of Conferences **22**, 01005 (2018) https://doi.org/10.1051/itmconf/20182201005
CMES-2018

Comparison of Encryption Algorithms in Wireless Sensor Networks

Murat Dener[1,*]

[1]Graduate School of Natural and Applied Sciences, Gazi University, Ankara, Turkey

Abstract. In general, Wireless Sensor Networks occur when sensor nodes are randomly left in an unreliable environment. Sensor node has limited processor, limited memory, limited radio capacity and low cost. In sensor network applications, security mechanisms must be used, because of unsafe environments, excessive number of sensor nodes, and wireless communication environments. Ensuring confidentiality, the primary goal of security, is one of the most important problems to be solved in order to realize time and vital objectives. While ensuring security, it is also necessary to consider other important criteria such as memory usage, energy and latency of Sensor Networks. In this study, encryption is described in Wireless Sensor Networks and Skipjack, XXTEA and AES encryption algorithms are compared using TOSSIM simulation program in TinyOS operating system considering memory usage, energy and delay criteria. The study is considered to be useful academicians who study security in Wireless Sensor Networks.

1 Introduction

Compared to classic computer networks, wireless sensor networks show many unique features. Therefore, security methods that can be used in conventional networks cannot be used in wireless sensor networks [1]. In sensitive wireless sensor networks applications such as monitoring enemy lines or monitoring border zones, security protocols that provide confidential data transmission from the sensor nodes to the base station must be used. However, low processor and radio capacities of sensor nodes do not allow the implementation of traditional security protocols in sensor networks [2]. To make sensor networks practical, many improvements are currently being made in software and hardware. However, the basis of every work done is the reduction of the energy consumption per unit and hence the extension of the life of the sensor nodes. The main reason for this is that replacing or refilling sensor nodes and used energy sources in the working environment is often impossible and costly [3]. Security methods designed for traditional networks and widely used in many applications today cannot be implemented directly in wireless sensor networks because sensor nodes have limited energy resources, insufficient memory capacities and limited processing capabilities. Because of this, the security methods developed for wireless sensor networks take into account safety, energy,

* Corresponding author: muratdener@gazi.edu.tr

memory usage and latency. A security method in which all resources are actively used will be ideal for sensor networks [4]. In sensor networks, encryption algorithms are used to provide data confidentiality. However, it should not be forgotten that sensor networks have limited hardware resources and that the first objective is energy efficiency while providing requirements. Otherwise, a protocol that consumes a lot of energy, even if it provides all security measures, would be useless for sensor networks [5]. This work describes encryption in wireless sensor networks and uses Skipjack [6], XXTEA [7] and AES [8] encryption algorithms to examine memory usage, energy and latency criteria using the TOSSIM simulation program in TinyOS.

2 TinyOS&TOSSIM

The TinyOS Operating System is installed on the sensor nodes that compose the Wireless Sensor Networks. TinyOS is encoded in the programming language NesC, a variant of the C programming language. With this coding, nodes can be given new features. Designed algorithms or protocols can be seen primarily in the TOSSIM simulation program in TinyOS. TOSSIM is designed to simulate sensor network applications. TinyDB is a small database that is used to collect data and to read the data during simulation. TinyViz is a graphical visualization tool associated with TOSSIM, which is used to view the results of any NesC application. TinyViz interacts with TOSSIM, providing an extensible graphical user interface for debugging and visualization. TOSSIM is a simulation tool designed for TinYOS sensor networks. NesC codes written in TOSSIM also work on sensor node hardware without any change. Then, it can be loaded into the nodes without making any changes. The TinyOS operating system is designed to support the needs of wireless sensor networks. TinyOS is an embedded operating system distributed as completely free and open source for use in wireless sensor networks. TinyOS has a component-based architecture. In TinyOS, sending messages is via the AMStandard component. The encryption can be written as a module in a separate file. After the message package is prepared in the AMStandard component, it is transmitted wirelessly to the receiver via radio. The node that receives the message can still see the contents of the message package in the AMStandard component. This means that users who want to use different encryption algorithms or encryption modes will have to write their own algorithms or modes in this partition. The code written in this way is both modular and easy to use. The diagram showing the message transmission in TinyOS is given in Figure 1.

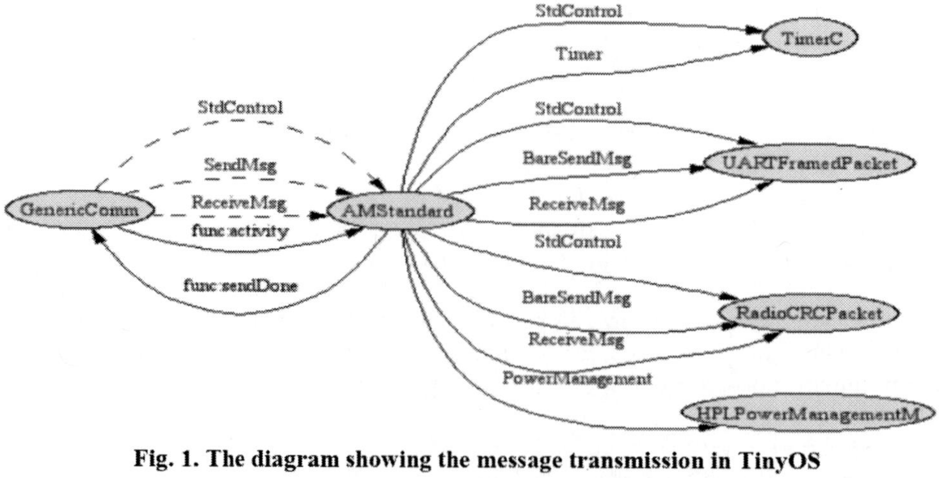

Fig. 1. The diagram showing the message transmission in TinyOS

GenericComm and AMStandard components are definitely used when sending / receiving messages in any application in TinyOS.

2 Simulation Works

First, Skipjack, XXTEA and AES encryption algorithms are written in C language. The encryption of the message values used in the future TOSSIM simulation results and the calculation of the MAC value are performed in this program. Then, after verifying the codes of the encryption algorithms, this written code is translated into NesC code.

Simulated results were obtained on 30 nodes with the help of the TOSSIM simulation program to test whether the NesC code written after this operation was performed correctly before loading the node. In this chapter, The message generated by the node, Message encryption and MAC calculation, Sending a message, Receiving message, Decryption of the password, Computation and comparison of MAC, If the MACs are equal, the process of accepting the message is shown. The simulated display is shown in Figure 2.

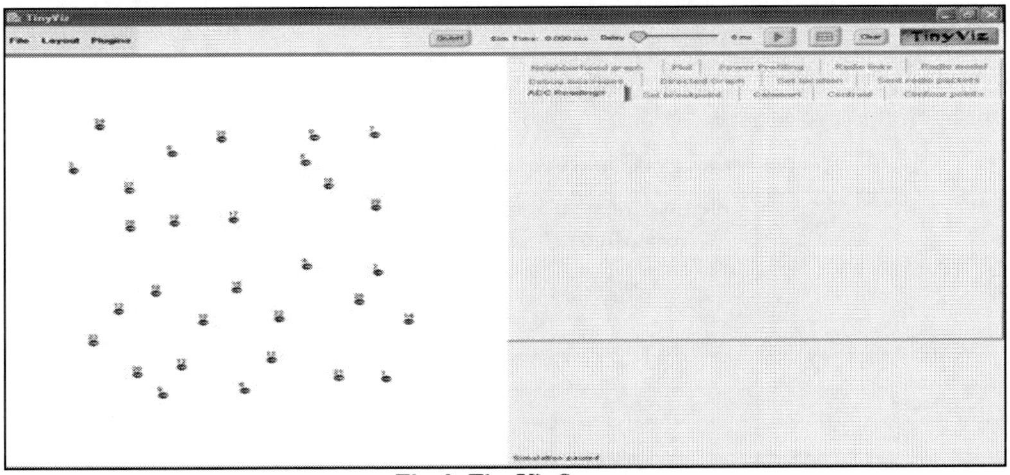

Fig. 2. TinyViz Screen

According to the simulation program, node 19 will send message to node 15. First, node 19 generates a message after performing node detection. Then, it performs the encryption process before sending the message. The message to be sent on the MAC calculated with the encrypted message is added to the package. The encrypted message sent by node 19 is received by node 15. So even if the node is seized on the road, it will not get a meaning for the aggressor. The decryption of the message received by node 15 is resolved and the MAC value is calculated. When the calculated MAC value is the same as the incoming MAC value, the message is processed.

It is seen that the NesC code for the encryption algorithm written by TinyViz runs smoothly on the nodes. After this, the same code can be loaded in the sensor nodes without making any changes.

3 Evaluation Results

The performance of a cryptographic algorithm can be determined according to the length of the system breakable period, the time spent in ciphering and deciphering processes, and the amount of memory required for encryption and decryption operations [9]. For a 40-bit key $n = 2^{40}$ or $n = 1\ 099\ 511\ 627\ 776$ (one trillion ninety nine billion five hundred ten ten million six hundred twenty seven thousand seven hundred seventy six) is a possible key word. In a competition, a credit card transaction on the Internet, encrypted with a 40-bit key with the RC4 algorithm, was solved in 3 and a half hours by a student who have just a modest computer lab in 1995 [10]. Therefore, the key length used must be carefully selected.

It is concluded that a 64-bit algorithm can be used up to 1994, an 80-bit algorithm up to 2013, and a 128-bit algorithm up to 2076 can be used in the proportions made by knowing that the 56-bit DES algorithm introduced in 1977 is broken in 1982 [11]. The comparison of the encryption algorithms used in the study according to the block length, key length and number of cycles are given in Table 1.

Table 1. Properties of the encryption algorithms used

Encryption Algorithms	Release date	Block Length (bit)	Key Length (bit)	Number of Cycles
Skipjack	1993	64	80	32
AES	2002	128	128,192,256	10,12,14
XXTEA	1998	64	128	32

In TOSSIM environment, the energy, delay and memory usage of the algorithms are compared in Tables 2. In comparisons, the key length of the AES algorithm is 128 bits and the number of cycles is 10.

Table 2. Energy comparison of the encryption algorithms used

	Increased amount of memory usage compared to TinyOS (%)		Increased amount of latency compared to TinyOS (%)	Increased amount of energy compared to TinyOS (%)
	ROM	RAM		
Skipjack	4.32	14.23	4.7	5
XXTEA	4.34	14.32	5.1	6
AES	4.64	45.22	8.4	13

As shown in Tables 2, although the key length of the XXTEA encryption algorithm is 128 bits, there is a negligible increase according to the Skipjack encryption algorithm according to the memory usage, energy and delay criteria. It would be better to use the XXTEA encryption algorithm instead of the Skipjack encryption algorithm, which has a key length of 80 bits, according to studies done in the literature. The AES encryption algorithm is more costly than other algorithms because of the specific methods it uses for encryption. However, the AES algorithm can be preferred in high-level applications that require much security.

4 Conclusions

When developing a security approach, it is necessary to consider the capacities of wireless sensor node resources (memory, processor, power supply). Sensor networks are expected to increase node energy consumption quantities and average end-to-end latency of encryption mechanisms added to improve security in applications. It is very important here that the needs of the applications are well established. In a simple large-scale environment or industrial sensor networks application, safety is of utmost importance, while energy consumption is a major factor. On the other hand, while security is of great importance in military and health applications, node energy consumption can be relatively neglected. Therefore, it is very important to select the encryption algorithm, encryption mode appropriate to the security solutions developed for military and health applications. Developed security solutions must be modular. That is, the newly introduced encryption algorithm and encryption modes of the literature should be integrated directly into the developed security solution if security, energy, memory usage, latency are better.

In this work, encryption is described in wireless sensor networks and Skipjack, XXTEA and AES encryption algorithms are compared using TOSSIM simulation program in TinyOS operating system considering memory usage, energy and delay criteria. XXTEA encryption algorithm is a negligible increase according to Skipjack encryption algorithm, but the result is that XXTEA encryption algorithm should be used instead of using Skipjack encryption algorithm with 80 bit key length. At the same time, it is concluded that the more costly AES encryption algorithm can be used in applications that require high security.

References

1. C-Y. Chong, S.P. Kumar, "*Sensor Networks : Evolution, opportunities, and challenges*", Proc IEEE, **91(8)**, 1247-1256, (2003).
2. M. Dener, "*Security Analysis in Wireless Sensor Networks*", International Journal of Distrubuted Sensor Networks, **2014**, Article ID 303501, 1-9, (2014).
3. R. Lin, Z. Wang, Y. Sun, "*Energy Efficient Medium Access Control Protocols for Wireless Sensor Networks and Its State-of-Art*", IEEE, pp 669-674, (2004).
4. I.F. Akyıldız, W. Su, Y. Sankarasubramaniam, E. Çayırcı, "*A survey on sensor networks", IEEE Communications Magazine*, **40(8)**, 102-114, (2002).
5. T. Kavitha, D. Sridharan, "*Security Vulnerabilities in Wireless Sensor Networks: A Survey*", Journal of Information Assurance and Security, **5**, 31-44, (2010).
6. C. Karlof, N. Sastry, D. Wagner, "*Tinysec: a link layer security architecture for wireless sensor networks*", In n: SenSys '04: Proceedings of the 2nd international conference on Embedded Networked Sensor Systems, **1008**, 162– 175, ACM, New York, USA, (2004).
7. D. Wheeler, R.M. Needham, "*XXTEA: Corrections to XTEA*", Technical report, Computer Laboratory, University of Cambridge, (1998).
8. D. Joan, R. Vincent, "*The Design of Rijndael*", Springer-Verlag New York, Inc., Secaucus, NJ, USA, (2002).
9. M. Tektaş, F. Baba, E.M. Çalışkan, "*Classification of Cryptographic Algorithms and a Credit Card Application*", Third International Advanced Technologies Symposium, Ankara, (2003).
10. H. Kodaz, "*Encryption for Security in Data Transmission*", Selçuk University, Graduate School of Natural and Applied Sciences, Master Thesis, (2002).
11. C.J. Devesh, R.P. Dhiren, S.D. Kankar, "*Investigating and Analyzing the Light-weight ciphers for Wireless Sensor Networks*", Journal of Computer Science, (2009).

A Fresh Look To Exact Solutions of Some Coupled Equations

Berat Karaagac[1,*] *Nuri Murat* Yagmurlu[2,**] *Alaattin* Esen[2,***] and *Selcuk* Kutluay[2,****]

[1]Adıyaman University, Department of Mathematics Education,Adıyaman, TURKEY
[2]Inonu University, Department of Mathematics, Malatya, TURKEY

Abstract. This manuscript is going to seek travelling wave solutions of some coupled partial differential equations with an expansion method known as Sine-Gordon expansion method. Primarily, we are going to employ a wave transformation to partial differential equation to reduce the equations into ordinary differential equations. Then, the solution form of the handled equations is going to be constructed as polynomial of hyperbolic trig or trig functions. Finally, with the aid of symbolic computation, new exact solutions of the partial differentials equations will have been found.

1 Introduction

The exact solutions of both ordinary and partial differential equations help scientists to comprehend the physical mechanism of the natural phenomena clearly. Thus, being able to obtain the exact solutions of several equations including but not limited to solitary wave solutions has become one of the main concerns of many scientists. For the sake of obtaining exact solutions, several methods and techniques have been developed can be seen in Refs. [1–7].

In the present study, we are going to seek the exact solutions of variant Boussinesq equation and coupled Klein Gordon equation via Sine-Gordon expansion method [8, 9]. Those equations have been solved by several authors using different methods and techniques in [10] and therein.

2 Description of the Sine-Gordon expansion method

In this section, we will introduce Sine-Gordon expansion method which is a well known and popular method [11, 12]. Let us take into consideration the following widely known Sine-Gordon equation

$$u_{xx} - u_{tt} = m^2 \sin u, \tag{1}$$

where $u = u(x, t)$ is the desired function and m is a constant. When the transformation $u(x, t) = U(\xi)$ in which $\xi = kx + lt$ is introduced, the Eq. (1) is reduced to a nonlinear

*e-mail: bkaraagac@adiyaman.edu.tr
**e-mail: murat.yagmurlu@inonu.edu.tr
***e-mail: alaattin.esen@inonu.edu.tr
****e-mail: selcuk.kutluay@inonu.edu.tr

ordinary differential equation given in the following form

$$U'' = \frac{m^2}{k^2 - l^2} \sin(U). \tag{2}$$

When the both sides of Eq. (2) is multiplied by U' and integrated once, the following form is obtained

$$\left[\left(\frac{U}{2}\right)'\right] = \frac{m^2}{k^2 - l^2} \sin^2\left(\frac{U}{2}\right) + K \tag{3}$$

in which the integration constant is denoted by K. When $K = 0$, $\frac{U}{2} = w(\xi)$, and $a^2 = \frac{m^2}{k^2-l^2}$ are taken, the new form of the equation (3) is obtained as follows

$$w' = a \sin(w). \tag{4}$$

When $a = 1$ is taken, the Eq. (4) , the most fundamental form of Eq. (1) is finally obtained as follows

$$w' = \sin(w). \tag{5}$$

It is obvious that the Eq. (5) is obtained from the original Sine-Gordon equation by some simplification processes. Now, Eq. (5) can be solved more easily than the original Sine-Gordon equation and its solutions can be obtained as follows

$$\sin(w) = \sin(w(\xi)) = \frac{2pe^{\xi}}{p^2e^{2\xi} + 1} |_{p=1} = \sec h(\xi), \quad \cos(w) = \cos(w(\xi)) = \frac{p^2e^{2\xi} - 1}{p^2e^{2\xi} + 1} |_{p=1} = \tanh(\xi) \tag{6}$$

where the integration constant is taken as $p \neq 0$. It is high time to generalize this approach to a nonlinear partial differential equation given as

$$F(u, u_x, u_t, u_{xt}, u_{tt}, ...) = 0. \tag{7}$$

When, again, the transformation $u(x, t) = U(\xi)$ in which $\xi = kx + lt$ is introduced, Eq. (7) is converted into the following nonlinear ordinary differential equation

$$G(U, U', U'', ...) = 0. \tag{8}$$

In a similar way followed above, it is assumed that the solution $U(\xi)$ of the nonlinear equation given in Eq. (8) can be formed as follows

$$U(\xi) = \sum_{i=1}^{N} \tanh^{i-1}(\xi) [B_i \sec h(\xi) + A_i \tanh(\xi)] + A_0. \tag{9}$$

When the Eqs. (6) are taken into consideration, Eq. (9) can also be written as follows

$$U(w) = \sum_{i=1}^{N} \cos^{i-1}(w) [B_i \sin(w) + A_i \cos(w)] + A_0. \tag{10}$$

When the value of N is determined via the homogenous balance principle, Eq. (10) is substituted into Eq. (8) and they are compared term by term; a nonlinear algebraic system is acquired. Solving this new system leads to obtain the travelling wave solutions of Eq. (7).

3 Applications

3.1 Variant Boussinesq equation

Let us consider the variant Boussinesq equations handled by Wang [13] in 1995 in the following form

$$u_{t+}(uv)_x + v_{xxx} = 0, \qquad v_t + u_x + vv_{x\cdot} = 0 \tag{11}$$

In this system, subscripts t and x denote partial derivatives with respect to time and space, respectively. In Eq. (11) when

$$u(x,t) = u(\xi), \qquad v(x,t) = v(\xi) \tag{12}$$

are taken, where $\xi = \alpha x - ct$, under these conditions, the following equations are obtained

$$-cu' + \alpha(uv)' + \alpha^3 v''' = 0, \qquad -cv' + \alpha u' + \alpha v v_x = 0.$$

By integrating the equations in Eq. (13) once, the following equations are found

$$-cu + \alpha uv + \alpha^3 v'' = r, \qquad -cv + \alpha u + \frac{\alpha}{2}v^2 = s_1$$

where r and s_1 are integration constants. When $u = \frac{c}{\alpha}v - \frac{1}{2}v^2 + \frac{s_1}{\alpha}$ are taken, and $s = s_1/\alpha$ is assumed, the following equation is obtained. When the some required simplifications are made, we obtain the following equations

$$\alpha^3 v'' - \frac{\alpha}{2}v^3 + \frac{3c}{2}v^2 + \left(\alpha s_1 - \frac{c^2}{\alpha}\right)v - (cs_1 + r) = 0. \tag{13}$$

Then using the homogenous balance condition we obtain, $N = 1$. Thus, the solution is formed as

$$u(\xi) = \frac{c}{\alpha}v(\xi) - \frac{1}{2}v(\xi)^2 + \frac{s_1}{\alpha}, \qquad v(\xi) = A_0 + A_1\tanh(\xi) + B_1 \sec h(\xi)$$

Substituting the solution given above in reduced equation given in Eq. (13), collecting the coefficient power of $\tanh(\xi)$, $\sec h(\xi)$ and $\sec h(\xi)\tanh^i(\xi)$ then equalizing each of coefficient to zero, we get the following algebraic equation systems in terms of the parameters A_0, A_1, B_1, c, s, r and α

$$\begin{aligned}
&(\sec h(\xi)\tanh(\xi))^0 : -r - \tfrac{c^2 A_0}{\alpha} + \tfrac{3cA_0^2 - \alpha A_0^3 + 3cB_1^2 - 3\alpha A_0 B_1^2}{2} - cs_1 + \alpha A_0 s_1, \quad (\tanh(\xi))^3 : 2\alpha^3 A_1 - \tfrac{\alpha A_1^3}{2} + \tfrac{3\alpha A_1 B_1^2}{2}\\
&(\sec h(\xi))^1 : -\tfrac{c^2 B_1}{\alpha} - \alpha^3 B_1 + 3cA_0B_1 - \tfrac{3\alpha A_0^2 B_1 - \alpha B_1^3}{2} + \alpha B_1 s_1, \qquad (\sec h(\xi)\tanh(\xi)) : 3cA_1B_1 - 3\alpha A_0 A_1 B_1\\
&(\tanh(\xi))^1 : \tfrac{-c^2 A_1}{\alpha} - 2\alpha^3 A_1 + 3cA_0A_1 - \tfrac{3\alpha A_0^2 A_1 + 3\alpha A_1 B_1^2}{2} + \alpha A_1 s_1, \qquad \left(\sec h(\xi)\tanh^2(\xi)\right) : \alpha^3 B_1 - \tfrac{3}{2}\alpha A_1^2 B_1\\
&(\tanh(\xi))^2 : \tfrac{3cA_1^2 - 3\alpha A_0 A_1^2 - 3cB_1^2 + 3\alpha A_0 B_1^2}{2} \qquad\qquad + \alpha^3 B_1 + \tfrac{\alpha B_1^3}{2}.
\end{aligned} \tag{14}$$

Solving the equation system simultaneously, we get the required coefficients for constructing desirable solution as follows:

$$\begin{aligned}
&\left\{A_1 = \mp 2\alpha, B_1 = 0, c = \alpha A_0, s = \tfrac{1}{2}\left(4\alpha^2 - A_0^2\right), r = \tfrac{1}{2}\left(4\alpha^3 A_0 - \alpha A_0^3 - 2cs_1\right)\right\}\\
&\left\{A_1 = 0, B_1 = 2 \mp i\alpha, c = \alpha A_0, s = \tfrac{1}{2}\left(-2\alpha^2 - A_0^2\right), r = \tfrac{1}{2}\left(-2\alpha^3 A_0 - \alpha A_0^3 - 2cs_1\right)\right\}\\
&\left\{A_1 = \mp\alpha, B_1 = -i\alpha, c = \alpha A_0, s = \tfrac{\alpha^2 - A_0^2}{2}, r = \tfrac{-2cs + \alpha^3 A_0 - \alpha A_0^3}{2}\right\}
\end{aligned} \tag{15}$$

and the solution is formed as

$$\begin{aligned}
&u_1(x,t) = 2\alpha^2 \operatorname{sech}^2(\alpha(x - A_0 t)), \qquad v_1(x,t) = A_0 \mp 2\alpha\tanh(\alpha(x - A_0 t))\\
&u_2(x,t) = \alpha^2\left(2\operatorname{sech}^2(\alpha(x - A_0 t)) - 1\right), \qquad v_2(x,t) = A_0 \mp 2i\alpha \operatorname{sech}(\alpha(x - A_0 t))\\
&u_3(x,t) = \frac{i\alpha^2}{i \mp \sinh\left(\alpha\left(x - tA_0\right)\right)}, \qquad v_3(x,t) = A_0 \mp \alpha\tanh(x\alpha - \alpha A_0 t) - i\alpha \sec h(x\alpha - \alpha A_0 t)
\end{aligned}$$

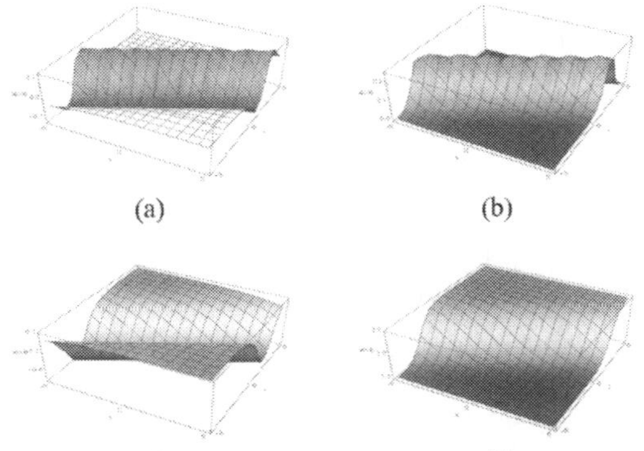

(a) (b)

(c) (d)

Figure 1: Solution of Variant Boussinesq equation for $\alpha = c = 0.5, A_0 = 2$

Additionally, the graphical representation of u_3 and v_3 is illustrated in Figure 1 for selected values of some parameters. In the literature; different types of solutions can be found [10, 14] and there in. However, within the knowledge of the authors, u_3 and v_3 solutions are new exact solutions of the Variant Boussinesq equation.

3.2 Coupled Klein Gordon equation

Now, we are going to consider coupled Klein Gordon equation in the form

$$u_{xx} - u_{tt} - u + 2u^3 + 2uv = 0, \qquad v_x - v_t - 4uu_t = 0 \tag{16}$$

where $u(x,t) = u(\xi)$ and $v(x,t) = v(\xi)$ and $\xi = \alpha x - ct$. After these assumptions we can write the following equations,

$$\alpha^2 u'' - c^2 u'' - u + 2u^3 + 2uv = 0, \qquad \alpha v' + cv' + 4cu\,u' = 0 \tag{17}$$

If we take the integral of the second equation and make some arrangements, we obtain $v = \frac{r-2cu^2}{(\alpha+c)}$ where r is an integral constant. If we write this newly obtained equation in the first equation in Eq. (17), we obtain $(\alpha^2 - c^2)u'' + (2 - \frac{4c}{\alpha+c})u^3 + (\frac{2r}{\alpha+c})u = 0$. Then using the homogenous balance condition, we obtain $n = 1$. Following the same procedure presented in the first example, we get algebraic equations after solving the system, required coefficients are given, respectively, by

$$\begin{array}{l} \left\{A_0 = 0, A_1 = -i(\alpha + c), B_1 = 0, r = \frac{1}{2}\left(2\alpha^3 + \alpha - 2c^3 - 2\alpha c^2 + 2\alpha^2 c + c\right)\right\} \\ \left\{A_0 = 0, A_1 = 0, B_1 = -\alpha - c, r = \frac{1}{2}\left(-\alpha^3 + \alpha + c^3 + \alpha c^2 - \alpha^2 c + c\right)\right\} \\ \left\{A_0 = 0, A_1 = \frac{1}{2}i(\alpha + c), B_1 = \frac{1}{2}(-\alpha - c), r = \frac{1}{4}\left(\alpha^3 + 2\alpha - c^3 - \alpha c^2 + \alpha^2 c + 2c\right)\right\} \end{array} \tag{18}$$

Thus, the solutions for $u\,(x,t)$ and $v\,(x,t)$ are obtained as follows

$$u_1(x,t) = i(\alpha + c)\tanh(ct - \alpha x)$$
$$v_1(x,t) = \alpha^2 - c^2 + 2c(\alpha + c)\tanh^2(ct - \alpha x) + \frac{1}{2}$$

$$u_2(x,t) = -(\alpha + c)\text{sech}(ct - \alpha x)$$
$$v_2(x,t) = \frac{1}{2}\left(-\alpha^2 + c^2\right) - 2c(\alpha + c)\text{sech}^2(ct - \alpha x) + \frac{1}{2}$$

$$u_3(x,t) = -\frac{1}{2}(\alpha + c)(\text{sech}(ct - \alpha x) - i\tanh(ct - \alpha x))$$
$$v_3(x,t) = \frac{1}{4}\left(2 + (\alpha + c)\left(\alpha + \frac{4ic}{\sinh(ct-\alpha x) - i} + c\right)\right)$$

Figure 2 shows selected solutions graphically. In the literature; different types of solutions related with Variant Boussinesq and Coupled Klein Gordon equations can be found [10, 14] and there in. However, within the knowledge of the authors, u_3 and v_3 solutions are new exact solutions of the considered equations.

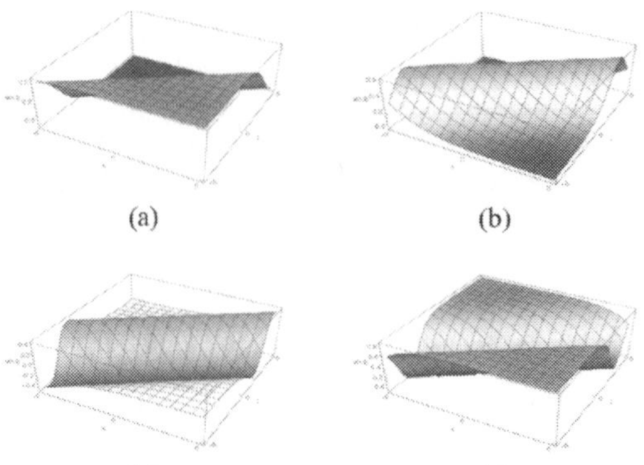

(a) (b)

(c) (d)

Figure 2: Solution of Coupled Klein equation for $\alpha = 0.5, c = 0.75$

4 Conclusion

The Sine-Gordon expansion method has been successfully applied to Variant Boussinesq equation and Klein Gordon equation. We have constructed the solutions in terms of hyperbolic functions. Also, the behavior of the newly obtained solutions are illustrated for some special values of the parameters. The method is quite efficient and practically well suited for use in obtaining the exact solutions of various partial differential equations.

References

[1] H. M. Baskonus, H. Bulut, Waves in Random and Complex Media, **26(4)**, 613-625, (2016)

[2] , M. J. Ablowitz, X. D. Luo, and Z. H. Musslimani, Journal of Mathematical Physics, **59.1**: 011501.(2018)

[3] X. H. Zhao, B. Tian, X.Y. Xie, X.-Y. Wu, Y. Sun and Y.-J. Guo, Waves in Random and Complex Media, **28.2**, 356-366,(2018)

[4] H. M. Baskonus, G. Yel, and H. Bulut , AIP Conference Proceedings, **1863: 1**, 2017.

[5] M. T. Darvishi, S. Ahmadian, S. B. Arbabi,M. Najafi , Optical and Quantum Electronics **50.1:32**,1-20,(2018)

[6] B. Zengin, H. Yaraneri, and S. Korunur, The European Physical Journal B **85.11**, 388-392, (2012).

[7] H. M.Baskonus, Nonlinear Dynamics, **86(1)**, 177–183, (2016)

[8] H. Bulut, T. A. Sulaiman and H. M. Baskonus, Optical and Quantum Electronics, **48, 12(564)**, 1-14, (2016)

[9] H. M. Baskonus, T. A. Sulaiman and H. Bulut,Optik- International Journal for Light and Electron Optics, **131**, 1036–1043, (2017)

[10] K. Khan, M. A. Akbar, SpringerPlus, **3: 324**, 1-17,(2014)

[11] H. M. Baskonus, T. A. Sulaiman, H. Bulut, Indian Journal of Physics,**91(10)**,1237–1243,(2017)

[12] C. Cattani, T. A. Sulaiman, H. M. Baskonus, H. Bulut, European Physical Journal Plus, **133(228)**, 1-12, (2018)

[13] , M. L. Wang, Physics Letters A, **199:3-4**, 169-172, (1995).

[14] M. Chen, Applied mathematics letters, **11.5**, 45-49.(1998)

ON CHARACTERIZATION OF OPEN SETS IN EUCLIDEAN SPACES

Firudin Kh. Muradov*
*Department of Mathematics, Near East University,Nicosia, North Cyprus, Mersin-10, Turkey

Abstract

A ternary semigroup is a nonempty set T together with a ternary operation $[abc]$ satisfying the associative law $[[abc]\,de] = [a\,[bcd]\,e] = [ab\,[cde]]$ for all $a, b, c, d, e \in T$. A map f between topological spaces X and Y is called open if the image of each set open in X is open in Y. The purpose of this paper is to give an abstract characterization of the ternary semigroups of open maps defined on open sets in Euclidean n-spaces.

1 Introduction

A ternary semigroup is a nonempty set T together with a ternary operation $[abc]$ satisfying the associative law $[[abc]\,de] = [a\,[bcd]\,e] = [ab\,[cde]]$, $\forall a, b, c, d, e \in T$. The theory of ternary algebraic systems was introduced by Lehmer in [1]. Sioson [2] studied ternary semigroups with special reference to ideals and radicals. Dutta [4], Santiago [5] developed regular and completely regular ternary semigroups.

2 An Abstract Charcterization of Ternary Semigroups of Open Maps

Let R^n be the Euclidean n-space. The collection of all open n-cubes in R^n forms a basis for a topology on R^n called the standard topology on R^n. Let X_1 and X_2 be two open subsets of R^n, $(n > 1)$ and $\varphi : X_1 \to X_2$, $\phi : X_2 \to X_1$. Denote the set of all pairs of (φ, ϕ) of open maps between X_1 and X_2 by $O\,(X_1, X_2)$. The set $O\,(X_1, X_2)$ is a ternary semigroup with respect to the ternary operation $[(\varphi_1, \phi_1)\,(\varphi_2, \phi_2)\,(\varphi_3, \phi_3)] = (\varphi_1\phi_2\varphi_3, \phi_1\varphi_2\phi_3)$.

Theorem 1 *Let X_1 and X_2 be open subsets of R^n and let Y_1 and Y_2 be open subsets of R^m, $(n, m > 1)$. The ternary semigroups $O\,(X_1, X_2)$ and $O\,(Y_1, Y_2)$ are isomorphic if and only if the spaces X_i and Y_i are homeomorphic $(i = 1, 2)$.*

*firudin.muradov@neu.edu.tr

Proof. It is obvious that if X_i and Y_i are homeomorphic $(i = 1, 2)$ then the ternary semigroups $O(X_1, X_2)$ and $O(Y_1, Y_2)$ are isomorphic. Specifically, if h_i is a homeomorphism from X_i to Y_i $(i = 1, 2)$ and $(\varphi, \phi) \in O(X_1, X_2)$ then the map $\theta(\varphi, \phi) \in O(Y_1, Y_2)$ defined by $\theta(\varphi, \phi) = (h_2 \circ \varphi \circ h_1^{-1}, h_1 \circ \phi \circ h_2^{-1})$ is an isomorphism from $O(X_1, X_2)$ to $O(Y_1, Y_2)$. The proof of the necessary condition follows from Lemmas 2-6. Throughout this paper, the symbol θ denotes the isomorphism from $O(X_1, X_2)$ onto $O(Y_1, Y_2)$. ■

Lemma 2 *Let X_1 and X_2 be two open subsets of R^n and let z be any point in X_1. Then, there exist open maps $f, g : X_1 \to X_2$ such that $f(z) \neq g(z)$ but $f(x) = g(x)$ for all $x \in X_1 \backslash \{z\}$.*

Proof. Let Z_1 and Z_2 be open n-cubes in R^n. The open n-cube is the set of all points $(z_1, ..., z_n) \in R^n$ such that $a_i < z_i < b_i$ for $i = 1, 2, ..., n$, i.e. Z_1 has the form $B_1 \times B_2 \times ... \times B_n$, where $B_1, B_2, ..., B_n$ are open intervals in R. For $z_1 \in B_1, z_2 \in B_2, ..., z_n \in B_n$ the sets of the form $z_1 \times B_2 \times ... \times B_n$, $B_1 \times z_2 \times ... \times B_n$,..., $B_1 \times B_2 \times ... \times z_n$ are called coordinate planes through the point $(z_1, ..., z_n)$. Analogously to the one dimensional case we use the procedure of constructing an open map f from n-cube to n-cube. Divide each coordinate interval $B_i, (i = 1, 2, ..., n)$ into 3 equal parts and the open n-cube bounded by the middle thirds map homeomorphically onto Z_2. Then divide the coordinate intervals of the remaining n-cubes into 3 equal parts and map the open n-cubes bounded by the middle thirds homeomorphically onto Z_2. Continuing this process infinitely we map countable number of open n-cubes onto Z_2. The constructed map is open.

Now let X_1 be an open subset in $R^n$and let $z \in X_1$. Assume without loss of generality that X_1 and X_2 are bounded. Let Z_1 be an open n-cube containing X_1 and let Z_2 be an open n-cube that is contained in X_2. Next we define the maps f and g at the point $z = (z_1, ..., z_n) \in X_1$ arbitrarily such that $f(z) \neq g(z)$. Determine the values of f and g on the coordinate planes $z_1 \times B_2 \times ... \times B_n$, $B_1 \times z_2 \times ... \times B_n$,..., $B_1 \times B_2 \times ... \times z_n$ through the point z that divide Z_1 at the point z into 2^n n-subcubes, to be a fixed point in Z_2. If we determine the values of f and g on the open n-subcubes applying the above procedure, then the maps f and g are open and we have $f(z) \neq g(z)$ but $f(x) = g(x)$ for all $x \in Z_1 \backslash \{z\}$. The restrictions of the maps f and g to X_1 are open and we have $f(z) \neq g(z)$ but $f(x) = g(x)$ for all $x \in X_1 \backslash \{z\}$. ■

Lemma 3 *Let $(\varphi_1, \phi_1), (\varphi_2, \phi_2)$ be arbitrary elements of $O(X_1, X_2)$. For every $f, g : X_1 \to X_2$ and $h_1, h_2 : X_2 \to X_1$ the condition*

$$[(f, h_1)(f, h_2)(\varphi_1, \phi_1)] = [(g, h_2)(g, h_1)(\varphi_1, \phi_1)] \tag{1}$$

implies

$$[(f, h_1)(f, h_2)(\varphi_2, \phi_2)] = [(g, h_2)(g, h_1)(\varphi_2, \phi_2)]$$

is necessary and sufficient for $\varphi_2(X_1) \subseteq \varphi_1(X_1)$ and $\phi_2(X_2) \subseteq \phi_1(X_2)$.

Proof. If the condition $\varphi_2(X_1) \subseteq \varphi_1(X_1)$ is satisfied, then for every $x \in X_1$ there exists a point $z \in X_1$ such that $\varphi_2(x) = \varphi_1(z)$. Then

$$fh_2\varphi_2(x) = fh_2\varphi_1(z) = gh_1\varphi_1(z) = gh_1\varphi_2(x).$$

If the condition $\phi_2(X_2) \subseteq \phi_1(X_2)$ is satisfied, then for every $x \in X_2$ there exists a point $z \in X_2$ such that $\phi_2(x) = \phi_1(z)$. Then

$$h_1f\phi_2(x) = h_1f\phi_1(z) = h_2g\phi_1(z) = h_2g\phi_2(x).$$

So the condition 1 holds.

Conversely, let the condition 1 hold for some $(\varphi_1, \phi_1), (\varphi_2, \phi_2) \in O(X_1, X_2)$. Suppose that $\varphi_2(X_1) \setminus \varphi_1(X_1)$ is not empty. It follows from Lemma 2 that for any point $y = \varphi_2(x_1)$ in $\varphi_2(X_1) \setminus \varphi_1(X_1)$ there exist open maps $h_1, h_2 : X_2 \to X_1$ such that $h_1(y) \neq h_2(y)$ but $h_1(x) = h_2(x)$ for all $x \in X_2 \setminus \{y\}$. Then for every $x \in X_1$ we have $h_1\varphi_1(x) = h_2\varphi_1(x)$ and $fh_1\varphi_1(x) = fh_2\varphi_1(x)$ for each open map $f : X_1 \to X_2$. But for $y = \varphi_2(x_1)$ in $\varphi_2(X_1) \setminus \varphi_1(X_1)$ we have $h_1\varphi_2(x_1) = h_1(y) \neq h_2(y) = h_2\varphi_2(x_1)$ and $fh_1\varphi_2(x_1) \neq fh_2\varphi_2(x_1)$ for some open map $f : X_1 \to X_2$ which contradicts to 1. ∎

Lemma 4 *Let X_1 and X_2 be open subsets of R^n and let Y_1 and Y_2 be open subsets of R^m, $(n, m > 1)$. Suppose that θ is an isomorphism from $O(X_1, X_2)$ to $O(Y_1, Y_2)$ and $(\varphi_1, \phi_1), (\varphi_2, \phi_2) \in O(X_1, X_2)$. If $\varphi_1(X_1) = \varphi_2(X_1)$ and $\phi_1(X_2) = \phi_2(X_2)$ then $(\theta\varphi_1)(Y_1) = (\theta\varphi_2)(Y_1)$ and $(\theta\phi_1)(Y_2) = (\theta\phi_2)(Y_2)$.*

Proof. Suppose that $\varphi_2(X_1) \subseteq \varphi_1(X_1)$. If $f'h_2'(\theta\varphi_1)) = g'h_2'(\theta\varphi_1)$ for some elements $f', g', h_1', h_2' \in O(Y_1, Y_2)$ then there exist $f, g, h_1, h_2 \in O(X_1, X_2)$ such that $f' = \theta f$, $g' = \theta g$, $h_1' = \theta h_1$ and $h_2' = \theta h_2$. Then $(\theta f)(\theta h_2)(\theta\varphi_1) = (\theta g)(\theta h_2)(\theta\varphi_1)$ and since θ is an isomorphism, $\theta(fh_2\varphi_1) = \theta(gh_2\varphi_1)$ and $fh_2\varphi_1 = gh_2\varphi_1$. We have $fh_2\varphi_2 = gh_2\varphi_2$, by Lemma 3. Again, since θ is an isomorphism, then $(\theta f)(\theta h_2)(\theta\varphi_2) = (\theta g)(\theta h_2)(\theta\varphi_2)$ and therefore $f'h_2'(\theta\varphi_2)) = g'h_2'(\theta\varphi_2)$. Because $f'h_2'(\theta\varphi_2)) = g'h_2'(\theta\varphi_2)$ is true for every $f', g', h_1', h_2' \in O(Y_1, Y_2)$ satisfying the condition $f'h_2'(\theta\varphi_1)) = g'h_2'(\theta\varphi_1)$ it follows from Lemma 3 that $(\theta\varphi_2)(Y_1) \subseteq (\theta\varphi_1)(Y_1)$. In the same way, we could show that if $\varphi_1(X_1) \subseteq \varphi_2(X_1)$ then $(\theta\varphi_1)(Y_1) \subseteq (\theta\varphi_2)(Y_1)$. ∎

Lemma 5 *Let X_1 and X_2 be open subsets of R^n and let A and B be open subsets of X_2 and X_1, respectively. Then there exist open maps $(\varphi, \phi) \in O(X_1, X_2)$ such that $\varphi(X_1) = A$ and $\phi(X_2) = B$.*

Proof. Let X_1 and X_2 be open subsets of R^n and let A be an open subset of X_2. Assume without loss of generality that X_1 and X_2 are bounded. Let R_2 be a closed n-cube contained in A and let R_1 be a closed n-cube contained in X_1. If E_1 is a closed n-ball containing $X_1 \setminus R_1$ and if E_2 is a closed n-ball that is contained in $A \setminus R_2$, then there exists a homeomorphism from E_1 onto E_2. Denote by g_1 the restriction of this homeomorphism to $X_1 \setminus R_1$. Clearly the map g_1 is open. Then construct an open map from $Int R_1$ onto $Int R_2$ using the same procedure as in Lemma 2. Denote by g_2 the extension of this map from $IntR_1$ to R_1, obtained by mapping the boundary of R_1 to a fixed point in $IntR_2$. We next define the map $f : X_1 \to A$ by

$$f(x) = \begin{cases} g_2(x), & \text{if } \ x \in R_1 \\ g_1(x), & \text{if } \ x \in X_1 \setminus R_1 \end{cases}$$

Suppose that V is an open set in X_1. If V is in $X_1 \setminus R_1$ then $f(V) = g_1(V)$ and therefore $f(V)$ is an open set of X_2. If $V \subset R_1$ then $V \subset IntR_1$ and then $f(V) = g_2(V)$ and therefore $f(V)$ is the open set of X_2. Now suppose that V has nonempty intersections with R_1 and $X_1 \setminus R_1$, i.e., $V \cap R_1 \neq \varnothing$ and $V \cap (X_1 \setminus R_1) \neq \varnothing$. Since the set $V \cap (X_1 \setminus R_1)$ is open in X_1 and is contained in $X_1 \setminus R_1$, the set $f(V \cap (X_1 \setminus R_1))$ is the open set of X_2. If the set $V \cap R_1$ does not contain boundary points of R_1, then $V \cap R_1 \subset IntR_1$. By the definition of f we have $f(V \cap R_1) = g_2(V \cap R_1)$ and therefore $f(V \cap R_1)$ is the open set of X_2. If the set $V \cap R_1$ contains boundary points of R_1, then at each neighborhood of these points there is an open n-cube that is homeomorphically mapped onto $IntR_2$. So, $f(V \cap R_1) = IntR_2$, that is $f(V \cap R_1)$ is open and therefore the map f is open.

Now, let us change f such that the image of the new constructed map φ was A, i.e. $\varphi(X_1) = A$. Denote by $\widetilde{R_1}$ the sequence of the decreasing n-cubes in R_1 converging to some vertex of R that are mapped homeomorphically onto $IntR_2$. Since A has a countable base, we can map these n-cubes homeomorpically onto base sets of A. Define $\varphi : X_1 \to A$ as follows. On the sets $X_1 \setminus R_1$ and $R_1 \setminus \widetilde{R_1}$, the values of φ coincide with the values of f. On the set $\widetilde{R_1}$ the image of φ is A, that is we map the n-cubes of $\widetilde{R_1}$ homeomorpically onto base sets of A . Show that the map φ is an open map. Suppose that V is an open set in X_1. If V is in $X_1 \setminus R_1$ then $\varphi(V) = f(V)$ and therefore $\varphi(V)$ is an open set of X_2. If $V \subset R_1 \setminus \widetilde{R_1}$ then $\varphi(V) = f(V)$ and therefore $\varphi(V)$ is the open set of X_2. Now suppose that V contains a point from boundary of $\widetilde{R_1}$. Then at each neighborhood of this point there is an open n-cube that is homeomorphically mapped onto $IntR_2$. Since the set V is open, there exists a neighborhood of this boundary point contained in V and therefore this neighborhood contains an n-cube from $R_1 \setminus \widetilde{R_1}$ that is homeomorphically mapped onto $IntR_2$. In this case $\varphi(V) = IntR_2 \cup \widetilde{A}$, where $\widetilde{A}$ is an open subset of A. Finally, if V is in $\widetilde{R_1}$, then because the n-cubes of $\widetilde{R_1}$ are mapped homeomorpically onto base sets of A it follows that $\varphi(V)$ is the open set of X_2. Consequently the map φ is open and $\varphi(X_1) = A$. In the same way we can show that there is an open map $\phi : X_2 \to B$ such that $\varphi(X_2) = B$. ■

Let X be a topological space. Let $C(X)$ denote the family of all open sets of X. The family $C(X)$ is a complete distributive lattice if set inclusion is taken as the ordering. The supremum is given by the union of open sets and the infimum by the interior of the intersection. Two topological spaces X and Y are said to be lattice-equivalent if there is a bijective map from $C(X)$ to $C(Y)$ which together with its inverse is order-preserving. In his paper [3], Thron proved that for T_D-spaces any lattice-isomorphism can be induced by a homeomorphism.

Lemma 6 *Let X_1 and X_2 be open sets of R^n and let Y_1 and Y_2 be open sets of R^m, $(n, m > 1)$. The lattices $C(X_i)$ and $C(Y_i)$ are lattice-isomorphic, $(i = 1, 2)$.*

Proof. Let U_1 and U_2 be open subsets of X_1 and X_2, respectively. By Lemma 5 there exists $(\varphi, \phi) \in O(X_1, X_2)$ such that $\varphi(X_1) = U_2$ and $\phi(X_2) = U_1$. Since the ternary semigroups $O(X_1, X_2)$ and $O(Y_1, Y_2)$ are isomorphic we can find $(\varphi', \phi') \in O(Y_1, Y_2)$ such that $\theta\varphi = \varphi'$ and $\theta\phi = \phi'$. Suppose that $\varphi'(Y_1) = U_2'$ and $\phi'(Y_2) = U_1'$. Let us define maps $\xi_i : C(X_i) \to C(Y_i)$, $(i = 1, 2)$ by mapping to each open set $U_i \subset X_i$ the set $U_i' \subset Y_i$. The maps ξ_i do not depend on the choice of $(\varphi, \phi) \in O(X_1, X_2)$. Indeed , if $\varphi_1(X_1) = U_2$ and $\phi_1(X_2) = U_1$ for some $(\varphi_1, \phi_1) \in O(X_1, X_2)$ then Lemma 4 says that $(\theta\varphi)(Y_1) = (\theta\varphi_1)(Y_1) = U_2'$ and $(\theta\phi)(Y_2) = (\theta\phi_1)(Y_2) = U_1'$. Let U and V be any two different open subsets of X_2. By Lemma 5 there are two open maps φ_1 and φ_2 from X_1 to X_2 such that $\varphi_1(X_2) = U$ and $\varphi_2(X_2) = V$. Since the ternary semigroups $O(X_1, X_2)$ and $O(Y_1, Y_2)$ are isomorphic the maps $\theta\varphi_1$ and $\theta\varphi_2$ are different and it follows from Lemma 4 that $(\xi_2\varphi_1)Y_2 \neq (\xi_2\varphi_2)Y_2$. Hence ξ_2 is a one-to-one map between $C(X_2)$ and $C(Y_2)$. In the same way we can show that the map ξ_1 is a one-to-one map between $C(X_1)$ and $C(Y_1)$.Now suppose that U_i' is an arbitrary open set in Y_i for $i = 1, 2$. From isomorphism between $O(X_1, X_2)$ and $O(Y_1, Y_2)$ and Lemma 4 there exists an open set $U_i \subset X_i$ such that $\xi_i(U_i) = U_i'$. Again using Lemma 4 we can conclude ξ_i saves order, that is, if $U_i \subseteq V_i$ then $\xi_i(U_i) \subseteq \xi_i(V_i)$ for $i = 1, 2$. ■

From Theorem 2.1 of [3] it follows that the open sets X_i and Y_i, $(i = 1, 2)$ are homeomorphic.

3 Conclusion

In this paper we consider ternary semigroups and give an abstract characterization of ternary semigroups of open maps defined on open sets of Euclidean n-spaces.

References

[1] D.H.Lehmer, *A ternary analogue of Abelian groups*, Am .J. Math. 54, (1932), 329-338

[2] F.M.Sioson, *Ideal theory in ternary semigroups*, Math. Jpn. 10 (1965), 63-84.

[3] W.J. Thron, *Lattice-equivalence of topological spaces*, Duke Math. J. Volume 29, No. 4 (1962), 671-679.

[4] T.K.Dutta, S.Kar, *On regular ternary semirings*, Advances in Algebra, Proceedings of the ICM Satellite conference in algebra and related topics, World Scientific, (2003), 343-355.

[5] M.L.Santiago, S.Sri Bala, *Ternary semigroups*, Semigroup Forum vol. 81, 380-388 (2010)

Novel Recursive Approximation for Fractional Nonlinear Equations within Caputo-Fabrizio Operator

Mehmet Yavuz[1,*]

[1] Department of Mathematics-Computer Sciences, Necmettin Erbakan University, Konya, Turkey

Abstract. This study displays a novel method for solving time-fractional nonlinear partial differential equations. The suggested method namely Laplace homotopy method (LHM) is considered with Caputo-Fabrizio fractional derivative operator. In order to show the efficiency and accuracy of the mentioned method, we have applied it to time-fractional nonlinear Klein-Gordon equation. Comparisons between obtained solutions and the exact solutions have been made and the analysis shows that recommended solution method presents a rapid convergence to the exact solutions of the problems.

1 Introduction

In the last decades, fractional calculus were applied to various fields of mathematical and physical analysis such as modelling some common viruses, estimation financial processes, diffusion, relaxation processes and so on [1-11]. Especially, modelling the nonlinear problems in these areas with fractional differential equations was extensively used. In order to find out the efficiency and superiority of the fractional differential equations, a lot of studies were emerged by some scientists [12-17]. On the other hand, some important fractional derivative operators have been developed such as Caputo-Fabrizio [18] and Atangana-Baleanu [19]. These operators are very important to model the complex nonlinear fractional dynamical systems and to solve them. In recent years, some scientists have preferred these operators to other fractional operators which have a singularity [11, 20-23]. In this study, we also use the Caputo-Fabrizio fractional operator to solve nonlinear Klein-Gordon equation by considering the Laplace homotopy method.

2 Caputo-Fabrizio Operator and its Main Properties

Definition 1. The Caputo-Fabrizio fractional derivative in Caputo sense (CFC) is defined by [18]

* Corresponding author: mehmetyavuz@konya.edu.tr

$$ {}^{CFC}_{\ 0}D_t^{\beta} f(t) = \frac{M(\beta)}{1-\beta} \int_0^t f'(\tau) \exp\left[-\frac{\beta(t-\tau)}{1-\beta}\right] d\tau, \quad 0 < \beta \le 1, \tag{1} $$

where $M(\beta)$ is a normalization function such that $M(0) = M(1) = 1$. The CF definition can be applied for the functions which do not belong to $H^1(a,b)$ and the kernel does not have singularity for $t = \tau$ unlike the Caputo fractional derivative.

Definition 2. The Laplace transform (LT) of ${}^{CFC}_{\ 0}D_t^{\beta} f(t)$ is given by [18, 24]:

$$ \begin{aligned} \mathcal{L}\left\{{}^{CFC}_{\ 0}D_t^{(\beta+n)} f(t)\right\}(s) &= \frac{1}{1-\beta} \mathcal{L}\left\{f^{(\beta+n)}(t)\right\} \mathcal{L}\left\{\exp\left[-\frac{\beta t}{1-\beta}\right]\right\} \\ &= \frac{s^{n+1} L\{f(t)\} - s^n f(0) - s^{n-1} f'(0) - \cdots - f^{(n)}(0)}{s + \beta(1-s)}. \end{aligned} \tag{2} $$

Definition 3. Let $0 < \beta < 1$. The CFC fractional integral of order β is defined by [25, 26]:

$$ {}^{CFC}_{\ 0}I^{\beta} f(t) = \frac{2(1-\beta)}{(2-\beta)M(\beta)} f(t) + \frac{2\beta}{(2-\beta)M(\beta)} \int_0^t f(\tau) d\tau, \quad t \ge 0. \tag{3} $$

3 Role of the Suggested Fractional Operator in the Fundamental Method

Consider the following nonlinear FPDE:

$$ {}^{CFC}D_t^{\beta} \phi(x,t) + \mu(\phi(x,t)) + \omega(\phi(x,t)) = v(x,t), \quad (x,t) \in [0,1] \times [0,T], \quad 0 < \beta \le 1, \tag{4} $$

with initial condition

$$ \phi(x,0) = k(x), \tag{5} $$

where ${}^{CFC}D_t^{\beta}$ shows the Caputo-Fabrizio-Caputo fractional derivative of order β, $\omega(\phi(x,t))$ is a non-linear operator and $v(x,t)$ is a known function and $\phi(x,t)$ is an unknown function. Using the LT of the CF operator we define $\mathcal{L}\{\phi(x,t)\}(s) = \Delta(x,s)$. Then we have the following homotopies:

$$ \Delta(x,s) = \left(\frac{\beta + s(1-\beta)}{s}\right)\left[\tilde{v}(x,s) - \mathcal{L}\left\{\mu(\phi(x,t)) + \omega(\phi(x,t))\right\}\right] + \frac{1}{s}(\phi_0(x)). \tag{6} $$

where $\Delta(x,s) = \mathcal{L}\{\phi(x,t)\}$ and $\tilde{v}(x,s) = \mathcal{L}\{v(x,t)\}$. Applying the suggested method we obtain the approximate solution as

$$ \Delta(x,s) = \sum_{m=0}^{\infty} p^m \Delta_m(x,s), \quad m = 0,1,2,..., \tag{7} $$

where $\Delta_m(x,s)$ values are known functions. The nonlinear term $\omega(\phi(x,t))$ in Eq. (4) can be evaluate as

$$ \omega(\phi(x,t)) = \sum_{m=0}^{\infty} p^m \eta_m(x,t), \tag{8} $$

and the polynomials $\eta_m(x,t)$ are given in [27] as

$$\eta_m(\phi_0,\phi_1,\ldots,\phi_m)=\frac{1}{m!}\frac{\partial^m}{\partial p^m}\left[\omega\left(\sum_{i=0}^{\infty}p^i\phi_i\right)\right]_{p=0},\quad m=0,1,2,\ldots. \tag{9}$$

Substituting Eqs.(8) and (9) into Eq.(6) we construct the solution series for the main problem.

$$\sum_{m=0}^{\infty}p^m\Delta_m(x,s)=-p\left[\left(\frac{\beta+s(1-\beta)}{s}\right)\left(\mathcal{L}\left\{\sum_{m=0}^{\infty}p^m\Delta_m(x,t)+\sum_{m=0}^{\infty}p^m\eta_m(x,t)\right\}\right)\right]$$
$$+\frac{1}{s}(\phi_0(x))+\left(\frac{\beta+s(1-\beta)}{s}\right)\tilde{v}(x,s). \tag{10}$$

By equating the same powers of p, we construct the homotopy steps of as

$$p^0:\Delta_0(x,s)=\frac{1}{s}(\phi_0(x))+\left(\frac{\beta+s(1-\beta)}{s}\right)\tilde{v}(x,s),$$
$$p^1:\Delta_1(x,s)=-\left(\frac{\beta+s(1-\beta)}{s}\right)\mathcal{L}\{\mu(\phi_0(x,t))+\omega(\phi_0(x,t))\},$$
$$p^2:\Delta_2(x,s)=-\left(\frac{\beta+s(1-\beta)}{s}\right)\mathcal{L}\{\mu(\phi_1(x,t))+\omega(\phi_1(x,t))\}, \tag{11}$$
$$\vdots$$
$$p^{n+1}:\Delta_{n+1}(x,s)=-\left(\frac{\beta+s(1-\beta)}{s}\right)\mathcal{L}\{\mu(\phi_n(x,t))+\omega(\phi_n(x,t))\}.$$

Considering $p\to 1$, we can see Eq.(11) gives the approximate solution for problem (10) and the solution is given by

$$\Im_n(x,s)=\sum_{j=0}^{n}\Delta_j(x,s). \tag{12}$$

Taking the inverse Laplace transform of the sum (12), we have the approximate solution

$$\phi_{approx}(x,t)\cong\phi_n(x,t)=\mathcal{L}^{-1}\{\Im_n(x,s)\}.$$

4 Solution of Nonlinear Fractional Klein-Gordon Equation

Consider the fractional nonlinear homogeneous Klein-Gordon equation (KGE) [28]

$${}^{CFC}D_t^{\beta}\phi(x,t)-\phi_{xx}(x,t)+\phi^2(x,t)=0,\quad x\in R,\quad t>0,\quad 0<\beta\le 1, \tag{13}$$

subject to the initial condition

$$\phi(x,0)=1+\sin x,\quad x\in R. \tag{14}$$

Firstly, applying the mentioned method in Section 3, to Eq.(13) with the initial condition (14), we obtain the approximate solution as

$$\Delta(x,s)=\left(\frac{\beta+s(1-\beta)}{s}\right)\mathcal{L}\{\phi_{xx}(x,t)-\phi^2(x,t)\}+\frac{1}{s}[\phi(x,0)]. \tag{15}$$

Then, we consider the inverse Laplace transform of Eq.(15), we have

$$\phi(x,t)=\phi(x,0)+\mathcal{L}^{-1}\left\{\left(\frac{\beta+s(1-\beta)}{s}\right)\mathcal{L}\{\phi_{xx}(x,t)-\phi^2(x,t)\}\right\}. \tag{16}$$

The nonlinear polynomials $\eta_m(x,t)$ are given in [27] as

$$\begin{aligned}
&\eta_0(\phi)=(1+\sin x)^2,\\
&\eta_1(\phi)=-2(1+\sin x)(\sin^2 x+3\sin x+1)(\beta t-\beta+1),\\
&\eta_2(\phi)=\left((\sin^2 x+3\sin x+1)(\beta t-\beta+1)\right)^2+2(1+\sin x)(3\sin x-2\cos 2x)(\beta t-\beta+1) \qquad (17)\\
&\qquad +4(1+\sin x)^2(\sin^2 x+3\sin x+1)\left(\beta^2-2\beta+1+\frac{\beta^2t^2}{2}-2(\beta^2-\beta)t\right),\\
&\vdots
\end{aligned}$$

Moreover, we use the series solution in Eq.(10), we can write

$$\sum_{m=0}^{\infty}p^m\Delta_m(x,s)=-p\left[\left(\frac{\beta+s(1-\beta)}{s}\right)\mathcal{L}\left\{\sum_{m=0}^{\infty}p^m\Delta_m(x,t)+\sum_{m=0}^{\infty}p^m\eta_m(x,t)\right\}\right]+\frac{1}{s}(\phi_0(x)). \quad (18)$$

By comparing the coefficients of powers of p, we construct the homotopy steps as in Eq.(11) and taking the inverse LT of obtained values, we have the iteration steps as

$$\begin{aligned}
&\phi_0(x,t)=1+\sin x,\\
&\phi_1(x,t)=-(\sin^2 x+3\sin x+1)(\beta t-\beta+1),\\
&\phi_2(x,t)=2(1+\sin x)(\sin^2 x+3\sin x+1)\left(\beta^2-2\beta+1+\frac{\beta^2t^2}{2}-2(\beta^2-\beta)t\right)\\
&\qquad +(3\sin x-2\cos 2x)(\beta t-\beta+1), \qquad (19)\\
&\vdots\\
&\phi_{n+1}(x,t)=\mathcal{L}^{-1}\left\{\left(\frac{\beta+s(1-\beta)}{s}\right)\mathcal{L}\{\mu(\phi_n(x,t))+\omega(\phi_n(x,t))\}\right\}.
\end{aligned}$$

Finally, considering the sum (19), we have the approximate solution

$$\begin{aligned}
\phi_{approx}(x,t)\cong\phi_n(x,t)&=1+\sin x-(\sin^2 x+2\cos 2x+1)(\beta t-\beta+1)\\
&+2(1+\sin x)(\sin^2 x+3\sin x+1)\left(\beta^2-2\beta+1+\frac{\beta^2t^2}{2}-2(\beta^2-\beta)t\right)+\ldots.
\end{aligned}$$

Following figures show the solution functions for some different values of x,t and fractional parameter β.

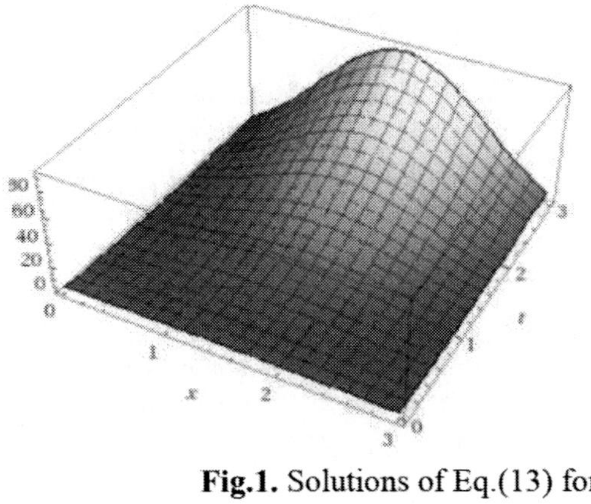

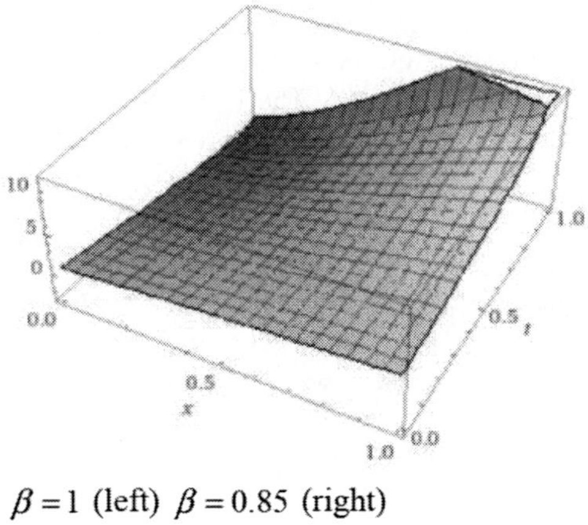

Fig.1. Solutions of Eq.(13) for $\beta=1$ (left) $\beta=0.85$ (right)

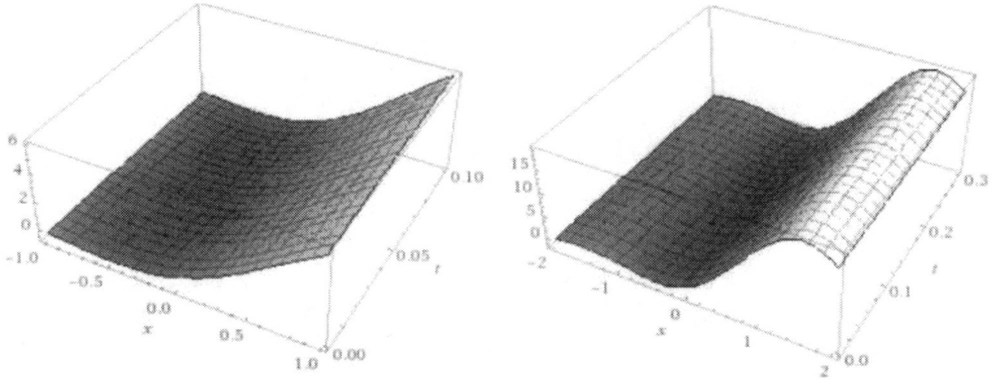

Fig.2. Solutions of Eq.(13) for $\beta = 0.5$ (left) $\beta = 0.1$ (right)

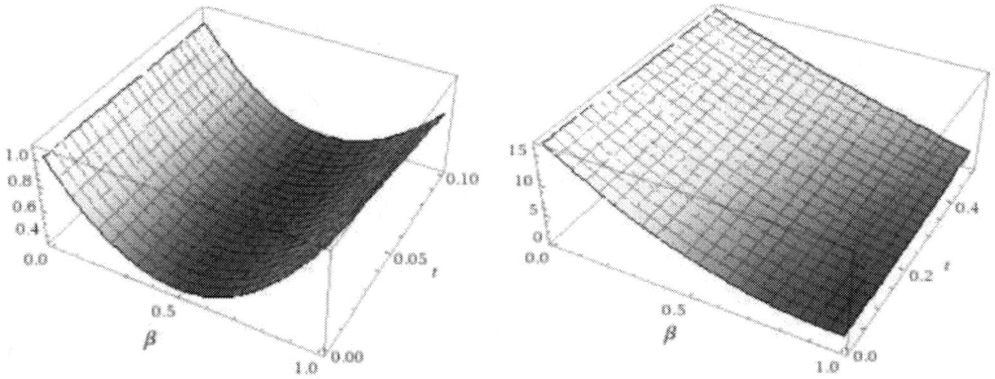

Fig.3. Solutions of Eq.(13) for $x = 0.1$ (left) $x = 1$ (right)

4 Conclusions

In this study, Laplace homotopy method which is defined with a new fractional operator is considered. This method presents an efficient solution for nonlinear FDEs without discretization. Therefore, this method can be applied to nonlinear FDEs easily. The nonlinear terms in the equation are evaluated by using Adomian polynomials. In this context, we obtain the approximate-analytical solution of nonlinear KGE of fractional order by using the LHM. The results obtained in this study carried out that the method is very efficient and accurate method for solving the fractional nonlinear homogeneous KGE. Maple software has been used for obtaining the numerical computations and presenting the solution plots with respect to various values of the variables.

References

1. T. Abdeljawad, D. Baleanu, *Integration by parts and its applications of a new nonlocal fractional derivative with Mittag-Leffler nonsingular kernel.* Journal of Nonlinear Sciences and Applications, 10(2017), pp. 1098-1107, (2017).

2. D. Avci, B.B. Iskender Eroglu, N. Ozdemir, *Conformable heat equation on a radial symmetric plate*. Thermal Science, 21(2), pp. 819-826, (2017).
3. E. Bonyah, A. Atangana, M.A. Khan, *Modeling the spread of computer virus via Caputo fractional derivative and the beta-derivative*. Asia Pacific Journal on Computational Engineering, 4(1), p. 1, (2017).
4. Y. Çenesiz, D. Baleanu, A. Kurt, O. Tasbozan, *New exact solutions of Burgers' type equations with conformable derivative*. Waves in Random and Complex Media, 27(1), pp. 103-116, (2017).
5. F. Evirgen, *Analyze the optimal solutions of optimization problems by means of fractional gradient based system using VIM*. An International Journal of Optimization and Control, 6(2), pp. 75-83, (2016).
6. J.F. Gomez-Aguilar, H. Yépez-Martínez, J. Torres-Jiménez, T. Córdova-Fraga, R.F. Escobar-Jiménez, V.H. Olivares-Peregrino, *Homotopy perturbation transform method for nonlinear differential equations involving to fractional operator with exponential kernel*. Advances in Difference Equations, 2017(1), p. 68, (2017).
7. J. Hristov, *Steady-state heat conduction in a medium with spatial non-singular fading memory: derivation of Caputo-Fabrizio space-fractional derivative with Jeffrey's kernel and analytical solutions*. Thermal science, 21(2), pp. 827-839, (2017).
8. M. Yavuz, *Novel solution methods for initial boundary value problems of fractional order with conformable differentiation*. An International Journal of Optimization and Control: Theories & Applications (IJOCTA), 8(1), pp. 1-7, (2018).
9. M. Yavuz, N. Ozdemir, *Numerical inverse Laplace homotopy technique for fractional heat equations*. Thermal Science, 22(2), pp. 185-194, (2018).
10. M. Yavuz, N. Ozdemir *A different approach to the European option pricing model with new fractional operator*. Mathematical Modelling of Natural Phenomena, 13(1), pp. 1-12, (2018).
11. M. Yavuz, N. Ozdemir, H.M. Baskonus, *Solutions of partial differential equations using the fractional operator involving Mittag-Leffler kernel*. The European Physical Journal Plus, 133(6), p. 215, (2018).
12. N. Özdemir, M. Yavuz, *Numerical Solution of Fractional Black-Scholes Equation by Using the Multivariate Padé Approximation*. Acta Physica Polonica A, 132(3), pp. 1050-1053, (2017).
13. H. Bulut, T.A. Sulaiman, H.M. Baskonus, *Dark, bright optical and other solitons with conformable space-time fractional second-order spatiotemporal dispersion*. Optik, 163, pp. 1-7, (2018).
14. H.M, Baskonus, T. Mekkaoui, Z. Hammouch, H. Bulut, *Active control of a chaotic fractional order economic system*. Entropy, 17(8), pp. 5771-5783, (2015).
15. A. Esen, Y. Ucar, N. Yagmurlu, O. Tasbozan, *A Galerkin finite element method to solve fractional diffusion and fractional diffusion-wave equations*. Mathematical Modelling and Analysis, 18(2), pp. 260-273, (2013).
16. A. Yokus, H.M. Baskonus, T.A. Sulaiman, H. Bulut, *Numerical simulation and solutions of the two-component second order KdV evolutionarysystem*. Numerical Methods for Partial Differential Equations, 34(1), pp. 211-227, (2018).
17. D. Kaya, S. Gülbahar, A. Yokuş, M. Gülbahar, *Solutions of the fractional combined KdV–mKdV equation with collocation method using radial basis function and their geometrical obstructions*. Advances in Difference Equations, 2018(1), p. 77, (2018).
18. M. Caputo, M. Fabrizio, *A new definition of fractional derivative without singular kernel*. Progress in Fractional Differentiation and Applications, 1(2), pp. 1-13, (2015).
19. A. Atangana, D. Baleanu, *New fractional derivatives with nonlocal and non-singular kernel: Theory and application to heat transfer model*. Thermal Science, 20(2), pp. 763-769, (2016).
20. N.A. Sheikh, F. Ali, M. Saqib, I. Khan, S.A.A. Jan, *A comparative study of Atangana-Baleanu and Caputo-Fabrizio fractional derivatives to the convective flow of a generalized Casson fluid*. The European Physical Journal Plus, 132(1), p. 54, (2017).
21. J.F. Gómez-Aguilar, R.F. Escobar-Jiménez, M.G. López-López, V.M. Alvarado-Martínez, *Atangana-Baleanu fractional derivative applied to electromagnetic waves in dielectric media*. Journal of Electromagnetic Waves and Applications, 30(15), pp. 1937-1952, (2016).

22. A. Atangana, D. Baleanu, *Caputo-Fabrizio derivative applied to groundwater flow within confined aquifer.* Journal of Engineering Mechanics, 143(5), p. D4016005, (2017).
23. M. Yavuz, N. Özdemir, *European vanilla option pricing model of fractional order without singular kernel.* Fractal and Fractional, 2(1), p. 3, (2018).
24. A. Atangana, B.S.T. Alkahtani, *New model of groundwater flowing within a confine aquifer: application of Caputo-Fabrizio derivative.* Arabian Journal of Geosciences, 9(1), p. 8, (2016).
25. J. Losada, J.J. Nieto, *Properties of a new fractional derivative without singular kernel.* Progress in Fractional Differentiation and Applications, 1(2), pp. 87-92, (2015).
26. D. Baleanu, B. Agheli, M.M. Al Qurashi, *Fractional advection differential equation within Caputo and Caputo–Fabrizio derivatives.* Advances in Mechanical Engineering, 8(12), p. 1687814016683305, (2016).
27. A. Ghorbani, *Beyond Adomian polynomials: he polynomials.* Chaos, Solitons & Fractals, 39(3), pp. 1486-1492, (2009).
28. A.K. Golmankhaneh, A.K. Golmankhaneh, D. Baleanu, *On nonlinear fractional Klein-Gordon equation.* Signal Processing, 91(2011), pp. 446-451, (2011).

An Alternative Approach for Nonlinear Optimization Problem with Caputo - Fabrizio Derivative

Fırat Evirgen[1], *and Mehmet* Yavuz[2,*]

[1] Department of Mathematics, Balikesir University, Balikesir, Turkey

[2] Department of Mathematics-Computer Sciences, Necmettin Erbakan University, Konya, Turkey

Abstract. In this study, a fractional mathematical model with steepest descent direction is proposed to find optimal solutions for a class of nonlinear programming problem. In this sense, Caputo-Fabrizio derivative is adapted to the mathematical model. To demonstrate the solution trajectory of the mathematical model, we use the multistage variational iteration method (MVIM). Numerical simulations and comparisons on some test problems show that the mathematical model generated using Caputo-Fabrizio fractional derivative is both feasible and efficient to find optimal solutions for a certain class of equality constrained optimization problems.

1 Introduction

Optimization theory and fractional calculus are the essential elements of the mathematical modeling process. Because of this reason, they are the most attractive two research areas studied by scientists in the word.

The optimization concept is applied for representing real-world problems, which aims to minimize and maximize some parameters, using mathematical models. Transportation systems, production processes, creating portfolios, routing of traffic, a system controlling and etc. al. all are examples of the optimization concept in our life. In particular, incredibly many new methods have been introduced in the literature for solving these type of problems, within the past twenty years. Gradient-based dynamic system method is one of these methods and was first introduced by Arrow and Hurwicz in 1958 [1]. In this method, the solution process is generated with the help of a differential equation system which includes optimality conditions. In recent past, Wang et al. [2], Jin et al. [3], Shikhman and Stein [4] and, Özdemir and Evirgen [5] have considered a gradient-based system for the different type of optimization problems.

* Corresponding author: mehmetyavuz@konya.edu.tr

On the other hand, the fractional calculus topics are drawn increasing attention from the researchers as it has a capability of representing real-world problems better than regular derivative. In this sense, various types of fractional derivatives, integrals, and applications are introduced by scientists. The Riemann-Liouville's fractional derivative and the Caputo fractional derivative are the most popular definitions in the literature. Podlubny [6] and Baleanu et al. [7] provided comprehensive discussions on these definitions. In their recent work, Caputo and Fabrizio [8] introduced a new derivative of fractional order with a non-singular kernel. Some interesting features and implementations of the new fractional derivative operator are introduced by Caputo and Fabrizio [9] and Losada and Nieto [10].

In the last two years, many authors used Caputo-Fabrizio fractional derivative operator in many interesting models and with many known methods, see [11-22].

In the present work, we will investigate the solution procedure of the optimization problems with the constructed mathematical model by Caputo-Fabrizio fractional derivative operator. To demonstrate the solution, we use the multistage variational iteration method (MVIM).

In this sense, the remainder paper is organized as follows. We present in Section 2 some basic notion and theoretical aspects of the optimization theory, Caputo-Fabrizio fractional derivative and variational iteration method. The newly defined mathematical model with Caputo-Fabrizio fractional derivative and the solution procedure by MVIM are given in Section 3, followed by the presentation of the solutions obtained by the proposed mathematical model and MVIM for well-known benchmarks.

2 Some Basic Definitions and Theorems

2.1 Fundamentals of optimization theory

To explain basic notions of the optimization theory let consider the following optimization problem which contains equality constraints,

$$\begin{aligned} &\min_{x\in\Re^n} \quad f(x), \\ &\text{s.t.} \qquad h(x)=0, \end{aligned} \tag{1}$$

where $f:\Re^n \to \Re$, $h:\Re^n \to \Re^p$ are functions. With the aid of the penalty function concepts, the equality constrained optimization problem is converted into the following optimization problem without constrained

$$\min F(x,\eta) = f(x) + \underbrace{\eta \frac{1}{\gamma}\sum_{i=1}^{p}(h_i(x))^{\gamma}}_{Penalty\ function} \quad s.t.\ x \in \Re^n. \tag{2}$$

where $\eta > 0$ is a sufficiently large auxiliary penalty variable and $\gamma > 0$ is a constant [23]. When $\gamma = 2$ is taken, it is named as quadratic penalty function method. The following theorem shows us the convergence between the problem (1) and the problem (2) solutions.

Theorem 1 Let $\{x^{(n)}\}$ be a solution sequence of the unconstrained optimization problems generated by the penalty function method (2) as $\eta \to \infty$. Then any accumulation points of the sequence are the solution of the original constrained problem (1).

2.2 Caputo-Fabrizio fractional derivative

Basic tools of the Caputo-Fabrizio fractional derivative of order α can be summarized as follow [8-10].

Definition 1 Let $f:[a, b] \to \Re$ be a continuous function. The left and the right Caputo fractional derivatives of order $\alpha > 0$ are defined, respectively, by

$$ {}^{C}D_{a^+}^{\alpha} f(t) = \frac{1}{\Gamma(m-\alpha)} \int_a^t (t-\tau)^{m-\alpha-1} f^{(m)}(\tau) d\tau, \ t \in (a,b], $$

and

$$ {}^{C}D_{b^-}^{\alpha} f(t) = \frac{1}{\Gamma(m-\alpha)} \int_t^b (-1)^n (t-\tau)^{m-\alpha-1} f^{(m)}(\tau) d\tau, \ t \in [a,b), $$

where α is the order of derivative, and $n = [\alpha]+1$ with $[\alpha]$ being the integer part of α

According to the above classical Caputo fractional derivative definition, novel fractional derivative definition consisting of the non-singular kernel can be given by the following definition.

Definition 2 Let $\alpha \in (0, 1)$, $f(x)$ be a function such that $f(x) \in H^1(a,b)$, $a < b$ and $M(\alpha)$ be a normalization function such that $M(0) = M(1) = 1$. The left Caputo-Fabrizio derivative of order α is defined as

$$ {}^{CF}D_{a^+}^{\alpha} f(t) = \frac{M(\alpha)}{1-\alpha} \int_a^t f'(\tau) \exp\left[-\frac{\alpha(t-s)}{1-\alpha}\right] ds. $$

where $M(\alpha)$ is normalization function such that $M(0) = M(1) = 1$ and, $0 < \alpha \le 1$.

Remark 1 The Caputo-Fabrizio fractional derivative is obtained by changing the kernel $(t-\tau)^{-\alpha}$ in Caputo fractional derivative formula with the function $\exp\left[-\frac{\alpha(t-\tau)}{1-\alpha}\right]$ and $\frac{1}{\Gamma(1-\alpha)}$ with $\frac{M(\alpha)}{1-\alpha}$.

Remark 2 The Caputo-Fabrizio fractional derivative, is zero where f as constant, as in the usual Caputo derivative. But, contrary to the usual Caputo fractional derivative, the kernel does not have singularity for $t = \tau$

In (2015), Losada and Nieto have redefined the Caputo-Fabrizio fractional derivative by taking $M(\alpha) = 1$ as follow:

$$ {}^{CF}D^{\alpha}_{a^+} f(t) = \frac{1}{1-\alpha} \int_a^t f'(\tau) \exp\left[-\frac{\alpha(t-s)}{1-\alpha} \right] ds $$

Furthermore, by using the Laplace transform for the general form of fractional differential equation involving Caputo-Fabrizio fractional derivative operator, they are stated the Caputo-Fabrizio fractional integral definition for $0 < \alpha \le 1$ as follow:

$$ {}^{CF}I^{\alpha}_{t} f(t) = \frac{1-\alpha}{M(\alpha)} f(t) + \frac{\alpha}{M(\alpha)} \int_a^t f(\tau) d\tau. $$

Remark 3 When $0 < \alpha \le 1$, the property below convinces

$$ \left({}^{CF}I^{\alpha}_{t} \right) \left({}^{CF}D^{\alpha}_{t} \right) f(t) = f(t) - f(0). $$

2.3 Variational iteration method

Let us consider the following general nonlinear differential equation to express the variational iteration method (VIM),

$$ L(u(t)) + N(u(t)) = g(t) \tag{3} $$

where L is a linear operator, N is a nonlinear operator and $g(t)$ is a known analytical function [24].

We can construct a correction functional for (4) as follows,

$$ u_{i,k+1}(t) = u_{i,k}(t) + \int_{t_0}^{t} \lambda(\tau) \left\{ L(u_{i,k}(\tau)) + N(\tilde{u}_{i,k}(\tau)) - g(\tau) \right\} d\tau, \tag{4} $$

where λ is a general Lagrange multiplier, and $\tilde{u}_{i,k}$ is a restricted variation with $\delta \tilde{u}_{i,k} = 0$. Thereby the exact solution can be obtained as

$$ u_i(t) = \lim_{k \to \infty} u_{i,k}(t) . $$

3 A mathematical model involving Caputo-Fabrizio derivative

Let consider the equality constrained optimization problem described by (1). As mentioned above, the optimization problem (1) is converted to the unconstrained optimization problem (2) by using the quadratic penalty function for $\gamma = 2$.

In this work, a mathematical model, which is constructed by using the Caputo-Fabrizio fractional derivative, is handled for determining the optimal solutions of (2), and also of the equality constrained optimization problem (1) with the help of the MVIM. This kind of fractional mathematical model for solving optimization problems was first introduced by Evirgen and Özdemir [25-28].

The mathematical model involving the Caputo-Fabrizio derivative can be defined by the following expression:

$$\begin{aligned} {}^{CF}D_t^{\alpha} x(t) &= -\nabla_x F(x,\eta) \quad ,0<\alpha \leqslant 1, \\ x_i(0) &= x_i 0 \qquad\qquad\quad ,i=1...n. \end{aligned} \tag{5}$$

where $\nabla_x F(x,\eta)$ is the gradient vector of (2) with respect to the $x \in \Re^n$.

Definition 4 A point x^* is referred to as an equilibrium point of (5) if it satisfies the right-hand side of Equation (5).

Without loss of generality, the mathematical model (5) can be rearranged by the following formula as follow

$${}^{CF}D_t^{\alpha} x_i(t) = g_i(t,\eta,x_1,x_2,...,x_n), \ i=1,2,...,n. \tag{6}$$

In this stage, the MVIM algorithm is applied to acquire the stable equilibrium point of the fractional model (6). The MVIM can be described in terms of certain modifications to VIM. In this approach, the integral interval $[t_0,t)$ is divided into the equal subinterval such as $[t_0,t_1)$, $[t_1,t_2)$,..., $[t_{j-1},t_j=t)$ to ensure the validity of the approximations of VIM in the large interval. The multistage technique is first adapted to the VIM by Batiha and et al. [29].

The correction functional for (6) according to the MVIM can be constructed as

$$x_i^{k+1}(t) = x_i^k(t) + I_t\left(\lambda_i(\tau)\left({}^{CF}D_{i\ \alpha}^{\alpha} x_i^k(\tau) - g_i\left(\tilde{x}_1^k(\tau),...,\tilde{x}_n^k(\tau)\right)\right)\right)d\tau \tag{7}$$

where λ_i $(i=1,2,...,n)$ are general Lagrange multiplier and $\tilde{x}_1,\tilde{x}_2,...,\tilde{x}_n$ denote restricted variations such that $\delta\tilde{x}_i = 0$.

To identify the multiplier, we take the usual Riemann integral instead of Caputo-Fabrizio fractional integral and taking variation with respect to the independent variable x_i, $i = 1,2,...,n$ with $\delta x_i(t^*) = 0$,

$$\delta x_i^{k+1}(t) = \delta x_i^k(t) + \delta \int_{t^*}^{t} \lambda_i(\tau)\left(x'^k_i(\tau) - gi\left(\tilde{x}_1^k(\tau),...,\tilde{x}_n^k(\tau)\right)\right)d\tau$$

and consequently we get the following stationary conditions:

$$\lambda_i'(\tau)\big|_{\tau=t} = 0$$
$$1 + \lambda_i(\tau)\big|_{\tau=t} = 0$$

for $i = 1,2,...,n$. Thus the Lagrange multipliers can be determined by the following equality

$$\lambda_i = -1, \ i = 1,2,...,n. \quad (8)$$

In this way, we acquire the MVIM formula by substituting the Lagrange multipliers (8) into the correctional functional (7) as follow:

$$x_i^{k+1}(t) = x_i^k(t) - I_t\left({}^{CF}D_t^{\alpha} x_i^k(\tau) - g_i\left(\tilde{x}_1^k(\tau),...,\tilde{x}_n^k(\tau)\right)\right)d\tau, \ i = 1,2,...,n. \quad (9)$$

If we begin with initial conditions $x_i(t^*) = x_i(t_0) = x_i(0)$, the iteration formula of the MVIM (9) can be carried out within every subinterval of equal length Δt, and so all solutions $x_i^k(t)$, $(i = 1,2,...,n; k = 1,2,...)$ are completely determined.

4 Numerical simulations

In this section, we give two examples to illustrate the effectiveness and applicability of the proposed fractional mathematical model involving the Caputo-Fabrizio derivative (6). The MVIM is applied and compared with the VIM and fourth order Runge-Kutta method (RK4).

Example 1 Consider the following nonlinear programming problem [30],

$$\begin{aligned} &\text{minimize} \quad f(x) = 100\left(x_1^2 - x_2\right)^2 + \left(x_1 - 1\right)^2 \\ &\text{subject to} \quad h(x) = x_1\left(x_1 - 4\right) - 2x_2 + 12 = 0. \end{aligned} \quad (10)$$

Firstly, the problem (10) is transformed to the unconstrained optimization problem (2) with the quadratic penalty function and so we obtain

$$F(x,\eta) = 100\left(x_1^2 - x_2\right)^2 + \left(x_1 - 1\right)^2 + \frac{1}{2}\eta\left(x_1\left(x_1 - 4\right) - 2x_2 + 12\right)^2$$

where $\eta \in \Re^{+} \left(\eta \to \infty\right)$ is an auxiliary penalty variable. Thereafter, the mathematical model from (5) is defined as

$$\left.\begin{aligned} {}^{CF}D_t^{\alpha} x_1(t) &= -400(x_1^2 - x_2)x_1 - 2(x_1 - 1) \\ &\quad - \eta(2x_1 - 4)(x_1^2 - 4x_1 - 2x_2 + 12), \\ {}^{CF}D_t^{\alpha} x_2(t) &= 200(x_1^2 - x_2) + 2\eta(x_1^2 - 4x_1 - 2x_2 + 12), \\ x_1(0) &= 0, x_2(0) = 0 \end{aligned}\right\} \tag{11}$$

where $0 < \alpha \leqslant 1$.

Finally, in order to obtain the solutions of (11), the MVIM and VIM are implemented with auxiliary penalty variable $\eta = 800$ and step size $\Delta T = 0.1 \times 10^{-4}$ by using the MVIM recursive formula (9).

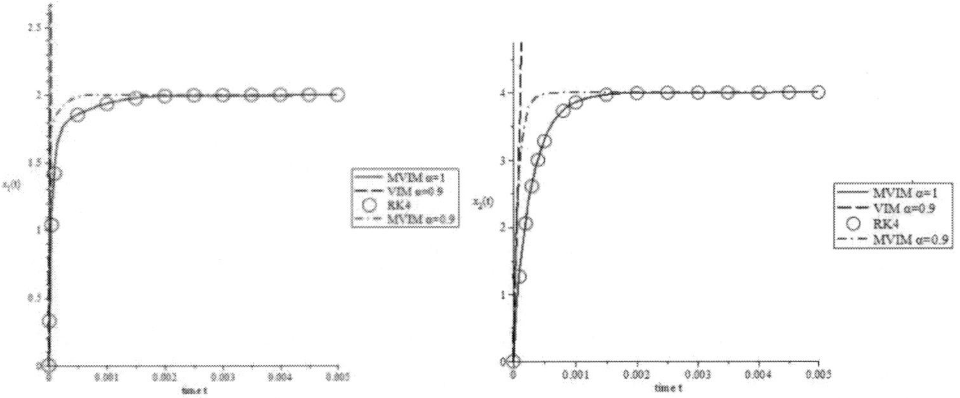

Fig.1. Numerical evaluation of $x(t)$ for problem (10) with MVIM, VIM and RK4.

In Figure 1, it can clearly be seen that the fractional model with the Caputo-Fabrizio fractional operator reaches the expected optimal solutions of (10) only one iteration using the MVIM. It can also be verified that MVIM with the fractional order has obtained the optimal solutions earlier than the VIM and RK4 with the integer order version.

Example 2 Let us consider the equality constrained problem in [30] as follow,

$$\begin{aligned} &\text{minimize} \quad f(x) = \left(x_1 - 20\right)^2 + \left(x_2 + 20\right)^2, \\ &\text{subject to} \quad h\left(x\right) = \frac{x_1^2}{100} + \frac{x_2^2}{4} - 1 = 0. \end{aligned} \tag{12}$$

This is a practical problem and the best known optimal solution is $x_1^* = 9.395$ and $x_2^* = -0.6846$. Thereafter, the corresponding mathematical model from (5) is defined as

$$\left.\begin{aligned}
&{}^{CF}D_t^{\alpha}x_1(t)=2x_1-40+\eta\left(\frac{1}{5000}x_1^3+\frac{1}{200}x_1x_2^2-\frac{1}{50}x_1\right),\\
&{}^{CF}D_t^{\alpha}x_2(t)=2x_2+40+\eta\left(\frac{1}{200}x_2x_1^2+\frac{1}{8}x_2^3-\frac{1}{2}x_2\right)\\
&x_1(0)=0,\ x_2(0)=0,
\end{aligned}\right\}\quad(13)$$

where $0<\alpha\leqslant 1$ is the order of the Caputo-Fabrizio fractional derivative. The optimal solution of the problem (12) is obtained by utilizing the MVIM with the auxiliary penalty variable $\eta=6\times10^5$ and step size $\Delta T=0.1\times10^{-5}$. Hence, the solutions can be summarized in Tables 1.

Table 1. The numerical evaluation of $x(t)$ for problem (12) with MVIM, VIM and RK4.

	MVIM ($\alpha=0.9$)		MVIM ($\alpha=1$)		RK4 ($\alpha=1$)	
t	$x_1(t)$	$x_2(t)$	$x_1(t)$	$x_2(t)$	$x_1(t)$	$x_2(t)$
0	0	0	0	0	0	0
0.1	9.3330	−0.7186	3.4920	−1.8741	3.4919	−1.8741
0.2	9.3966	−0.6847	6.0764	−1.5885	6.0764	−1.5885
0.3	9.3967	−0.6846	7.8531	−1.2383	7.8530	−1.2383
0.4	9.3967	−0.6846	8.8690	−0.9242	8.8690	−0.9242
0.5	9.3967	−0.6846	9.2690	−0.7510	9.2690	−0.7510

In the numerical simulation, we see that the VIM solutions for $\alpha=0.9$ are divergent. Furthermore, as we saw in the previous example, although the fractional order MVIM approaches the optimal solution at $t=0.5$, other methods have not yet been approached the desired optimal solution.

4 Conclusion

In this paper, a fractional mathematical model with the newly defined Caputo-Fabrizio fractional derivative is designed successfully for solving a certain class of equality constrained optimization problem. For the purpose of obtaining the optimal solutions, the steepest descent direction is integrated into the mathematical model. In addition to this, the solution trajectory of the generated fractional mathematical model is shown by applying the MVIM. Numerical comparisons between the RK4 and MVIM, VIM with the fractional and integer order are shown that the Caputo-Fabrizio fractional gradient based mathematical model is more reliable and stable than the integer order one for obtaining the optimal solutions of this type optimization problem (1).

References

1. K.J. Arrow, L. Hurwicz and H. Uzawa, *Studies in Linear and Non-Linear* Programming, Stanford University Press, California (1958).
2. S. Wang, X.Q. Yang and K.L. Teo, *A unified gradient flow approach to constrained nonlinear optimization problems*, Comput. Optim. Appl., 25, pp. 251-268, (2003).
3. L. Jin, L.-W. Zhang and X. Xiao, *Two differential equation systems for equality-constrained optimization*, Appl. Math. Comput., 190, pp. 1030-1039, (2007).
4. V. Shikhman and O. Stein, *Constrained optimization: projected gradient flows*, J. Optim. Theory Appl., 140, pp. 117-130, (2009).
5. N. Özdemir and F. Evirgen, *A dynamic system approach to quadratic programming problems with penalty method*, Bull. Malays. Math. Sci. Soc., 33, pp. 79-91, (2010).
6. I. Podlubny, *Fractional Differential Equations*, Academic Press, New York (1999).
7. D. Baleanu, K. Diethelm, E. Scalas and J.J. Trujillo, *Fractional calculus: models and numerical methods*, Series on complexity, nonlinearity and chaos, World Scientific, Singapore, (2012)
8. M. Caputo and M. Fabrizio, *A new definition of fractional derivative without singular kernel*, Progr. Fract. Differ. Appl., 1, pp. 73-85, (2015).
9. M. Caputo and M. Fabrizio, *Applications of new time and spatial fractional derivatives with exponential kernels*, Progr. Fract. Differ. Appl., 2, pp. 1-11, (2016).
10. J. Losada and J.J. Nieto, *Properties of a new fractional derivative without singular kernel*, Progr. Fract. Differ. Appl., 1, pp. 87-92, (2016).
11. A. Atangana, B.S.T. Alkahtani, *New model of groundwater flowing within a confine aquifer: Application of Caputo–Fabrizio derivative*, Arab. J. Geosci., 9, pp. 1-8, (2016)
12. J. Singh, D. Kumar, J.J. Nieto, *Analysis of an El Nino-Southern Oscillation model with a new fractional derivative*, Chaos Solitons Fractals, 99, pp. 109-115, (2017).
13. I. Koca, A. Atangana, *Solutions of Cattaneo–Hristov model of elastic heat diffusion with Caputo–Fabrizio and Atangana–Baleanu fractional derivatives*, Therm. Sci., 21, pp. 2299-2305, (2017).
14. J. Hristov, *Steady-state heat conduction in a medium with spatial non-singular fading memory: Derivation of Caputo–Fabrizio space-fractional derivative with Jeffrey's kernel and analytical solutions*, Therm. Sci., 21, pp. 827-839, (2017).
15. M.S. Aydogan, D. Baleanu, A. Mousalou, and et al., *On high order fractional integro-differential equations including the Caputo–Fabrizio derivative*, Bound Value Probl., 2018: 90, pp. 1-15, (2018).
16. M. Yavuz, N. Özdemir, *A different approach to the European option pricing model with new fractional operator*. Mathematical Modelling of Natural Phenomena, 13(1), pp. 1-12, (2018).
17. S. Ullah, M. Altaf Khan and M. Farooq, *A new fractional model for the dynamics of the hepatitis B virus using the Caputo-Fabrizio derivative*, Eur. Phys. J. Plus, 133: 237, (2018).
18. N.A. Asif, Z. Hammouch, M.B. Riaz and et al., Analytical solution of a Maxwell fluid with slip effects in view of the Caputo-Fabrizio derivative, Eur. Phys. J. Plus, 133: 272, (2018).
19. M.A. Dokuyucu, E. Celik, H. Bulut and et al., Cancer treatment model with the Caputo-Fabrizio fractional derivative, Eur. Phys. J. Plus 133: 92, (2018).
20. M. Yavuz, N. Özdemir, *European vanilla option pricing model of fractional order without singular kernel,* Fractal and Fractional, 2(1), 3, (2018).
21. A. Yokus, *Numerical solutions of time fractional Korteweg-de vries equation and its stability analysis*, Commun. Fac. Sci. Univ. Ank. Ser. A1 Math. Stat., 68(1), pp. 353-361, (2019).
22. A. Yokus, D. Kaya, *Numerical and exact solutions for time fractional Burgers' equation*, J. Nonlinear Sci. Appl., 10(7), pp. 3419–3428, (2017).
23. D.G. Luenberger and Y. Ye, *Linear and Nonlinear Programming*, Third Edition, Springer, New York (2008).
24. J.H. He, *Variational iteration method for delay differential equations*, Commun. Nonlinear Sci. Numer. Simul., 2, pp. 235-236, (1997).

25. F. Evirgen and N. Özdemir, Multistage Adomain decomposition method for solving NLP problems over a nonlinear fractional dynamical system. J. Comput. Nonlinear Dyn., 6, 021003 (2011).
26. F. Evirgen and N. Özdemir, *A fractional order dynamical trajectory approach for optimization problem with HPM*, Fractional Dynamics and Control, Springer, Eds. Baleanu, D., Machado, J.A.T., Luo, A.C.J., pp. 145-155 (2012).
27. F. Evirgen, *Analyze the optimal solutions of optimization problems by means of fractional gradient based system using VIM*, An International Journal of Optimization and Control: Theories & Applications (IJOCTA), 6(2), pp. 75-83, (2016).
28. F. Evirgen, *Conformable fractional gradient based dynamic system for constrained optimization problem*, Acta Physica Polonica A, 132 (3), 1066-1069, (2017).
29. B. Batiha, M.S.M. Noorani, I. Hashim and E.S. Ismail, *The multistage variational iteration method for a class of nonlinear system of ODEs*, Phys. Scr., 76, pp. 388-392, (2007).
30. K. Schittkowski, *More test examples for nonlinear programming codes*, Springer, Berlin, (1987).

IMAGE CODING USING LAPLACE TRANSFORM

Muharrem Tuncay Gençoğlu[1] **Mehmet Vural**[2]

[1] Fırat University, Elazig/NationalDefenseUniversity, Ankara, Turkey, mtgencoglu23@gmail.com

[2] Fırat University,Elazığ,Turkey

Abstract

In this paper a different cryptographic method is introduced by using Power series transform. A new algorithm for cryptography is produced. The extended Laplace transform of the exponential function is used to encode an explicit text. The key is generated by applying the modular arithmetic rules to the coefficients obtained in the transformation. Here, ASCII codes used to hide the mathematically generated keys strengthen the encryption. Text steganography is used to make it difficult to break the password. The made encryption is reinforced by image steganography. To hide the presence of the cipher text, it is embedded in another open text with a stenographic method. Later, this text is buried in an image. For decryption it is seen that the inverse of the Power series transform can be used for decryption easily.

Experimental results are obtained by making a simulation of the proposed method. As a result, it is stated that the proposed method can be used in crypto machines.

Keywords: Cryptography, Power Series Transform, Data Encryption, Embedded Image

1.Introduction

Processed in this environment and protection of the transmitted information or security to provide is great importance. There are many threats such as unauthorized access, damage, etc. while data communication is being performed in the digital environment. In order to eliminate these threats many encryption techniques are developed [1-6].Cryptography is the whole mathematical technical work on information security. The cipher is often expressed with parameters called key which is as part of the external information. Decryption is almost impossible without an appropriate key. For the definition, theorem and mathematical transaction details used in this study, [3,7] can be looked at.

1.2. Our Contribution

We wrote a Power series transformation algorithm which can be used in the encryption methods existed in the literature. Next, we buried a hidden text into any image by cryptology and steganography. We suggested a hybrid method for crypto machines.

In this article, a new information security model based on the Taylor series for steganography is proposed. The proposed model is used for both cryptography and steganography. The characterization of the proposed model is given as follows.

- A new method in which steganography and cryptography reused together, is proposed.
- The proposed steganography method is a media independent steganography method. This method is used for both in image steganography and text steganography.
- A Taylor series based coding method is defined mathematically and practically.
- The method is simulated and the simulation results are shown in the experimental results clearly

2.The Proposed Method

Combining cryptography and steganography methods, the application stages that increase the data security and privacy of this proposed hybrid model are as follows.

2.1.Encryption

In this method, a new encryption method based on the Taylor series is proposed. The steps of the proposed method are as follows.

Step 1: The Taylor series is expanded withe^t. Then this value is multiplied by t^3 to generalize the mathematical relation to be used in the encryption algorithm.

Step 2: The number values corresponding to the letters in the alphabet is applied to the text to be encrypted.

Step 3: The numbers found are replaced in the generalized encryption algorithm.

Step 4: The Power series transform is applied to the function obtained from here.

Step 5: The obtained coefficients are found mod 28 values.

Step 6: Instead of these numbers, the encryption keys are found by taking the quotients in the mode operation.

Step 7: The text is encrypted by writing the letters corresponding to these keys.

Step 8: Encrypted text is converted to ASCII code and the corresponding numbers are found. Then these numbers are converted into a binary system.

Step 9: These numbers are hidden into any text by a method that the user will specify.

Step 10: The sender sends this embedded text along with the private key.

Example

Suppose that we want to send the message "FIRAT". Firstly we take into accord extended Power series withe^t:

$$f(x) = f(a) + \frac{f'(a)}{1!}(x-a) + \frac{f''(a)}{2!}(x-a)^2 + \cdots + \frac{f^n(a)}{n!}(x-a)^n + \cdots$$

$$= \sum_{n=0}^{\infty} \frac{f^n(a)}{n!}(x-a)^n. \quad (3.1)$$

Then, if we expand;

$$e^t = 1 + \frac{t}{1!} + \frac{t^2}{2!} + \frac{t^3}{3!} + \cdots = \sum_{n=0}^{\infty} \frac{t^n}{n!} \qquad (2.1)$$

With t³, then we get:

$$t^3 e^t = t^3 + \frac{t^4}{1!} + \frac{t^5}{2!} + \frac{t^6}{3!} + \cdots = \sum_{n=0}^{\infty} \frac{t^{n+3}}{n!} \qquad (2.2)$$

Therefore, we obtain:

$$f(t) = \sum_{n=0}^{\infty} K_n \frac{t^{n+3}}{n!}. \qquad (2.3)$$

If we enumerate letters of the alphabet from scratch "FIRAT" plain text be equal 6,9,19,0,22. If we write K_0 =6, K_1 =9, K_2 =19, K_3 =0, K_4 =22 in to (2.3), we get

$$f(t) = \sum_{n=0}^{\infty} K_n \frac{t^{n+3}}{n!}$$

$$= K_0 \frac{t^3}{0!} + K_1 \frac{t^4}{1!} + K_2 \frac{t^5}{2!} + K_3 \frac{t^6}{3!} + K_4 \frac{t^7}{4!} \qquad (2.4)$$

If we apply extended Taylor series transformation to both sides of (2.4), we get

$$T[f(t)](h) = T[\sum_{n=0}^{\infty} K_n \frac{t^{n+3}}{n!}](h)$$

$$= T[K_0 \frac{t^3}{0!} + K_1 \frac{t^4}{1!} + K_2 \frac{t^5}{2!} + K_3 \frac{t^6}{3!} + K_4 \frac{t^7}{4!}](h)$$

$$= 6.3!\,h^3 + 9.4!\,h^4 + 19.5!\frac{h^5}{2!} + 0.6!\frac{h^6}{3!} + 22.7!\frac{h^7}{4!}$$

$$\sum_{n=0}^{\infty} K_n (n+3)! \frac{h^{n+3}}{n!} = 36h^3 + 216h^4 + 1140\frac{h^5}{2!} + 0\frac{h^6}{3!} + 4620\frac{h^7}{4!}. \quad (2.5)$$

The equivalence of 36,216,1140,0,4620 in the modes (28) are(K_n) 8,20,20,0,0. If we write quotient in mode operation place of these numbers, we get the key (K_n') 1, 7,40,0,165. "FIRAT" plain text converts "HSSAA" by (2.2).

If we convert "HSSAA" encrypted text to 8-bit characters in the ASCII code we obtain 72, 83,83,65,65. If these codes are written in binary system we get the keys$(1001000)_2$,$(1010001)_2$,$(1010001)_2$,$(1000001)_2$, $(1000001)_2$. ASCII 8 bit keys are in the text as follows: A provision giving the binary number system in the space between each word of the text namely if the number between two words 1 then we get $(1)_2$ and we define 2 with $(0)_2$.

Sender also send this text clearly with (1, 7, 40, 0,165) secret key.

2.2. Decryption

Steps of the proposed decryption method are given below.

Step 1: Thereceiver writes hidden ASCII codes into the text.

Step 2: He finds the letters corresponding to these codes.

Step 3: He obtains the numbers corresponding to the letters.

Step 4: These numbers and the secret key are written in place of the inverse power series.

Step 5: The letters corresponding to the coefficients obtained here are written.

Step 6: The first clear text is obtained.

Example

The recipient receives a text message and by reading the spaces between words with software that will get the data buried create the necessary numerical equivalents. If these numbers are divided into 8-bit groups, the ASCII equivalence of the embedded data is obtained.

If we write H, S, S, A, A→8,20,20,0,0 andsecret key values (1,7,40,0,0,165) into

$$A_n = \frac{K_n - K_n'}{28}$$

$\mathbf{36} = 28x1 + 8$

$\mathbf{216} = 28x7 + 20$

$\mathbf{1140} = 28x40 + 20$

$\mathbf{0} = 28x0 + 0$

$\mathbf{4620} = 28x165 + 0$ are obtained.

If we apply these values 36,216,1140,0,4620 to the

$$\sum_{n=0}^{\infty} K_n (n+3)! \frac{h^{n+3}}{n!}$$

then, we get

$$\sum_{n=0}^{\infty} K_n (n+3)! \frac{h^{n+3}}{n!} = 36h^3 + 216h^4 + 1140\frac{h^5}{2!} + 0\frac{h^6}{3!} + 4620\frac{h^7}{4!}$$

$= 6.3!\,h^3 + 9.4!\,h^4 + 19.5!\frac{h^5}{2!} + 0.6!\frac{h^6}{3!} + 22.7!\frac{h^7}{4!}.$ (2.6)

If we apply inverse Extended Power Series Transformation to both sides of the (3.7), then we get

$$T^{-1}\left[\sum_{n=0}^{\infty} K_n (n+3)! \frac{h^{n+3}}{n!}\right] = T^{-1}[6.3!\,h^3 + 9.4!\,h^4 + 19.5!\frac{h^5}{2!} + 0.6!\frac{h^6}{3!} + 22.7!\frac{h^7}{4!}]$$

$$\sum_{n=0}^{\infty} K_n \frac{t^{n+3}}{n!} = 6.t^3 + 9.t^4 + 19.\frac{t^5}{2!} + 0.\frac{t^6}{3!} + 22.\frac{t^7}{4!}.$$

If we convert the K_n coefficients we will get the first plain text 6,9,19,0,22→F,I,R,A,T.

The operations performed in this section is shown in Figure 1 and Figure 2.

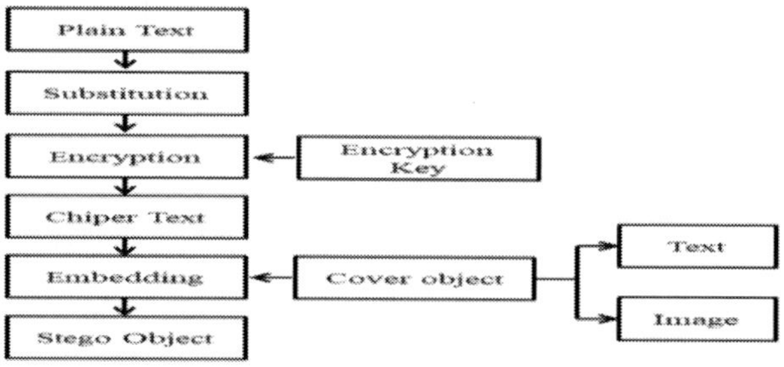

Figure-1. Flow Diagram of Encryption System

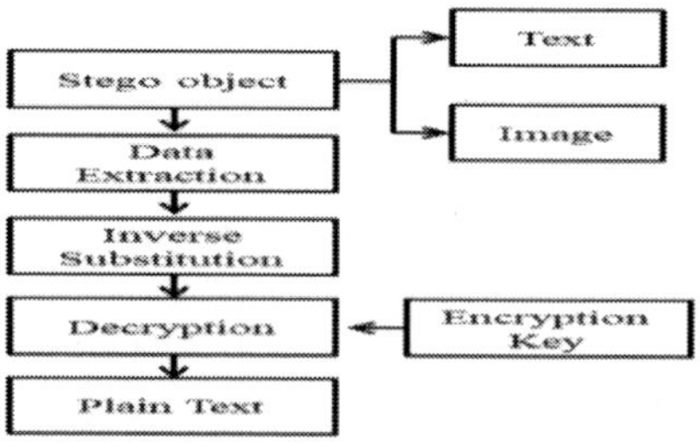

Figure-2. Flow Diagram of Decryption System

3.Experimental Results

In this article, recommended the method for both text and image media is applied. A simulation program is carried out to obtain experimental results. Text steganography and image steganography results are obtained in the performed simulation. Simulation of text steganography windows Fig.3 and4 'are shown.Fig.3 shows encryption method and data embedding window and Fig.4 shows the data extraction and decryption window, respectively.

Figure-3. Encryption window of the simulation

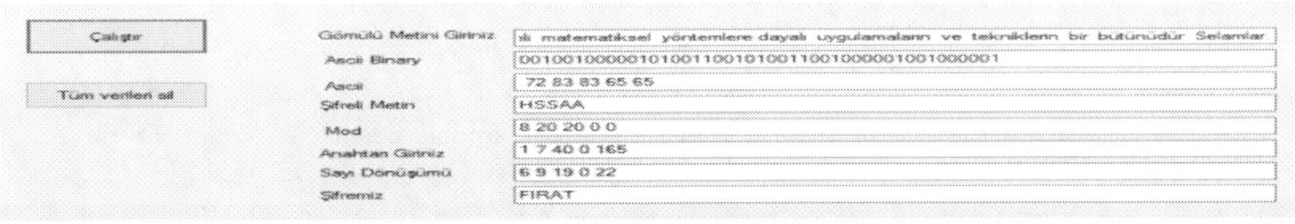

Figure-4. Decryption window of the simulation

In this part of the simulation, images are used as data hiding media.Data encryption and embedding window is shown in Fig. 5.

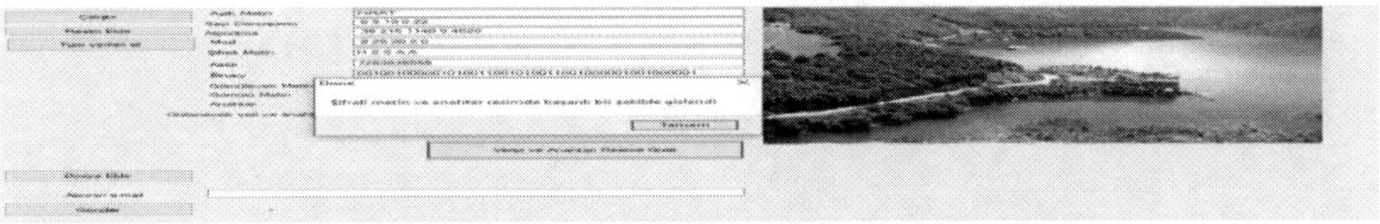

Figure-5. Data encryption and embedding windowfor image.

Data extraction window for images is shown in Fig. 7.

Figure-6. Data extraction window for image.

Data decryption window for images is shown in Fig. 7.

Figure-7. Data decryption window for image.

4. Conclusion and Recommendations

The proposed algorithm is created by using the power series transformation. Keys generated using this algorithm is applied to the method known as substitution method in the literature. In our practice, the keys obtained by the proposed method that aroused as an end result of digitization are used. A hybrid model is developed by using steganography to provide high security and explained in detail. The user can hide this massage which is obtained by taking the coefficient q_n instead of the coefficient $(K_n^{'})$

andalso presence of this message is hidden with ASCII code.Then, using another password, an encrypted text can be hidden into a text by the proposed method. In this way, by using the steganography approach, the security level of the data can be increased.It is verydifficulttofindtheprivatekeybythebrutefarceattackorbyanyotherattack. Finally; the written program can be used in crypto machine.

References

[1]Koç, Ç.K. (2009). Cryptographic Engineering, Springer. PP 125-128.

[2]Gençoğlu, M.T. (2016). Use of Integral Transform in Cryptology. Science and Eng. of Fırat Univ.,**28** (2), 217-220.

[3]Martin, K.M. (2012). Everyday Cryptography Fundamental Principles and Applications, Oxford University Press.

[4]Delfs, H.andKnebl, H. (2007). Introduction to Cryptography Principles and Applications, Springer.

[5]Paar, C.andPelzl, J. (2010). UnderstandingCryptography, Springer.

[6] Gençoğlu, M.T. (2017). Combining Cryptography with Steganography. ITM web of Conferences,**13**, 01010.

Four Point Implicit Methods for the Second Derivatives of the Solution of First Type Boundary Value Problem for One Dimensional Heat Equation

Suzan Cival Buranay[1,*] and *Lawrence Adedayo* Farinola [1]

[1]Department of Mathematics, Faculty of Arts and Sciences, Eastern Mediterranean University, Famagusta, North Cyprus, Via Mersin 10 Turkey

Abstract. We construct four-point implicit difference boundary value problem for the first derivative of the solution $u(x,t)$ of the first type boundary value problem for one dimensional heat equation with respect to the time variable t. Also, for the second derivatives of $u(x,t)$ special four-point implicit difference boundary value problems are proposed. It is assumed that the initial function belongs to the Hölder space $C^{8+\alpha}, 0<\alpha<1$, the heat source function given in the heat equation is from the Hölder space $C_{x,t}^{6+\alpha,3+\frac{\alpha}{2}}$, the boundary functions are from $C^{4+\frac{\alpha}{2}}$, and between the initial and the boundary functions the conjugation conditions of orders $q=0,1,2,3,4$ are satisfied. We prove that the solution of the proposed difference schemes converge uniformly on the grids of the order $O(h^2+\tau)$ (second order accurate in the spatial variable x and first order accurate in time t) where, h is the step size in x and τ is the step size in time. Theoretical results are justified by numerical examples.

1 Introduction

Many practical heat conduction questions lead to problems not conveniently solvable by classical methods such as separation of variables techniques or the use of Green's functions. As a result approximate procedures like finite difference methods have been proposed, (see [1] and the references there in). On the other hand, besides the solution of the heat conduction problem it is very important to provide information about some physical phenomena related with the derivatives. Such as, there are a number of problems in heat conduction in which one material is transformed into another with generation or absorption of heat. Examples are the melting or freezing of a solid and the process of a temperature-dependent chemical reaction through a solid [2], [3]. In melting problems to calculate the

* Corresponding author: suzan.buranay@emu.edu.tr

quantities as the rate of change in temperature $(\partial u/\partial t)$, the gradient of the rate of change in temperature $(\partial^2 u/\partial t^2)$, the rate of heat flow into the solid $(\partial u/\partial x)$, the gradient of the rate of heat flow into the solid $(\partial^2 u/\partial x^2)$ and the rate of melting $(\partial s/\partial t)$ where $x = s(t)$ is the thickness melted, $u(x,t)$ is the temperature of the solid are considerably important.

In this study we consider the first type boundary value problem for one dimensional heat equation of which the initial function belongs to $C^{8+\alpha}$ $0<\alpha<1$, the heat source function is from $C_{x,t}^{6+\alpha,3+\frac{\alpha}{2}}$, the boundary functions are from $C^{4+\frac{\alpha}{2}}$, and between the initial and the boundary functions the conjugation conditions of orders $q=0,1,2,3,4$ are satisfied. Denoting the exact solution of this problem by $u(x,t)$, difference problems of four point implicit schemes approximating $\frac{\partial u}{\partial t},\frac{\partial^2 u}{\partial x^2},\frac{\partial^2 u}{\partial t^2}$ and $\frac{\partial^2 u}{\partial x \partial t}$ are constructed. It is obtained that the solution of the constructed difference schemes converge uniformly to the exact values of $\frac{\partial u}{\partial t},\frac{\partial^2 u}{\partial x^2},\frac{\partial^2 u}{\partial t^2}$ and $\frac{\partial^2 u}{\partial x \partial t}$ respectively, on the grids of order $O(h^2+\tau)$.

2 Implicit Four-Point Difference Approximation of $\frac{\partial u}{\partial t},\frac{\partial^2 u}{\partial x^2},\frac{\partial^2 u}{\partial t^2}$ and $\frac{\partial^2 u}{\partial x \partial t}$

Let $\Omega=(0,b)$, $\sigma=(0,T)$, and $\bar{\Omega}$, $\bar{\sigma}_T$ be the closure of these sets respectively. Also denote by $Q_T=\{(x,t):0<x<b,0<t\le T\},\gamma_1=\{(0,t):t\in\bar{\sigma}_T\},\gamma_2=\{(x,0):x\in\bar{\Omega}\}$, $\gamma_3=\{(b,t):t\in\bar{\sigma}_T\}$. Let $\gamma=\bigcup_{i=1}^{3}\gamma_i$, represent the boundary of Q_T, and $\bar{Q}_T=Q_T\cup\gamma$. Let s be a noninteger positive number, $C_{x,t}^{s,s/2}(\bar{Q})$ and $C^s(\bar{\Omega})$ denote the Hölder spaces as defined in [4]. We consider:

Problem 1 : The first type boundary value problem for one dimensional heat equation

$$Lu=f(x,t) \text{ on } Q_T, \tag{1}$$

$$u(x,0)=u_0(x) \text{ on } \gamma_2, \tag{2}$$

$$u(0,t)=u_1(t) \text{ on } \gamma_1,\ u(b,t)=u_2(t) \text{ on } \gamma_3, \tag{3}$$

Where $L\equiv\frac{\partial}{\partial t}-a\frac{\partial^2}{\partial x^2}$, a is positive constant and

$$u_0(x)\in C^{8+\alpha}(\bar{\Omega}), f(x,t)\in C_{x,t}^{6+\alpha,3+\frac{\alpha}{2}}(\bar{Q}_T) \text{ and } u_j(t)\in C^{4+\frac{\alpha}{2}}(\bar{\sigma}_T), j=1,2, \tag{4}$$

also the conjugation conditions

$$u^{(q)}(0)=u_1^{(q)}(0), \tag{5}$$

$$u^{(q)}(b)=u_2^{(q)}(0), \tag{6}$$

of order $q=0,1,2,3,4$ between the initial and boundary conditions are satisfied (see [4]). Here,

$$u^{(0)}(x) = u_0(x), u^{(q)}(x) = \partial_t^q u(x,t)|_{t=0}, q = 1,2,..., \tag{7}$$
$$u_1^{(0)}(0) = u_1(t)|_{t=0}, u_2^{(0)}(0) = u_2(t)|_{t=0}, \tag{8}$$
$$u_1^{(q)}(0) = D_t^q u_1(t)|_{t=0}, u_2^{(q)}(0) = D_t^q u_2(t)|_{t=0}, q = 1,2,..., \tag{9}$$

$\partial_t^q u(x,t) = \dfrac{\partial^q u}{\partial t^q}$, $D_t^q u_i(t) = \dfrac{d^q u_i}{dt^q}$ $i = 1,2$ and $u^{(q)}(x)$, $q = 1,2,3,4$ in (7) are defined as

$$u^{(1)}(x) = aD_x^2 u_0(x) + f(x,0), \tag{10}$$
$$u^{(q)}(x) = a\partial_x^2 u^{(q-1)}(x) + f^{(q-1)}(x), \tag{11}$$

where, $f^{(0)}(x) = f(x,0)$ and $f^{(q)}(x) = \partial_t^q f(x,t)|_{t=0}$. Let $w = \partial_t u$ and $\phi = \partial_x^2 u$. The following boundary value problems are considered [5]

Problem 2

$$Lw = \partial_t f(x,t) = F(x,t) \text{ on } Q_T, \tag{12}$$
$$w(x,0) = u^{(1)}(x) = w_0(x) \text{ on } \gamma_2, \tag{13}$$
$$w(0,t) = D_t u_1(t) = w_1(t) \text{ on } \gamma_1, \tag{14}$$
$$w(b,t) = D_t u_2(t) = w_2(t) \text{ on } \gamma_3, \tag{15}$$

Problem 3

$$L\phi = \partial_x^2 f(x,t) = G(x,t) \text{ on } Q_T, \tag{16}$$
$$\phi(x,0) = \partial_x^2 u_0(x) = \phi_0(x) \text{ on } \gamma_2, \tag{17}$$
$$\phi(0,t) = \frac{1}{a}(D_t u_1(t) - f(0,t)) = \phi_1(t) \text{ on } \gamma_1, \tag{18}$$
$$\phi(b,t) = \frac{1}{a}(D_t u_2(t) - f(b,t)) = \phi_2(t) \text{ on } \gamma_3. \tag{19}$$

We represent the set of nodes on γ_i, $i = 1,2,3$ by

$$\omega_{0,\tau} = \{(0,t_j) : t_j = j\tau, \tau = \frac{T}{M}, j = 0,...,M\},$$
$$\bar{\omega}_{h,0} = \{(x_m,0) : x_m = mh, h = \frac{b}{N}, m = 0,...,N\},$$
$$\omega_{b,\tau} = \{(b,t_j) : t_j = j\tau, \tau = \frac{T}{M}, j = 0,...,M\},$$

respectively. To realize the numerical solution of the Problem 2, and Problem 3 we use implicit four-point difference scheme [6]. Let $\tilde{w}$ denote the solution of

$$\tilde{w}_{\bar{t},m}^{h,\tau} = a\Lambda\tilde{w}_m^{h,\tau} + \varphi_{\tilde{w},m} \text{ on } \omega_{h,\tau}, \tag{20}$$
$$\tilde{w}_m^0 = \tilde{w}(x_m,0) = w_0(x_m) \text{ on } \bar{\omega}_{h,0}, \tag{21}$$
$$\tilde{w}_0^j = \tilde{w}(0,t_j) = w_1(t_j) \text{ on } \omega_{0,\tau}, \quad \tilde{w}_N^j = \tilde{w}(b,t_j) = w_2(t_j) \text{ on } \omega_{b,\tau}, \tag{22}$$

where,

$$\tilde{w}_{\bar{t},m}^{h,\tau} = \frac{1}{\tau}(\tilde{w}_m^{j+1} - \tilde{w}_m^j), \tag{23}$$

$$\Lambda\tilde{w}_m^{h,\tau} = \frac{1}{h^2}\left(\tilde{w}_{m+1}^{j+1} - 2\tilde{w}_m^{j+1} + \tilde{w}_{m-1}^{j+1}\right), \tag{24}$$

$$\varphi_{\tilde{w},m} = F(x_m, t_{j+1}) = \partial_t f(x_m, t_{j+1}). \tag{25}$$

For the approximate solution of Problem 3 we give the next four point difference problem and denote the solution by $\tilde{\phi}$

$$\tilde{\phi}_{\bar{t},m}^{h,\tau} = a\Lambda\tilde{\phi}_m^{h,\tau} + \varphi_{\tilde{\phi},m} \text{ on } \omega_{h,\tau}, \tag{26}$$

$$\tilde{\phi}_m^0 = \tilde{\phi}(x_m, 0) = \phi_0(x_m) \text{ on } \bar{\omega}_{h,0}, \tag{27}$$

$$\tilde{\phi}_0^j = \tilde{\phi}(0, t_j) = \phi_1(t_j) \text{ on } \omega_{0,\tau}, \ \tilde{\phi}_N^j = \tilde{\phi}(b, t_j) = \phi_2(t_j) \text{ on } \omega_{b,\tau}, \tag{28}$$

where, $\tilde{\phi}_{\bar{t},m}^{h,\tau}$, $\Lambda\tilde{\phi}_m^{h,\tau}$ are formulae analogous to (23) and (24) respectively, and $\varphi_{\tilde{\phi},m} = G(x_m, t_{j+1}) = \partial_x^2 f\,|_{(x_m,t_{j+1})}$. We denote $q_i = \partial_t^2 u$ and $p_i = \partial_x\partial_t u$ on γ_i, $i = 1,2,3$, and construct the boundary value Problem 4 [5] for $z = \partial_t^2 u$, and Problem 5 for $y = \partial_x\partial_t u$.

Problem 4

$$Lz = \partial_t^2 f(x,t) \text{ on } Q_T, \tag{29}$$

$$z(x,0) = q_2 \text{ on } \gamma_2, \tag{30}$$

$$z(0,t) = q_1 \text{ on } \gamma_1, \ z(b,t) = q_3 \text{ on } \gamma_3, \tag{31}$$

where, $f(x,t)$ is the heat source function in (1). We take

$$q_{1\tau} = D_t^2 u_1(t) = D_t w_1(t) \text{ on } \omega_{0,\tau} \tag{32}$$

$$q_{2\tau}(\tilde{w}) = \frac{1}{\tau}\left(\tilde{w}(x,\tau) - w_0(x)\right) \text{ on } \bar{\omega}_{h,0}, \tag{33}$$

$$q_{3\tau} = D_t^2 u_2(t) = D_t w_2(t) \text{ on } \omega_{b,\tau}, \tag{34}$$

where, $w_0(x) = u^{(1)}(x)$ is as defined in (10) and $u_1(t), u_2(t)$ are given in (3). We propose the next difference problem for the numerical solution of Problem 4

$$\tilde{z}_{\bar{t},m}^{h,\tau} = a\Lambda\tilde{z}_m^{h,\tau} + \varphi_{\tilde{z},m} \text{ on } \omega_{h,\tau}, \tag{35}$$

$$\tilde{z}_m^0 = q_{2\tau}(\tilde{w}) \text{ on } \bar{\omega}_{h,0}, \tag{36}$$

$$\tilde{z}_0^j = q_{1\tau} \text{ on } \omega_{0,\tau}, \ \tilde{z}_N^j = q_{3\tau} \text{ on } \omega_{b,\tau}, \tag{37}$$

where $q_{i\tau}$, $i = 1,2,3$ are defined by (32)-(34) and $\varphi_{\tilde{z},m} = \partial_t^2 f\,|_{(x_m,t_{j+1})}$.

Problem 5

$$Ly = \partial_x\partial_t f(x,t) \text{ on } Q_T, \tag{38}$$

$$y(x,0) = p_2 \text{ on } \gamma_2, \tag{39}$$

$$y(0,t) = p_1 \text{ on } \gamma_1, \ y(b,t) = p_3 \text{ on } \gamma_3, \tag{40}$$

where, $f(x,t)$ is the given function in (1). We take

$$p_{1h}(\tilde{w}) = \frac{1}{2h}(-3w_1(t) + 4\tilde{w}(h,t) - \tilde{w}(2h,t)) \text{ on } \omega_{0,\tau} \quad (41)$$

$$p_{2h} = D_x w_0(x) \text{ on } \bar{\omega}_{h,0} \quad (42)$$

$$p_{3h}(\tilde{w}) = \frac{1}{2h}(3w_2(t) - 4\tilde{w}(b-h,t) + \tilde{w}(b-2h,t)) \text{ on } \omega_{b,\tau} \quad (43)$$

where, $w_0(x) = u^{(1)}(x)$ and $w_1 = D_t u_1(t), w_2 = D_t u_2(t)$.

For the numerical solution of Problem 5 we propose the next difference problem

$$\tilde{y}^{h,\tau}_{\bar{t},m} = a\Lambda\tilde{y}^{h,\tau}_m + \varphi_{\tilde{y},m} \text{ on } \omega_{h,\tau}, \quad (44)$$

$$\tilde{y}^0_m = \tilde{y}(x_m, 0) = p_{2h} \text{ on } \bar{\omega}_{h,0}, \quad (45)$$

$$\tilde{y}^j_0 = \tilde{y}(0,t_j) = p_{1h}(\tilde{w}) \text{ on } \omega_{0,\tau},\ \tilde{y}^j_N = \tilde{y}(b,t_j) = p_{3h}(\tilde{w}) \text{ on } \omega_{b,\tau}, \quad (46)$$

and denote the solution by $\tilde{y}$. Here, p_{ih} are defined by (41)-(43) and $\tilde{y}^{h,\tau}_{\bar{t},m}$, $\Lambda\tilde{y}^{h,\tau}_m$ are formulae analogous to (23) and (24) respectively, also $\varphi_{\tilde{y},m} = \partial_x\partial_t f\,|_{(x_m,t_{j+1})}$.

Theorem : *The solution* $\tilde{z}$ *of the finite difference problem (35)-(37) satisfies*

$$\max_{\bar{\omega}_{h,\tau}} |\tilde{z} - z| \le c_1\left(h^2 + \tau\right) \quad (47)$$

and the solution $\tilde{y}$ *of the finite difference problem (44)-(46) satisfies*

$$\max_{\bar{\omega}_{h,\tau}} |\tilde{y} - y| \le c_2\left(h^2 + \tau\right) \quad (48)$$

where, $z = \partial_t^2 u$ is the exact solution of Problem 4, and $y = \partial_x\partial_t u$ is the exact solution of Problem 5 also, c_1, c_2 are constants independent from h and τ.

Remark : *The extended theoretical analysis and numerical examples will be presented in the full research article.*

References

1. J.Douglas, H.H. Rachford, Trans. Amer. Math. Soc. 82, No.2. 421-439, (1956)
2. H.G. Landau, Quart. Appl, Math. 8, 81-94 (1950)
3. M. Lotkin, Quart. Appl, Math. 18(1), 79-85 (1960).
4. O.A. Ladyzenskaja, V.A. Solonnikov, N.N. Ural'ceva, Translation of Mathematical Monographs, American Mathematical Society, USA, (1967).
5. S.C. Buranay, L.A. Farinola, Proceedings of Fourth International Conference on Analysis and Applied Mathematics (ICAAM 2018), AIP publishing (accepted for publication).
6. A.A. Samarskii, Marcel Dekker, Inc. New York (2001).

A Soft Approach to Ring-Groupoids

Gülay Oğuz[1,*], *M.Habil* Gürsoy[1], and *Ilhan* Içen[1]

[1]Department of Mathematics, Faculty of Science and Arts, Inonu University, Malatya, TURKEY

Abstract. This study introduces a soft approach to the concept of ring-groupoid which is the one of the structured groupoids. Some properties and characterizations of soft ring-groupoids are established. Also, the category of soft ring-groupoids constructed by the homomorphism between two soft ring-groupoids is presented.

1 Introduction

Several theories have been developed to model the uncertainties appearing in engineering, economics, medicine, and many other field. Soft set theory is one of the most powerful theories that modeling the uncertainty and vagueness [1]. It, which was initiated by Molodtsov in 1999, has been worked on many areas including mathematics. The boundaries of this theory, which attracted a great interest of mathematicians, have been expanded in algebraic, topological and categorical sense [2-9, 12-14].

The first algebraic studies on this theory were presented by Maji and et.al. [2]. They presented different operators on soft sets and obtained several of their fundamental features. After that, Aktas and Cagman described the concept of soft group, which is a parametrized collection of subgroups of a group, over a group [3]. In [7], Acar and et.al. defined the notion of soft ring over a ring. On the other hand, the topological studies on this theory were first made by Shabir and Naz [4]. They examined the separation axioms on a soft topological space by defining soft topological space. Later, Cagman and et.al., and Min established some basic results by studying on soft topological spaces and algebraic structures [6, 9].

Recently, categorical studies on soft theory have been carried out. We refer to [12-14] for these studies. The definition of soft category was first proposed by Sardar and Gupta [12]. Also, they introduced the main notions of soft category theory and explained the relationship between soft category and fuzzy category.

Also, another concept in the focus of this study is groupoid. Although the notion of groupoid was first mentioned by Brandt, pioneering work on it was done by Ehresmann and Pradines [10, 20-21]. Subsequently, some algebraic structures were added to the groupoid, which the composition of groupoid is compatible with operators of these added structures. One of these is the ring groupoid which is defined as a groupoid endowed with a ring structure [17]. Afterwards, Gursoy extended the notion of ring groupoid to the generalized ring-groupoid [18].

In this manuscript, we describe the notion of soft ring-groupoid and discuss its features. Subsequently, by the definition of soft ring-groupoid homomorphism , we establish the cate-

*Corresponding author: gulay.oguz@inonu.edu.tr

gory of soft ring-groupoids. We also present the definition of soft subring-groupoid. Lastly, we complete this manuscript with a summary for further research.

In summary, we contribute to this expanding literature by offering a different approach to soft set theory.

2 Preliminaries

Here, we give some of the main notions such as soft set, soft ring, groupoid and ring-groupoid that we will use later. For more about these concepts, please refer [1 – 3, 7, 11, 15 – 17].

Let X be a set of parameters and let $P(U)$ denotes the power set of U for the initial universe set U.

Definition 2.1 [1] *Let A be a non-empty subset of X. A pair (F, A) is said to be a soft set over U, in which F is a mapping defined by*

$$F : A \longrightarrow P(U)$$

That is, a soft set over U can actually be considered as a parametrized collection of subsets of the universe U.

In this study, we will sometimes use the triplet (U, F, A) instead of soft set (F, A) over U.

Definition 2.2 [2] *A soft set (U, F, A) is called a null soft set, indicated by Φ, if $F(\alpha) = \emptyset$ for all $\alpha \in A$.*

Definition 2.3 [2] *The support of the soft set (F, A) over U is defined as the set $Supp(F, A) = \{\alpha \in A \ : \ F(\alpha) \neq \emptyset\}$.*

Definition 2.4 [2] *Let (U, F, A) and (U, F', B) be two soft sets. Then, (U, F', B) is said to be a soft subset of (U, F, A) if*
i. *$B \subset A$, and*
ii. *$F(\beta)$ and $F'(\beta)$ are identical approximations for all $\beta \in B$.*

We denote it as $(F', B)\widetilde{\subset}(F, A)$.

Now we proceed on to recall the concept of soft ring. Suppose that $\mathcal{R}$ is a commutative ring and A be a non-empty set. Then,

Definition 2.5 [7] *For the non-null soft set (F, A) over $\mathcal{R}$, (F, A) is said to be a soft ring over $\mathcal{R}$ if and only if $F(\alpha)$ is a subring of $\mathcal{R}$ for all $\alpha \in A$.*

From here, it is said that the soft ring (F, A) is a parametrized family of subrings of the ring $\mathcal{R}$. In what follows, the notation $(\mathcal{R}, F, A)$ stands for the soft ring (F, A) over $\mathcal{R}$.

Definition 2.6 [7] *Let $(\mathcal{R}, F, A)$ and $(\mathcal{R}, F', B)$ be two soft rings. Then, $(\mathcal{R}, F', B)$ is said to be a soft subring of $(\mathcal{R}, F, A)$ if*
i. *$B \subset A$,*
ii. *$F'(\beta)$ is a subring of $F(\alpha)$ for all $\beta \in Supp(F', B)$.*

Definition 2.7 [7] *Let $(\mathcal{R}, F, A)$ and $(\mathcal{R}', F', B)$ be two soft rings and let $\theta : \mathcal{R} \longrightarrow \mathcal{R}'$ and $\psi : A \longrightarrow B$ be two mappings. Then, the pair (θ, ψ) is said to be a soft ring homomorphism if*
i. *θ is a ring epimorphism,*
ii. *ψ is a surjection, and*
iii. *$\theta(F(\alpha)) = F'(\psi(\alpha))$ for all $\alpha \in A$.*

In this case, (F, A) is called soft homomorphic to (F', B) and this situation is indicated by $(F, A) \sim (F', B)$. Further, (F, A) is called softly isomorphic to (F', B) if θ is a ring isomorphism and ψ is bijective.

Here, the definitions of groupoid and ring-groupoid will be presented assuming that the foundation of the category concept is known.

Definition 2.8 [15] *A groupoid* $\mathcal{G}$ *is a category* $C = (Mor(C),\ Ob(C),\ s,\ t,\ m,\ i)$ *where each morphism is an isomorphism.*

Definition 2.9 [15] *Let* $\mathcal{G}$ *and* $\mathcal{G}'$ *be two groupoids. Then,* $\mathcal{G}'$ *is called a subgroupoid of* $\mathcal{G}$ *if* $\mathcal{G}'$ *consists of the subcollections of* $Mor(\mathcal{G})$ *and* $Ob(\mathcal{G})$*, where it is closed under the composition and inversion in* $\mathcal{G}$*.*

The following definition is due to Mucuk [17].

Definition 2.10 [17] *A ring-groupoid* $\mathcal{R}$ *is a groupoid endowed with a ring structure such that the following maps are groupoid homomorphisms:*
i. $m : \mathcal{R} \times \mathcal{R} \longrightarrow \mathcal{R}$, $(x, y) \longmapsto x + y$, *group operation,*
ii. $u : \mathcal{R} \longrightarrow \mathcal{R}$, $x \longmapsto -x$, *group inversion map,*
iii. $n : \mathcal{R} \times \mathcal{R} \longrightarrow \mathcal{R}$, $(x, y) \longmapsto xy$, *ring operation, and*
iv. $e : \{\star\} \longrightarrow \mathcal{R}$, *where* $\{\star\}$ *is a singleton.*

A ring-groupoid homomorphism between two ring-groupoids is a homomorphism of underlying groupoids which preserves the ring structure. Therefore, the category of ring-groupoids is established with ring-groupoids and their homomorphisms.

Example 2.1 [17] *For a ring* $\mathcal{R}$ *, the cartesian product* $\mathcal{R} \times \mathcal{R}$ *is a ring-groupoid over* $\mathcal{R}$ *with the ring operation described by* $(x, y)(x', y') = (xx', yy')$ *for all* $x, y, x', y' \in \mathcal{R}$.

Definition 2.11 [19] *Let* $\mathcal{R}$ *be a ring-groupoid and* $\mathcal{P} \subset \mathcal{R}$. *Then* $\mathcal{P}$ *is called a subring-groupoid of* $\mathcal{R}$ *If* $\mathcal{P}$ *is a ring-groupoid with the groupoid structure maps and ring maps induced from* $\mathcal{R}$.

3 Soft ring-groupoids

In this section, we will describe the notion of soft ring-groupoid.

Definition 3.1 *Let* $\mathcal{R}$ *be a ring-groupoid and let* $P(\mathcal{R})$ *denotes the set of all subring-groupoids of* $\mathcal{R}$ *. A pair* (F, A) *is said to be a soft ring-grupoid over* $\mathcal{R}$ *if each* $F(\alpha)$ *is a subring-grupoid of* $\mathcal{R}$ *for all* $\alpha \in A$*, where* F *is a mapping given by*

$$F : A \longrightarrow P(\mathcal{R})$$

and A *is a set of parameters.*

In fact, a soft ring-groupoid can be considered as a parameterized collection of subring-groupoids of the ring-groupoid $\mathcal{R}$. After that, we will occasionally write $(\mathcal{R}, F, A)$ for the soft ring-groupoid (F, A) over the ring-groupoid $\mathcal{R}$.

Example 3.1 *Let* (F, A) *be a soft ring over the ring* $\mathcal{R}$ *with identity. Then, since* (F, A) *is a soft ring, each* $F(\alpha)$ *is a subring of* $\mathcal{R}$ *for all* $\alpha \in A$*. From the viewpoint of groupoid theory, every ring is a groupoid with only one object in which all of the arrows are the elements of the ring and the composition of arrows is the multiplication operation of the ring and the addition of* *arrows is the addition of elements of the ring. Furthermore, every ring is a ring-groupoid. For this reason , the ring $\mathcal{R}$ and $F(\alpha)$ are ring-groupoids for all $\alpha \in A$. From here, it is easy to verify that $F(\alpha)$ is a subring-groupoid of $\mathcal{R}$ for all $\alpha \in A$. Therefore, (F, A) is a soft ring-groupoid over $\mathcal{R}$.*

In this example, it is not hard to see that each soft ring is a soft ring-groupoid as above. As a more specific example, the following can be given.

Example 3.2 *Let (F, A) be a soft ring over $\mathcal{R}$. Hence, the mapping F can be written as*

$$\begin{array}{rcl} F : A & \longrightarrow & P(\mathcal{R}) \\ \alpha & \mapsto & F(\alpha) \end{array}$$

Using this mapping, we can define a mapping F' as follows:

$$\begin{array}{rcl} F' : A & \longrightarrow & P(\mathcal{R} \times \mathcal{R}) \\ \alpha & \mapsto & F'(\alpha) = F(\alpha) \times F(\alpha) \end{array}$$

we know from the groupoid theory that If $\mathcal{R}$ is a ring, $\mathcal{R} \times \mathcal{R}$ is a ring-groupoid over $\mathcal{R}$ with the ring operation defined by $(x, y)(x', y') = (xx', yy')$ for all $x, y, x', y' \in \mathcal{R}$. Furthermore, each $F'(\alpha) = F(\alpha) \times F(\alpha)$ is a subring-groupoid of $\mathcal{R} \times \mathcal{R}$ with the groupoid structure maps and ring maps induced from $\mathcal{R} \times \mathcal{R}$ for all $\alpha \in A$. So (F', A) is a soft ring-groupoid over $\mathcal{R} \times \mathcal{R}$.

Therefore, it is important to note that a soft ring-groupoid can be obtained from each soft ring.

Definition 3.2 *For two soft ring-groupoids $(\mathcal{R}, F, A)$ and $(\mathcal{P}, F', B)$, define the pair $(\mathcal{K}, \psi)$, where $\psi : A \longrightarrow B$ is a surjection and $\mathcal{K} : \mathcal{R} \longrightarrow \mathcal{P}$ is a functor. Then, the pair $(\mathcal{K}, \psi)$ which satisfies the following conditions is called a soft ring-groupoid homomorphism if it is a soft ring homomorphism:*

i. *$\mathcal{K}$ is full.*

ii. *$\mathcal{K}(F(a)) = F'(\psi(\alpha))$ for all $\alpha \in A$.*

In this manner, we construct the category of softring-groupoids whose objects are soft ring-groupoids and whose morphisms are soft ring-groupoid homomorphisms. This category is written as $SRgd$.

Example 3.3 *For two soft rings $(\mathcal{R}, F, A)$ and $(\mathcal{P}, F', B)$, let the pair (θ, ψ) be a soft ring homomorphism between them as defined in Definition 2.7. It follows from Example 3.1 that the soft rings $(\mathcal{R}, F, A)$ and $(\mathcal{P}, F', B)$ are soft ring-groupoids. Also, the conditions **i.** and **ii.** in Definition 2.7 imply that the soft ring homomorphism (θ, ψ) is a soft ring-groupoid homomorphism.*

We now present a concept related to the soft ring-groupoids as follows:

Definition 3.3 *Let $(\mathcal{P}, F', B)$ and $(\mathcal{R}, F, A)$ be two soft ring-groupoids. Then $(\mathcal{P}, F', B)$ is called a soft subring-groupoid of $(\mathcal{R}, F, A)$ if the following conditions are satisfied:*

i. *$B \subset A$.*

ii. *$F'(\beta)$ is a subgroupoid of $F(\beta)$ for all $\beta \in Supp(F', B)$.*

iii. *$Ob(F'(\beta))$ and $Mor(F'(\beta))$ are subrings of $Ob(F(\beta))$ and $Mor(F(\beta))$ for all $\beta \in Supp(F', B)$, respectively.*

Example 3.4 *For the soft ring-groupoid $(\mathcal{R}, F, A)$, let us constitute*

$$\mathcal{P} = \{Id_x : x \in Ob(F(\alpha)), \forall \alpha \in A\}$$

. It is straightforward to see that $(\mathcal{P}, F, A)$ is a soft subring-groupoid of $(\mathcal{R}, F, A)$.

4 Conclusion

In this manuscript, the notion of ring-groupoid is examined in the context of soft set theory and some fundamental features are presented. As the notion of ring-grupoid is softfied , it would be interesting to develop the other concept of groupoid theory with this setting.

References

[1] D. A. Molodtsov, Soft set theory-First results, Comput. Math. Appl. **37**(4-5) (1999), 19-31.

[2] P. K. Maji, R. Biswas and A. R. Roy, Soft set theory, Comput. Math. Appl. **45**(4-5) (2003), 555-562.

[3] H. Aktas and N. Cagman, Soft sets and soft groups, Inform. Sci. **77**(13) (2007), 2726-2735.

[4] M. Shabir and M. Naz, On soft topological spaces, Comput. Math. Appl. **61**(7) (2011), 1786-1799.

[5] T. Shah and S. Shaheen, Soft topological groups and rings, Ann. Fuzzy Math. Inform. **7**(5)(2014), 725-743

[6] N. Cagman, S. Karatas and S. Enginoglu, Soft topology, Comput. Math. Appl. **62**(1) (2011), 351-358.

[7] U. Acar, F. Koyuncu , B. Tanay, Soft sets and soft rings, Comput. Math. Appl. **59** (2010), 3458- 3463.

[8] A. O. Atagun, A. Sezgin, Soft substructures of rings, fields and modules, Comput. Math. Appl. **61** (2011), 592-601.

[9] W.K. Min, A Note on Soft Topological Spaces, Comput. Math. Appl. **62** (2011), 3524-3528.

[10] H. Brandt, Uber eine Verallgemeinerungdes Gruppenbegriffes, Math. Ann. **96** (1926), 360-366.

[11] S. Mac Lane, Categories for working mathematician, Springer-Verlag, New York (1998)

[12] S.K. Sardar, S. Gupta, Soft category theory-an introduction, J. Hyperstruct. **2** (2013), 118-135.

[13] B. P. Varol, A. Shosttak, and H. Aygun, Categories related to topology viewed as soft sets, EUSFLAT-LFA 2011, **1**(1)2011, 883-890.

[14] O. Zahiri, Category of soft sets, An. Univ. Craiova Ser. Mat. Inform. **40**(2) (2013), 154–166.

[15] R. Brown, Topology and Groupoids, BookSurge LLC, North Carolina (2006).

[16] R. Brown and C. B. Spencer, G-groupoids, crossed modules and the fundamental groupoid of a topological group, Proc. Konn. Ned. Akad. v. Wet. **79** (1976), 296-302.

[17] O. Mucuk, Coverings and ring-groupoids, Georgian Math. J. **5**(5) (1998), 475-482.

[18] M H. Gursoy, Generalized ring-groupoids, An. Univ. Craiova Ser. Mat. Inform. **44**(1) (2017), 126–136.

[19] M. Q. Mann'a, Some properties of topological ring-groupoid, Int. J. Contemp. Math. Sci. **7** (11) 2012, 517 - 529.

[20] C. Ehresmann, Oéuvres complétes et commentées, Cahiers de topologie et géométrie différentielle, Parties I.1-I.2, Dunod, Paris (1950).

[21] J. Pradines, Théories de Lie pour les groupoides différentiables: relations entre propriétés locales et globales, C. R. Acad. Sei. Paris Sér. **263** (1966), 907-910.

Option of the Optical-mechanical Positioning of the Sensor and Sample in the Magnetometer. Determination of Magnetic Susceptibility of Powders and particles

Alexander Sandulyak[1], *Anna* Sandulyak[1,*], *Dmitriy* Kiselev[1], *Maria* Polismakova[1], and *Hasan* Bulut[2]

[1] Moscow technological university, Moscow, RF

[2] Department of Mathematics, University of Fırat, Elazığ, Turkey

Abstract. In Faraday's magnetometer it is recommended to apply remotely the located poles hemispheres. In this case (and similar cases) there is a need for expeditious and exact positioning of the measuring sensor (Hall) and the studied sample – for obtaining the most authentic characteristics of induction and its gradient, identification and use of a zone of stability. This problem is solved by the corresponding optical-mechanical system of positioning. It consists of the laser modules promoting aim positioning, the Web camera promoting final positioning. On the received concentration dependences of magnetic susceptibility of powder samples existence of limited line sections is confirmed that gives the chance to define susceptibility of particles. The critical relation to use for the similar purposes of samples (disperse) in the form of colloids and suspensions expresses.

1 Introduction

Faraday's ponderomotive method is preferable in a case of statement of the problem about determining a sample' s magnetic susceptibility χ with a small volume V. Herewith it could be both solid and disperse sample (for example, powder sample of ferro- or ferromagnetic particles, which are disperse phase of many technological mediums, including those which are affected by magnetophoresis and magnetic control [1-6]). A magnetic (ponderomotive) force F, acting on study "microsample", becomes basic measurement value when magnetic loop of Faraday balance of field with intensity H (induction $B = \mu_0 H$) and nonuniformity *gradH* (*gradB*) is creating between pole pieces. Considering that its functional view, as a rule, is determining by expression:

$$F = \mu_0 \cdot \chi \cdot V \cdot HgradH = \chi \cdot V \cdot BgradB / \mu_0, \quad (1)$$

* Corresponding author: anna.sandulyak@mail.ru

where μ_0 = $4 \cdot 10^{-7}$ H/m – magnetic constant, so having known (measured) values of expression' s parameters and using subsequent calculating formula:

$$\chi = \frac{F}{\mu_0 VHgradH} = \frac{F\mu_0}{VBgradB} \quad (2)$$

quantitative information about χ becomes available.

Placement of the sample in a certain (working) zone of the interpolar region, in which the field nonuniformity must be constant [7-13], is considered an important condition for the implementation of the method. Proper attention has not been paid to this problem at the literature. Existing recommendations about choosing a configuration of pole pieces are greatly general, still within question about the necessity for distinct sample positioning.

2 Discussion and results

The coordinate characteristic of parameter H or B is needed for objective estimation about acceptance of choosing pole pieces and for having an information about, for example, parameter's behavior of the *grad B* (*grad H*) in one x-direction or another at interpolar region (for example, in direction of action of the ponderomotive force). Hall sensor, in particular, is completely applicable for obtaining such characteristic, moreover essentially in detail, with small spacing (in choosing x-direction). It is this characteristic B, presented at graphic or/ and analytical form, being basic, lets getting coordinate characteristic, for example, parameter *grad B*. Then it can be inferred, at first, by presence (or absence) of a region of interest (i.e. region with constant *gradB* values) in a certain interpolar region and, in second, by coordinates of its location and thus coordinates of sample positioning.

So, if the case in hand is diagnostics of interpolar region using parameter B in one x-direction or another (for example, direction of force F action) for finding a region with a stable value of *gradB*, it is enough to identify a linear (close to linear) segment, even short (comparable with sample' s dimensions) on the obtained dependence of B on x. The presence of such a segment means that $gradB = dB/dx$ = Const at this location. In this regard it must be admitted that obtaining and analysis of dependence B on x (H on x) are outstanding necessary step for realization of the method.

Herewith it should not go unspoken that these dependences are nonlinear in whole for pole pieces with traditional and specific form. This fact disagrees an approach in finding linear segment of dependence of B (or H) on x in some way.

The essence of a simple and affordable approach for excluding this disagreement consists in selecting pole pieces shaped such that the key characteristic $B(x)$ or $H(x)$ is tortuous between them, i.e., has an inflection. Then, its segment (which is short relative to the entire unfolded nonlinear characteristic) in the vicinity of the inflection point can be considered close to the desired linear one – with almost stable value of *gradB* (*gradH*). In this case the corresponding derivative of the characteristic B (or H), i.e., the characteristic *gradB* (*gradH*), has an extremum here; it is the neighborhood of this extremum that is an indication of the presence of a zone with an invariable *gradB* (*gradH*) value.

Results of the theoretical and experimental researches about magnetizing (by field with intensity H_0) chain of balls point to acceptable (and easy-designed) form of pole pieces [14-17] (Fig. 1*a*). Precisely these nonlinear characteristics of intensity H ($H >> H_0$), having an inflection, are representative of region between opposite spherically shaped surfaces in the direction x (Fig. 1*a*) – from line between centers of balls and normal to it (along "bisecting line" of wedgegap between balls). So, there is a distinct extremum of characteristic dH/dx at the inflection point (Fig. 1*b*) [14-17]. Consequently, from the standpoint of the creation of a guaranteed working region, the using of spherically shaped pole pieces (Fig. 2) appears to be one of the most effective solutions to the problem of selecting the pole pieces for a Faraday balance [18-20].

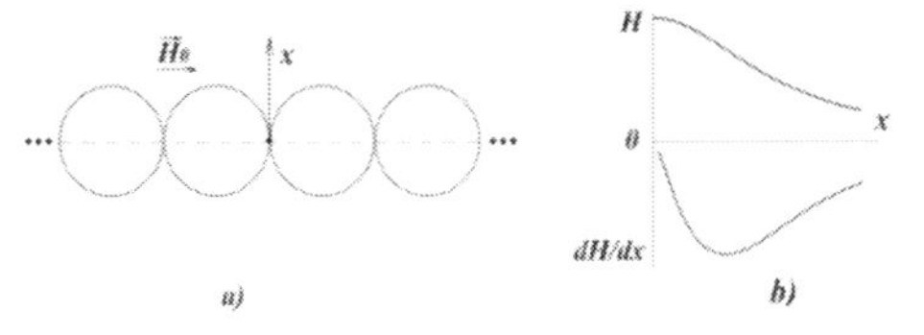

Fig.1. The chain of balls in field of solenoid with intensity H_0 (*a*) and corresponding illustration of coordinate (along *x*) characteristics [14-17] of intensity *H* and nonuniformity *dH*/*dx* of field at wedgegap between the spherical pole surfaces (*b*)

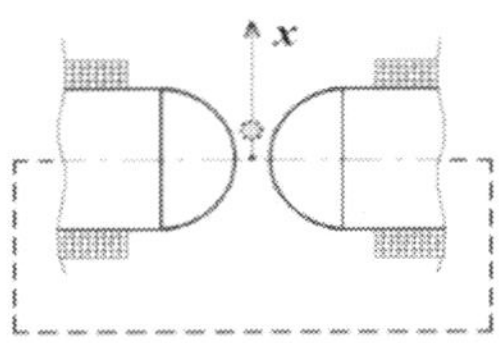

Fig.2. The variant of creation a working region (is located near the abscissa of the inflection of the coordinate characteristic of the magnetic field strength or induction and the abscissa of the extremum of the gradient, correspondingly) between the spherical pole pieces

Figure 3 (points) presents the measured values of induction *B* in the plane of symmetry of the region between the pole pieces–hemispheres 100 mm in diameter (separated by 10 mm) at different distances *x* from the axis and different values *I* of the winding current. The inflection of each characteristic *B*(*x*) is clearly seen. Identification of the inflection allows linear approximation of data in the vicinity of the inflection point (shown with dashed lines) that indicates the presence and location of the working region. An extremum view of characteristic *grad B* indicates the presence, too (Fig. 4).

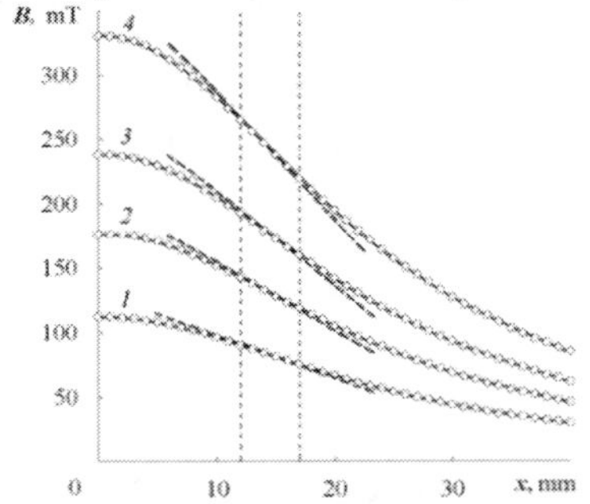

Fig.3. The characteristics of the field induction at the symmetry plane of region between the pole pieces shaped as hemispheres 100 mm in diameter with the distance between them is 10 mm for different values of the winding current *I*: *1* - *I* =4A, *2* – *I*=8A, *3* – *I* = 16A, *4* – *I* = 40A; the dots are the experimental results, the lines present the results of calculation by (3)

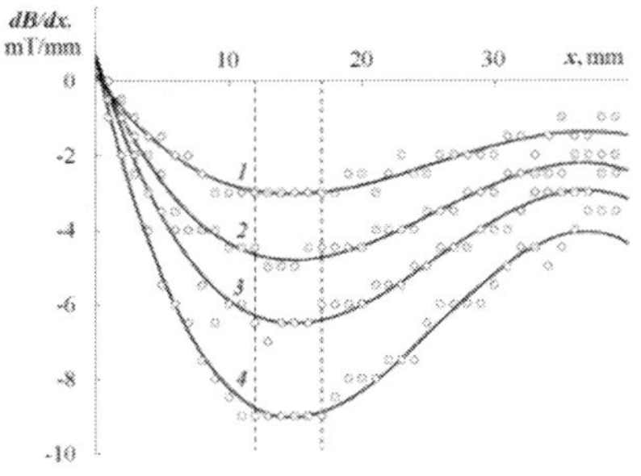

Fig.4. The characteristics of the induction gradient (by data of Fig. 3); dashed and solid lines – are the calculation by using equation (4), which is written for polynomial (3) of 5th and 4th degree correspondingly

It should be used analytical (phenomenological) approximation at analysis of characteristic *B* – with subsequent mathematical differentiation and obtaining *gradB*-characteristic herewith obvious identifies an extremum's absciss. In this case it is appropriate using, for example, an approximation function of polynomial [18-20]:

$$B = a_1 x + a_2 x^2 + a_3 x^3 + a_4 x^4 + ... + a_0 \quad (3)$$

with fourth, maximum fifth degree – it is enough.

Analytical dependences *B* on *x* such as (3) could be easy differentiate, i.e. *gradB*=*dB*/*dx* characteristics could be obtained:

$$dB/dx = a_1 + 2a_2x + 3a_3x^2 + 4a_4x^3 + \ldots \qquad (4)$$

Dependences (4), which are obtained for polynomial (3) of 5th and 4th degree, are obvious extremal (Fig. 4; dashed and solid lines corresponding), herewith they are in good agreement with each other. The mutual noticeable discrepancy of polynomial (3) of 5th and 4th degree is observed only at high and low x values (Fig. 4). However, taking the fact into account that these values B are not important here a simpler polynomial of the 4th degree is preferable for approximating the B data, and quadrinomial dependence correspondingly (4).

Moreover, it provides an analytical formula for calculating the abscissa of the extremum $x= x_{extr}$ of the *gradB* characteristic (Fig. 4) if we based on polynomial (3) of 4th degree. In fact, we could obtain a calculating formula, if we differentiate quadrinomial dependence (4) with subsequent nulling obtained equation of the 2nd degree, i.e. $a_2+3a_3x_{extr}+6a_4(x_{extr})^2=0$:

$$x_{extr} = \left(-3a_3 + \sqrt{9a_3^2 - 24a_2a_4}\right)\Big/(12\cdot a_4)\cdot \qquad (5)$$

Formula (5) allows calculating abscissas values of the extrema x_{extr} of characteristics *gradB*.

When the current load varies from experiment to experiment the sample location may remain unchanged in the course of the measurements because of abscissas of the extrema are mutually identical (Fig. 3, 4). In this case, the size of the working region in the interpolar area under investigation (between choosing pole pieces), i.e., the region of practically stable *gradB* values, is (in terms of x) 12−17 mm. This is clearly seen in Fig. 3, where the dashed straight lines, as noted above, show the possibility of linearly approximating short segments of dependences $B(x)$ in the vicinity of their inflection point (the corresponding interval of x is marked). This can be seen even more clearly in Fig. 4 on the x interval (it is also highlighted), where the dB/dx value near the extrema of dependences dB/dx on x is relatively stable.

It should be noted that when using pole-hemispherical pieces, additional complications (and with them errors) can arise in the implementation of precise positioning (in particular, in ensuring an accurate starting position), first of all, Hall-effect device. This can adversely affect the accuracy of obtaining the necessary coordinate characteristics of the field, and therefore – the accuracy of determining the working zone of a stable gradient and, as a consequence, the reliability of the result of the definition of susceptibility. We note that even with accurate determination of the working region, it is equally important to realize the exact positioning of the sample.

In this case, it is advisable to use the optomechanical positioning system of the measuring sensor and the sample in the interpolar region. It should include a bi-directional laser module, in particular with a crossed beam for initial (targeted) positioning of the sensor, and a Web camera (connected to a computer), with mutually perpendicular scales for initial and control positioning of the sensor and the sample. Herewith, initial the positioning of the Hall-effect device and its moving in the interpolar region is easy to provide with a three-coordinate table, rigidly connected with the Hall-effect device.

In order to derive the magnetic susceptibility χ of disperse phase' ferroparticles, with a certain degree of conditionality it is possible to use the famous relationship $\chi = \langle\chi\rangle/\gamma$ ($\langle\chi\rangle$ – magnetic susceptibility of disperse medium).

However, this expression is valid only for rather small values of volume fraction of ferroparticles γ (in the volume of studied dispersive medium) – for such volumes, where the mutual magnetic effect of the ferroparticles can be practically excluded. This is possible when the ferroparticles in a sample of the dispersive medium (powder is more preferable instead suspension or colloid in which ferroparticles are susceptible being formed in chains and aggregates [21]) are sufficiently mutually separated. In powder dispersive media, yet another dispersive phase from non-ferromagnetic particles (often obtained after pulverizing sand) is used to achieve a particular γ – values (a particular certain degree of dissociation of ferroparticles), which mixed with dispersive phase of ferroparticles.

Fig. 5 shows the family of dependences of $\langle\chi\rangle$ on γ (for various H) [22]. The dependences are well nigh linear throughout the studied range γ – up to γ =0.2–0.25. If one is guided by the results of linear approximation in this range of γ, then by relationship $\chi=\langle\chi\rangle/\gamma$ it is possible to determine the values of the susceptibility of particles χ as a coefficient of proportionality of each dependence from the family of $\langle\chi\rangle$ on γ dependences.

The values of χ as a function of H, obtained for the entire set of data, are presented on Fig. 6 in logarithmic coordinates. In such coordinates they are well-quasilinearized, thereby being represented in the specified range of H by a dependence similar to the power function ($\chi \sim 1/H^{0.75}$). With accuracy up to coefficient K = 11400, it takes the form $\chi = K/H_x^{0,75}$,where $H_x = H/1$ A/m is the field intensity converted to dimensionless form.

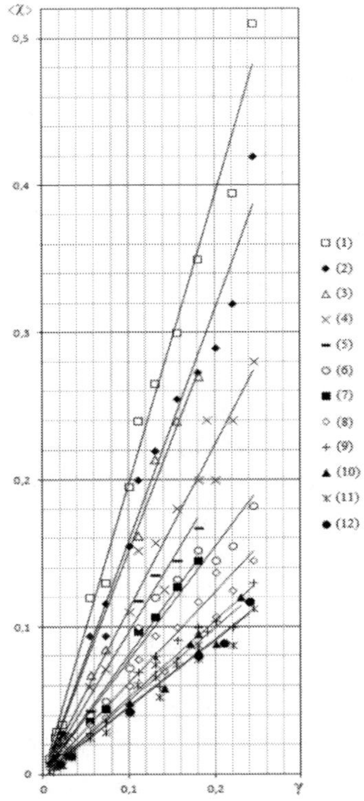

Fig.5. Effect of the volume fraction of magnetite particles in the sample of dispersive medium on its magnetic susceptibility for various values of field intensity H [22]; 1- H = 90*kA*/m, 2- 150, 3- 190, 4- 220, 5- 270, 6- 340, 7- 365, 8- 420, 9- 550, 10- 650, 11- 730, 12- 780

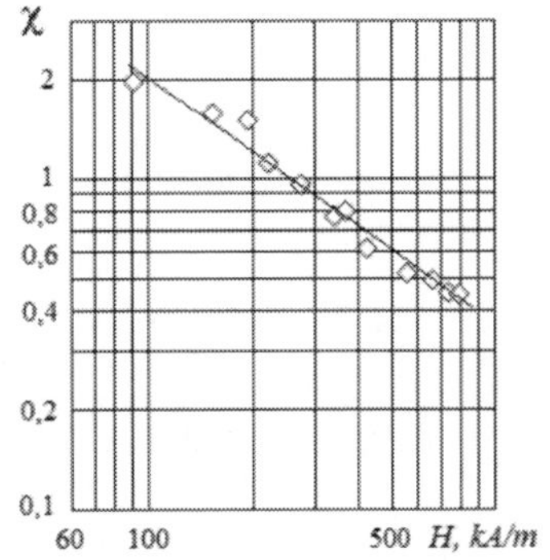

Fig.6. Dependence of the magnetic susceptibility of magnetite particles on magnetic field intensity

3 Conclusion

In this paper corresponding optical-mechanical system of positioning of the measuring sensor (Hall) and the studied sample is recommended. On the received concentration dependences of magnetic susceptibility of powder samples existence of limited line sections is confirmed that gives the chance to define susceptibility of particles.

Acknowledgement

The research is conducted with financial support from RFBR within the frame works of research project No 16-38-60034 mol_a_dk and from Russian Federation Ministry of Education and Science No 9.9626.2017.

References

1. S. Das, S. Chakraborty, S.K. Mitra. Microfluidics and Nanofluidics, **13(5)**, 799–807 (2012)
2. M. Kawano, H. Watarai, Analytical and Bioanalytical Chemistry. **403(9)**, 2645–2653 (2012)
3. K. Nandy, S. Chaudhuri, R. Ganguly, I.K. Puri. JMMM. **320**, 1398–1405 (2008)
4. V. Murariu, J. Svoboda. Phys. Separ. in Sci. and Engineering. **12(1),** 1–11 (2003)
5. A.A. Sandulyak, A.V. Sandulyak, B.M. Belgacem Fethi, D.O. Kiselev. Journal of Magnetism and Magnetic Materials. **401**, 902–905 (2016)
6. Patent USA №4492921. Sandulyak A.V., Garaschenko V.I., Korkhov O.J. Method of Determining the Quantity of Solid Fraction of Ferromagnetic Matter in a Fluid. 1985.
7. J.C.P. Klaasse. *The Faraday balance*, Van der Waals Zeeman Institute, November 1999, URL: https://iop.fnwi.uva.nl/cmp//klaasse/fdb.html.
8. V.I. Chechernikov. *Magnitnye izmereniya (Magnetic Measurements)*, Moscow: Izd. MGU, 1969, 388 p.
9. A.Y. Dorfman, S.K. Sidorov. Zh. Eksp. Teor. Fiz. **9(1)**, 25 (1939)
10. V.R. Karasik. *Fizika i tekhnika sil'nykh magnitnykh polei* (Physics and Technology of Strong Magnetic Fields), Moscow: Nauka, 1964, 348 p.
11. R.D. Heyding, J.B. Taylor, M.L. Hair. Rev. Sci. Instrum. **32(2)**, 161–163 (1961)
12. D. Govindarajan, R. Gopalakrishnan. International J. Minerals, Metallurgy and Materials. **16(3)**, 349–354 (2009)
13. R. Gopalakrishnan, S. Barathan, D. Govindarajan, American J. Materials Science. **2(1)**, 32–36 (2012)
14. A.V. Sandulyak. Magnetic and filtration purification of liquids and gases. Moscow: Chemistry, 1988, 136 p. https://dlib.rsl.ru/viewer/01001440011#?page=136
15. A.V. Sandulyak, A.A. Sandulyak, V.A. Ershova. Technical Physics. **54(5)**, 743–745 (2009)
16. A.A. Sandulyak, V.A. Ershova, D.V. Ershov, A.V. Sandulyak. Solid State Physics. **52(10)**, 1967–1974 (2010)
17. A.V. Sandulyak, A.A. Sandulyak, V.A. Ershova. Doklady Physics. **52(4)**. 179–181 (2007)
18. A.V. Sandulyak, A.A. Sandulyak, M.N. Polismakova, D.O. Kiselev, D.A. Sandulyak, V.A. Ershova. MATEC Web of Conferences 10, 01007 (2017)
19. A.V. Sandulyak, A.A. Sandulyak, M.N. Polismakova, V.A. Ershova, D.A. Sandulyak, D.O. Kiselev. Applied Physics, System Science and Computers, Lecture Notes in Electrical Engineering **428**, 77-83 (2016)
20. A.A. Sandulyak, A.V. Sandulyak, M.N. Polismakova, D.O. Kiselev, V.A. Ershova , D.A. Sandulyak. Instrum. and Experim. Tech. **61(1)**, 123-126 (2018)
21. A.A. Sandulyak, D.A. Sandulyak, M.N. Polismakova, A.V. Sandulyak, D.O. Kiselev, V.A. Ershova. J. Eng. Phys. and Thermophys. **90(4)**, 845-850 (2017)
22. D.A. Sandulyak, A.A. Sandulyak, D.O. Kiselev, A.V. Sandulyak, M.N. Polismakova, M.A. Kononov, V.A. Ershova. Meas. Tech. **60(9)**, 928-933 (2017)

ITM Web of Conferences **22**, 01014 (2018) https://doi.org/10.1051/itmconf/20182201014
CMES-2018

Measurement of the Actual Contact Area of Details by the Power of a Reflected Light Beam Directed to a Contact Surface Covered with a Film

Arkadiy Titov, *Anna* Sandulyak*, *Mikhail* Baharev, and *Alexander* Sandulyak

Moscow technological university, Moscow, RF

Abstract. Measurement of the actual area of details contact is important for control of friction and wear of such details. For this purpose, in particular, on a surface of one of details of couple covered with the thinnest layer of lubricant the thin film of coal is putted. On the raised dust surface impose the second detail of couple and squeeze them the application of normal loading. Light spots of contact are measured by planimetry. Instead of direct (labor-consuming) measurement of a large number of the areas of spots of contacts it is expedient to use integrated determination of the actual area of details – on the power of the beam of light reflected from ledges of a contact surface. The optical scheme of the device is developed, the efficiency of which for implementation is shown. Measurement of the actual area of details contact is important for control of friction and wear of such details. For this purpose, in particular, on a surface of one of details of couple covered with the thinnest layer of lubricant the thin film of coal is putted. On the raised dust surface impose the second detail of couple and squeeze them the application of normal loading. Light spots of contact are measured by planimetry. Instead of direct (labor-consuming) measurement of a large number of the areas of spots of contacts it is expedient to use integrated determination of the actual area of details – on the power of the beam of light reflected from ledges of a contact surface. The optical scheme of the device is developed, the efficiency of which for implementation is shown.

1 Introduction

At present, measuring the actual contact area of details are a very urgent task, since it largely determines the friction and wear of the details. Nowadays, the following methods for measuring the actual area of contact of surfaces are known: electrical; geometric; adhesive; optic; physicochemical [1]. Of the listed methods, the simplest and most convenient one should recognize adhesion, which is that a thin film of coal is deposited on the surface of one of the piece of a pair coated with a thin layer of lubricant, a second piece

* Corresponding author: anna.sandulyak@mail.ru

of a pair is impressed to the evaporated surface and compressed by applying a normal load, light blots of touch measured by planimetration [2]. The disadvantage of the method is low productivity, since it is necessary to measure a large number of areas of contact spots. To eliminate this shortcoming, we proposed an integral method for determining the actual contact area of details by the power of the light beam reflected from the projections of the contact area [3].

2 Calculations

To implement the recommended method, we proposed a device whose schematic structure is shown in Fig. 1. The device contains: 1 – frequency generator, 2 – power amplifier, 3 – LED, a 4 – collimator lens, 5 – detail coated by film, 6 – lens, a 7 – photodetector, 8 – amplifier, 9 – high frequency filter, 10 – computer. The device works as follows. From the generator 1 voltage through the amplifier 2 goes to the LED 3, the output of which produces an alternating luminous flux. LED 3 is in the focus of lens 4, forming a parallel light flux illuminating detail 5.

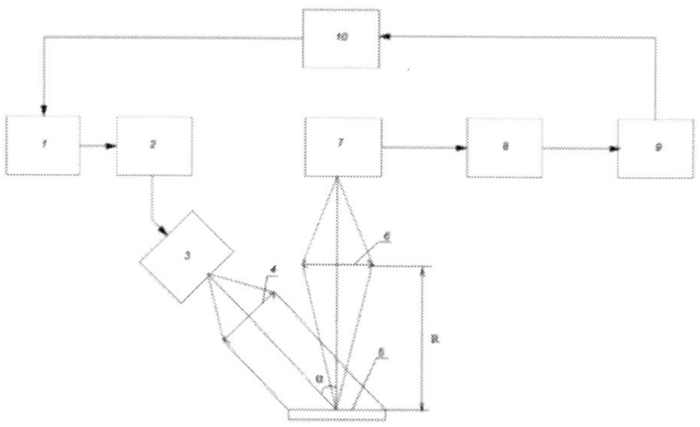

Fig.1. Schematic structure of the device for measuring the actual contact area of details

The light flux scattered from the surface of the detail 5 is collected by lens 6 on the photodetector 7, from the output of which the voltage through the amplifier 8 goes to the filter 9, which extinguishes the external illumination of the low frequency of 50 Hz. Then the signal goes to the computer 10, which processes the received information according to the developed algorithm, which will be considered later.

Define the basic characteristics of the device. To do this, it is necessary to determine the radiation power scattered from the surface of the detail 5. Suppose that the film and projections of part 5 reflect light according to Lambert's law (diffuse reflection). Then the power of the light beam reflected from the film entering the lens 6, taking into account [3, 4], can be found from the expression:

$$P_1 = \frac{IST_{пл}\tau D^2 \cos\alpha}{4R^2}, \quad (1)$$

where I - is the radiation intensity falling on the contact surface covered by the film; S - the area of the detail; $T_{пл}$ - reflection coefficient of the film; τ - loss of light in optics; D – diameter of lens 6; R - the distance from the contact surface of detail 5 to the lens 6; α - angle between the normal to the contact surface and the direction to the lens 4.

Then, the second detail is pressed against the contact surface of the first detail 5 and after deformation of the individual parts of the film, the second detail is removed. The contact surface is again illuminated with light. Then the radiation power reflected from the contact surface after deformation, taking into account [3], can be represented as:

$$P_2 = \frac{I[S_{пл}T_{пл} + S_в T_д]\tau D^2 \cos\alpha}{4R^2} \tag{2}$$

where $S_в$ - the area of the projections; $S_{пл}$ - area of the detail occupied by the film; $T_д$ - the reflection coefficient of the part corresponding to the reflection coefficient of the projections.

Taking into account that $S = S_в + S_{пл}$ and expressions (1) and (2), we find the area of the projections:

$$S_в = \left| \frac{4R^2 S_n (P_1 - P_2)}{PD^2 \cos\alpha (T_{пл} - T_д)\tau} \right| \tag{3}$$

where P - power of the radiator; S_n is the area of the beam illuminating the part.

In order for the light flux from the entire area of the contact surface to completely reach the sensitive area of the photodetector, it is necessary to determine the parameters of the lens and the distance between it and the photodetector. To do this, we determine the lens magnification:

$$\beta = -\frac{l_{фп}}{l_д} \tag{4}$$

where $l_{фп}$ is the size of the photodetector; $l_д$ is the largest detail size.

The distance R can be expressed through the lens magnification and its focal distance f' [4]:

$$R = \frac{(1-\beta)}{\beta} f' \tag{5}$$

The distance from the lens to the photodetector is:

$$a = (1-\beta) f' \tag{6}$$

Taking into account (5), expression (3) takes the form:

$$S_в = \left| \frac{4(1-\beta)^2 f'^2 S_n (P_1 - P_2)}{PD^2 \beta^2 \cos\alpha (T_{пл} - T_д)\tau} \right| \tag{7}$$

Let's find the sensitivity of the method, which is determined by the minimum area of projections that the device can detect. To do this, we use expression (7), in which the power difference P1¬P2 will correspond to the minimum power detected by the photodetector. Then using expression (7) taking into account [4] we obtain:

$$S_{в\min} = \left| \frac{4(1-\beta)^2 f'^2 S_n g w \sqrt{v}}{PD^2 \beta^2 \cos\alpha (T_{пл} - T_д)\tau} \right| \tag{8}$$

where g is the threshold sensitivity of the photodetector 7, w is the signal-to-noise ratio; v- Bandwidth of the device.

We estimate the accuracy of the measurement. To this end, we select in the expression (7) parameters that have systematic errors, to which S_n, α, $T_{пл}$, $T_д$, D can be assigned. These parameters can be measured when the instrument is adjusted. The remaining parameters included in expression (7) have random errors. The relative measurement error is found by differentiating expression (7) with respect to parameters that have random errors:

$$\frac{\Delta S_в}{S_в} = \sqrt{4\left(\frac{\Delta\beta}{(1-\beta)\beta}\right)^2 + 4\left(\frac{\Delta f'}{f'}\right)^2 + \left(\frac{\Delta P}{P}\right)^2} \tag{9}$$

We estimate the parameters entering into (9). As shown in [4], the errors occurring in (9) are respectively equal to: $\frac{\Delta P}{P} = 10^{-2}, \frac{\Delta f}{f} = 10^{-4}, \frac{\Delta \beta}{\beta} = 10^{-4}$.

Substituting the values of these parameters in (9), we obtain $\frac{\Delta S_в}{S_в} \approx 1\%$.

By expression (8), we determine the sensitivity of the device. For the experimental sample, elements with the following parameters were chosen. The source of radiation is a white LED with a power of 50 Lm. The photodetector has a sensitive area of 3x3 mm with a threshold sensitivity of 2.73 * 10^{-11} $Lm/Hz^{0.5}$. The condenser forms a beam of 21 mm in diameter on the investigated surface. The magnification of the lens 6 can be found from the expression (4) taking into account the size of the photodetector and beam diameter β = -0.142. The lens has a diameter of 20 mm and a focal length of 20 mm. The angle α=56°. The modulation frequency is chosen to be 5 kHz, and the signal-to-noise ratio w = 10. Coefficient of reflection from the black film $T_{пл}$ = 0.04, and from the detail $T_д$ = 0.8. Loss in optics τ = 0.7. Substituting the above parameters in (8), we obtain $S_{вmin} \approx 70\ \mu m^2$. When the emitter power is increased (using a laser), the sensitivity can be significantly increased.

Consider now how information is processed by a computer 10. As a medium for measuring the output voltage and generating a modulating one, we select the Lab View virtual programming environment and as a hardware part for receiving and transmitting data, a sound card.

Generation of the modulating signal is effected by the linear audio output of the computer sound card 10. The procedure for generating this signal passes in two stages. On the first stage, a sinusoidal signal is generated by a component representing itself as a virtual signal generator Simulate Signal. Figure 2 shows that the frequency is set to 5 kHz.

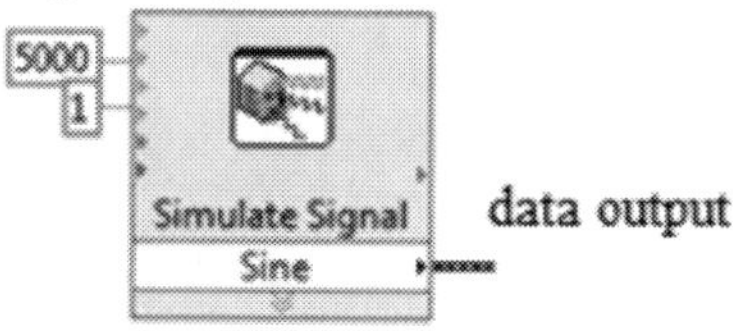

Fig.2. Virtual Generator

At the second stage, the signal is output via the line port of computer 10 via an element like Play Waveform. In this element, you can specify only the device number that will reproduce the signal.

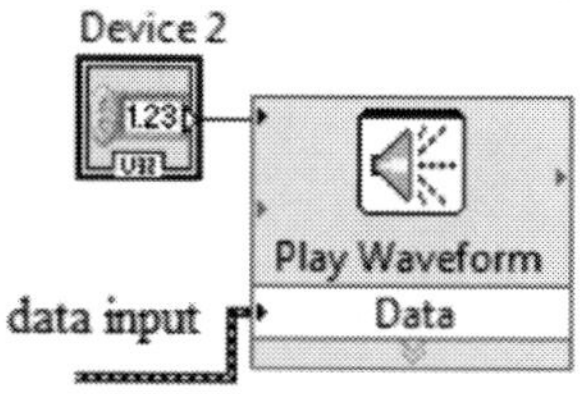

Fig.3. Element for signal output to the line output of the computer's audio card

The signal from the output of the filter 9 goes to the line input of the computer's audio card 10 and is entered into the program using the standard component Acquuire Sound, which accesses its audio card.

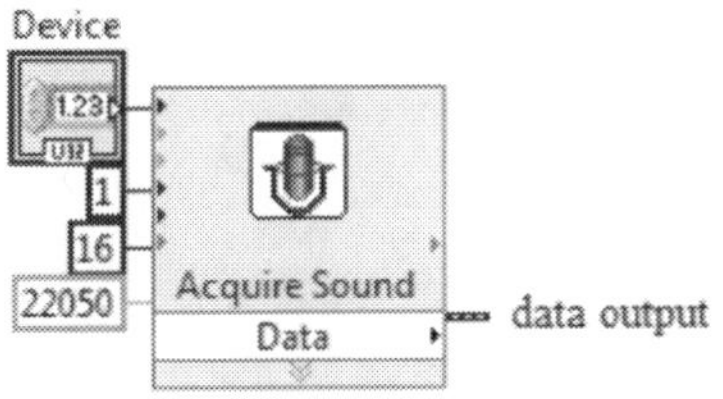

Fig.4. Data access element of the input line port

As you can see in Figure 4, on this component can be relocate device number (Device ID), sampling data per channel (Number of Samples/ch), sampling rate (95 kHz), bit data (16 bits) and the number of channels involved (1 channel).

Then the signal goes to the amplitude meter - the standard component Tone Measurements Express VI. Lab View. It determines the amplitude of the input signal. The tasks that this component solves, are the search for a single signal with the largest amplitude, or search for a specified frequency range to search for a single signal with the largest amplitude.

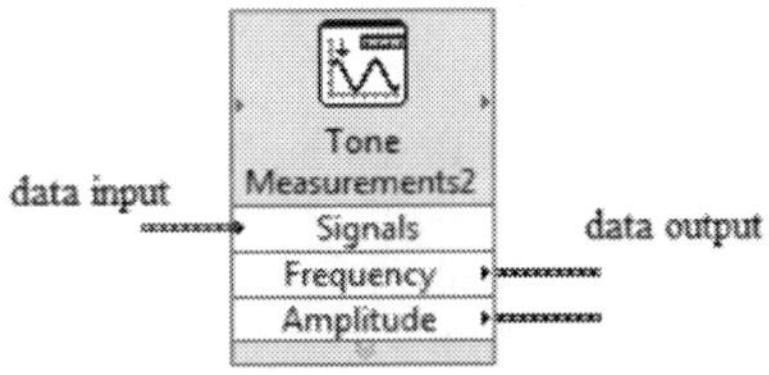

Fig.5. Tone Measurements element for analyzing the sinusoidal signal

To perform amplitude measurements, it is necessary to make a reference system. This function allows you to start measuring at any value of the initial signal. Therefore, the following circuit was shown, shown in Figure 6, performing the position setting to zero.

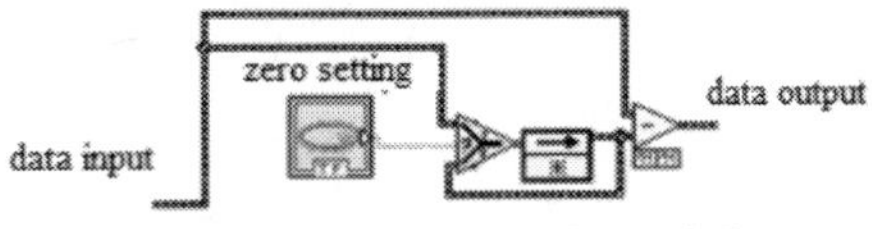

Fig.6. Link to set the origin to zero

Here, Lab View, by pressing a button, captures data from the bus coming from the amplitude analyzer and stores these values in the Feedback Node component (remembers the last value), and after disabling the button, the data is stored and subtracted from the signal coming from the amplitude analyzer. The overall system of program is shown in Fig. 7.

After the development and fabrication of the device model, experimental studies were carried out. To do this, samples were made with different contact areas.

These samples were in turn installed in the device instead of detail 5 (Fig. 1). Using computer 10, the dependence of the voltage of the signal on different contact areas was obtained. The graph of this relationship is shown in Fig. 8.

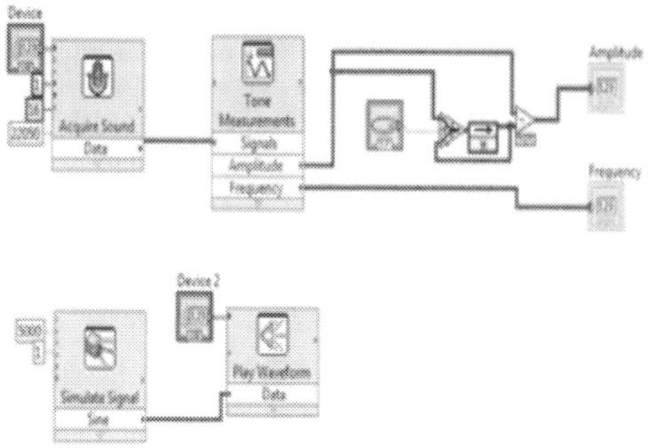

Fig.7. Program construction for the contact area meter

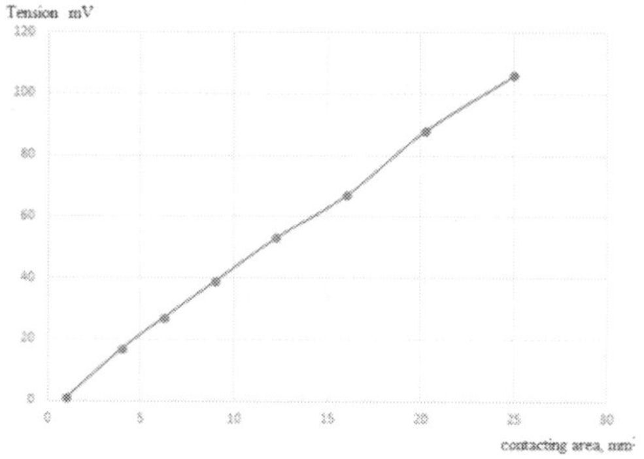

Fig.8. Voltage versus contact area

3 Conclusions

As you can see from Figure 8, the graph has a fairly linear character, which corresponds to the theoretical positions.

Thus, analytical expressions for calculating the basic parameters of the device were obtained in this paper, its sensitivity and measurement accuracy were determined, a schematic structure of the device and a special program for the computer were developed, a model was developed and the actual contact area of the details was measured.

4 Acknowledgements

The research is conducted with financial support from RFFI within the frameworks of research project № 16-38-60034 mol_a_dk and the support of the Ministry of Education and Science of the Russian Federation for the state task in the field of scientific activities no. 9.9626.2017.

References

1. V.M Gryazev. Izvestiya TulGU. Technical Science **10,** 26-32 (2013)
2. I.I. Berkovich, D.G. Gromakovskiy. *Tribology. Physical bases, mechanics and technical applications*: The textbook for higher education institutions. Samara (2000) p. 213.
3. Patent RU 2642915. Way of measurement of the actual area of contact of details. By Albagachiev A.Y., Titov A.A., Sandulyak A.A. (2018).
4. A.A. Titov. Measurement techniques. **56 (6),** 628-650 (2013).

ITM Web of Conferences **22**, 01015 (2018) https://doi.org/10.1051/itmconf/20182201015
CMES-2018

A highly accurate corrected scheme in solving the Laplace's equation on a rectangle

Adıgüzel A. Dosiyev[1,*] and *Hediye Sarıkaya* [1]

[1]Department of Mathematics, Near East University,Nicosia, TRNC, Mersin 10, Turkey

Abstract.A pointwise error estimation of the form $O(\rho h^8)$, h is the mesh size, for the approximate solution of the Dirichlet problem for Laplace's equation on a rectangular domain is obtained as a result of three stage (9-point, 5-point and 5-point) finite difference method; here $\rho = \rho(x, y)$ is the distance from the current grid point $(x, y) \in \Pi^h$ to the boundary of the rectangle Π.

1 Introduction

A highly accurate corrected scheme is one of the compact approximations for decreasing the number of unknowns, that gives the effective numerical solution of differential equations.

In [1] a two-stage difference method was given on the way of solving the Dirichlet problem for Laplace's equation on a rectangular parallelepiped. It was assumed that the sixth order derivative of the boundary functions which are given on the faces of a parallelepiped satisfy the Hölder condition, and on the edges, besides the continuity they satisfy the compatibility condition for second derivatives, which results from the Laplace equation. It was proved that by using a simple 7-point scheme in two stages the order of uniform error can be improved up to $O(h^4 ln\, h^{-1})$.

In this paper, a new three-stage difference method to solve the Dirichlet problem for Laplace's equation on the rectangular domain is given. By using at the first stage the 9-point, and at the second and the third stages the 5-point schemes for error of the approximate solution a pointwise estimation of order $O(\rho h^8)$ is obtained, where $\rho = \rho(x, y)$ is the distance from the current grid point $(x, y) \in \Pi^h$ to the boundary of the rectangle Π. Numerical experiment is given to support the theoretical analysis made.

The weighted estimates for the approximate solutions by the finite difference method of the Dirichlet problem for the Laplace and Poisson equations, at the rate as $(\rho + h)^\sigma h^2, 0 < \sigma \leq 1$, were obtained in [2]. A highly accurate finite difference approximations by correcting the right hand term using the high order differences of the numerical solution of differential equation without the weighted estimates was constructed in [3], [4].

2 Finite Difference Problem

Let $\Pi = \{(x,y): 0 < x < a, 0 < y < b\}$ be rectangle, a/b be rational, $\gamma_j(\gamma_j'), j = 1,2,3,4,$ be the sides, including (excluding) the ends, numbered counterclockwise begining by the left side ($\gamma_0 \equiv \gamma_4, \gamma_1 \equiv \gamma_5$) and let $\gamma = \cup_{j=1}^{4} \gamma_j$ be the boundary of Π. Denote by s the arclength, measured along γ, and by s_j the value of s at the beginning of γ_j. We say that $f \in C^{k,\lambda}(D)$, if f has kth derivatives on D satisfying a Hölder condition where $0 < \lambda < 1$

We consider the following boundary value problem

$$\Delta u = 0 \ \ on\ \Pi,\ \ u = \phi_j(s)\ on\ \gamma_j,\ \ j = 1,2,3,4, \tag{1}$$

where $\Delta \equiv \partial^2/\partial x^2 + \partial^2/\partial y^2$, ϕ_j are given functions of s. Assume that

---

[*]Corresponding author: adiguzel.dosiyev@emu.edu.tr

$$\phi_j \in C^{12,\lambda}(\gamma_j),\ 0 < \lambda < 1,\quad j = 1,2,3,4, \tag{2}$$

$$\phi_j^{(2q)}(s_j) = (-1)^q \phi_{j-1}^{(2q)}(s_j),\ q = 0,\ldots,5. \tag{3}$$

On Π, we introduce the functions $u^m,\ m = 4,8$ defined by

$$u^m = u^m(x,y) = \left(\frac{1}{2}\right)\left(\left(\frac{\partial^m u(x,y)}{\partial x^m}\right) + \left(\frac{\partial^m u(x,y)}{\partial y^m}\right)\right), \tag{4}$$

where u is the solution of Dirichlet problem defined by (1).

It is easy to check that for each of $m = 4,8$ the function (4) is the solution of following boundary value problem

$$\Delta u^m = 0 \ on\ \Pi, \tag{5}$$

$$u^m = \Psi_k^m = \Psi_k^m(y)\ on\ \gamma_k,\ k = 1,3, \tag{6}$$

$$u^m = \Psi_l^m = \Psi_l^m(x)\ on\ \gamma_l,\ l = 2,4, \tag{7}$$

where

$$\Psi_k^m(y) = \left(\frac{d^m\phi_k}{dy^m}\right), k = 1,3,\ \Psi_l^m(x) = \left(\frac{d^m\phi_l}{dx^m}\right), l = 2,4. \tag{8}$$

From (2), (3) and (8) it follows that the following $(12-m)-th$ order derivatives are bounded on Π:

$$\max_{0\le p\le\frac{12-m}{2}} \sup_{(x,y)\in\Pi} \left|\frac{\partial^{12-m}u^m}{\partial x^{2p}\partial y^{12-m-2p}}\right| < \infty, m = 4,8. \tag{9}$$

Let $h > 0,\ b/h \ge 8\ and\ a/h \ge 8$ where a/h and b/h are integers. We define Π^h a square grid net on Π with step size h, obtained by the lines $x, y = 0, h, 2h, \ldots$. Let γ_j^h be a set of nodes on the interior of γ_j, and let

$$\gamma^h = \bigcup_{j=1}^{4}\gamma_j^h, \dot{\gamma}_j = \gamma_{j-1}\cap\gamma_j,\qquad \bar{\gamma}^h = \bigcup_{j=1}^{4}(\gamma_j^h \cup \dot{\gamma}_j),\qquad \bar{\Pi}^h = \Pi^h \cup \bar{\gamma}^h.$$

Let the operator $B^p, p \in \{0,1\}$ be defined as follows:

$$B^p u(x,y) = \left(\frac{1}{5-p}\right)\{u(x+h,y) + u(x-h,y) + u(x,y+h) + u(x,y-h)\}$$
$$+\left(\frac{1-p}{20}\right)\{u(x+h,y+h) + u(x-h,y+h) + u(x+h,y-h)$$
$$+u(x-h,y-h)\}. \tag{10}$$

We consider for the approximation of problems (5)-(7) by the following system of difference equations:

$$u_h^4 = B^0 u_h^4\ on\ \Pi^h,\ u_h^4 = \Psi_j^4\ on\ \gamma_j^h \cup \dot{\gamma}_j,\ j = 1,2,3,4; \tag{11}$$

$$u_h^8 = B^1 u_h^8\ on\ \Pi^h, u_h^8 = \Psi_j^8\ on\ \gamma_j^h,\ j = 1,2,3,4. \tag{12}$$

By the maximum principle (see[5], Chap.4), problems (11) and (12) have the unique solution.

Consider the following systems:

$$q_h = B^p q_h + g_h, on\ \Pi^h,\ q_h = 0\ on\ \gamma_h, \tag{13}$$

$$\bar{q}_h = B^p \bar{q}_h + \bar{g}_h, on\ \Pi^h, \bar{q}_h \ge 0\ on\ \gamma_h, \tag{14}$$

where $B^p,\ p \in \{0,1\}$ is the operator defined by (10), g_h and $\bar{g}_h$ are given functions, and $|g_h| \le \bar{g}_h$ on Π^h.

Lemma 1. The solutions q_h and $\overline{q}_h$ of systems (13) and (14) hold the inequality

$$|q_h| \leq \bar{q}_h \; on \; \overline{\Pi}^h.$$

Proof. Lemma 1 follows by the Comparison theorem (see Chapter 4 in [5]).■

In what follows and for simplicity $c, c_0, c_1, ...$ defined the positive constants which are independent of h and the nearest factor, the similar notation is eonvenient for varied constants.

Let ϖ be a solution of the problem

$$\Delta\varpi = 0 \;\; on \; \Pi, \;\; \varpi = \theta_j(s) \;\; on \; \gamma_j, \;\; j = 1,2,3,4, \tag{15}$$

where θ_j, $j = 1,2,3,4$are given functions and

$$\theta_j \in C^{2^{3-p},\lambda}(\gamma_j), \;\; 0 < \lambda < 1, \;\; j = 1,2,3,4, \;\; p \in \{0,1\} \tag{16}$$

$$\theta_j^{(2q)}(s_j) = (-1)^q \theta_{j-1}^{(2q)}(s_j), \;\; q = 0,..,3 - 2p. \tag{17}$$

According to Taylor formula, we have

$$\max_{(x,y)\in\overline{\pi}} |B^p\varpi - \varpi| \leq ch^{2^{3-p}}, where \; p \in \{0,1\} \tag{18}$$

where ϖ is the solution of problem (15)-(17) on Π^h.

Lemma 2. For the solution of problem

$$q_h = B^p q_h + h^\ell \; on \; \Pi^h, \;\; q_h = 0 \; on \; \gamma_h, \;\; where \; p \in \{0,1\} \tag{19}$$

the following inequality holds:

$$q_h \leq \Bbbk\rho d h^{\ell-2} \; on \; \Pi^h,$$

where $\Bbbk$ is a positive constant defined by follows,

$$\Bbbk = \begin{cases} \frac{5}{3}, & if \; p = 0, \\ 2, & if \; p = 1, \end{cases}$$

$\ell > 2$ is an integer, $d = max\{a, b\}, and \; \rho = \rho(x,y)$ is the distance between the boundary of the rectangle Π and the current point $(x, y) \in \overline{\Pi}^h$.

3 Approximation of the solution using three stage method

Lemma 3. The estimation holds

$$|u_h^m - u^m| \leq c\rho h^{10-m}, \;\; where \; m \in \{4,8\} \tag{20}$$

where u_h^m is the solution of system (11) when $m = 4$, and of system (12) when $m = 8$, u^m is the trace of exact solution of problem (5)-(7) on $\overline{\Pi}^h$.

Proof. Let

$$\varepsilon_h = u_h^m - u^m \; on \; \overline{\Pi}^h. \tag{21}$$

It is obvious that

$$\varepsilon_h = B^p\varepsilon_h + (B^p u^m - u^m) \; on \; \Pi^h, \;\; \varepsilon_h = 0 \; on \; \gamma^h. \tag{22}$$

By virtue of Lemma 2, for the solution of problem (22), we obtain

$$|\varepsilon_h| \leq cph^{10-p}. \tag{23}$$

From equations (21) and (23) follows the proof of Lemma 3.■

Let u_h^4 and u_h^8 be the solutions of difference problem (11) and (12), respectively. The solution of the problem (1) on the grid $\bar{\Pi}^h$ is approximated by a solution u_h of the following difference problem

$$u_h = B^1 u_h - \left(\frac{h^4}{4!}\right) u_h^4 - \left(\frac{h^8}{8!}\right) u_h^8 \ on\ \Pi^h,\ u_h = \phi_j\ on\ \gamma_j^h,\ j = 1,2,3,4. \tag{24}$$

Theorem 1. On Π^h, the pointwise estimation is true

$$|u_h - u| \leq cph^8, \tag{25}$$

where u is the trace of the problem defined by (1) on Π^h, u_h is a solution of system (24).

Proof. By Taylor's formula, we have

$$u(x,y) = B^1 u(x,y) - \left(\frac{h^4}{4!}\right) u^4(x,y) - \left(\frac{h^8}{8!}\right) u^8(x,y) - r(x,y) \tag{26}$$

where $(x,y) \in \bar{\Pi}^h$, u^4 and u^8 are the functions defined by (4),

$$r(x,y) = \left(\frac{h^{12}}{2\times 12!}\right)\left(\left(\frac{\partial^{12} u(x+\theta_1,y)}{\partial x^{12}}\right) + \left(\frac{\partial^{12} u(x,y+\theta_2)}{\partial y^{12}}\right)\right),\ |\theta_i| < 1, i = 1,2. \tag{27}$$

By (9), we have

$$|r(x,y)| \leq ch^{12}. \tag{28}$$

We put

$$\varepsilon_h = u_h - u\ on\ \bar{\Pi}^h \tag{29}$$

where u is the solution of Dirichlet problem (1) and u_h is the solution to the system (24).

In the view of (24), (26) and (29), we get

$$\varepsilon_h = B^1 \varepsilon_h + \left(\frac{h^4}{4!}\right)\left(u_h^4(x,y) - u^4(x,y)\right) + \left(\frac{h^8}{8!}\right)\left(u_h^8(x,y) - u^8(x,y)\right) + r\ on\ \Pi^h, \tag{30}$$

$$\varepsilon_h = 0\ on\ \gamma_j^h. \tag{31}$$

By virtue of Lemma 3 and (28) it follows that

$$\left(\frac{h^4}{4!}\right)|u^4 - u_h^4| + \left(\frac{h^8}{8!}\right)|u^8 - u_h^8| + |r| \leq c_1 ph^{10} + c_2 h^{12} \leq c_3 h^{10}\ on\ \Pi^h. \tag{32}$$

Now, from (30)-(32) we have

$$\varepsilon_h = B^1 \varepsilon_h + g_h\ on\ \Pi^h,$$

$$\varepsilon_h = 0\ on\ \gamma^h,$$

where

$$|g_h| \leq c_3 h^{10}.$$

Let ε_h be a solution of the problem

$$\varepsilon_h = B^1 \varepsilon_h + c_3 h^{10}\ on\ \Pi^h,$$

$\varepsilon_h = 0 \; on \; \gamma^h.$

Then, by Lemma 1 and Lemma 3 , we obtain

$$|\varepsilon_h| \leq |\varepsilon_h| \leq 2c_3 dh^8 \rho$$

or

$$|\varepsilon_h| \leq ch^8 \rho,$$

where $c = 2c_3 d,\; d = max\{a, b\}$, and $\rho = \rho(x, y)$ is the distance between the boundary of the rectangle Π and the current point $(x, y) \in \overline{\Pi}^h$.

Remark 1. The proposed method can be generalised for the mixed boundary value problem for Laplace's equation.Furthermore, the idea of the proposed method can be applied to increase an accuracy of the finite difference approximation of KdV equation used in [6].

References

1. E.A. Volkov, Comp. Maths Math. Phys.**49(3)**, 496-501(2009)
2. E.A. Volkov, Proc. Steklov Inst. Math., **117,** 119–134, (1972)
3. E.A. Volkov, Differ. Equations, **1(7)**, 946-960, (1965)
4. G. K. Berikelashvili and B. G. Midodashvili, Differential Equations, **51(1)**,107–115, (2015)
5. A.A. Samarskii,The theory of difference schemes, Marcel Dekker, Inc.(2001)
6. A. Yokus and H. Bulut, Indian Journal of Physics, **(5)**, 1-5, (2018)

An Approximate Grid Solution of a Nonlocal Boundary Value Problem with Integral Boundary Condition for Laplace's Equation

Adıgüzel A. Dosiyev[1,*] and *Rifat Reis* [1]

[1]Department of Mathematics, Near East University ,Nicosia, TRNC, Mersin 10, Turkey

Abstract. A new method for the solution of a nonlocal boundary value problem with integral boundary condition for Laplace's equation on a rectangular domain is proposed and justified. The solution of the given problem is defined as a solution of the Dirichlet problem by constructing the approximate value of the unknown boundary function on the side of the rectangle where the integral boundary condition was given. Further, the five point approximation of the Laplace operator is used on the way of finding the uniform estimation of the error of the solution which is order of $O(h^2)$, where hi s the mesh size. Numerical experiments are given to support the theoretical analysis made.

1 Introduction

In [1]-[5], a new constructive method that reduces the multilevel nonlocal problem for Poisson's equation to local Dirichlet problem, was given.

In this paper, the idea of the method in [5] is applied for the approximate solution of the Laplace's equation with integral boundary condition. By applying trapeizodal rule for the integral boundary condition, the problem is approximated by the multilevel nonlocal boundary value problem which is solved as the sum of two classical 5-point finite-difference Dirichlet problems. Finally, the numerical experiments are illustrated to support to obtain theoretical results.

On the convergence of difference schemes for problems with integral boundary conditions without reducing to the local problems see [6],[7] and references therein.

2 Nonlocal Boundary Value Problem

Let

$$R = \{(x, y): 0 < x < 1, 0 < y < 2\} \tag{1}$$

be an open rectangle, $\gamma^m, m = 1,2,3,4,$ be its sides including the endpoints, numbered in the clockwise direction, begining with the side lying on the y-axis and let $\gamma = \cup_{m=1}^{4} \gamma^m$ be the boundary of R.

Let C^0 denote the linear space of continous functions of one variable x on the interval $[0,1]$ of x-axis, and vanish at the points $x = 0$ and $x = 1$. For the function $f \in C^0$ we define the norm

$$\|f\|_{C^0} = \max_{0\le x\le 1} |f(x)|.$$

*Correspondingauthor: adiguzel.dosiyev@emu.edu.tr

Consider the following nonlocal boundary value problem on R :

$$\Delta u = 0 \; on \; R, \qquad u = 0 \; on \; \gamma^1 \cup \gamma^3, \qquad u = \tau \;\; on \; \gamma^2, \tag{2}$$

$$u(x,0) = \alpha \int_{\xi}^{2} u(x,y)dy + \mu(x), \quad 0 \le x \le 1, \qquad 0 < \xi < 2, \tag{3}$$

where $\Delta \equiv \partial^2/\partial x^2 + \partial^2/\partial x^2$ is the Laplacian, $\tau = \tau(x) \in C^0$ is a given function, and $|\alpha| < \frac{1}{2-\xi}$ is a given constant.

By replacing the integral in condition (3) with its approximation using trapezoidal rule, we have

$$u(x_i,0) = \alpha \sum_{k=1}^{M} \rho_k . u(x_i,\eta_k) + \mu_i, \qquad i = 1,2,\dots,N-1.$$

where $\rho_1 = \rho_M = \frac{h}{2}$, $\rho_j = h$ for $j = 2,3,\dots,M-1$, $\eta_j = \xi + (j-1)h$, $j = 1,2,\dots,M$, $h = \frac{1}{N}$,

$(M-1)h + \xi = 2$ and $\frac{\xi}{h}$ is a fixed integer number.

It follows that

$$|\alpha| \sum_{k=1}^{M} \rho_k = q_0 < 1. \tag{4}$$

We consider the following multilevel nonlocal boundary value problem

$$\Delta U = 0 \; on \; R, \qquad U = \tau \; on \; \gamma^2, \qquad U = 0 \; on \; \gamma^1 \cup \gamma^3, \tag{5}$$

$$U(x,0) = \alpha \sum_{k=1}^{M} \rho_k . U(x,\eta_k) + \mu(x), \;\; 0 \le x \le 1. \tag{6}$$

By the maximum principle and the error estimate of trapeziodal rule, we have

$$\max_{(x,y)\in \bar{R}} |u - U| \le c_1 h^2, \tag{7}$$

where c_1 is a constant independent of h.

Let V be a solution of the Dirichlet problem,

$$\Delta V = 0 \; on \; R, \qquad V = \tau \; on \; \gamma^2, \qquad V = 0 \; on \; \gamma/\gamma^2,$$

and denote by

$$\phi_k(x) = V(x,\eta_k) \; for \; k = 1,2,\dots,M, \tag{8}$$

$$\phi = \alpha \sum_{k=1}^{M} \rho_k \phi_k. \tag{9}$$

We consider the Dirichlet problem

$$\Delta W = 0 \; on \; R, \qquad W = 0 \; on \; \gamma/\gamma^4, \qquad W = f \;\; on \; \gamma^4,$$

where f be an unknown function from C^0.

We define the operator $B_i: C^0 \to C^0$ as

$$B_i f(x) = W(x,\eta_i) \in C^0, \qquad i = 1,2,\dots,M. \tag{10}$$

Since W is a harmonic function on R, by the maximum principle for the norm of operator B_i, we have

$$|B_i| < 1 - \frac{\xi + (i-1)h}{2}, \quad i = 1,2,\dots,M \tag{11}$$

and

$$0 < |B_M| < |B_{M-1}| <\dots< |B_1| < 1.$$

Then the following inequality holds

$$|B_1|q_0 = q < 1, \tag{12}$$

where q_0 is defined in (4).

It is obivious that,

$U(x,0) = f(x),\ 0 \le x \le 1.$

Relying on (8), (9) and (10), the funtion f satisfies relation

$$f = \phi + \mu + \alpha \sum_{k=1}^{M} \rho_k B_k f. \tag{13}$$

On the basis of (11), by analogy with the proof of Theorem 1 and 2 in [4] the following Theorem is proved.

Theorem 1. There is a unique function $f \in C^0$ for which relation (13) holds.

3 Finite Difference Approximation

We say that $F \in C^{k,\lambda}(E)$, if F has k-th derivatives on E satisfying the Hölder condition with exponent λ. We assume that $\tau(x) \in C^{2,\lambda}(\gamma^2)$ *and* $\mu(x) \in C^{2,\lambda}(\gamma^4), 0 < \lambda < 1$, in (2) and (3), respectively.

We define a square mesh with the mesh size $h = 1/N, N > 2$ is an integer, constructed with the lines $x, y = h, 2h, \dots$. Let D_h be the set of nodes of this square grid, $R_h = R \cap D_h$,where R is the rectangle (1), and $\gamma_h^m = \gamma^m \cap D_h,\ m = 1,2,3,4$. We denote by $\gamma_h = \cup_{m=1}^{4} \gamma_h{}^m, \bar{R}_h = R_h \cup \gamma_h$.

Let

$$[0,1]_h = \left\{x = x_i\colon\ x_i = ih,\ i = 0,1,\dots,N,\ h = \tfrac{1}{N}\right\}$$

be the set of points divided by the step size h on $[0,1]$.

Let C_h^0 be the linear space of grid functions defined on $[0,1]_h$ that vanish at $x = 0$ and $x = 1$. The norm of a function $f_h \in C_h^0$ is defined as

$$\|f_h\|_{C_h^0} = \max_{x\in[0,1]_h} |f_h|.$$

Let A be the operator defined by

$$Au_h \equiv (u_h(x+h,y) + u_h(x-h,y) + u_h(x,y+h) + u_h(x,y-h))/4.$$

Consider the system of grid equations

$$v_h = Av_h \text{ on } R_h,\quad v_h = \tau_h \text{ on } \gamma_h^2,\quad v_h = 0 \text{ on } \gamma_h/\gamma_h^2,$$

where τ_h is the trace of τ on γ_h^2 and for each $\eta_i = \xi + (i-1)h$, we define

$$\widetilde{\phi}_{i,h}(x) = v_h(x,\eta_i),\qquad i = 1,2,\dots,M. \tag{14}$$

Let w_h be a solution of the finite difference problem

$$w_h = Aw_h \; on \; R_h, \quad w_h = 0 \; on \; \gamma_h/\gamma_h^4, \quad w_h = \widetilde{f}_h \; on \; \gamma_h^4,$$

where $\widetilde{f}_h \in C_h^0$, is an arbitrary function.

Let B_i^h be a linear operator from C_h^0 to C_h^0 defined by

$$B_i^h f_h(x) = w_h(x, \eta_i), \quad i = 1,2,\dots,M.$$

The following inequality holds

$$\left\|B_i^h f_h(x)\right\|_{C_h^0} \le \|f_h\|_{C_h^0}\left(1 - \frac{\xi + (i-1)h}{2}\right), \tag{15}$$

We define

$$\tilde{\phi}_h = \alpha \sum_{k=1}^{M} \rho_k \tilde{\phi}_{k,h}(x), \quad x \in [0,1]_h, \tag{16}$$

where $\tilde{\phi}_{k,h}(x)$ defined by (14).

Now, we introduce the function

$$\tilde{f}_h = \tilde{\phi}_h + \mu_h + \alpha \sum_{k=1}^{M} \rho_k \tilde{\psi}_{k,h},$$

where μ_h is the trace of μ on $[0,1]_h$ and $\tilde{\psi}_{k,h} \in C_h^0, k = 1,2,\dots M,$ are the solution of the system of the equations

$$\tilde{\psi}_{i,h} = B_i^h(\tilde{\phi}_h + \mu_h + \alpha \sum_{k=1}^{M} \rho_k \tilde{\psi}_{k,h}), \quad i = 1,2,\dots,M. \tag{17}$$

We investigate the solution of system (17) by the the following fixed point iteration

$$\tilde{\psi}_{i,h}^0 = 0, \quad \tilde{\psi}_{i,h}^n = B_i^h(\tilde{\phi}_h + \mu_h + \alpha \sum_{k=1}^{M} \rho_k \tilde{\psi}_{k,h}^{n-1}), i = 1,2,\dots,M; \quad n = 1,2,\dots. \tag{18}$$

Define

$$\tilde{f}_h^{\,n} = \tilde{\phi}_h + \mu_h + \alpha \sum_{k=1}^{M} \rho_k \tilde{\psi}_{k,h}^n, \tag{19}$$

where $\tilde{\psi}_{k,h}^n$ is the n-th iteration of (18).

Theorem 2. The following inequality is true.

$$\left\|\tilde{f}_h^{\,n} - f_h\right\|_{C_h^0} \le c_2 h^2 + q_0\left(\frac{q_1^{n+1}}{1 - q_1}\right)(\|\phi\|_{C^0} + \|\mu\|_{C^0}), \tag{20}$$

where f_h is the trace of f on $[0,1]_h$, q_0 is a number defined by (4), $q_1 = max\{1 - \frac{\xi}{2}, q\}$, q is a number given in (12), c_2 is a constant independent of h.

Consider the actual finite difference problem for the approximate solution of problem (7) and (8),

$$\tilde{u}_h^n = A\tilde{u}_h^n \; on \; R_h, \quad \tilde{u}_h^n = \tau_h \; on \; \gamma_h^2, \quad \tilde{u}_h^n = 0 \; on \; \gamma_h^1 \cup \gamma_h^3, \tag{21}$$

$$\tilde{u}_h^n = \tilde{f}_h^{\,n} \; on \; \gamma_h^4, \tag{22}$$

where $\tilde{f}_h^{\,n}$ is computed function which approximates to f.

On the basis of (7) and Theorem 2, the following Theorem is proved:

Theorem 3. Let the boundary function $\tau(x)$ and $\mu(x)$ in (5), (6) be from the class $C^{2,\lambda}(\gamma^2)$. The estimation holds

$$\max_{(x,y)\in\bar{R}_h} |\tilde{u}_h^n - u| \leq c_3 h^2 + q_0 \left(\frac{q_1^{n+1}}{1-q_1}\right)(\|\phi\|_{C^0} + \|\mu\|_{C^0}), \tag{23}$$

where $\tilde{u}_h^n$ is a solution of the problem (21), (22), u is the solution of problem (2), (3) and c_3 is a constant independent of h.

In the estimation (23)

$$n = max\left\{\left[\left(\frac{ln\ h^{-2}(1-q^1)^{-1}q_0}{ln\ q_1^{-1}}\right)\right], 1\right\}.$$

4 Numerical Experiments

The following problem is solved:

$$\Delta u = 0 \ on\ R, \quad u(0,y) = u(1,y) = 0, \quad 0 \leq y \leq 2,$$

$$u(x,2) = e^{2\pi} sin(\pi x), \quad 0 \leq x \leq 1,$$

$$u(x,0) = \alpha \int_{\xi}^{2} u(x,y)dy + \mu(x), \ 0 \leq x \leq 1,$$

where $u(x,y) = e^{\pi x} sin(\pi x)$ is the exact solution, $\mu(x) = sin(\pi x)\left[1 + \left(\frac{\alpha}{\pi}\right)(1 - e^{2\pi})\right]$ and $\alpha = \frac{1}{100}$ and $\xi = \frac{1}{16}$.

By using the present method, the approximate solutions found and showed on line $y = 0$ according to the decreasing mesh sizes h. This shows that the error of approximation has order of $O(h^2)$. To attain this accuracy just 4 iterations in (19) are applied. The comparisons of the CPU times of the proposed method and the method without reducing to the local problem show the efficiency of our approach.

References

1. E.A. Volkov, Comput. Math. Phys. **53(8),** 1128-1138 (2013)
2. E.A. Volkov, A.A. Dosiyev, S.C. Buranay, Comput. Math. Appl. **66,** 330-338(2013)
3. E.A. Volkov, Comput. Math. Math. Phys. **53**(10), 1494-1498 (2013)
4. E.A. Volkov, A.A. Dosiyev, Mediterr. J. Math. **13**, 3589-3604 (2016)
5. A.A. Dosiyev, J.Comput. Appl. Math. (2018), https://doi.org/10.1016/j.cam.2018.04.046
6. G. Berikelashvili, N. Khomeriki, Lith. Math. J. 52 (4), 353-362 (2012)
7. L.Zhou, H.Yu, Advances in Difference Equations, **2018:225**, 1-11 (2018)

Exact Solutions with Lie Symmetry Analysis for Nano-Ionic Currents along Microtubules

Zühal Küçükarslan Yüzbaşı[1,*], *Ebru* Cavlak Aslan[1], and *Mustafa* Inc[1]

[1]Fırat University, Faculty of Science, Department of Mathematics, 23119 Elazig / Turkey

Abstract. In this work, the Lie symmetry analysis is tested by its applications to nano-solitons of ionic waves propagation along microtubules in living cells and nano-ionic currents of MTs.

1 Introduction

The main aim of this work is to study the executive equation defining nano-ionic currents along microtubules with the Lie symmetry analysis. This method is one of the most important and efficient techniques to find the exact solutions [1]-[7].

The nano ionic currents are elaborated in [8] take the form

$$\frac{L^2}{2}q_{xxx}+\frac{Z^{3/2}}{L}(\chi G_0-2\delta C_0)qq_t+2q_x+\frac{ZC_0}{L}q_t+\frac{1}{L}(RZ^{-1}-G_0Z)q=0, \tag{1}$$

where $R = 0.34\times10^9\Omega$ the resistance of the elementary rings (ER), $L = 8\times10^{-9}m$, $C_0 = 1.8\times10^{-15}F$ is the total maximal capacitance of the ER. $G_0 = 1.1\times10^{-13}Si$ is the conductance of pertaining nano-pores (NPS) and $Z = 5.5\times10^{10}\Omega$ is the characteristic impedance of the system. The parameter δ and χ describe the nonlinearity of ER capacitor and conductance of NPS in ER, respectively.

2 Lie Symmetry Analysis

We will now consider the one parameter group of point transformation of the form

$$x\to x+\varepsilon\xi(x,t,q)\,,$$
$$t\to t+\varepsilon\tau(x,t,q)\,,$$
$$q\to q+\varepsilon\phi(x,t,q)\,,$$

where $\varepsilon < 1$ is a group parameter [9]. The vector field associated with the above group of transformations can be written as

$$V=\xi(x,t,q)\frac{\partial}{\partial x}+\tau(x,t,q)\frac{\partial}{\partial t}+\phi(x,t,q)\frac{\partial}{\partial q}. \tag{2}$$

*Corresponding Author: zuhal2387@yahoo.com.tr

In order to study symmetries, from Eq. (1), it can be shown by a well established procedure as following

$$\xi = C_2,$$
$$\tau = C_1,$$
$$\phi = 0,$$

where C_1 and C_2 are constants. The Lie symmetry algebra of Eq. (1) is spanned by the following two generators.

$$V_1 = \frac{\partial}{\partial x},$$
$$V_2 = \frac{\partial}{\partial t}.$$

Now we consider the above mentioned vector fields for Eq. (1).

Case 1: For the generator $V_1 = \frac{\partial}{\partial x}$, the similarity variables are $\eta = t$, $q = f(\eta)$. Then from Eq. (1), we get the following reduction equation

$$\frac{Z^{3/2}}{L}(\chi G_0 - 2\delta C_0)ff' + \frac{ZC_0}{L}f' + \frac{1}{L}(RZ^{-1} - G_0Z)f = 0. \tag{3}$$

Solution of this equation

$$f(\eta) = \frac{C_0}{Z^{1/2}(\chi G_0 - 2\delta C_0)} ProductLog[\frac{Z^{1/2}(\chi G_0 - 2\delta C_0)}{C_0}e^{\alpha\eta + C}], \tag{4}$$

where $\alpha = (RZ^{-1} - G_0Z)/ZC_0$.

Case 2: For the generator $V_2 = \frac{\partial}{\partial t}$, the similarity variables are $\eta = x$, $q = f(\eta)$. Then from Eq. (1), we get the following reduction linear equation

$$\frac{L^2}{3}f''' + 2f' + \frac{1}{L}(RZ^{-1} - G_0Z)f = 0. \tag{5}$$

Case 3: For the generator $V_3 = V_1 + V_2 = \partial/\partial x + \partial/\partial t$, we have $q = f(\eta)$ that $\eta = x - t$. If we substitute $q = f(\eta)$ into the governing equation, we obtain following NODE;

$$\frac{L^2}{3}f''' - \frac{Z^{3/2}}{L}(\chi G_0 - 2\delta C_0)ff' + (\frac{2L - ZC_0}{L})f' + \frac{1}{L}(RZ^{-1} - G_0Z)f = 0. \tag{6}$$

Case 4: For the infinitesimal generator $V_4 = \frac{1}{\lambda}\frac{\partial}{\partial x} + \frac{1}{\rho}\frac{\partial}{\partial t}$, we have $q = f(\eta)$, $\eta = \lambda x - \rho t$ where λ and ρ are constants. If we substitute $q = f(\eta)$ into Eq. (1), so that

$$\frac{L^2}{2}\lambda^3 f''' - \frac{Z^{3/2}}{L}(\chi G_0 - 2\delta C_0)\rho ff' + (2\lambda - \frac{ZC_0}{L}\rho)f' + \frac{1}{L}(RZ^{-1} - G_0Z)f = 0. \tag{7}$$

In this, we find the solution of Eq. (7) using the "Generalized Projective Riccati Equations Method" (GPREM) [10]. So, balancing f''' and ff' gives $N = 2$. Therefore

$$f(\eta) = A_0 + \sum_{i=1}^{2} \sigma^{i-1}[A_i\sigma(\eta) + B_i\nu(\eta)], \tag{8}$$

where A_j and B_j, $j = 0, 1, 2$ are constants.

In Eq. (8), $\sigma(\eta)$ and $\nu(\eta)$ satisfy the ODE

$$\sigma' = \epsilon\sigma(\eta)\nu(\eta), \tag{9}$$

$$\nu' = R + \epsilon\nu^2(\eta) - \mu\sigma(\eta), \epsilon = \pm 1, \tag{10}$$

where

$$\nu^2 = \epsilon(R - 2\mu\sigma(\eta)) + \frac{\mu^2 + r}{R}\sigma^2(\eta), \tag{11}$$

r, R and μ are non zero constants.

So, if Eq. (8) are written in Eq. (7) for $\epsilon = -1$ and $r = 1$, we get solving the system of the algebraic equation.

Consequently, the solution of f(z) is obtained as following

$$f(z) = -\frac{3\lambda^2}{2\alpha_2\rho}\frac{\csc h\left[z\sqrt{\frac{-i}{\lambda\alpha_1}}\right]}{\left(1 + i\csc h\left[z\sqrt{\frac{-i}{\lambda\alpha_1}}\right]\right)} + \frac{2\lambda - i\lambda^2 - \rho\alpha_3}{\alpha_2\rho},$$

where $\alpha_1 = \frac{L^2}{2}\lambda^3$, $\alpha_2 = \frac{Z^{3/2}}{L}(\chi G_0 - 2\delta C_0)\rho$, $\alpha_3 = (2\lambda - \frac{ZC_0}{L}\rho)$ and finally we get $\alpha_4 = \frac{1}{L}(RZ^{-1} - G_0 Z)$.

3 Conclusion

In this work, the Lie symmetry analysis has been successfully applied to find the solution of the nonlinear partial differential equation which plays an important role in biology. Some of the equations obtained with the Lie symmetry are solved and we found new exact solutions for these equations. Also, while finding these new exact solutions, Generalized Projective Riccati method was applied to the reduced equation by Lie analysis.

References

[1] M. Younis and S. Ali, Solitary wave and shock wave solitons to the transmission line model for nano-ionic currents along microtubules, Appl. Math. Comput., 246, 460-463, 2014.

[2] M. Younis and S. T. R. Rizvi, Analytical and soliton solutions: Nonlinear model of nanobioelectronics transmission lines, Appl. Math. Comput., 265, 994-1002, 2015.

[3] D. Baleanu, M. Inc, A. Yusuf and A. I. Aliyu, Lie symmetry analysis, exact solutions and conservation laws for the time fractional modified Zakharov Kuznetsov equation, Nonlinear Anal. Model. Control, 22, 6, 861-876, 2017.

[4] M. Inc, A. I. Aliyu and A. Yusuf, Solitons and conservation laws to the resonance nonlinear Shrödinger equation with both spatio-temporal and inter-modal dispersions, Optik, 142, 509-522, 2017.

[5] D. Baleanu, M. Inc, A. Yusuf and A. I. Aliyu, Lie symmetry analysis, exact solutions and conservation laws for the time fractional Caudrey- Dadd- Gibbon- Sawada- Kotera equation, Commun. Nonlinear Sci. Numer. Simul., 59, 222-234, 2018.

[6] D. Baleanu, M. Inc, A. Yusuf and A. I. Aliyu, Time Fractional Third-Order Evolution Equation: Symmetry Analysis, Explicit Solutions, and Conservation Laws, J. Comput. Nonlinear Dynam, 13(2), 021011, 2017.

[7] M. Inc, A. I. Aliyu, A. Yusuf and D. Baleanu, Optical solitons and modulation instability analysis with (3 + 1)-dimensional nonlinear Shrödinger equation, Superlattices and Microstructures, 112, 296-302, 2017.

[8] D. L. Sekulic, M. V. Sataric and M. B. Zivanov, Symbolic computation of some new nonlinear partial differential equations of nanobiosciences using modified extended tanh-function method, Appl. Math. Comput., 218, 3499-3506, 2011.

[9] T. Ozis and I. Aslan, Simirity Solutions to Burgers' Equation in terms of Special Functions of Mathematical Physics, Acta Phy. Pol. B, 48, 1349-1369, 2017.

[10] M. Inc, A. I. Aliyu and A. Yusuf, Solitons and conservation laws to the resonance nonlinear Shrdingers equation with both spatio-temporal and inter-modal dispersions, Optik, 142, 59-522, 2017.

Risk Analysis of Ports in Maritime Industry in Turkey using FMEA Based Intuitionistic Fuzzy TOPSIS Approach

Mine Şenel[1], *Bilgin Şenel*[1], *Celal Alpay Havle*[2,*]

[1] Munzur University, Department of Industrial Engineering, Tunceli, Turkey

[2]Özyeğin University, Department of Pilot Training, İstanbul, Turkey

Abstract. Today, international trade is extremely important for countries' economies. It is possible to show the ship's transport as one of the most important execution channels of this trade. International commercial maritime transport subject to many rules and regulations carries many risks in processes such as loading, handling and unloading. Commercial ports are the places where these risks are seen intensely. From this point, many risk components in terms of the process in the port, one of the international commercial port in Turkey, based on a risk analysis is tried to be evaluated in this study. Especially, loading and unloading processes are focused with the directions of industrial experts. The main dimensions of the risks in the port are determined via expert opinions, and the sub-criteria of these dimensions are revealed. In this way, a generic model is proposed based on failure mode and effect analysis and the model is digitized using intuitionistic fuzzy TOPSIS. Interpretations are made in the direction of the obtained analysis results.

1 Introduction

Seaborne trade in world is vital for country economies in terms of balance of import and export. Nowadays, with 90% contribution to world trade, the importance of maritime industry has increased and to meet customer demands in the aspect of this, countries increase their number of vessels and facilities [1]. World marine trade fleet has reached 1.7 billion DWT [8] and maritime trade volume has reached 10.3 billion tons [1] and an estimated $ 500 billion in revenue is generated annually from maritime trade. The vitality of well-structured and well-functioning international ports for industrial activities such as trade, globalized production processes should always be taken into consideration because ports have key role in global transport and trade chains [1].

Place of ports in global trade creates increased competitive forces and these influence the need for higher performance levels. These levels are not only related with cost

* Corresponding author: celal.alpay.havle@gmail.com

reduction, optimization of operations, time efficiency and trade promotions but also security, safety, resource conservation, and environmental protection [1]. By considering the importance of international trade ports with the aspects mentioned above, port risks are focused in this study.

A port is a place that includes a terminal and the usual places where vessels are loaded and / or evacuated, loads and ships wait for their order, or wait or queue their order and it usually has an interface similar to other forms of transport and provides linking services in this way [5-6]. Within the context of increased transport concepts, port location is very important in the transport chain and to illustrate this concept in general terms, the diagram in Figure 1 is given [5-6]:

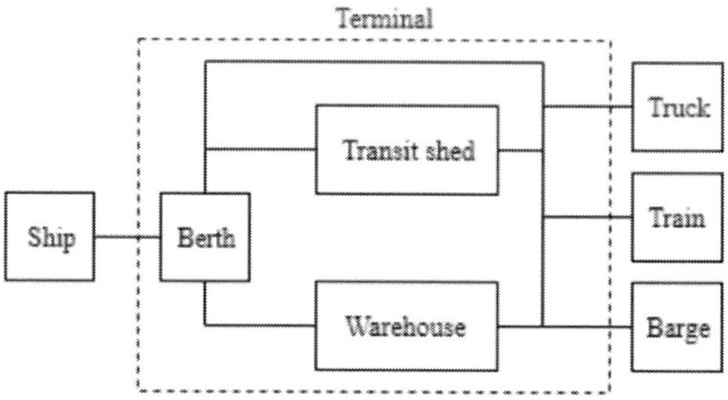

Figure 1. Port operations [5]

In this study, risks in the processes between ship and berth parts of port operations shown in Figure 1 are focused. Loading and unloading processed are examined with the industrial experts from Turkey's maritime industry. The study aims to determine the most important risk factors and failures in ports. To perform this failure mode and effect analysis is used to develop o model to construct a base for analyze. After construction of the model intuitionistic fuzzy TOPSIS approach is used.

The study contains following sections: Section 2 gives information about utilized methodology, section 3 gives numerical illustration of application of proposed methodology in Turkey maritime industry. By following this, section 4 highlights the obtained results before the conclusion.

2 Methodology

Failure mode and effect analysis (FMEA) is a technique to identify and potential failures, or errors in a system, process or design [14]. FMEA was developed as a formal design methodology by the aerospace industry for reliability and safety requirements in the 1960s [4-12].

The traditional FMEA determines the risk priorities of failure modes using the risk priority numbers (RPNs), which is defined as the probability of occurrence (O), severity (S) and detection (D) of a failure [17]. To calculate RPNs; O, S, and D of each failure should be multiplicated but, multiplication is not the right aggregation operator and FMEA factors must be aggregated in a nonlinear rather than linear manner [9]. The traditional FMEA has been well-accepted risk/criticality assesment method, however, it has been criticized for many drawbacks/shortcomings [4-7-10-11-12-13-16-17]. Significant criticisms include to the following:

- The relative importance among O,S,D is not taken into consideration in determining the priority of the failure modes. Three elements are assumed to have the same importance. However, this may not always be the case.
- Various sets of O,S and D may produce an identical value of RPN, however, the risk implication may be totally different. For example, two different failures with the O, S and D values of 1,4,6 and 2,3,4,respectively, have the same RPN value of 24. However, the risk implication is different.
- It is usually difficult or even impossible to give exact numerical evaluations of practically intangible quantities associated with the risk factors.
- The RPN scale is not continuous, with many holes because many numbers between 1 and 1000 cannot be obtained from the product of O, S, and D.
- The three risk factors are difficult to be precisely evaluated. Much information in FMEA can be expressed linguistically in terms such as *Likely, Important or Very high* and so on.

To overcome obscurity during the evaluation process of risks and to address the above listed drawbacks, in this study Intuitionistic Fuzzy FMEA is conducted to assess risks in ports of Turkey. Intuitionistic fuzzy sets (IFS) are based on fuzzy set theory [19] but firstly introduced by Atanassov to eliminate obscurity [2]. Obscurity can be seen due to lack information in some cases [3-15]. In this study, FMEA model and TOPSIS method are augmented to IFSs [15] and an integrated intuitionistic fuzzy approach is applied to Turkey's maritime industry to analyze ports' risks.

Suppose that there are n failure modes and they are evaluated by experts $E_k(1,2,...,k)$ opinions based FMEA model. The model has risk factors and they are indicated as $C=\{O,S,D\}$ [15]. These FMEA risk factors based on failure modes are evaluated with intuitionistic fuzzy sets. To perform this intuitionistic fuzzy TOPSIS (IF-TOPSIS) is applied. Steps of proposed approach are given as following [3-15]:

Step 1: Calculate the weight of experts: Suppose that there are k experts in the decision makers team. To calculate the weights intuitionistic fuzzy numbers are used and they are shown with Table 1. Let $E_k=\{\mu_k,\upsilon_k,\pi_k\}$ be an intuitionistic fuzzy number (IFN) of kth expert. To determine the weight of the expert formula (1) is used as follows:

$$\lambda_k=\frac{\left(\mu_k+\pi_k\left(\frac{\mu_k}{\mu_k+\upsilon_k}\right)\right)}{\sum_{k=1}^{n}\left(\mu_k+\pi_k\left(\frac{\mu_k}{\mu_k+\upsilon_k}\right)\right)} \quad \text{where } \sum_{k=1}^{n}\lambda_k=1 \tag{1}$$

Table 1. Linguistic terms for importance of criteria and experts [15]

Linguistic terms	IFNs
Very important	[0.9, 0.1]
Important	[0.75, 0.2]
Medium	[0.5, 0.45]
Unimportant	[0.35, 0.6]
Very unimportant	[0.1, 0.9]

Step 2: Construct the aggregated intuitionistic fuzzy decision matrix based on experts' opinion: Let $R^{(k)} = \left(r_{ij}^{(k)}\right)_{m\times n}$ be the intuitionistic fuzz decision matrix of kth experts and $\lambda = \{\lambda_1, \lambda_2, \lambda_3, ..., \lambda_k\}$ is the set of weights of each expert where $\sum_{k=1}^{l} \lambda_k = 1$ and $\lambda_k \in [0,1]$. At the beginning of aggregation process, decision matrix of each experts should be constructed. To perform this, linguistic variables shown in Table 2 are used. To aggregate the decision matrices, IFWA operator [18] is used:

$$r_{ij} = IFWA_{\lambda}(r_{ij}^{(1)}, r_{ij}^{(2)}, ..., r_{ij}^{(l)}) = \lambda_1 r_{ij}^{(1)} \oplus \lambda_2 r_{ij}^{(2)} \oplus \lambda_3 r_{ij}^{(3)} \oplus \cdots \oplus \lambda_l r_{ij}^{(l)}$$
$$= \left[1 - \prod_{k=1}^{l}\left(1-\mu_{ij}^{(k)}\right)^{\lambda_k}, \prod_{k=1}^{l}\left(\upsilon_{ij}^{(k)}\right)^{\lambda_k}, \prod_{k=1}^{l}\left(1-\mu_{ij}^{(k)}\right)^{\lambda_k} - \prod_{k=1}^{l}\left(\upsilon_{ij}^{(k)}\right)^{\lambda_k}\right] \quad (2)$$

where $r_{ij} = \left(\mu_{A_i}(x_j), \upsilon_{A_i}(x_j), \pi_{A_i}(x_j)\right)$ and $(i = 1,2,3,...,m, j = 1,2,3,...,n)$.

Table 2. Intuitionistic fuzzy linguistic variables [15]

Detectability		Failure severity probability		Failure occurrence probability	
Linguistic terms	**IFNs**	**Linguistic terms**	**IFNs**	**Linguistic terms**	**IFNs**
Absolutely impossible	[1, 0]	Risky without warning	[1, 0]	Very high	[0.9, 0.1]
Highly unlikely	[0.9, 0.1]	Risky with warning	[0.9, 0.1]	High	[0.75, 0.1]
Unlikely	[0.8, 0.1]	Very high	[0.8, 0.1]	Medium	[0.5, 0.45]
Very low	[0.7, 0.2]	High	[0.7, 0.2]	Low	[0.35, 0.6]
Low	[0.6, 0.3]	Medium	[0.6, 0.3]	Very low	[0.1, 0.9]
Medium	[0.5, 0.4]	Low	[0.5, 0.4]		
Relatively high	[0.4, 0.5]	Very low	[0.4, 0.5]		
High	[0.25, 0.6]	Insignificant	[0.25, 0.6]		
Very high	[0.1, 0.75]	Very insignificant	[0.1, 0.75]		
Absolutely possible	[0.1, 0.9]	None	[0.1, 0.9]		

Aggregated intuitionistic fuzzy decision matrix is obtained by following this aggregation process as follows:

$$R=\begin{bmatrix} (\mu_{FM_1}(O),\upsilon_{FM_1}(O),\pi_{FM_1}(O)) & (\mu_{FM_1}(S),\upsilon_{FM_1}(S),\pi_{FM_1}(S)) & (\mu_{FM_1}(D),\upsilon_{FM_1}(D),\pi_{FM_1}(D)) \\ (\mu_{FM_2}(O),\upsilon_{FM_2}(O),\pi_{FM_2}(O)) & (\mu_{FM_2}(S),\upsilon_{FM_2}(S),\pi_{FM_2}(S)) & (\mu_{FM_2}(D),\upsilon_{FM_2}(D),\pi_{FM_2}(D)) \\ & \ddots & \\ (\mu_{FM_n}(O),\upsilon_{FM_n}(O),\pi_{FM_n}(O)) & (\mu_{FM_n}(S),\upsilon_{FM_n}(S),\pi_{FM_n}(S)) & (\mu_{FM_n}(D),\upsilon_{FM_n}(D),\pi_{FM_n}(D)) \end{bmatrix}$$

$$=\begin{bmatrix} r_{1O} & r_{1S} & r_{1D} \\ r_{2O} & r_{2S} & r_{2D} \\ \vdots & \ddots & \vdots \\ r_{nO} & r_{nS} & r_{nD} \end{bmatrix} \tag{3}$$

Step 3: Calculate the weight of risk factors: Assume that $w_j^k=\left(\mu_j^{(k)},\upsilon_j^{(k)},\pi_j^{(k)}\right)$ is an intuitionistic fuzzy number assigned by kth expert to jth criterion, then the weights of risk factors are determined via IFWA operator [18] as follows:

$$w_j=IFWA_\lambda(w_j^{(1)},w_j^{(2)},\ldots,w_j^{(l)})=\lambda_1 w_j^{(1)}\oplus\lambda_2 w_j^{(2)}\oplus\lambda_3 w_j^{(3)}\oplus\cdots\oplus\lambda_l w_j^{(l)}$$

$$=\left[1-\prod_{k=1}^{l}\left(1-\mu_j^{(k)}\right)^{\lambda_k},\prod_{k=1}^{l}\left(\upsilon_j^{(k)}\right)^{\lambda_k},\prod_{k=1}^{l}\left(1-\mu_j^{(k)}\right)^{\lambda_k}-\prod_{k=1}^{l}\left(\upsilon_j^{(k)}\right)^{\lambda_k}\right] \tag{4}$$

where $W=\left[w_1,w_2,w_3,\ldots,w_j\right]$, $w_j=\left(\mu_j,\upsilon_j,\pi_j\right)$ and $(j=1,2,\ldots,n)$.

Step 4: Determine the weighted aggregated intuitionistic fuzzy decision matrix: Determination of the weight of each factor leads to aggregated decision matrix. To obtain this matrix following formulas (5) and (6) is used

$$R\otimes W=\left\{\left\langle c,\mu_{FM_i}(c).\mu_W(c),\upsilon_{FM_i}(c)+\upsilon_W(c)-\upsilon_{FM_i}(c).\upsilon_W(c)\mid x\in X\right\rangle\right\} \tag{5}$$

$$\pi_{FM_i}(c)=1-\upsilon_{FM_i}(c)-\upsilon_W(c)-\mu_{FM_i}(c).\mu_W(c)+\upsilon_{FM_i}(c).\upsilon_W(c) \tag{6}$$

The weighted aggregated decision matrix is obtained as follows:

$$R=\begin{bmatrix} (\mu_{FM_1W}(O),\upsilon_{FM_1W}(O),\pi_{FM_1W}(O)) & (\mu_{FM_1W}(S),\upsilon_{FM_1W}(S),\pi_{FM_1W}(S)) & (\mu_{FM_1W}(D),\upsilon_{FM_1W}(D),\pi_{FM_1W}(D)) \\ (\mu_{FM_2W}(O),\upsilon_{FM_2W}(O),\pi_{FM_2W}(O)) & (\mu_{FM_2W}(S),\upsilon_{FM_2W}(S),\pi_{FM_2W}(S)) & (\mu_{FM_2W}(D),\upsilon_{FM_2W}(D),\pi_{FM_2W}(D)) \\ & \ddots & \\ (\mu_{FM_nW}(O),\upsilon_{FM_nW}(O),\pi_{FM_nW}(O)) & (\mu_{FM_nW}(S),\upsilon_{FM_nW}(S),\pi_{FM_nW}(S)) & (\mu_{FM_nW}(D),\upsilon_{FM_nW}(D),\pi_{FM_nW}(D)) \end{bmatrix} \tag{7}$$

where $r'_{ij}=\left(\mu'_{ij},\upsilon'_{ij},\pi'_{ij}\right)=(\mu_{FMW}(c),\upsilon_{FMW}(c),\pi_{FMW}(c))$.

Step 5: Determine the positive and negative ideal solutions based on IFNs: Let J_1 and J_2 are profit and cost criteria respectively. Then, positive and negative ideal solutions can be obtained by formulas (8) and (9) respectively:

$$FM^+=(\mu_{FM^+W}(c_j),\upsilon_{FM^+W}(c_j)) \tag{8}$$

$$FM^-=(\mu_{FM^-W}(c_j),\upsilon_{FM^-W}(c_j)) \tag{9}$$

where

$$\mu_{FM^+W}(c_j)=\left(\left\langle\max_i\,\mu_{FM_i W}(c_j)\mid j\in J_1\right\rangle,\left\langle\min_i\,\mu_{FM_i W}(c_j)\mid j\in J_2\right\rangle\right) \tag{10}$$

$$\upsilon_{FM^{+}W}(c_j)=\left(\left\langle \min_i \upsilon_{FM_i W}(c_j) \mid j\in J_1\right\rangle,\left\langle \max_i \upsilon_{FM_i W}(c_j) \mid j\in J_1\right\rangle\right) \quad (11)$$

$$\mu_{FM^{-}W}(c_j)=\left(\left\langle \min_i \mu_{FM_i W}(c_j) \mid j\in J_1\right\rangle,\left\langle \max_i \mu_{FM_i W}(c_j) \mid j\in J_2\right\rangle\right) \quad (12)$$

$$\upsilon_{FM^{-}W}(c_j)=\left(\left\langle \max_i \upsilon_{FM_i W}(c_j) \mid j\in J_1\right\rangle,\left\langle \min_i \upsilon_{FM_i W}(c_j) \mid j\in J_1\right\rangle\right) \quad (13)$$

Step 6: Calculate the distances from positive and negative ideal solutions: To calculate separation measures normalized Euclidean distance is used as follows [3]:

$$S^{+}=\sqrt{\frac{1}{2n}\sum_{j=1}^{n}\left[\left(\mu_{FM_i W}(c_j)-\mu_{FM^{+}W}(c_j)\right)^2+\left(\upsilon_{FM_i W}(c_j)-\upsilon_{FM^{+}W}(c_j)\right)^2+\left(\pi_{FM_i W}(c_j)-\pi_{FM^{+}W}(c_j)\right)^2\right]} \quad (14)$$

$$S^{-}=\sqrt{\frac{1}{2n}\sum_{j=1}^{n}\left[\left(\mu_{FM_i W}(c_j)-\mu_{FM^{-}W}(c_j)\right)^2+\left(\upsilon_{FM_i W}(c_j)-\upsilon_{FM^{-}W}(c_j)\right)^2+\left(\pi_{FM_i W}(c_j)-\pi_{FM^{-}W}(c_j)\right)^2\right]} \quad (15)$$

Step 7: Obtain the relative closeness coefficients based on separation measures:

$$C_i^{+}=\frac{S_i^{-}}{S_i^{-}+S_i^{+}} \quad (16)$$

3 Numerical Illustration

FMEA based IF-TOPSIS approach is performed in maritime industry in Turkey to analyze risks and their failure modes. Scope area is selected as ports loading and unloading areas with the directions of highly qualified industrial experts. Risk factors and their failure modes are determined with face to face interview and group decision. Constructed model and its sub-criteria are given in Table 3:

Table 3. FMEA model

Dimensions and sub-criteria		
Crane (C_1)	FM_{11}	Overloaded crane load limits
	FM_{12}	Shaking of crane control mechanisms
	FM_{13}	Hold damage with ropes
Vessel Stress (C_2)	FM_{21}	Discharging sequence
	FM_{22}	Discrepancy of aft/forward drafts
	FM_{23}	Load positioning failures
Vessel Performance (C_3)	FM_{31}	Hull (chest) condition
	FM_{32}	Misuse of fuel
	FM_{33}	Lack of maintenance
Loading /Unloading (C_4)	FM_{41}	Wing tank damage
	FM_{42}	Fuel tank damage
	FM_{43}	Hatchway damage
	FM_{44}	Equipment failures
Personnel (C_5)	FM_{51}	Gang coordination failures

	FM_{52}	Knowledge level of crew
	FM_{53}	Neglect of technical personnel
Weather (C_6)	FM_{61}	Separation
	FM_{62}	Manoeuver risks
	FM_{63}	Degradation of vision fields
	FM_{64}	Swell condition

Following the construction of FMEA model proposed methodology steps are performed. Weights of experts are given in Table 4, importance of risk factors are shown with Tables 5 and 6 as follows:

Table 4. Weights of experts

Decision makers		Exp_1			Exp_2			Exp_3	
Linguistic terms		VI			I			M	
IFNs	[0.9	0.1	0]	[0.75	0.2	0.05]	[0.5	0.45	0.05]
Weights		0.406			0.356			0.238	

Table 5. Importance of O, S, D with linguistic variables

FMEA Factors	Exp_1	Exp_2	Exp_3
O	VI	I	I
S	M	I	M
D	I	VI	I

Table 6. Importance of O, S, D with IFNs

FMEA Dimensions		Exp_1			Exp_2			Exp_3	
O	[0.900	0.100	0.000]	[0.750	0.200	0.050]	[0.750	0.200	0.050]
S	[0.500	0.450	0.050]	[0.750	0.200	0.050]	[0.500	0.450	0.050]
D	[0.750	0.200	0.050]	[0.900	0.100	0.000]	[0.750	0.200	0.050]

Experts' opinions are aggregated via IFWA operator and given with Table 7:

Table 7. Aggregated importance of O, S, D based on experts' opinion

	O			S			D	
[0.828	0.151	0.021]	[0.609	0.337	0.054]	[0.819	0.156	0.025]

From aggregated intuitionistic fuzzy decision matrix, positive and negative ideal solutions are obtained and shown in Table 8:

Table 8. Ideal solutions

Ideal Solutions		O			S			D	
FM^+	[0.516	0.358	0.126]	[0.609	0.337	0.054]	[0.558	0.330	0.112]
FM^-	[0.175	0.799	0.026]	[0.331	0.573	0.096]	[0.082	0.883	0.035]

Separation measures are calculated and used to obtained relative closeness coefficients and these are given in Table 9 below.

Table 9. Ideal solutions

FMs	Si^{+}	Si^{-}	C_i
FM_{11}	0.126	0.037	0.172
FM_{12}	0.128	0.045	0.208
FM_{13}	0.120	0.041	0.117
FM_{21}	0.119	0.047	0.049
FM_{22}	0.127	0.058	0.176
FM_{23}	0.082	0.080	0.363
FM_{31}	0.055	0.120	0.658
FM_{32}	0.094	0.109	0.620
FM_{33}	0.045	0.123	0.687
FM_{41}	0.103	0.082	0.649
FM_{42}	0.115	0.074	0.669
FM_{43}	0.124	0.095	0.180
FM_{44}	0.125	0.081	0.183
FM_{51}	0.094	0.074	0.214
FM_{52}	0.107	0.060	0.470
FM_{53}	0.049	0.136	0.793
FM_{61}	0.079	0.090	0.853
FM_{62}	0.115	0.082	0.369
FM_{63}	0.136	0.036	0.076
FM_{64}	0.121	0.090	0.064

4 Results

FMEA model in international trade ports is analyzed with IF-TOPSIS. Performing the steps of proposed methodology leads to determine the priority of failure modes. Determined relative closeness coefficients which are shown in Table 9, are used to prioritize these failures. This process is shown in Table 10:

Table 10. Ranked failure modes

Ranked Order	Failure Modes (Sub-criteria)		C_i
1	FM_{53}	Neglect of technical personnel	0.736
2	FM_{33}	Lack of maintenance	0.732
3	FM_{31}	Hull (chest) condition	0.686
4	FM_{32}	Misuse of fuel	0.538
5	FM_{61}	Separation	0.531
6	FM_{23}	Load positioning failures	0.493

7	FM_{41}	Wing tank damage	0.443
8	FM_{51}	Gang coordination failures	0.442
9	FM_{43}	Hatch coaming (Hatchway) damage	0.434
10	FM_{64}	Swell condition	0.426
11	FM_{62}	Manoeuver risks	0.417
12	FM_{44}	Equipment failures	0.394
13	FM_{42}	Fuel tank damage	0.391
14	FM_{52}	Knowledge level of crew	0.359
15	FM_{22}	Discrepancy of aft/forward drafts	0.314
16	FM_{21}	Discharging sequence	0.284
17	FM_{12}	Shaking of crane control mechanisms	0.258
18	FM_{13}	Hold damage with ropes	0.255
19	FM_{11}	Overloaded crane load limits	0.226
20	FM_{63}	Degradation of vision fields	0.211

According to analysis results, first 5 most prone failure modes with higher priority in the failure of loading/unloading process at the ports are revealed as follows:

- Neglect of technical personnel,
- Lack of maintenance,
- Hull (chest) condition,
- Misuse of fuel,
- Separation.

Neglecting the known and applicable procedures negatively affects navigation safety. It affects the safety of the cargo carried and unintentional accidents cause injury and loss of personnel. Lack of maintenance of auxiliary machinery, generators, and other technical equipment can cause performance degradation and accidents. Maintenance deficiencies may cause the system to interfere with loading and unloading processes. Coverage of the bottom of the ship with moss and mussel, reduces the cruising speed and increases the fuel consumption. The reduction of cruising speed results in negative time in the planned cruising process. In addition, water-borne filters cause failure and failure of the required circulation. Like in land vehicles, marine fuels have different gravity values. It is also forbidden to use fuels with high sulfur content in some ports. Improper use of fuel increases consumption, causes pollution and reduces performance. The cargoes placed in the ship's warehouses are not properly separated, causing the cargo to slip under bad sea conditions. Loads are stacked on one side, risking the navigation and safety of the ship. It can even cause the ship to sink.

Conclusion

FMEA is a useful technique to analyze risks among different approaches. In this study, international trade ports in maritime industry is analyzed in terms of risks. To perform this, an FMEA model is constructed based in risk factors. Sub-criteria of these factors and their main dimensions are determined with experts' opinions. To eliminate obscurity and hesitancy during evaluation process intuitionistic fuzzy sets are used. IF-TOPSIS analytical tool is used to prioritize the failure modes and most important ones are determine. Study shows that FMEA based IF-TOPSIS methodology is useful for this application area.

Constructed model is also proved that risks are evaluated in a hierarchical structure, so as a future work, analytic hierarchy process (AHP) can be used or different multi-criteria decision-making tools can be combined to reach more sensitive results.

Acknowledgements

Authors would like to give deep gratitude to industrial experts.

References

1. http://unctad.org/en/PublicationsLibrary/rmt2017_en.pdf
2. Atanassov, K. T. (1983). Intuitionistic fuzzy sets, VII ITKR's Session, Sofia deposed in Central Sci. Technical Library of Bulg. Acad. of Sci, 1697, 84.
3. Boran, F. E., Genç, S., Kurt, M., & Akay, D. (2009). A multi-criteria intuitionistic fuzzy group decision making for supplier selection with TOPSIS method. Expert Systems with Applications, 36(8), 11363-11368.
4. Bowles, J. B., & Pelaez, C. E. (1995). Fuzzy logic prioritization of failures in a system failure mode, effects and criticality analysis. Reliability Engineering and System Safety, 50, 203–213.
5. Branch, A.E. Elements of Port Operation and Management. Chapman and Hall. London (1986).
6. Branch, A.E. Elements of port operation and management. Springer Science & Business Media (2012).
7. Gilchrist, W. (1993). Modelling Failure Modes and Effects Analysis. International Journal of Quality& Reliability Management, 10(5), 16-23.
8. IMEAK Deniz Ticaret Odası, Deniz Ticareti Dergisi, Şubat 2017, 1-64
9. Kutlu, A. C.,& Ekmekçioğlu, M. (2012). Fuzzy failure modes and effects analysis by using fuzzy TOPSIS-based fuzzy AHP. Expert Sys. with Applications, 39(1), 61-67.
10. Liu, H.C., Liu, L. and Liu, N. (2013). Risk evaluation approaches in failure mode and effects analysis: A literature review, Expert Sys. with Applications, 40(2), 828-838.
11. Pillay, A., & Wang, J. (2003). Modified failure mode and effects analysis using approximate reasoning. Reliability Engineering & System Safety, 79, 69–85.
12. Sankar, N. R., & Prabhu, B. S. (2001). Modified approach for prioritization of failures in a system failure mode and effects analysis. International Journal of Quality and Reliability Management, 18(3), 324–335.
13. Sharma,R.K., Kumar, D., & Kumar, P. (2005). Systematic failure mode effect analysis (FMEA) using fuzzy linguistic modelling, International Journal of Quality & Reliability Management, 22(9). 986–1004.
14. Stamatis, D. H. (1995). Failure mode and effect analysis: FMEA from theory to execution. Milwaukee, WI: ASQC Quality Press.
15. Tooranloo, H. S. &Ayatollah, A.S. (2016). A model for failure mode and effects analysis based on intuitionistic fuzzy approach. Applied Soft Computing, 49, 238-247.
16. Ung, S.T., Williams, V. Bonsall, S. and Wang, J. (2009). The Risk Assessment and Management of Port Security Using Fuzzy Modeling, Marine Tech., 46(2). 61-73.
17. Wang, Y. M., Chin, K. S., Poon, G. K. K., & Yang, J. B. (2009). Risk evaluation in failure mode and effects analysis using fuzzy weighted geometric mean. Expert Systems with Applications, 36, 1195–1207.
18. Xu, Z. (2007). Intuitionistic fuzzy aggregation operators. IEEE Transactions on fuzzy systems, 15(6), 1179-1187.
19. Zadeh, L. A. (1996). Fuzzy sets. In Fuzzy Sets, Fuzzy Logic, And Fuzzy Systems: Selected Papers by Lotfi A. Zadeh (pp. 394-432).

On New Knot Tables

Ceren Sultan ELMALI[1*], Tamer UĞUR[2], Tuğçe KUNDURACI[3]

[1]*Department of Mathematics, Faculty of Science, Erzurum Technical University, 25100 Erzurum-TURKEY*

[2]*Department of Mathematics, Faculty of Science, Atatürk University, 25200 Erzurum-TURKEY*

[3]*Graduate School of Natural and Applied Science, Erzurum Technical University, 25100 Erzurum-TURKEY*

Abstract. The quasi-pseudo metrics on the vertices of a digraph induces a unique bitopology. In this work, we obtained that a bitopology is associated with any knot k_m, where k is crossing points of knot and $m = 1,2$ by using quasi-pseudo metrics on the vertices of a digraph which is called knot digraph by us, when we consider relation between the knot and its graph (knot digraph notation). Also some topological properties of bitopology obtained by using quasi-pseudo metric and associated with the knot digraphs are investigated.

1.Introduction

The mathematical notion of bitopological space (X, τ_1, τ_2), namely, a set X equipped with two arbitrary topologies τ_1, τ_2. For the first time, J.C.Kelley formulated this space [4]. Girija and Plakkat showed that bitopology is get from a digraph which has orientation by using quasi-pseudo metric entitled with 'Bitopological space associated with digraphs' [3] in 2013. A quasi pseudo metric on nonempty set X is a nonnegative real valued function ρ on $X \times X$ such that $\rho(x,x) = 0, \forall x \in X$ and $\rho(x,z) \leq \rho(x,y) + \rho(y,z), \forall x,y,z \in X$associated with ρ there is another quasi pseudometric q such that $\rho(x,y) = q(y,x)$.In this study the authors is called digraph if the vertex of graph is reachable from edge of it. Let be V the vertex set of a digraph. The function $p: V \times V \to \mathbb{R}$ and $q: V \times V \to \mathbb{R}$ defined by

$$p(x,y) = \begin{cases} 0, & If\ x\ is\ reachable\ from\ y \\ 1, & otherwise \end{cases}, \quad q(x,y) = \begin{cases} 0, & If\ y\ is\ reachable\ from\ x \\ 1, & otherwise \end{cases}$$

respectively. These functions are quasi-pseudo metrics on V and induces a unique bitopology on V with $S_p(x,\varepsilon) = \{y: p(x,y) < \varepsilon\}$ and $S_q(x,\varepsilon) = \{y: q(x,y) < \varepsilon\}$ as a basis. All open ρ-balls collection forms a base for a topology on X. We show this topology with τ_1. Similarly all open q-balls collection forms a base for a topology on X. We show this topology with τ_2. So, we get the bitopology (X, τ_1, τ_2). A bitopological space (X, τ_1, τ_2) is a set X with two topologies, τ_1 and τ_2, on it.

We have seen that this method can be applied for the oriented knot graph which is obtained form regular diagram of a knot. We can get a bitopology matched with any knot by using knot digraph when we consider relation between the knot and its graph.

In this way, our purpose contributes the problem of classification the most important problem of mathematic by using bitopology. We plan knot table of prime knot, which has at least 11 crossing by using this method. In our day, there isn't any table of all prime knots. So, we aim to contribute knot table of all prime knot in future period.

2.Knot and Knot Graphs

A *knot* is a simple closed polygonal curve in R^3 [1]. A knot projection is called a *regular projection* if no three points on the knot project to the same point, and no vertex projects to the same point as any other point on the knot. It divides the plane into several regions. Each of this regions is homeomorphic to an open disk. It is useful tool for separating regions (domains) into two classes, white-black or shaded-unshaded or shadow-unshadow regions, in the study of the knot theory.

A regular projections of knot K divides plane into $n+2$ regions where n is the number of double point A ($A_1, A_2, \ldots A_n$). The Figure 1 show that the number of double point of Trefoil knot is 3. For this reason the number of regions is 5 for Trefoil knot.

* Corresponding Author: ceren.elmali@erzurum.edu.tr

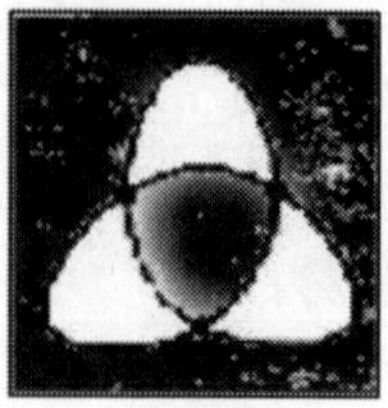

Figure 1

Starting with the outer most region, the regions can be colored black or white where neighboring regions are never this same color. White and black regions are labelled, in order of, such that $\alpha = \{\alpha_1, \alpha_2, \ldots, \alpha_k\}$ and $\beta = \{\beta_1, \beta_2, \ldots, \beta_l\}$. The point at infinity is contained in a region of black. $P_i \in \alpha_i$ $(i = 1,2,\ldots,k)$ points are taken. We connect this points by n- non-intersection arcs $d_1, d_2, \ldots, d_n$ in such a way each d_t corresponds to A_t $(t = 1,2,\ldots,n)$ and P_i and P_j connected by d_t, if and only if α_i, α_j have a common double point A_t on their boundaries.Thus, a plane graph is obtained from regular projection of the knot G. From same consideration about class β, another graph G^d is get. We call G^d the dual graph of G. This method is called as Tait's method. Figure 2(a) and 2(b) show that graph and dual graph of Trefoil knot, respectively,

(a) (b)

Figure 2

The regions of class α in all can be considered as a projection of a surface spanning K which is twisted 180°at each double point of projection. From this consideration, we define the sign of d_t in such a way that it is + or – according as the surface is twisted at A_t as the right screw or the left one, respectively (Figure 3).

\+ Figure 3 -

It is well known that every deformation of a knot is equivalent to a combination of the fundamental deformation (Reidemeister Movies) Ω_1, Ω_2 and Ω_3 in [7].

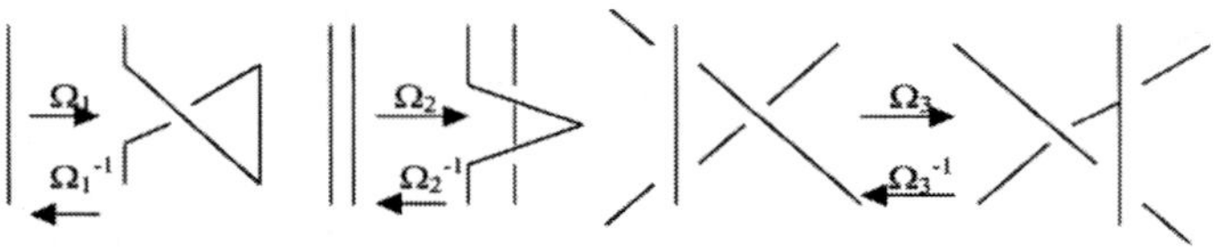

Figure 4

These fundamental deformations can be translated into the operations on graphs as follows [8].

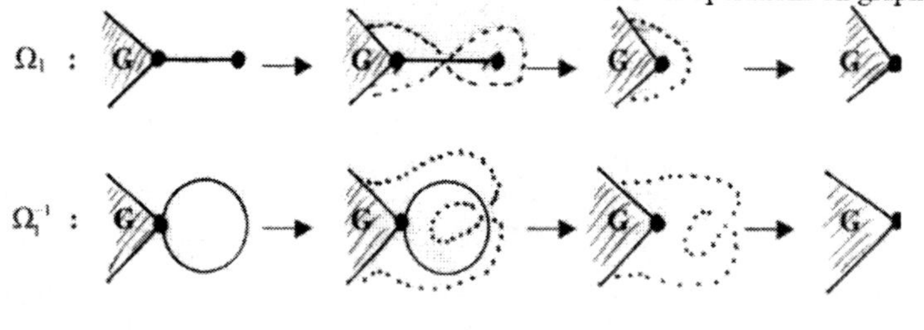

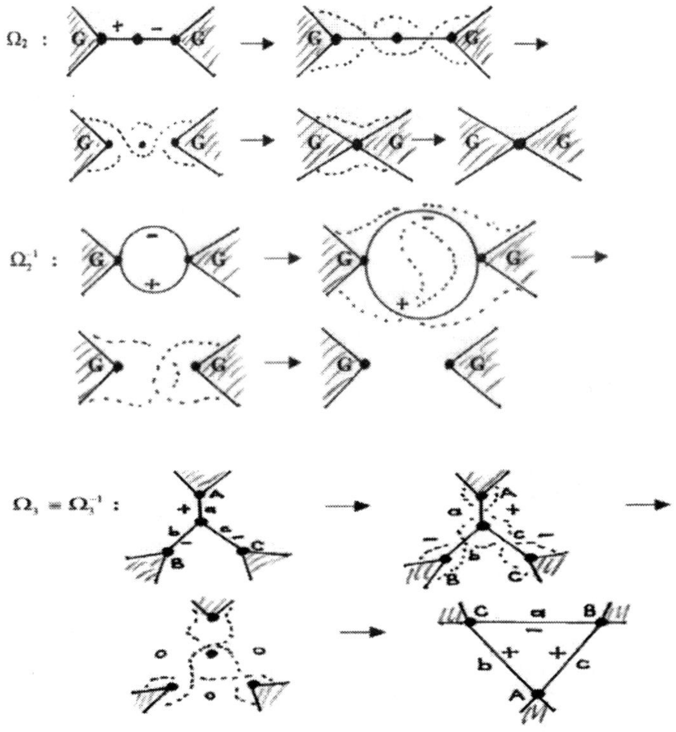

Figure 5

We recommend [2,5,6] for details about this section.

3. Knotdigraph Notation

Firstly, signed graph is obtained from regular diagram of a knot by using Tait methods explained in Section 2.Secondly, orientation is given depending on the signed of the signed knot graph as Figure 6.If the signed of edge is +, orientation of this edge is given from a to b. If the signed of edge is -, orientation of this edge is given from b to a. So we can get two directed graphs, when we consider graph and dual graph of a knot. We called this directed graph as knotdigraph.

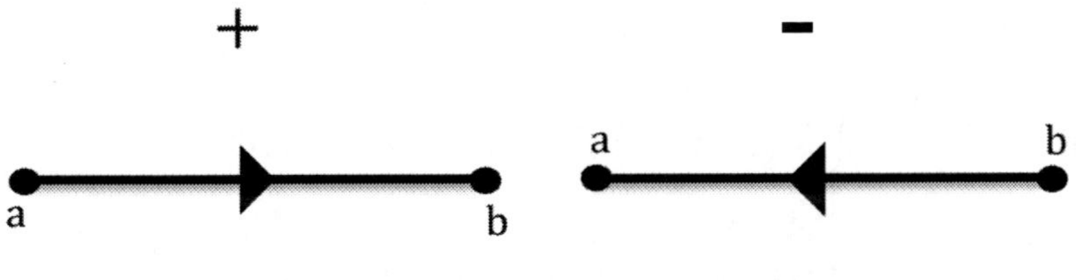

Figure 6. Orientation of knot graph depend on the signed of the signed knot graph

After that, every vertices of knot digraph are labelled by lowercases. Finally, the base is obtained from knot digraph by using quasi-pseudo metric for bitopology. Thus, every knot is matched with bitopology obtained by using this method.We called this method as Knotdigraph Notation

We can apply this method to dual graph of trefoil. Firstly we draw signed dula graph of oriented right-hand trefoil. Secondly, we draw knot digraph and label every vertices of knot digraph by 1,2 as a Figure 7

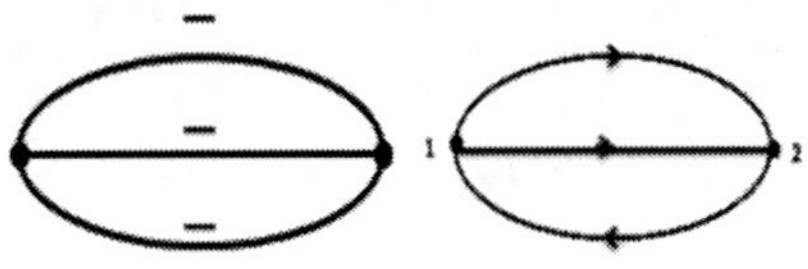

Figure 7

After that we apply Reidemester deformation to labelled knot digraph as a Figure 8. Adding point denoted by ε_1^{12} depending on edge of knot

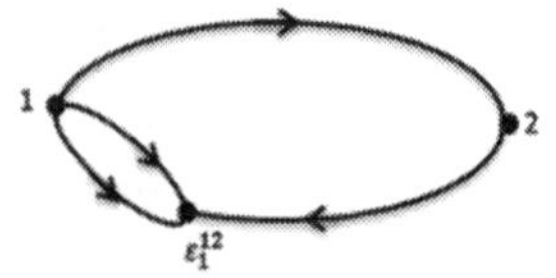

Figure 8

Finally, we obtained the base by using quasi-pseudo metrics mentioned in Section 3.Then, bitopology of dual graph of oriented right-hand trefoil is $\tau_1 = \{\emptyset, X, \{1\}, \{1,2\}\}$, $\tau_2 = \{\emptyset, X, \{\varepsilon_1^{12}\}, \{\varepsilon_1^{12}, 2\}\}$.
We get a bitopology by applying this method to knots k_m ,where n is crossing points of knot and $m = 1,2$. We get some tables as follows.

$k_1 (k\ is\ odd)$	Bitopology associated with knot digraph
3_1	$\tau_1 = \{\emptyset, X, \{1\}, \{1,2\}\}, \tau_2 = \{\emptyset, X, \{\varepsilon_1^{12}\}, \{\varepsilon_1^{12}, 2\}\}$
5_1	$\tau_1 = \{\emptyset, X, \{1\}, \{1,2\}\}, \tau_2 = \{\emptyset, X, \{\varepsilon_1^{12}, \varepsilon_2^{12}\}, \{\varepsilon_1^{12}, \varepsilon_2^{12}, 2\}\}$
7_1	$\tau_1 = \{\emptyset, X, \{1\}, \{1,2\}\}, \tau_2 = \{\emptyset, X, \{\varepsilon_1^{12}, \varepsilon_2^{12}, \varepsilon_3^{12}\}, \{\varepsilon_1^{12}, \varepsilon_2^{12}, \varepsilon_3^{12}, 2\}\}$
9_1	$\tau_1 = \{\emptyset, X, \{1\}, \{1,2\}\}, \tau_2 = \{\emptyset, X, \{\varepsilon_1^{12}, \varepsilon_2^{12}, \varepsilon_3^{12}, \varepsilon_4^{12}\}, \{\varepsilon_1^{12}, \varepsilon_2^{12}, \varepsilon_3^{12}, \varepsilon_4^{12}, 2\}\}$
⋮	⋮
k_1	$\tau_1 = \{\emptyset, X, \{1\}, \{1,2\}\}, \tau_2 = \left\{\emptyset, X, \left\{\varepsilon_1^{12}, \varepsilon_2^{12}, \dots, \varepsilon_{\frac{k-1}{2}}^{12}\right\}, \left\{\varepsilon_1^{12}, \varepsilon_2^{12}, \dots, \varepsilon_{\frac{k-1}{2}}^{12}, 2\right\}\right\}$

$k_1 (k\ is$ even)	Bitopology associated with knot digraph
4_1	$\tau_1 = \{\emptyset, X, \{1\}, \{1,2\}, \{1,2,3\}\}, \tau_2 = \{\emptyset, X, \{\varepsilon_1^{13}\}, \{\varepsilon_1^{13}, 3\}, \{\varepsilon_1^{13}, 2,3\}\}$
6_1	$\tau_1 = \{\emptyset, X, \{1\}, \{1,2\}, \{1,2,3\}, \{1,2,3,4\}, \{1,2,3,4,5\}\}$ $\tau_2 = \{\emptyset, X, \{\varepsilon_1^{15}\}\{\varepsilon_1^{15}, 5\}\{\varepsilon_1^{15}, 4,5\}, \{\varepsilon_1^{15}, 3,4,5\}, \{\varepsilon_1^{15}, 2,3,4,5\}\}$
8_1	$\tau_1 = \left\{\begin{matrix}\emptyset, X, \{1\}, \{1,2\}, \{1,2,3\}, \{1,2,3,4\}, \{1,2,3,4,5\}, \\ \{1,2,3,4,5,6\}, \{1,2,3,4,5,6,7\}\end{matrix}\right\}$ $\tau_2 = \left\{\begin{matrix}\emptyset, X, \{\varepsilon_1^{17}\}, \{\varepsilon_1^{17}, 7\}, \{\varepsilon_1^{17}, 6,7\}, \{\varepsilon_1^{17}, 5,6,7\}\{\varepsilon_1^{17}, 4,5,6,7\}, \\ \{\varepsilon_1^{17}, 3,4,5,6,7\}, \{\varepsilon_1^{17}, 2,3,4,5,6,7\}\end{matrix}\right\}$
10_1	$\tau_1 = \left\{\begin{matrix}\emptyset, X, \{1\}, \{1,2\}, \{1,2,3\}, \{1,2,3,4\}, \{1,2,3,4,5\}, \{1,2,3,4,5,6\}, \\ \{1,2,3,4,5,6,7\}, \{1,2,3,4,5,6,7,8\}, \{1,2,3,4,5,6,7,8,9\}\end{matrix}\right\}$ $\tau_2 = \left\{\begin{matrix}\emptyset, X, \{\varepsilon_1^{19}\}, \{\varepsilon_1^{19}, 9\}, \{\varepsilon_1^{19}, 8,9\}, \{\varepsilon_1^{19}, 7,8,9\}, \\ \{\varepsilon_1^{19}, 6,7,8,9\}\{\varepsilon_1^{19}, 5,6,7,8,9\}, \\ \{\varepsilon_1^{19}, 4,5,6,7,8,9\}\{\varepsilon_1^{19}, 3,4,5,6,7,8,9\}, \{\varepsilon_1^{19}, 2,3,4,5,6,8,97\}\end{matrix}\right\}$
⋮	⋮
k_1	$\tau_1 = \{\emptyset, X, \{1\}, \{1,2\}, \dots, \{1,2, \dots, k-1\}\}$ $\tau_2 = \left\{\begin{matrix}\emptyset, X, \{\varepsilon_1^{1k-1}\}, \{\varepsilon_1^{1k-1}, k-1\}, \{\varepsilon_1^{1k-1}, k-2, k-1\}, \\ \{\varepsilon_1^{1k-1}, k-3, k-2, k-1\}, \dots, \{\varepsilon_1^{1k-1}, 2, \dots, k-2, k-1\}\end{matrix}\right\}$

$k_2 (k\ is\ odd)$	Bitopology associated with knot digraph
5_2	$\tau_1 = \{\emptyset, X, \{1\}, \{1,2\}, \{1,2,3\}, \{1,2,3,4\}\}$ $\tau_2 = \{\emptyset, X, \{\varepsilon_1^{14}\}, \{\varepsilon_1^{14}, 4\}, \{\varepsilon_1^{14}, 3,4\}, \{\varepsilon_1^{14}, 2,3,4\}\}$

7_2	$\tau_1 = \{\emptyset, X, \{1\}, \{1,2\}, \{1,2,3\}, \{1,2,3,4\}, \{1,2,3,4,5\}, \{1,2,3,4,5,6\}\}$ $\tau_2 = \left\{\begin{matrix}\emptyset, X, \{\varepsilon_1^{16}\}, \{\varepsilon_1^{16}, 6\}, \{\varepsilon_1^{16}, 5,6\}, \{\varepsilon_1^{16}, 4,5,6\}, \\ \{\varepsilon_1^{16}, 3,4,5,6\}, \{\varepsilon_1^{16}, 2,3,4,5,6\}\end{matrix}\right\}$
9_2	$\tau_1 = \left\{\begin{matrix}\emptyset, X, \{1\}, \{1,2\}, \{1,2,3\}, \{1,2,3,4\}, \{1,2,3,4,5\}, \\ \{1,2,3,4,5,6\}, \{1,2,3,4,5,6,7\}, \{1,2,3,4,5,6,7,8\}\end{matrix}\right\}$ $\tau_2 = \left\{\begin{matrix}\emptyset, X, \{\varepsilon_1^{18}\}, \{\varepsilon_1^{18}, 8\}, \{\varepsilon_1^{18}, 7,8\}, \{\varepsilon_1^{18}, 6,7,8\}, \{\varepsilon_1^{18}, 5,6,7,8\}, \\ \{\varepsilon_1^{18}, 4,5,6,7,8,\}, \{\varepsilon_1^{18}, 3,4,5,6,7,8\}, \{\varepsilon_1^{18}, 2,3,4,5,6,7,8\}\end{matrix}\right\}$
⋮	⋮
k_2	$\tau_1 = \{\emptyset, X, \{1\}, \{1,2\}, \{1,2,\dots,k-1\}\}$ $\tau_2 = \left\{\begin{matrix}\emptyset, X, \{\varepsilon_1^{1k-1}\}, \{\varepsilon_1^{1k-1}, k-1\}, \{\varepsilon_1^{1k-1}, k-2, k-1\}, \\ \{\varepsilon_1^{1k-1}, k-3, k-2, k-1\}, \dots, \{\varepsilon_1^{1k-1}, 2, \dots, k-2, k-1\}\end{matrix}\right\}$

k_2(k *is* even)	Bitopology associated with knot digraph
6_2	$\tau_1 = \{\emptyset, X, \{1\}, \{1,2\}, \{1,2,3\}\}$ $\tau_2 = \{\emptyset, X, \{\varepsilon_1^{23}, \varepsilon_1^{13}\}, \{\varepsilon_1^{23}, \varepsilon_1^{13}, 3\}, \{\varepsilon_1^{23}, \varepsilon_1^{13}, 2,3\}\}$
8_2	$\tau_1 = \{\emptyset, X, \{1\}, \{1,2\}, \{1,2,3\}\}$ $\tau_2 = \{\emptyset, X, \{\varepsilon_1^{23}, \varepsilon_2^{23}, \varepsilon_1^{13}\}, \{\varepsilon_1^{23}, \varepsilon_2^{23}, \varepsilon_1^{13}, 3\}, \{\varepsilon_1^{23}, \varepsilon_2^{23}, \varepsilon_1^{13}, 2,3\}\}$
10_2	$\tau_1 = \{\emptyset, X, \{1\}, \{1,2\}, \{1,2,3\}\}$ $\tau_2 = \left\{\begin{matrix}\emptyset, X, \{\varepsilon_1^{23}, \varepsilon_2^{23}, \varepsilon_3^{23}, \varepsilon_1^{13}\}, \{\varepsilon_1^{23}, \varepsilon_2^{23}, \varepsilon_3^{23}, \varepsilon_1^{13}, 3\}, \\ \{\varepsilon_1^{23}, \varepsilon_2^{23}, \varepsilon_3^{23}, \varepsilon_1^{13}, 2,3\}\end{matrix}\right\}$
⋮	⋮
k_2	$\tau_1 = \{\emptyset, X, \{1\}, \{1,2\}, \{1,2,3\}\}$ $\tau_2 = \left\{\begin{matrix}\emptyset, X, \left\{\varepsilon_1^{23}, \varepsilon_2^{23}, \dots, \varepsilon_{\frac{k-4}{2}}^{23}, \varepsilon_1^{13}\right\}, \left\{\varepsilon_1^{23}, \varepsilon_2^{23}, \dots, \varepsilon_{\frac{k-4}{2}}^{23}, \varepsilon_1^{13}, 3\right\}, \\ \left\{\varepsilon_1^{23}, \varepsilon_2^{23}, \dots, \varepsilon_{\frac{k-4}{2}}^{23}, \varepsilon_1^{13}, 2,3\right\}\end{matrix}\right\}$

References

1. Alexander, J.W., G.B., *On types of knotted curve.* Ann. Math., 28, 562-586, 1956.
2. Bankwitz, C., *Uber die Torsionszahlen der alternierden Knoten.* Math. Ann., 103, 145-161, 1930.
3. Girija B., Pilakkat R., *Bitopological spaces associated with digraphs,* South Asian Journal of Mathematics, Vol.3 (1):56-65, 2013.
4. Kelley J.C., *Bitopological Spaces,* Proc. London Math.(3), 13, 71-89, 1963.
5. Murasugi, K., *On invariants of graphs with applications to knot theory.* Trans. Amer. Math. Soc., 314, 1-49, 1989.
6. Murasugi, K., *Knot Theory and Its Application.*Boston,1993.
7. K. Reidemeister: *Knotentheorie,* Julius Springer, 1932.
8. Yajima T., Kinoshita S., *On the graphs of knots,* Osaka Math. J., 9, 155-163, 1957.

ITM Web of Conferences **22**, 01020 (2018) https://doi.org/10.1051/itmconf/20182201020
CMES-2018

Representation of the Matrix for Conversion between Triangular Bezier Patches and Rectangular Bezier Patches

Pembe Sabancıgil[1,*] and *Mustafa Kara*[1]

[1]Department of Mathematics, Eastern Mediterranean University, Gazimağusa, TRNC, Mersin 10, Turkey

[1] Department of Mathematics Eastern Mediterranean University, Gazimağusa, TRNC, Mersin 10, Turkey

Abstract. In this paper we studied Bezier surfaces that are very famous techniques and they are widely used in the area of Computer Aided Geometric Design. Mainly there are two kinds of Bezier surfaces which are classified as rectangular and triangular Bezier patches. In this paper we will give a simple representation for the conversion matrix which converts one type to another type in one step.

1 Introduction

The theory of Bezier curves has an important role and they are numerically the most stable among all polynomial bases currently used in CAD systems. On the other hand in these days Bezier surfaces are very famous techniques and widely used in Computer Aided Geometric Design (see [1] and [2]). Mainly there are two kinds of Bezier surfaces which are triangular and rectangular Bezier patches and their definition is given by the uni-variate Bernstein polynomials $B_i^n(s)=\binom{n}{i}s^i(1-s)^{n-i}$ and the bi-variate Bernstein polynomials

$B_{i,j,k}^n(\alpha,\beta,\eta)=\binom{n}{i,j,k}\alpha^i\beta^j\eta^k$ where the sum $\alpha+\beta+\eta=1$. Definition of an n-th degree triangular Bezier patch with control points $D_{i,j,k}$ is given by

$$D(\alpha,\beta,\eta)=\sum_{i+j+k=n} D_{i,j,k}B_{i,j,k}^n(\alpha,\beta,\eta),\quad \alpha,\beta,\eta\ge 0,\quad \alpha+\beta+\eta=1.$$

On the other hand $n\times m$ degree rectangular Bezier patch with the control points $P_{i,j}$ is defined by

$$P(\alpha,\beta)=\sum_{i=0}^{n}\sum_{j=0}^{m}P_{i,j}B_i^n(\alpha)B_j^m(\beta),\ 0\le\alpha,\beta\le 1,\ \text{(see [3])}$$

* Corresponding author: pembe.sabancigil@emu.edu.tr

Since these two patches have different geometric properties, it is not easy to deal with both of them in one Computer Aided Design system (see [4] and [5]). Thus from this reason we have to convert one type to another.

2 Construction of the Conversion Matrices

In the following theorem (see [3]), it is given that a triangular Bezier patch can be converted to an equivalent degenerate rectangular Bezier patch, whose control points are computed by degree elevating some Bezier curves. Here a degenerate rectangular patch means a rectangular patch in which one of its edges collapsed into a point.

Theorem 1. (see [3]) *A degree n triangular Bézier patch $D(\alpha,\beta,\eta)$ can be represented as a degenerate rectangular Bézier patch of degree $n\times n$:*

$$P(\alpha,\beta)=\sum_{i=0}^{n}\sum_{j=0}^{n}P_{i,j}B_i^n(\alpha)B_j^n(\beta),\ 0\le\alpha,\beta\le 1$$

where $P_{i,j}$ are the control points and they are determined as

$$\begin{pmatrix} P_{i,0}\\ P_{i,1}\\ \vdots\\ P_{i,n}\end{pmatrix}=A_1A_2\ldots A_i\begin{pmatrix} D_{i,0}\\ D_{i,1}\\ \vdots\\ D_{i,n-i}\end{pmatrix},\qquad i=0,1,\ldots,n.$$

and A_i where $i=0,1,\ldots n$, are matrices (or operators) of degree elevation which are represented in the following form:

$$A_i=\begin{bmatrix} 1 & 0 & 0 & \ldots & 0 & 0\\ \frac{1}{n+1-i} & \frac{n-i}{n+1-i} & 0 & \ldots & 0 & 0\\ 0 & \frac{2}{n+1-i} & \frac{n-i-1}{n+1-i} & \ldots & 0 & 0\\ \vdots & \vdots & \vdots & \ddots & \vdots & \vdots\\ 0 & 0 & 0 & 0 & \frac{n-i}{n+1-i} & \frac{1}{n+1-i}\\ 0 & 0 & 0 & 0 & 0 & 1\end{bmatrix}_{(n-i+2)\times(n-i+1)}$$

Here in Theorem 2 we give a simple representation for the matrix product $A_1 A_2 ... A_i$.

Theorem 2. *The product of the degree elevation matrices that converts a degree n triangular Bezier patch to a degenerate rectangular Bezier patch of degree $n \times n$ is given as follows :*

$$A_1 A_2 ... A_k = \overline{A}_k = \left[\overline{a}_{i,j}^{(k)} \right]_{(n+1)\times(n-k+1)}$$

where

$$\overline{a}_{i,j}^{(k)} = \frac{\binom{i-1}{j-1} (k)_{i-j} (n-k)_{j-1}}{(n)_{i-1}},$$

$$(k)_n = k(k-1)...(k-n+1) = \prod_{j=1}^{n} (k-(j-1)) \quad \text{and}$$

$$A_k = \begin{bmatrix} 1 & 0 & 0 & \cdots & 0 & 0 \\ \frac{1}{n+1-k} & \frac{n-k}{n+1-k} & 0 & \cdots & 0 & 0 \\ 0 & \frac{2}{n+1-k} & \frac{n-k-1}{n+1-k} & \cdots & 0 & 0 \\ \vdots & \vdots & \vdots & \ddots & \vdots & \vdots \\ 0 & 0 & 0 & 0 & \frac{n-k}{n+1-k} & \frac{1}{n+1-k} \\ 0 & 0 & 0 & 0 & 0 & 1 \end{bmatrix}_{(n-k+2)\times(n-k+1)}$$

Now in the following theorem we will consider the inverse process which will convert a rectangular Bezier patch of degree $n \times n$ to a triangular Bezier patch of degree n :

Theorem 3 *A rectangular Bezier patch $P(\alpha, \beta)$ of degree $n \times n$ can be represented as a triangular Bezier patch $D(\alpha, \beta, \eta)$ of degree n :*

$$D(\alpha,\beta,\eta) = \sum_{i+j+k=n} D_{i,j,k} B^n_{i,j,k}(\alpha,\beta,\eta), \quad \alpha,\beta,\eta \geq 0, \quad \alpha+\beta+\eta = 1$$

where $D_{i,j,k}$ *are the control points and they are defined as*

$$\begin{pmatrix} D_{i,0} \\ D_{i,1} \\ \vdots \\ D_{i,n-i} \end{pmatrix} = B_i B_{i-1} \ldots B_1 \begin{pmatrix} P_{i,0} \\ P_{i,1} \\ \vdots \\ P_{i,n} \end{pmatrix} \quad i = 0,1,\ldots,n$$

and $B_i (i = 0,1,\ldots,n)$ *are matrices (or operators) of degree reduction which are given in the following form*

$$B_k = \begin{bmatrix} 1-t & t & 0 & 0 & \cdots & 0 & 0 \\ 0 & 1-t & t & 0 & \cdots & 0 & 0 \\ 0 & 0 & 1-t & t & \cdots & 0 & 0 \\ \vdots & \vdots & \vdots & \ddots & \ddots & \vdots & \vdots \\ 0 & 0 & 0 & 0 & 1-t & t & 0 \\ 0 & 0 & 0 & 0 & & 1-t & t \end{bmatrix}_{(n-k+1)\times(n-k+2)}$$

Theorem 4 *The product of the matrices in the above theorem* $B_k B_{k-1} \ldots B_1$ *that converts a rectangular Bezier patch of degree* $n \times n$ *to a triangular Bezier patch of degree* n *can be generalized as follows:*

$$Z^k = B_k B_{k-1} \ldots B_1 = \begin{bmatrix} b_{k,0} & b_{k,1} & b_{k,2} & \cdots & b_{k,k-1} & b_{k,k} & 0 & 0 & 0 & \cdots & 0 \\ 0 & b_{k,0} & b_{k,1} & b_{k,2} & \cdots & b_{k,k-1} & b_{k,k} & 0 & 0 & \cdots & 0 \\ 0 & 0 & b_{k,0} & b_{k,1} & b_{k,2} & \cdots & b_{k,k-1} & b_{k,k} & 0 & \cdots & 0 \\ 0 & 0 & 0 & b_{k,0} & b_{k,1} & b_{k,2} & \cdots & b_{k,k-1} & b_{k,k} & \cdots & 0 \\ \vdots & \vdots & \vdots & \ddots & \ddots & \ddots & \ddots & \ddots & \ddots & \ddots & 0 \\ 0 & 0 & 0 & 0 & 0 & b_{k,0} & b_{k,1} & b_{k,2} & \cdots & b_{k,k-1} & b_{k,k} \end{bmatrix}_{(n-k+1)\times(n+1)}$$

where $b_{k,j} = \binom{k}{j} t^j (1-t)^{k-j}$.

Remark *The proofs of Theorems 2, 3 and 4 will be presented later in the complete paper.*

3 Conclusion

In this paper, we studied on the conversion matrix to convert triangular Bezier patch to a rectangular Bezier patch and a rectangular Bezier patch to a triangular Bezier patch. We found simple representations for these two matrices which will allow the conversion in one step.

References

1. Farin, G., Curve and Surface for CAGD : A Practical Guide. Academic Press, New York (1990)
2. Farin, G., Computer Aided Design, 21, 293-296 (1989)
3. Shi-Min Hu, Computer Aided Geometric Design, 18, 667-671 (2001)
4. Goldman, R., Filip, D., Computer Aided Design, 19, 25–27 (1987)
5. Shi-Min Hu, Computer Aided Geometric Design, 13, 219–226 (1996)

Use And Comparison of Topis And Electre Methods In Personnel Selection

Bilgin Şenel[1],*Mine* Şenel[1,*] ,*Gizem* Aydemir[2]

[1]Department of Industrial Engineering, University of Munzur, Tunceli, Turkey

[2]Project Manager, OBSS

Abstract. One of the most important function of human resources is personnel selection process. This process should be done professionally, in a short time and with minimum cost. After personnel selection process, performance of the hired person is very important for the permanence and success of the company. From this point of view, the aim of this study is to select a personnel among the candidates efficiently, with minimum cost and within a short time in one of the leading companies of Turkey in automotive sector. In order to select the right personnel all criterias which has great impact on blue collar worker selection was decided and these criterias are weighted. From the candidate pool of automotive company, appropriate candidates were selected by using TOPSIS AND ELECTRE method which are multi-criteria decision making methods

1 Introduction

Personnel selection is an important part of human resources management policy in any enterprise. Personnel selection process is aimed at choosing the best candidate to fill the defined vacancy in a company. It determines the input quality of personnel and thus plays an important role in human resource management [2].

Especially today, with the rapid development of the increasing globalization processes, increasing competition environment necessitates development in personnel selection processes. What makes this compulsory is that with the diversity of candidates in the human resources market, the employer wants to know which candidates qualify which candidates are best suited to perform the defined job [6].

Many businesses that understand the importance of this are now allocating a lot of funding to be one step ahead of the competition. These funds are used by many scientists to develop new decision-making techniques to make the ideal choice [10]. The objective of a selection process depends mainly on assessing the differences among candidates and predicting future performance [3]. For the solution of the complexity inherent in the selection of personnel, many scientists have tried to achieve the result using multi-criteria decision making (MCDM) in methods. [1-2-5- 6-7-10-11].

* Corresponding author: mines@anadolu.edu.tr

MCDM methods deal with problems of compromise selection of the best solutions from the set of available alternatives according to objectives. Usually neither of the alternatives satisfies all the objectives therefore satisfactory decision is made instead of optimal one [1].

In this study, an evaluation was conducted on the ideal staff selection in a pioneering company operating in the automotive sector in Turkey and in accordance with the wishes of the company's senior management, TOPSIS and ELECTRE methods, which are multi criteria decision making techniques, were used to determine which method gave better results.

2 Fundamental Properties of the Method

2.1 TOPSIS Method

Using the TOPSIS method, it is necessary to compare the alternative options according to certain criteria and between the maximum and minimum values that the criteria can take against the ideal situation. A multi-criteria decision problem with *n* alternatives and *m* criterions can be represented by *n* points in *m*-dimensional space. Yoon and Hwang (1980) have constructed the TOPSIS method based on the assumption that the solution alternative is the shortest distance to the ideal solution point and the farthest distance to the negative ideal solution point in some geometrical sense [9].The TOPSIS method consists of following steps [9]:

Step 1 Construct the decision matrix (A) :

The decision matrix is a matrix that must be created by the decision maker. This generated matrix will be a *mxp* dimensional matrix. Decision-making lines show decision points while columns contain factors.

$$A_{ij}=\begin{bmatrix} a_{11} & a_{12} & . & a_{1p} \\ a_{21} & a_{22} & . & a_{2p} \\ . & . & . & . \\ . & . & . & . \\ . & . & . & . \\ a_{m1} & a_{m2} & . & a_{mp} \end{bmatrix}$$

Step 2 Construct the normalized decision matrix (R):

After the decision matrix is formed, an element r_{ij} of the normalized decision matrix R is calculated as follows;

$$R_{ij}=\begin{bmatrix} r_{11} & r_{12} & . & r_{1n} \\ r_{21} & r_{22} & . & r_{2n} \\ . & . & . & . \\ . & . & . & . \\ . & . & . & . \\ r_{m1} & r & . & r_{mn} \end{bmatrix}$$

$$r_{ij}=\frac{a_{ij}}{\sqrt{\sum_{i=1}^{m} a_{ij}^2}} \quad (i = 1,2, \dots\dots m \;\; and \;\; j = 1,2, \dots\dots n) \qquad (1)$$

Step 3 Calculate the weighted normalized decision matrix (V_{ij}):

$$V_{ij} = \begin{vmatrix} w_1 r_{11} & w_2 r_{12} & \cdot & w_n r_{1p} \\ w_1 r_{21} & w_2 r_{22} & \cdot & w_n r_{2p} \\ \cdot & \cdot & \cdot & \cdot \\ \cdot & \cdot & \cdot & \cdot \\ \cdot & \cdot & \cdot & \cdot \\ w_1 r_{m1} & w_2 r_{m2} & \cdot & w_n r_{mp} \end{vmatrix} \rightarrow V_{ij} = \begin{vmatrix} v_{11} & v_{12} & \cdot & v_{1p} \\ v_{21} & v_{22} & \cdot & v_{2p} \\ \cdot & \cdot & \cdot & \cdot \\ \cdot & \cdot & \cdot & \cdot \\ \cdot & \cdot & \cdot & \cdot \\ v_{m1} & v_{m2} & \cdot & v_{mp} \end{vmatrix}$$

Each value of the normalized matrix is weighted with a value such as w_{ij}.

$v_{ij} = r_{ij}.w_{ij}$ (2) The weighted normalized value v_{ij} is calculated in the left way;

Where w_j is the weight of the j-th criterion, $\sum_{j=1}^{n} w_j = 1$

Step 4 Determine Ideal (A^+) and Negative ideal (A^-) solutions :

Once the weighted normalized matrix (V_{ij} matrix) is obtained, the maximal values of each column are determined, provided that our goal is maximization, provided that it depends on the structure of the problem. These maximum values are our ideal solution values. Then, minimum values for each column are obtained again. This is the negative ideal solution value. If our goal is minimization, the values obtained will be the exact opposite. The notation for obtaining ideal and negative ideal solution values is shown below.

$$\left\{ {}^{max}_{\ \ i} \ on\ condition\ that \right\} \rightarrow A^+ = \{v_1^+, v_2^+, \dots \dots \dots v_n^+\} max\ values\ for\ each\ column \quad (3)$$

$$\left\{ {}^{min}_{\ \ i} \ on\ condition\ that \right\} \rightarrow A^- = \{v_1^-, v_2^-, \dots \dots \dots v_n^-\} min\ values\ for\ each\ column \quad (4)$$

Step 5 Calculate the separation measures from the positive and the negative ideal solution :

The n-dimensional Euclidean distance method is applied to measure the separation distances of each alternative from the ideal solution and negative-ideal solution. In order to be able to calculate the distance of the ideal and negative ideal point, following formulas are used :

Ideal Distance

$$S_i^+ = \sqrt{\sum_{j=1}^{N} \left(V_{ij} - V_j^+\right)^2} \quad (5)$$

Negative Ideal Distance

$$S_i^- = \sqrt{\sum_{j=1}^{N} \left(V_{ij} - V_j^-\right)^2} \quad (6)$$

Step 6 Calculate the relative closeness to the positive ideal solution :

Ideal and negative ideal solution is used to calculate the ideal resolving relative closeness of each decision point. The ideal solution is symbolized by the relative closeness C_i^+. Where C_i^+ value takes a value in the range $0 \leq C_i^+ \leq 1$ and while $C_i^+ = 1$ represents the absolute solution closeness to the ideal solution of the corresponding decision point, C_i^-represents the absolute solution closeness to the negative ideal solution of the corresponding decision point.

$$C_i^+ = \frac{S_i^-}{S_i^- + S_i^-} \quad (7)$$

Step 7 Rank the preference order or select the alternative closest to 1 :

A set of alternatives now can be ranked by descending order of value of C_i^+.

2.2 ELECTRE Method

ELECTRE (Elemination and Choice Expressing Reality) method is a multi-decision method first proposed by Benayoun and Roy in 1966. The method is based on binary superiority comparisons between alternative decision points for each rating factor [4]. In the ELECTRE Method concordance and discordance indexes are defined as measurements of satisfaction and dissatisfaction for decision maker in choosing one alternative over anothor. These indexes are then used to analyze the outranking relations among the alternatives [8]. The method is solved by the following steps [9]. The first three steps (Preparation of the decision matrix, the normalized decision matrix and weighted matrix) are not included here because they are the same as TOPSIS method.

Step 4 Ascertainment of Concordance $C_{kl}$to Discordance D_{kl} set :

In determining the concordance set, y matrix is used. The decision points for the evaluation factor are compared with another one and the sets are determined using the relationship shown in the following formula;

$$C_{kl} = \{J, y_{kj} \geq y_{lj}\} \qquad j=1,2,3,\ldots..n$$

The formula is basically based on comparing the sizes of the line elements relative to each other. Each concordance set (C_{kl}) corresponds to an discordance set (D_{kl}). The elements of the discordance set consist of J values that do not belong to the concordance set. On complimentation of (C_{kl}), we obtain the discordance interval set (D_{kl}) using;

$$D_{kl} = \{J, y_{kj} < y_{lj}\} = J - C_{kl} \qquad j=1,2,3,\ldots..n$$

Step 5 Calculation of The Concordance (C) and Discordance (D) Matrices

The relative value of the elements in the concordance matrix C is calculated by means of the concordance index. The concordance index C_{kl} is the sum of the weights associated with the criteria contained in the concordance set. That is;

$$C_{kl} = \sum_{J \in C_{kl}} W_j \qquad (8)$$

$$C = \begin{bmatrix} - & c_{12} & \cdot & c_{1m} \\ c_{21} & - & \cdot & c_{2m} \\ \cdot & \cdot & \cdot & \cdot \\ \cdot & \cdot & \cdot & \cdot \\ \cdot & \cdot & \cdot & \cdot \\ c_{m1} & c_{m2} & \cdot & c_{mn} \end{bmatrix}$$

The concordance matrix is defined as left ;

The discordance matrix D_{kl} elements is defined by formula.

$$D_{kl} = \frac{\max\limits_{j \in D_{kl}} |y_{kj} - y_{lj}|}{\max\limits_{j} |y_{kj} - y_{lj}|} \qquad (9)$$

$$D = \begin{bmatrix} - & d_{12} & . & d_{1m} \\ d_{21} & - & . & d_{2m} \\ . & . & . & . \\ . & . & . & . \\ . & . & . & . \\ d_{m1} & d_{m2} & . & d_{mn} \end{bmatrix}$$

Then, using discordance index set, we can obtain discordance matrix as left;

Step 6 Determine The Concordance (F) and Discordance Dominance (G) Matrix :

For finding matrix (F), it is needed to compute threshold value $(\bar{c})$ as follow:

$$\bar{c} = \frac{1}{m(m-1)} \sum_{k=1}^{m} c_{kl} \qquad (m\ is\ dimension\ of\ matrix) \tag{10}$$

$$\begin{cases} f_{kl} = 1, & if\ \ c_{kl} \geq \bar{c} \\ f_{kl} = 0 & if\ \ c_{kl} < \bar{c} \end{cases}$$

The left inequalities mean that if each element of matrix C_{kl}, is greater than or equal to $\bar{c}$, than 1 would be set in matrix (F). (corresponding element)

To determine discordance dominance matrix, we calculate matrix of (G).

$$\bar{d} = \frac{1}{m(m-1)} \sum_{k=1}^{m} \sum_{l}^{m} d_{kl} \qquad (\bar{d}: Discordance\ Threshold\ Value) \tag{11}$$

$$\begin{cases} g_{kl} = 1 & if\ \ d_{kl} \geq \bar{d} \\ g_{kl} = 0 & if\ \ d_{kl} < \bar{d} \end{cases}$$

The elements of the matrix (G). take the value 0 or 1. There is no value on the diagonal of the matrix (G) as it shows the same decision points.

Step 7 Determine the aggregate dominance matrix (E):

Matrix E is performed by multiplying corresponding elements of F and G.

$$e_{kl} = f_{kl} \cdot g_{kl} \tag{12}$$

The matrix (E) is dimensioned according to the matrix (F) and (G) and consists of the values 0 and 1.

Step 8 Eliminate the less favourable alternative and rank them:

We must scan the line of matrix (E), line that has the highest score should be chosen as the best one.

3 Application

Since the first 3 steps of the TOPSIS and ELECTRE Methods to be used in practice are the same (a decision matrix, a normalized decision matrix and a weighted normalized decision matrix) 3 steps for both methods were created in the same way. Afterwards, the analysis were carried out according to the steps of both methods.

3.1 TOPSIS

At this step, the first question is which criteria are important for recruitment. As a result of interviews with business managers, it has been determined that the following criteria are taken into account in recruitment. Then, a total of 65 applications were made and each candidate was asked to answer the previously determined criteria. A matrix was prepared based on the answers received.

In determining the weight of the factors, business managers, team leaders, and recruitment staff were asked which criteria weighted the recruitment process. The points awarded are distributed in such a way that the sum of the criterion values is 1 point. The weights of the criteria are shown in table 1.

Table 1. Criteria and Weights

CRITERIA	WEIGHTS	CRITERIA	WEIGHTS
C1. Education Status	0,099535124	C20. Task Responsibility	0,019917012
C2. Graduated Section	0,086828512	C21. Mobility	0,02127499
C3. Birthyear	0,01300578	C22. Anger	0,019917012
C4. Gender	0,011823437	C23. Uncontrolled Behavior	0,022632968
C5. Marital status	0,010641093	C24. Self-discipline	0,023990946
C6. Number of children	0,007094062	C25. Sociability	0,017201056
C7. Homeland	0,007882291	C26. Strength Resistance	0,022180309
C8. Smoking	0,016158696	C27. Harmony	0,023538287
C9. Driver's license	0,013399895	C28. Attention and Concentration	0,040685543
C10. Heavy and Dangerous Work Certificate	0,012217551	C29. Visual Comparison	0,037555886
C11. the work is registered	0,012217551	C30. Follow Visual Instruction	0,0383383
C12. Criminal record	0,016946926	C31. Mechanical-Technical Tendency	0,035991058
C13. Military Status for Male Candidates	0,014976353	C32. Verbal Instructions following	0,0383383
C14. Computer Skill	0,033849129	C33. General Image	0,026194145
C15. Course / Certificate Information	0,029819471	C34. General Attitude and Behavior	0,03077812
C16. Foreign Language Information	0,020148291	**35.** Willingness	0,032087827
C17. Work Experience Period (years)	0,029819471	C36. Listening and Understanding	0,033397535
C18. Experience Clarity	0,02127499	C37. Reliability	0,032087827
C19. Layout	0,023538287		

After the decision matrix formed by the data taken in the direction of the criterion has been put forward, Eq.(1) has been used to form the normalized decision (A) matrix. Then, the weighted normalized decision matrix was formed by Eq.(2). After the weighted normalized decision matrix is constructed, two solution sets have been established by using the Eq.(3) for the positive ideal solution and Eq.(4) for the negative ideal solution. In the next step, the deviations of the evaluation factor values for each decision point from the positive ideal and negative ideal solution sets were found. while the Eq.(5) formula is used to find positive ideal deviations, the Eq.(6) formula is used to find negative ideal deviations.

Positive ideal relative closeness values (S_i^+) and negative ideal relative closeness values (S_i^-) were obtained in the direction of the obtained results. By using the obtained differences in Eq.(7), the relative closeness to the ideal solution of each decision point was found and the results are given in Table 2.

Table 2. Relative closeness to the positive ideal solution

1	0,537037	**14**	0,408097	**27**	0,462341	**40**	0,454777	**53**	0,451602
2	0,509735	**15**	0,555171	**28**	0,477382	**41**	0,463658	**54**	0,305041
3	0,478116	**16**	0,082199	**29**	0,600987	**42**	0,526335	**55**	0,390352
4	0,553469	**17**	0,488091	**30**	0,454299	**43**	0,476687	**56**	0,354146
5	0,441776	**18**	0,519921	**31**	0,355324	**44**	0,494757	**57**	0,292422
6	0,402958	**19**	0,45544	**32**	0,445327	**45**	0,497137	**58**	0,542106
7	0,429644	**20**	0,465373	**33**	0,459771	**46**	0,428475	**59**	0,464441
8	0,495668	**21**	0,448104	**34**	0,458169	**47**	0,517146	**60**	0,498541
9	0,5489	**22**	0,528596	**35**	0,375747	**48**	0,548075	**61**	0,477263
10	0,433854	**23**	0,395076	**36**	0,097782	**49**	0,532101	**62**	0,528715
11	0,444864	**24**	0,492628	**37**	0,483895	**50**	0,45384	**63**	0,538521
12	0,524402	**25**	0,518692	**38**	0,557004	**51**	0,382064	**64**	0,323209
13	0,484628	**26**	0,381532	**39**	0,423482	**52**	0,216875	**65**	0,460184

As a result of Table 2, the most suitable candidates for the business were determined to be the ranking result from small value to the large value. Candidates with the highest score were candidates for 29, 38, 15 and 4 respectively, while 16th candidate is the lowest scored candidate.

3.2 ELECTRE

In this method, as in the TOPSIS method, 65 candidates were examined. We tried to make the best choice by using the Electre method in the direction of the data received from these candidates. While the first 3 steps of this method are performed as done in TOPSIS, the other steps will require different operations. Therefore, firstly, Concordance (C_{kl}) and Discordance (D_{kl}) sets were found by using the decision matrix, normalized decision matrix and weighted decision matrices obtained in the TOPSIS method.

Candidates for the same criteria as the values obtained from the weighted normalized matrix are compared with each other and the row number of the greatest value is determined as Concordance and the row number of the smallest name is determined as Discordance.

After this step, matrices of Concordance and Discordance were established. The Eq.(8) is used to construct the concordance matrices. On the other hand, the Eq.(9) is used to construct Discordance matrices. In this step, after the matrices of concordance and discordance are established, matrices of concordance and discordance dominance are established. When these two matrices are generated, the threshold values in Eq.(10)-Eq (11) for both matrices are first calculated.

After two calculated threshold values are obtained, the concordance matrix equal value is compared to each concordance matrix element. In this comparison, if a concordance matrix element is greater than the computed threshold value, it takes on the value 1, while if it is smaller, it takes on the value 0. In this way, a concordance dominance matrix is formed.

On the other hand, after the discordance threshold value is calculated with the Eq (11), it is compared with each element of the discordance matrix. In this comparison, if the value of each element of the discordance matrix is greater than the discordance threshold, it takes 1 and if it is smaller, the value of 0 is taken. Total dominance matrix consisting of 0 and 1 is formed as a result of mutual multiplication of elements of concordance and discordance dominance matrices as indicated by Eq (12).

Table 3. The Aggregate Dominance Matrix (E)

P.ID																																																								TOPLAM
29	0	1	1	1	1	1	1	1	1	1	1	1	1	1	1	1	0	1	1	1	1	1	1	1	1	1	1	1	1	0	1	1	1	1	1	1	1	1	1	1	1	1	1	0	1	1	1	1	1	1	1	1	1	1	1	1	60
22	0	1	0	0	1	1	1	1	0	1	1	0	1	1	0	1	1	1	1	1	0	1	0	1	1	1	1	1	0	1	1	1	1	1	1	1	1	1	0	1	1	1	1	0	1	0	0	1	0	1	1	1	0	0	1	1	44
9	0	0	1	0	1	1	1	1		1	1	0	1	0	0	1	1	0	1	0	1	0	0	1	1	1	1	1	0	1	1	1	0	0	1	1	1	1	0	1	0	1	1	0	1	1	1	1	0	1	0	1	1	0	1	1	43
15	0	1	1	0	1	1	1	1	0	1	1	0	1	0		1	1	1	1	0	1	0	0	1	1	1	1	1	0	0	1	1	0	0	1	1	1	1	0	1	0	0	1	0	1	1	1	1	1	1	0	1	1	0	1	1	43
53	0	1	1	0	1	1	1	1	1	1	1	1	0	0	0	1	0	1	1	0	1	0	0	1	1	1	1	1	0	1	0	0	1	1	1	1	1	0	1	0	0	1	1		1	1	1	1	1	0	0	1	1	0	1	1	42
25	0	1	0	0	1	1	1	0	0	1	1	0	1	1	0	1	1	0	1	0	1	0	1	1		1	0	1	0	0	1	1	0	1	1	1	1	1	0	0	1	1	1	0	1	0	0	1	0	1	0	1	0	0	1	1	39
48	0	0	1	0	1	1	0	1	1	1	1	0	1	0	0	1	0	0	1	0	1	0	0	1	0	1	1	1	0	0	1	1	0	1	1	1	1	0		0	0	0	1	0	1	1	1	1	1	1	0	1	1	0	0	1	37
38	0	0	0	0	1	1	1	0	0	1	1	1	0	1	0	1	0	0	1	1	1	0	1	1	0	1	0	1	0	1	1	0	0	1	0	0	1	0	0	0	0	1	1	0	1	0	1	1	0	1	0	1	1	0	1	1	35
42	0	0	0	0	1	1	0	0	0	1	1	0	1	0	0	1	1	0	1	0	1	0	0	1	0	1	0	1	0	1	0	1		1	1	1	1	1	0	0	0	1	1	0	1	0	0	1	0	1	0	1	0	0	1	1	33
49	0	0	0	0	1	1	1	0	0	1	1	0	0	1	0	1	0	0	1	1	1	0	1	1	0	0	0	1	0	0	1	0	0	0	0	0	1	0	0		0	1	1	0	1	0	1	1	0	1	0	1	0	0	1	1	31
4	0	0	1		1	1	1	1	0	1	1	0	1	0	0	1	0	0	1	0	1	0	1	1	0	1	1	1	0	1	0	0	0	0	1	0	1	0	0	0	0	0	1	0	0	1	1	1	0	1	0	0	0	0	1	0	30
18	0	0	0	0	1	1	1	0	0	1	1	0	0	1	0	0	0		1	1	1	0	1	1	0	0	0	1	0	0	1	1	0	1	0	0	1	0	0	0	0	1	1	0	1	0	0	1	0	1	0	1	0	0	1	1	30
60	0	0	0	0	1	1	1	0	0	1	1	0	0	1	0	0	1	1	1	1	1	0	1	1	0	1	0	1	0	0	0	1	0	0	1	0	1	0	0	0	0	0	1	0	0	0	0	0	0	1		0	0	0	1	1	30
13	0	0	0	0	1	1	0	0	0	1	1	0		1	0	1	0	0	1	0	1	0	0	1	0	1	0	1	0	1	0	1	0	0	1	0	1	0	0	0	1	1	1	0	1	0	0	1	0	1	0	1	0	0	1	1	29
2	0		0	0	1	1	0	0	0	1	1	0	1	0	0	1	0	0	1	0	1	0	1	1	0	1	0	1	0	1	1	1	0	1	0	1	1	0	0	0	0	0	1	0	1	0	0	1	0	1	0	1	0	0	1	1	28
17	0	0	0	0	1	1	1	0	0	1	1	0	0	1	0	0		0	1	0	1	0	1	1	0	1	0	1	0	0	0	0	0	0	1	0	1	0	0	0	0	0	1	0	0	0	0	0	0	1	0	0	0	0	1	1	25
47	0	0	0	0	1	1	0	0	0	0	1	0	0	1	0	1	1	0	0	0	1	0	0	1	0	1	0	1	0	0	0	1	0	0	1	0	1		0	0	1	0	0	0	1	0	0	1	0	1	0	1	0	0	1	1	25
12	0	0	0	0	1	1	1	0	0	1	1		0	1	0	0	0	0	1	1	1	0	1	0	0	1	0	1	0	1	0	0	0	1	0	0	0	0	0	0	0	0	1	0	1	0	1	0	0	0	0	0	0	0	1	1	24
45	0	0	0	0	1	1	0	0	0	1	0	0	1	0	0	0	0	0	1	0	1	0	0	1	0	1	0	1	0	0	0	1	0	0	1		1	0	0	0	0	0	1	0	0	0	0	0	0	1	1	0	0	0	0	1	24
30	0	0	0	0	1	1	1	0	0	1	1	0	0	1	0	1	0	0	1	0	1	0	1	1	0	1	0	1	0		0	0	0	0	0	0	0	0	0	0	0	0	0	1	0	1	0	0	1	0	0	0	0	0	1	1	23
51	0	0	0	0	0	1	1	0	0	1	1	0	0	1	0	1	0	0	1	0	1	0	0	1	0	0	0	0	0	0	0	1	0	0	0	0	1	0	0	1	0		1	0	1	0	1	0	0	1	0	0	0	0	1	1	23
58	0	0	0	0	1	1	0	1	0	1	0	0	1	0	0	1	0	0	1	0	1	0	0	1	1	1	0	1	0	0	1	0	0	0	0	0	1	0	0	0	0	0	1	0	1	0	0	1		0	0	1	0	0	0	1	21
43	0	0	0	0	0	1	1	0	0	1	1	0	0	1	0	0	0	0	1	0	1	0	1	0	0	0	0	0	0	0	1	0		0	0	1	0	0	0	0	0	1	0	0	0	0	1	0	1	0	0	0	0	0	1	1	20
55	0	0	0	0	1	1	0	0	0	1	1	0	0	0	0	1	0	0	1	0	1	0	0	1	0	0	0	1	0	0	1	1	0	0	0	0	1	0	0	0	0	0	1	0	1	0	0	0	0	1	0	0	0	0	0	1	20
63	0	0	1	1	1	1	1	0	0	0	1	0	0	0	0	0	0	0	0	0	0	1	0	0	1	1	0	0	1	0	0	0	0	0	0	0	0	0	0	0	0	0	0	0	0	1	1	0	0	0	0	0	0		1	0	19
20	0	0	0	0	1	0	1	0	0	0	1	0	0	1	0	0	0	0	1		0	0	1	0	0	0	0	0	0	0	1	0	0	0	0	0	0	0	0	0	0	0	1	0	0	0	0	0	0	1	0	0	0	0	1	0	16
27	0	0	0	0	1	1	0	0	0	0	0	0	0	0	0	0	0	0	0	0	1	0	0	1	0	1		0	0	0	0	0	0	0	0	0	0	0	0	0	0	0	1	0	1	1	1	1	0	0	0	0	1	0	0	0	15
16	0	0	0	0	0	0	0	0	0	1	0	0	0	0	0		0	0	0	0	0	0	0	0	0	0	0	0	0	0	1	0	0	0	0	1	0	0	0	0	0	1	0	0	0	0	1	0	0	0	0	0	0	0	1	1	13
8	0	0	1	0	0	1	1		0	0	0	0	0	0	1	0	0	0	0	0	1	0	0	0	0	1	0	1	0	0	0	0	0	0	0	1	0	0	0	0	0	0	0	1	0	0	0	0	0	0	0	0	0	0	0	0	12
41	0	0	0	0	0	0	0	0	0	1	0	0	0	0	0	0	0	0	0	0	0	1	0	0	0	0	0	0	0	1		0	0	0	0	1	0	0	0	0	0	1	0	0	0	0	1	0	0	0	0	0	0	0	1	1	12
50	0	0	0	0	0	0	1	0	0	0	1	0	0	1	0	0	0	0	0	0	0	0	0	1	0	0	0	0	0	0	0	0	0	0	0	0	0	0	0		0	0	0	0	0	0	0	0	1	0	0	0	0	0	1	1	12
11	0	0	0	0	0	1	0	0	0	0	0	0	0	1	0	0	0	0	0	0	0	1	0	0	0	0	0	0	0	0	0	0	0	0	0	0	0	0	0	0	0	0	0	1	0	0	0	0	1	0	0	0	0	0	1	0	11
33	0	0	0	0	0	1	0	0	0	0	0	0	0	1	0	0	0	0	1	0	0	0	1	0	0	0	0	0	0	0	0	0	0	0	0	1	0	0	0	0	0	1	0	0	0	0	0	0	1	0	0	0	0	0	1	0	11
44	0	0	0	0	0	0	0	0	0	1	0	0	0	0	0	0	0	0	0	0	0	0	0	0	0	0	0	0	0	0	0	0	0		0	1	0	0	0	0	0	1	0	0	0	1	1	1	1	0	0	0	0	0	0	1	11
37	0	0	0	0	1	1	0	0	0	0	0	0	0	0	0	0	0	0	0	0	0	0	0	0	0	1	0	0	0	0	0	0	0	0	0	0	0	0	0	0	0	0	0	0	0	0	0	1	0	0	0	0	0	0	1	0	10
62	0	0	0	0	1	1	0	0	0	0	0	0	0	0	0	0	0	0	1	0	0	0	0	0	0	0	0	0	1	0	0	0	0	0	0	0	0	0	0	0	0	0	0	1	0	0	1	0	0	0	0	0		0	0	0	10
24	0	0	0	0	0	1	0	0	0	0	1	0	0	0	0	0	0	0	0	0	1	0	0		0	0	0	0	0	0	0	0	0	0	0	0	0	1	0	0	0	0	0	0	0	1	0	0	0	0	1	0	0	0	0	0	9
34	0	0	0	0	1	1	1	0	0	1	0	0	0	0	0	0	0	0	0	0	0	0	0	1	0	0	0	0	0	0	0	0	0	0	0	0	0	0	0	0	0	0	0	0	0	0	0	0	0	0	0	0	0	0	1	0	8
61	0	0	0	0	0	0	0	0	0	0	0	0	0	0	0	0	0	0	0	0	1	0	0	0	0	0	0	0	0	0	0	0	0	0	0	0	1	0	0	0	0	0	1	0	0	0	0	0	0	0	0		0	0	1	0	8
28	0	0	0	0	0	1	0	0	0	0	0	0	0	0	0	0	0	0	1	0	0	0	0	0	0	0	0		0	0	0	0	0	0	0	0	0	0	0	0	0	0	1	0	0	0	0	0	0	0	1	0	0	0	0	0	7
32	0	0	0	0	0	0	0	0	0	1	0	0	0	0	0	0	0	0	0	0	0	0	0	0	0	0	0	0	0	0	0	0	0	0	0	0	1	0	0	0	0	0	1	0	0	0	0	1	0	0	0	0	0	0	0	0	7
56	0	0	0	0	1	1	0	0	0	0	0	0	0	0	0	0	0	0	1	0	0	0	0	0	0	0	0	0	0	0	0	0	0	0	0	0	0	0	0	0	0	0	0	0	0		0	0	0	0	0	0	0	0	0	0	7
5	0	0	0	0		1	0	0	0	0	0	0	0	1	0	0	0	0	1	0	0	0	0	0	0	0	0	0	0	0	0	0	0	0	0	0	0	0	0	0	0	0	1	0	0	0	0	0	0	0	0	0	0	0	0	0	6
7	0	0	0	0	0	1		0	0	0	0	0	0	1	0	0	0	0	0	0	0	1	0	0	0	0	0	0	0	0	0	0	0	0	0	0	0	0	0	0	0	0	0	0	0	0	0	0	0	0	0	0	0	0	0	1	6
65	0	0	0	0	0	0	0	0	0	0	0	0	0	1	0	0	0	0	0	0	0	0	0	0	0	0	0	0	0	0	0	0	0	0	0	0	1	0	0	0	0	0	0	0	0	0	0	0	0	0	0	0	0	0	0		6

Table 3. shows the rows and columns of the aggregate dominance matrix generated. In the line of the matrix, the candidate with the highest sum of all criteria has become the most suitable candidate. The best candidate for the job is number 29.

Subsequently, candidates numbered 22, 9 and 15, respectively, are followed candidate number 29 in terms of their suitability for work. The candidate with the lowest suitability is 36.

4 CONCLUSION

As a result of this study, both methods applied were compared with each other and evaluations were made about which one would be the most correct result in terms of usage in the sector.

The Electre method compares the criterion values of the candidates and makes the conformity order. This method performs the evaluation by comparing whether the values are greater than or equal to the concordance threshold and whether the values is less than the discordance threshold. However, it does not take any action regarding how large the values are from each other, it also place candidates whose values are equal to each other in the same concordance set as the larger ones.

The Topsis method decides by looking at the Euclidean distance approach how far candidates are away from ideal. Even those values which are slightly different in magnitude among each other can be reflected in the order. Thus, the Topsis method gives the opportunity to analyze how close the candidates are to the ideal. For this reason, there are differences in the order of suitability of the candidates between the outputs of the two methods. When the results obtained in both methods are compared, the most suitable candidate is number 29 for business. However, since the methods calculate differently from each other, there is no similarity in the intermediate values.

Moving from here, it seems more appropriate to use the Topsis method in the problem of personnel selection. It is thought that the topsis method gives more reliable and precise results than Electre method and it is more accurate to use the Topsis method in the problem of personnel selection.

REFERENCES

1. Balezentis, A,. Balezentis, T & Brauers, W, K., (2012). Personnel selection based on computing with words and fuzzy MULTIMOORA, Expert Syst. Appl. 39, pp:7961–7967.
2. Dursun, M., & Karsak, E. E. (2009). A fuzzy MCDM approach for personnel selection.Expert Systems with Applications. doi:10.1016/j.eswa.2009.11.067
3. Güngör, Z., Serhadlioglu, G., & Kesen, S. E. (2009). A fuzzy AHP approach to personnel selection problem. Applied Soft Computing, 9, 641–646.
4. Ishizaka, A., Nemery, P.(2013). Multi-Criteria Decision Analysis Methods and Software, John Willey&Sons,Ltd. United Kingdom
5. Kabak, M., Burmaoglu, S & Kazancoglu, Y., (2012). A fuzzy hybrid MCDM approach for professional selection, Expert Syst. Appl. 39, pp: 3516–3525.
6. Kelemenis, A., & Askounis, D. (2010). A new TOPSIS-based multi-criteria approach to personnel selection. Expert Systems with Applications. doi:10.1016/j.eswa.2009.12.013
7. Sang, X., Lui, X & Qin., (2015). An analytical solution to fuzzy TOPSIS and its application in personnel selection for knowledge-intensive enterprise. Applied Soft Computing 30, pp:190-204.

8. Sarı, U. Irem, Öztayşi, B., Kahraman, C. (2013). Fuzzy Analytic Hierarch Process Using Type-2 Fuzzy Set: An Application to Warehouse Location Selection. Multicriteria Decision Aid and Artificial Intelligence Links, Theory and Applications. John Willey&Sons,Ltd. United Kingdom
9. Triantaphyllou, E. (2000) Multi-Criteia Decision Making Methods: A Comparative Study, Kluwer Academic Publishers, Boston, USA
10. Zavadskas, E. K., Turskis, Z., Tamosaitiene, J, Marina, V (2008). Multicriteria Selection of Project Managers by Applying Grey Criteria. Ukio Technologinis ir Ekonominis Vystmas 14. (4): 462-477.
11. Zhang, X., & Liu, P. D. (2010). Method for aggregating triangular fuzzy intuitionistic fuzzy information and its application to decision making. Technological and Economic Development of Economy, 16(2), 280–290.

Novel Exact Solutions of the Extended Shallow Water Wave and the Fokas Equations

Serbay DURAN[1,*], *Berat KARAAGAC*[1] and *Alaattin ESEN*[2]

[1]Adiyaman University, Faculty of Education, 020040 Adıyaman, Turkey
[2]Inonu University, Department of Mathematics, 44280 Malatya, Turkey

Abstract. In this study, a Sine-Gordon expansion method for obtaining novel exact solutions of extended shallow water wave equation and Fokas equation is presented. All of the equations which are under consideration consist of three or four variable. In this method, first of all, partial differential equations are reduced to ordinary differential equations by the help of variable change called as travelling wave transformation, then Sine Gordon expansion method allows us to obtain new exact solutions defined as in terms of hyperbolic trig functions of considered equations. The newly obtained results showed that the method is successful and applicable and can be extended to a wide class of nonlinear partial differential equations.

1 Introduction

Over the run of humanity history; people are taken their attention on sciences or engineering because of their deeply curious nature. They believe that patterns in all of nature can be discovered with the help of using intellect. Where at the laws of the natural and physical world are generally modeled in the form of evolution equations, searching the solutions of evolution equations is become one of the main areas of scientific works [1]. During the years, various new techniques are developed for solving evolution equation [2-5].

The aim of this study, to investigate solution of extended Shallow Water Wave equation and Fokas equation with the help of the Sine Gordon expansion method which is a useful and popular tool for constructing exact solutions.

2 Sine-Gordon expansion method

This section is cover a simple definition of the Sine-Gordon expansion method. Firstly, let us take into consideration the general Sine-Gordon equation given as

$$u_{xx} - u_{tt} = m^2 \sin u, \tag{2.1}$$

* Corresponding author: sduran@adiyaman.edu.tr

where $u = u(x,t)$ is the desired function and m is a constant. The using of wave transformation $u(x,t) = U(\xi)$ ($\xi = kx + lt$) to the Eq. (2.1) leads to reduce Sine-Gordon equation to a nonlinear ordinary differential equation given in the form of

$$U'' = \frac{m^2}{k^2 - l^2}\sin(U), \tag{2.2}$$

where k and l refer the number of wave in a unit distance and the wave speed, respectively. When the both sides of Eq. (2.2) is multiplied by U' and integrated once, the following form is found

$$\left(\frac{U}{2}\right)' = \frac{m^2}{k^2 - l^2}\sin^2\left(\frac{U}{2}\right) + K \tag{2.3}$$

where $K = 0$. When $\frac{U}{2} = w(\xi)$ and $a^2 = \frac{m^2}{k^2 - l^2}$ are taken, the new form of the equation (2.3) is found as

$$w' = a\sin(w). \tag{2.4}$$

When $a = 1$ is taken, the Eq. (2.4), the basic form of Eq (2.1) is finally found as follows

$$w' = \sin(w). \tag{2.5}$$

Then, the solutions of the Eq. (2.5) can be found as follows

$$\sin(w(\xi)) = \frac{2pe^{\xi}}{p^2e^{2\xi} + 1}\Big|_{p=1} = \sec h(\xi) \tag{2.6}$$

$$\cos(w(\xi)) = \frac{p^2e^{2\xi} - 1}{p^2e^{2\xi} + 1}\Big|_{p=1} = \tan h(\xi) \tag{2.7}$$

Where the integration constant is taken as $p \neq 0$. Thus, the general form is expressed of nonlinear PDE with two variables as following

$$P(u, u_x, u_t, u_{xx}, u_{xt}, u_{tt}, \ldots, x, t) = 0 \tag{2.8}$$

and using the transformation given as

$$u(x,t) = U(\xi), \quad \xi = kx + lt. \tag{2.9}$$

we obtain a nonlinear ordinary differential equation (NODE) for $U(\xi)$

$$G(U, U', U'', \ldots) = 0. \tag{2.10}$$

In a similar way followed above, it is assumed that the solution $U(\xi)$ of the nonlinear equation given in Eq. (2.10) can be formed as follows

$$U(\xi) = \sum_{i=1}^{m} \tanh^{i-1}(\xi)\left[B_i \sec h(\xi) + A_i \tanh(\xi)\right] + A_0. \tag{2.11}$$

Assuming that the wave solutions of Eq. (2.11) can be written in the following form:

$$U(w) = \sum_{i=1}^{m} \cos^{i-1}(w)\left[B_i \sin(w) + A_i \cos(w)\right] + A_0. \tag{2.12}$$

Where $A_i, B_i, (0 \le i \le m)$ are desired coefficients and m is positive integers. m will be determine with homogeneous balance principle. After, using the Eq. (2.12) is substituted into Eq. (2.10), collecting all the terms according to powers of function $(\sec h(\xi)\tanh(\xi))^i$ or $(\sin(\xi)\cos(\xi))^i$ and at the last equalizing coefficients to zero, an algebraic equation system is obtained. When this new system is solved, the travelling wave solutions of Eq. (2.8) is obtained.

3 Exact Solitary Wave Solutions of the Nonlinear Differential Equations

3.1 The (4+1) Dimensional Fokas equation

Consider The (4+1) dimensional Fokas equation [6] given by:

$$4u_{xt} - u_{xxxy} + u_{xyyy} + 12u_x u_y + 12uu_{xy} - 6u_{zw} = 0 \tag{3.1}$$

$$u(x,y,z,w,t) = u(\phi), \quad \phi = \alpha x + \beta y + \gamma z + \lambda w + ct \tag{3.2}$$

where α, β, γ and λ are real parameters. Using the transformation given in Eq. (3.2) into Eq. (3.1), we get the following NODE;

$$(\alpha\beta^3 - \alpha^3\beta)\tilde{u}^{(4)} + (4\alpha c - 6\gamma\lambda)\tilde{u}'' + 12\alpha\beta(\tilde{u}\tilde{u}')' = 0. \tag{3.3}$$

And integrating twice with respect to ξ Eq. (3.3) yields following equation,

$$(\alpha\beta^3 - \alpha^3\beta)\tilde{u}'' + (4\alpha c - 6\gamma\lambda)\tilde{u} + 6\alpha\beta\tilde{u}^2 = 0. \tag{3.4}$$

where $\tilde{u}^2$ is the highest order nonlinear term and $\tilde{u}''$ is the highest order derivative. Balancing them leads us to obtain $m = 2$. Thus the solution is obtained as

$$\tilde{u}(\phi) = A_0 + A_1 \tanh(\phi) + A_2 \tanh^2(\phi) + B_1 \sec h(\phi) + B_2 \tanh(\phi)\sec h(\phi) \tag{3.5}$$

We insert Eq. (3.5) into Eq. (3.4), a polynomial of $\tanh(\phi)$, $\sec h(\phi)$, and also $\alpha, \beta, \gamma, \lambda$ is come out. Summing all the terms of the polynomial according to power of $\tanh(\phi)$, $\sec h(\phi)$, and equating whole summation to zero with the help of Wolfram Mathematica 9, gives us desired coefficients as

$$A_0 = \frac{1}{2}(-\alpha^2 + \beta^2),\ A_1 = B_1 = 0,\ A_2 = \frac{1}{2}(\alpha^2 - \beta^2), B_2 = \frac{-i}{2}(\alpha^2 - \beta^2),\ c = \frac{\alpha^3\beta - \alpha\beta^3 + 6\gamma\lambda}{4\alpha}$$

substituting these coefficients into Eq. (3.5), produces the following solution:

$$\begin{aligned} u(x,y,z,w,t) = \frac{1}{2}(-\alpha^2 + \beta^2) + \tanh\left(\alpha x + \beta y + \gamma z + \lambda w + \frac{(\alpha^3\beta - \alpha\beta^3 + 6\gamma\lambda)t}{4\alpha}\right) \\ \left(-\frac{i}{2}(\alpha^2 - \beta^2)\sec h\left(\alpha x + \beta y + \gamma z + \lambda w + \frac{(\alpha^3\beta - \alpha\beta^3 + 6\gamma\lambda)t}{4\alpha}\right)\right. \\ \left. + \frac{1}{2}(\alpha^2 - \beta^2)\tanh\left(\alpha x + \beta y + \gamma z + \lambda w + \frac{(\alpha^3\beta - \alpha\beta^3 + 6\gamma\lambda)t}{4\alpha}\right)\right). \end{aligned} \tag{3.6}$$

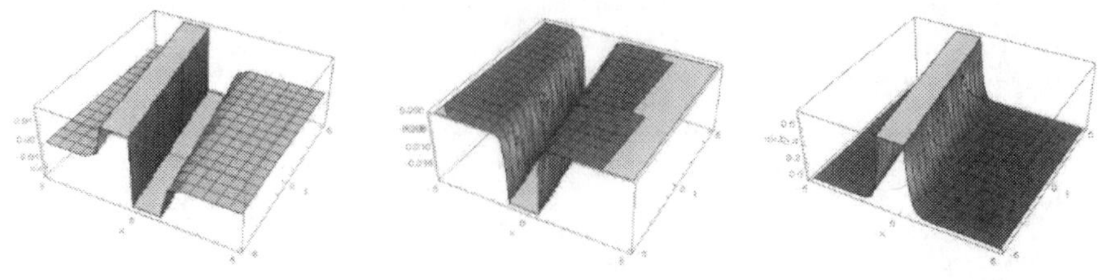

Fig. 1. The imaginary part, real part and Absolute solutions of Fokas equation by substituting the values $\alpha = 5,\ \beta = 0.1, \gamma = \mu = 0.5\ \ -5 < x, t < 5,$ respectively.

3.2 Extended Shallow Water Wave Equation

Extended shallow water wave equation reads [7]:

$$u_{xzt} + u_{xxxyz} - 2\left(u_{xx}u_{yz} + u_y u_{xxz}\right) - 4\left(u_x u_{xyz} + u_{xz}u_{xy}\right) + \alpha u_{xyz} = 0 \tag{3.7}$$

$$u(x,y,z,t) = \tilde{u}(\phi), \quad \phi = \beta x + \gamma y + \lambda z - ct \tag{3.8}$$

where α, β, γ and λ are real parameters, c denotes wave velocity. Transformed form of Eq. (3.8) is

$$\beta^3\gamma\lambda\tilde{u}''' + \left(\alpha\beta\gamma\lambda - \beta\gamma c\right)\tilde{u}' - 3\beta^2\gamma\lambda\left(\tilde{u}'\right)^2 = 0. \tag{3.9}$$

With the help of the balancing principle, we can found $m = 1$. Thus, solution is obtained

$$\tilde{u}(\phi) = A_0 + A_1 \tanh(\phi) + B_1 \operatorname{sech}(\phi) \tag{3.10}$$

When the Sine- Gordon expansion method is applied, we get following coefficients and desired solutions given as

$$A_1 = -\frac{\sqrt{-\alpha\beta\lambda + c\lambda}}{\sqrt{\gamma\lambda}}, \quad B_1 = \frac{\sqrt{\alpha\beta\lambda - c\lambda}}{\sqrt{\gamma\lambda}}, \quad \beta = \frac{\sqrt{-\alpha\beta\lambda + c\lambda}}{\sqrt{\gamma\lambda}}$$

and

$$\begin{aligned} u(x,y,z,t) = A_0 &+ \frac{\sqrt{-\alpha\gamma\lambda + c\lambda}}{\sqrt{\gamma\lambda}} \sec h\left(\beta x + \gamma y + \lambda z - ct\right) \\ &- \frac{\sqrt{-\alpha\gamma\lambda + c\lambda}}{\sqrt{\gamma\lambda}} \tan h\left(\beta x + \gamma y + \lambda z - ct\right) \end{aligned} \tag{3.11}$$

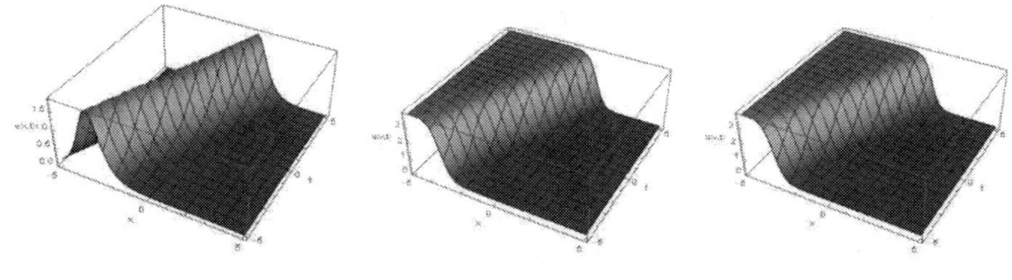

Fig. 2. The imaginary part, real part and Absolute solutions of shallow water wave equation by substituting the values
$\alpha = 0.5,\ \lambda = 2,\ \gamma = 0.25, \mu = 0.5\ \ c = 0.75,\ A_0 = 2,\ -5 < x, t < 5,$ respectively.

4 Conclusions

Under investigation in the present study, we investigate new solutions to the well-known extended Shallow Water Wave equation and Fokas equation. Sine-Gordon expansion method is adopted to carry out this goal in a straightforward manner. With the assistance of the considered method, we construct novel and general solutions including hyperbolic trig functions. 3-dimensional of each solution obtained in the paper is illustrated in the study. As a result, the effectiveness and helpfulness of the Sine Gordon method deal with the nonlinear higher order equation is proved. The method can be used for constructing various nonlinear evolution equations in future studies.

References

1. A. Yokus and D. Kaya, Numerical and exact solutions for time fractional Burgers' equation, Journal Of Nonlinear Sciences And Applications 10.7, 3419-3428, (2017).
2. I.E. Inan, B. Kiliç, and S. Duran, G′/G expansion method and its applications to the 2 - Dimensional Burgers equation and coupled Burgers type equation, World Appl. Sci. J. (2011).
3. E. M. E. Zayed and SA Hoda Ibrahim, Exact solutions of nonlinear evolution equations in mathematical physics using the modified simple equation method, Chinese Physics Letters 29.6 (2012).
4. A. Yokus, T. A. Sulaiman, H. Bulut, On the analytical and numerical solutions of the Benjamin–Bona–Mahony equation. Opt. Quant. Electron. 50, 31, (2018).
5. H. M. Baskonus and H. Bulut, On the complex structures of Kundu-Eckhaus equation via improved Bernoulli sub-equation function method, Waves in Random and Complex Media 25.4, 720-728 (2015).
6. A. M. Wazwaz, Multiple-soliton solutions for extended shallow water wave equations, Studies in Mathematical Sciences 1.1, 21-29 (2010).
7. H. Yinghui, Exact Solutions for-Dimensional Nonlinear Fokas Equation Using Extended F-Expansion Method and Its Variant, Mathematical Problems in Engineering 1-14, (2014).

Risk Assessment with Failure Mode and Effects Analysis and Grey Relational Analysis Method in Plastic Injection Process

Ceren Ünlükal[1*], *Mine Şenel*[1] , *Bilgin Şenel*[1]

[1]Munzur University, Department of Industrial Engineering, Tunceli, Turkey

Abstract. This study aims to evaluate the risks that may arise during the production process in a plastic injection manufacturing enterprise with traditional Failure Mode and Effect Analysis (FMEA) and Grey Relational Analysis (GRA). Although it is a widely used analytical technique that helps to identify and reduce the risks of failure in a process, the Failure Mode and Effects Analysis (FMEA) has some drawbacks that the different risks can have the same risk priority values and the weight of risk factors is not take into consideration. This situation has been tried to be eliminated by integrating the FMEA with the GRA. As a result, it is seen that the order of risk priority values of the identified failure changes according to both methods.

1 Introduction

All businesses in the service or manufacturing sector face risks and hazards from internal and external factors. Businesses that want to take part and be permanent at the markets where the hard competition have to manage their risks well. Risks must be identified and analyzed in order to be managed.

In businesses producing high volume and fast, a process that starts with a failure can result in very large losses if precaution is not taken in time. Failures that occur during production are difficult and high cost to compensate after they reach the customer [1]. For this reason, especially mass production businesses need to keep their risks under control.

The failure rates that occur in the production step in the mass production businesses like plastic injection are very high and this situation leads up both economic inefficiency and loss of confidence in customers for businesses. This study aims that in order to reduce the damages to the business by eliminating the failures that occur in a plastic injection production business, in addition to the Failure Mode and Effects Analysis technique, the Grey Relational Analysis technique has been also applied and the risk priorities have been compared according to two techniques and the effects of the methods on the results have been investigated.

* Corresponding author: senelbilgin@gmail.com

This study is arranged as follows: In Section 2, Failure Mode and Effects Analysis is introduced and its drawbacks are given shortly. Grey Relational Analysis and its steps are given in Section 3. Section 4 is about application of risk assessment for plastic injection process. Conclusions are made in Section 5.

2 Failure Mode and Effects Analysis

The purpose of the Failure Mode and Effects Analysis (FMEA), which was first introduced into the literature by the application of the American Army and NASA, is to detect and correct the potential problems in the design and production processes [2,3]. FMEA is one of the most established methods used in engineering discipline for risk analysis and preventing failure [4]. In this method, which is frequently used to by businesses seeking to maintain the continuity of high quality products and services, it is very important that the root causes of the risks are identified, and their effects reduced [5,6]. Early response and prevention of possible failures contribute to reducing system losses.

FMEA, which helps to prioritize risks and failures, consists of three factors: Occurrence (O), Detectability (D) and Severity (S). Risk Priority Numbers (RPN) are calculated by multiplying the numerical values of the risk factors according to the expressions given in Table 1.

$$RPN = O \times D \times S \quad (1)$$

Table 1. Scales used in traditional FMEA [3]

(a) Chance of occurrence		
Criteria	**Score**	**Possible failure occurrence rates**
Remote chance of failure	1 2	0 1/20000
Low failure rate	3 4	1/10000 1/2000
Moderate failure rate	5 6	1/1000 1/200
High failure rate	7 8	1/100 1/20
Very high failure rate	9 10	1/10 1/2
(b) Chance of non-detection		
Chance of not detecting fault	**Score**	**Probability of a defect reaching the customer**
Remote	1 2	0-5 6-15
Low	3 4	16-25 26-35
Moderate	5 6	36-45 46-55
High	7 8	56-65 66-75
Very high	9 10	76-85 86-100
(c) Severity		
Severity		**Score**
Customer will probably not notice		1

Slight annoyance	2,3
Customer dissatisfaction	4,5,6
High degree of dissatisfaction	7,8
Safety/regulatory consequences	9,10

The higher the RPN that can be at any value between 1 and 1000, the more critical it is. The highest RPN value has been highest priority for the potential failure. Corrective action should be planned for each failure by prioritizing the RPN value from highest to lowest. It is not necessary to plan corrective action for failures that are less than 100 RPN value.

Although FMEA has gained huge acceptance and applications in a wide range of industries and helps the analysts prioritize the identified failure modes and work out corrective actions, it has been some limitations and criticized for various reasons [7-13]:

- Some failure modes that have different risk factor values may have the same RPN value, but real risk implications may be completely different. For example, two different failures with the O, S and D values of 2,5,4 and 8,5,1, respectively, have the same RPN value of 40 (Eq.1). However, the risk implication is not same.
- In traditional FMEA, O, S, and D risk factors are considered as equally important. However, for each business, these factors may not have equal importance.
- The mathematical formula for calculating RPN is questionable and debatable. There is no rationale as to why O, S and D should be multiplied to produce the RPN.
- The RPN scale is not continuous with many holes, because many numbers between 1 and 1000 cannot be obtained from the product of O, S and D.
- The three factors are difficult to numerically detection. Much information in FMEA can be expressed linguistically such as likely, important or very high and so on.

In this study, the Grey Relational Analysis method will be used and the results will be compared in order to eliminate all these problems and prioritize the failures easier and more accurately.

3 Grey Relational Analysis

Grey Relational Analysis (GRA) is one of the subheadings of Grey Theory which was introduced by Julong Deng in 1982. [14,15]. The grey system which is between the white system, where information is completely known, and the black system, where information is not completely known, expresses poor, incomplete and uncertain information [16,17].

GRA which is an interdisciplinary analysis technique investigates the factors affecting the system and determines the dependence of these factors on each other. In addition to determining the interrelationships between multiple factors, it also detects which factor become prominent [18]. It has been successfully applied on cluster analysis, path planning, selection problems, prediction analysis, performance evaluation, factor effect evaluation, decision making and also FMEA [19-21].

In the FMEA method, GRA is applied according to the following steps [3]:
Step 1. Establish comparative series: The information series consisting of O, D and S can be expressed as follows.

$$x_i' = (x_i'(1), x_i'(2), ..., x_i'(k)) \in X \ , \ i = 1,2,...,n \quad (2)$$

k is number of the risk factors and n is number of the failure modes. $x_i(k)$ denotes the kth factors of x_i and the n information series can be described as the following matrix [3]:

$$x = \begin{bmatrix} x_1 \\ x_2 \\ . \\ . \\ x_n \end{bmatrix} = \begin{bmatrix} x_1(1) \; x_1(2) \; \ldots \; x_1(k) \\ x_2(1) \; x_2(2) \; \ldots \; x_2(k) \\ . \\ . \\ x_n(1) \; x_n(2) \; \ldots \; x_n(k) \end{bmatrix} \tag{3}$$

Equation 2 is a matrix that express O, D and S values for each failure mode.

Step 2. Establish the standard series: The purpose of defining standard series is to determine the degree of relation [22]. Standard series is established according to the lowest score of the risk factor scale.

$$x_0 = (x_0(1), x_0(2), \ldots, x_0(k)) = (1, 1, \ldots, 1) \tag{4}$$

Step 3. Obtain the difference between the comparative series and standard series: To obtain the difference matrix, the difference between the comparative series and standard series must to be calculated.

$$\Delta_{0i}(k) = \begin{bmatrix} \Delta_{01}(1) \; \Delta_{01}(2) \; \ldots \; \Delta_{01}(k) \\ \Delta_{01}(1) \; \Delta_{01}(2) \; \ldots \; \Delta_{01}(k) \\ . \\ . \\ \Delta_{0n}(1) \; \Delta_{0n}(2) \; \ldots \; \Delta_{0n}(k) \end{bmatrix} \text{and where } \Delta_{0i}(k) = |x_0(k) - x_i(k)| \tag{5}$$

Step 4. Compute the grey relational coefficient: To calculate grey relational coefficient, we use the following equation:

$$\gamma\,(x_0(k), x_i(k)) = \frac{\Delta_{\min} + \zeta\,\Delta_{\max}}{\Delta_{0i}(k) + \zeta\,\Delta_{\max}} \tag{6}$$

$\zeta \in (0,1)$ is an identifier coefficient and only affects the relative value of the risk without changing the priority [22].

Step 5. Determine the degree of relation: There are two approaches to determine the degree of relation. Firstly, we assume that all of risk factors have equal importance;

$$\tau_i(k) = \frac{1}{n}\sum_{k=1}^{n} \Delta_i(k) \tag{7}$$

If the risk factors do not have equal importance;

$$\tau_i(k) = \sum_{k=1}^{n} \Delta_i(k)\beta(k) \text{ and } \sum_{k=1}^{n} \beta(k) = 1 \tag{8}$$

$\beta(k)$ is the weights of the risk factors and determined by the experts work in plastic injection business.

Step 6. Obtain the relational series to rank the priority of risk: The lower degree of relation represents the higher risk priority. As the degree of relation is closer to 1, it means the failure mode is closer to the optimal value [22].

4 Application

This study aims to reduce the risks by determining the failures that can occur during the period from production to delivery in a company that produces plastic injection for the white goods, bathroom accessories and complementary wares sector in Eskisehir. The risk analysis studies are first performed with traditional FMEA and then Grey Relational Analysis is applied to determine the risk priorities according to these methods.

A team consisting of a quality control manager, a quality control engineer and a production manager is made. The experts are identified problems that could arise during and after the production process. Table 2 lists failures occurring in the firm.

Table 2. Failure modes and definitions

Failure Mode	Failure	Failure Definition
FM1	Warpage	Warpage can be defined as a dimensional distortion in a molded product after it is ejected from the mold at the end of the injection molding process [23].
FM2	Flash/Burrs	Flash is a molding defect that occurs when some molten plastic escapes from the mold cavity [24].
FM3	Short shot/Non-fill	Short shot can be defined as an incomplete molded part caused by insufficient material injected into the mold [23].
FM4	Label defect	Incorrect filling of card containing information about product
FM5	Assembly/Grouping defects	Incorrect grouping of multiple parts or incorrect selection of parts to be grouped (Manual assembly)
FM6	Brittleness	Brittleness can be defined as the tendency of a molded plastic part to break or crack under conditions in which it would not normally do so [23].
FM7	Dimensional defects	Incorrect product size (out of tolerance values)
FM8	Cracked/Broken product	Some parts of the product are cracked or broken.
FM9	Oily, dirty and stained product	It means that oiling and contamination of product after remove from the mold.
FM10	Discoloration	Discoloration can be defined as a change in the original color of a plastic material [23].
FM11	Burn marks	Burn marks can be defined as small dark brown or black discolorations on the surface of a molded part [23].
FM12	Shipment of wrong product	Sending wrong product that is not suitable for customer order
FM13	Afterflame	Afterflame is the duration of the material burning after the combustion source is removed.
FM14	Surface defects	Surface defects express the scratches, gouges, specks and blisters on the product surface.

O, D and S risk factors for each failure modes listed in Table 2 are evaluated by experts and arithmetic mean values are calculated. According to Equation 1, O, D and S mean values are multiplied and obtained RPN. Similar calculations are made with GRA, too. There are two different approaches in GRA. The first is that the risk factors have been equal importance. The other is that the weights of risk factors are determined by experts. The weights determined with Pair-Wise Comparison method are given in Table 3. And then the risk priorities are determined and ordered according to the RPN values.

Table 3. The weights of risk factors

Risk Factors	Weights
Occurrence	0,23
Detectability	0,15
Severity	0,62
Consistency Rate: 0,01	

The results are shown in Table 4 and the RPN values are listed according to the GRA-FMEA method, which is obtained both by traditional FMEA and by the approach of equal and different weights.

Table 4. Comparison of priorities determined by different methods

Failure mode	Traditional RPN	Rank	Grey RPN (Equal weight)	Rank	Grey RPN (Unequal weight)	Rank
FM1	114	9	0,64	9	0,64	7
FM2	199	5	0,58	4	0,61	6
FM3	222	3	0,56	3	0,58	5
FM4	30	14	0,82	14	0,83	14
FM5	346	2	0,51	2	0,54	3
FM6	164	6	0,61	6	0,56	4
FM7	487	1	0,47	1	0,48	1
FM8	51	12	0,75	11	0,69	10
FM9	62	10	0,73	10	0,75	11
FM10	148	7	0,62	8	0,65	8
FM11	36	13	0,80	13	0,76	12
FM12	144	8	0,62	7	0,66	9
FM13	203	4	0,59	5	0,53	2
FM14	61	11	0,75	12	0,77	13

It is seen that the highest risk is FM7 (Dimensional defects) and the lowest risk is FM4 (Label defect) according to all of methods. It means corrective actions should be planned for FM7 firstly. And then for each failure mode exceeding 100 RPN value, corrective action should be planned.

The main reason for changing the rest of the ranks is that the importance level of the risk factors changes according to experts' opinions. Because importance level of O, D and S risk factors changes according to business sector and it differs according to firms and experts.

Conclusion

Traditional FMEA is practical and helpful for risk assessment, but it may not be sufficient when used alone. In this study, FMEA and GRA methods are used together to assess and prioritize the risks more effectively and realistic. Weights of risk factors determined according to experts' opinions are also included in the calculation and all the results are compared. It has been observed that the order of priority differs according to all the methods. It has been determined that the FMEA based GRA method which takes the firm priorities into account is useful. This study will be helpful for other plastic injection firms to assess their risks. Fuzzy logic and multi-criteria decision-making methods can be used in future work and the results can be compared with FMEA based GRA method.

Acknowledgements

Authors would like to thank to Şefik Tanrısevdi, who works as the quality control manager at the firm, for his great contributions.

References

1. M. Zerenler, K. Karaboğa, "Müşteri Memnuniyetinin Sağlanmasında Hataların Önlenmesine Yönelik Üretim Odaklı Bir Bakış Açısı: Poka-Yoke Sistemleri", Selcuk University Journal of Institute of Social Sciences, Dr. Mehmet YILDIZ Special Edition, 263-275, 2014.
2. P. M. Özfırat, "Bulanık Önceliklendirme Metodu ve Hata Türü ve Etkileri Analizini Birleştiren Yeni Bir Risk Analizi Yöntemi", Journal of the Faculty of Engineering and Architecture of Gazi University, Vol 29, No 4, 755-768, 2014.
3. C. L. Chang, C. C. Wei, Y. H. Lee, "Failure Mode and Effects Analysis Using Fuzzy Method and Grey Theory", Kybernetes, Vol. 28 Issue: 9, pp.1072-1080, 1999.
4. R. E. McDermott, R. J. Mikulak, M. R. Beauregard, "The Basics of FMEA", Productivity Press 2nd edition, ISBN 9781563273773, 2008.
5. W. C. Ng, S. Y. Teh, H. C. Low, P. C. Teoh, "The Integration of FMEA with Other Problem Solving Tools: A Review of Enhancement Opportunities", Journal of Physics: Conference Series 890/012139, 2017.
6. C. Kahraman, İ. Kaya, Ö. Şenvar, "Healthcare Failure Mode and Effects Analysis Under Fuzziness", Human and Ecological Risk Assessment, 19:2, 538–552, 2013.
7. K. S. Chin, Y. M. Wang, G. K. K. Poon, J. B. Yang, "Failure Mode and Effects Analysis Using A Group-Based Evidential Reasoning Approach", Computers & Operations Research, 36, 1768-1779, 2009.
8. J. B. Bowles, "An Assessment of PRN Prioritization in A Failure Modes Effects and Criticality Analysis", Journal of the IEST, 6, 47-51, 2004.
9. N. R. Sankar, B. S. Prabhu, "Modified Approach for Prioritization of Failures in A System Failure Mode and Effects Analysis", International Journal of Quality & Reliability Management, 18(3):324–35, 2001.
10. M. Ben-Daya, A. Raouf, "A Revised Failure Mode and Effects Analysis Model", International Journal of Quality & Reliability Management, 13(1):43–7, 1996.
11. M. Braglia, M. Frosolini, R. Montanari, "Fuzzy Criticality Assessment Model for Failure Modes and Effects Analysis", International Journal of Quality & Reliability Management, 20(4):503–24, 2003.
12. W. Gilchrist, "Modelling Failure Modes and Effects Analysis", International Journal of Quality & Reliability Management, 10(5):16–23, 1993.

13. A. Pillay, J. Wang, "Modified Failure Mode and Effects Analysis Using Approximate Reasoning", Reliability Engineering & System Safety, 79:69–85, 2003.
14. N. Uçkun, N. Girginer, "Türkiye'deki Kamu ve Özel Bankaların Performanslarının Gri İlişki Analizi ile İncelenmesi", Akdeniz University Faculty of Economics & Administrative Sciences Faculty Journal, 21: 46-66, 2011.
15. D. Julong, "Control Problems of Grey Systems, Systems and Control Letters", 5,288-94, 1982.
16. N. Tosun, "Determination of Optimum Parameters for Multi-Performance Characteristics in Drilling by Using Grey Relational Analysis", The International Journal of Advanced Manufacturing Technology, 28, 450–455, 2006.
17. D. Julong, "Introduction to Grey System Theory", The Journal of Grey System, 1, 1-24, 1989.
18. C. Erden, E. Ceviz, "Gri Sistem Teorisi Kullanılarak Türkiye'nin Büyüme Oranı Faktörlerinin Analizi", Sakarya University Journal of Science, 19:3, 361-369, 2015.
19. K.C. Chang and M.-F. Yeh, "Grey-Relational Analysis Based Approach for Data Clustering", IEEE Proceedings - Vision Image and Signal Processing, 152(2):165–172, 2005.
20. H. C. Lu, M. F. Yeh, "Robot Path Planning Based on Modified Grey Analysis", Cybern. Syst., 33(2):129-159, 2002.
21. M.-F. Yeh, H. C. Lu, "Evaluating Weapon Systems Based on Grey Relational Analysis and Fuzzy Arithmetic Operations", J. Chinese Inst. Eng., 23(2):211-221, 2000.
22. Ç. Sofyalıoğlu, Ş. Öztürk, "Application of Grey Relational Analysis with Fuzzy AHP to FMEA Method", Doğuş University Journal, 13(1):114-130, 2012.
23. http://www.plastictroubleshooter.com/ThePlasticTroubleshooter/defects.htm (23.05.2018)
24. https://www.creativemechanisms.com/blog/what-cause-injection-molding-defects-and-how-to-fix-them (23.05.2018)

ITM Web of Conferences **22**, 01024 (2018) https://doi.org/10.1051/itmconf/20182201024
CMES-2018

On Intuitionistic Fuzzy 2-Metric Spaces

Elif Güner[1,*], *Vildan* Çetkin[1] and *Halis* Aygün[1]

[1]Department of Mathematics, University of Kocaeli, Kocaeli, Turkey

Abstract In this study, we first recall the notion of an intuitionistic fuzzy 2-metric space and fundamental definitions with several illustrative examples. Then we define the notion of δ −chainable space and (δ, λ) −uniform locally contractive mapping between intuitionistic fuzzy 2-metric spaces. After that, by using the proposed concepts, we obtain a few fixed point theorems of self-mappings defined on this spaces.

1 Introduction

Fuzzy set theory which is one of the most important and useful branch of mathematics, was introduced by Zadeh [13] in 1965. The range of applications of this theory is wide and starting from artificial intelligence and also touching on a lot of scientific disciplines such as engineering, economics, etc. Many researchers have obtained a lot of interesting results and applications concerning the fuzzy set theory. Kramosil and Michalek [5] described the concept of fuzzy metric space applying the idea of fuzziness to the clasical notions of metric and metric spaces. Park [8] defined the concept of intuitionistic fuzzy metric space and proved some well-known theorems of metric spaces for this spaces.

The notion of 2-metric which was proposed in Euclidean space by the area function, was initiated by Gähler [2]. In 2002, Sharma [11] gave the definition of fuzzy 2-metric space and obtained some common fixed point theorems. Mursaleen and Lohani [6, 7] launched the notion of intuitionistic fuzzy 2-metric space which is the generalization of the intuitionistic fuzzy metric space. They studied some topological properties of this spaces such as convergence, completeness, completion. Bakry [1] studied the existence and uniqueness of a common fixed point theorem in complete intuitionistic fuzzy 2-metric spaces. Shrivastava et al. [12] gave the definition of the weak compatible mappings in intuitionistic fuzzy 2-metric spaces and investigated a common fixed point theorems with the help of these mappings.

2 Preliminaries

Definition 2.1. [10] If a binary operation $*$ $(\diamond)$ on the interval $[0,1]$ satisfies the following conditions, then $*$ $(\diamond)$ is said to be a t-norm (t-conorm):
(a) $*$ $(\diamond)$ is associative and commutative.
(b) $*$ $(\diamond)$ is continuous.

* Corresponding author: elif.guner@kocaeli.edu.tr

(c) $x * 1 = x$ $(x \diamond 0 = x)$ for all $x \in [0,1]$.
(d) $x * y \leq z * u$ $(x \diamond y \leq z \diamond u)$ whenever $x \leq z$ and $y \leq u$ for all $x, y, z, u \in [0,1]$.
Definition 2.2. [2, 4] Let $X \neq \emptyset$ be a set. A function $d: X \times X \times X \to \mathbb{R}$ is called a $2-$metric if d satisfies the following conditions :
(i) to each pair of points $x, y \in X$ $(x \neq y)$, there is a point $z \in X$ such that $d(x, y, z) \neq 0$,
(ii) $d(x, y, z) = 0$ when at least two of x, y, z are equal,
(iii) $d(x, y, z) = d(x, z, y) = d(y, z, x)$ for all $x, y, z \in X$,
(iv) $d(x, y, z) \leq d(x, y, w) + d(x, w, z) + d(w, y, z)$ for all $x, y, z, w \in X$.
The ordered pair (X, d) is called a $2-$metric space when d is a $2-$metric on X. We may think of this as meaning that $d(x, y, z)$ is a family of distance-like functions of x and y indexed by $z \in X$.
Example 2.3. [9] Let $X = \{0, 1, \frac{1}{2}, \frac{1}{3}, \ldots\}$. The mapping $d : X \times X \times X \to [0, \infty)$ defined by

$$d(x,y,z) = \begin{cases} 1, & x, y, z \text{ are distinct and } \left\{\frac{1}{n}, \frac{1}{n+1}\right\} \subseteq \{x, y, z\} \\ 0, & otherwise \end{cases}$$

for all $x, y, z \in X$, is a 2-metric on X.
Definition 2.4. [1, 6] Let $X \neq \emptyset$ be an arbitrary set, $*$ be a continuous t-norm, $\diamond$ be a continuous t-conorm and $\mu, \nu : X \times X \times X \times [0, \infty) \to [0,1]$ be the mappings. If the following conditions are satisfied, then a 5-tuple $(X, \mu, \nu, *, \diamond)$ is called an intuitionistic fuzzy 2-metric space or shortly if2ms:
(IFM1) $\mu(x, y, z, t) + \nu(x, y, z, t) = 1$.
(IFM2) $\mu(x, y, z, 0) = 0$.
(IFM3) $\mu(x, y, z, t) = 1$ for all $t > 0$ when at least two of x, y, z are equal.
(IFM4) $\mu(x, y, z, t) = \mu(x, z, y, t) = \mu(y, z, x, t) = \mu(z, x, y, t)$.
(IFM5) $\mu(x, y, z, t_1 + t_2 + t_3) \geq \mu(x, y, u, t_1) * \mu(x, u, z, t_2) * \mu(u, y, z, t_3)$.
(IFM6) $\mu(x, y, z, \cdot): [0, \infty) \to [0,1]$ is left continuous.
(IFM7) $\nu(x, y, z, 0) = 1$.
(IFM8) $\nu(x, y, z, t) = 0$ for all $t > 0$ when at least two of x, y, z are equal.
(IFM9) $\nu(x, y, z, t) = \nu(x, z, y, t) = \nu(y, z, x, t) = \nu(z, x, y, t)$.
(IFM10) $\nu(x, y, z, t_1 + t_2 + t_3) \leq \nu(x, y, u, t_1) \diamond \nu(x, u, z, t_2) \diamond \nu(u, y, z, t_3)$.
(IFM11) $\nu(x, y, z, \cdot): [0, \infty) \to [0,1]$ is right continuous.
for all $x, y, z \in X$ and $t, t_1, t_2, t_3 > 0$. The values $\mu(x, y, z, t)$ and $\nu(x, y, z, t)$ may be interpreted as the degrees of nearness and non-nearness that the area of triangle enlarged x, y, z with respect to t.
Example 2.5. Let (X, d) be a 2-metric space, $a * b = ab$ and $a \diamond b = min\{1, a + b\}$. Let $\mu_d, \nu_d : X^3 \times [0, \infty) \to [0,1]$ be two mappings defined by

$$\mu_d(x,y,z,t) = \begin{cases} \frac{t}{t + d(x,y,z)}, & t > 0 \\ 0, & t = 0 \end{cases}, \quad \nu_d(x,y,z,t) = \begin{cases} \frac{d(x,y,z)}{t + d(x,y,z)}, & t > 0 \\ 1, & t = 0 \end{cases}$$

for all $x, y, z \in X$. Then $(X, \mu_d, \nu_d, *, \diamond)$ is an if2ms and said to be the standard if2ms induced by the 2-metric d. Hence, we have an if2ms when a 2-metric d is given.
Definition 2.6. [6] Let (x_n) be a sequence in a given if2ms $(X, \mu, \nu, *, \diamond)$.
(1) $(x_n) \subseteq X$ is said to converge to $x \in X$ if $\lim_{n\to\infty} \mu(x_n, x, z, t) = 1$ and $\lim_{n\to\infty} \nu(x_n, x, z, t) = 0$ for all $z \in X$ and $t > 0$. It is denoted by $\lim_{n\to\infty} x_n = x$ or $x_n \to x$.
(2) $(x_n) \subseteq X$ is called a Cauchy sequence if $\lim_{n,m\to\infty} \mu(x_n, x_m, z, t) = 1$ and $\lim_{n,m\to\infty} \nu(x_n, x_m, z, t) = 0$ for all $z \in X$ and $t > 0$.

(3) If every Cauchy sequence in if2ms $(X, \mu, \nu, *, \diamond)$ is convergent, then this space is called complete.

Note 2.7. [1] From now, we will assume that $(X, \mu, \nu, *, \diamond)$ is an if2ms with the condition (IFM12) $\lim_{t\to\infty} \mu(x, y, z, t) = 1$ and $\lim_{t\to\infty} \nu(x, y, z, t) = 0$ for all $x, y, z \in X$.

Lemma 2.1. If (x_n) is a sequence in a given if2ms $(X, \mu, \nu, *, \diamond)$, then the following inequalities hold for all $z \in X, t > 0$ and $p > 0$:

$$\mu(x_n, x_{n+p}, z, t) \geq \mu\left(x_n, x_{n+1}, x_{n+p}, \frac{t}{2(p-1)+1}\right) * \mu\left(x_{n+1}, x_{n+2}, x_{n+p}, \frac{t}{2(p-1)+1}\right)$$
$$* \dots * \mu\left(x_{n+p-2}, x_{n+p-1}, x_{n+p}, \frac{t}{2(p-1)+1}\right) * \mu\left(x_n, x_{n+1}, z, \frac{t}{2(p-1)+1}\right)$$
$$* \dots * \mu\left(x_{n+p-1}, x_{n+p}, z, \frac{t}{2(p-1)+1}\right) * \mu\left(x_{n+p-1}, x_{n+p}, z, \frac{t}{2(p-1)+1}\right) \text{ and}$$
$$\nu(x_n, x_{n+p}, z, t) \leq \nu\left(x_n, x_{n+1}, x_{n+p}, \frac{t}{2(p-1)+1}\right) \diamond \nu\left(x_{n+1}, x_{n+2}, x_{n+p}, \frac{t}{2(p-1)+1}\right)$$
$$\diamond \dots \diamond \nu\left(x_{n+p-2}, x_{n+p-1}, x_{n+p}, \frac{t}{2(p-1)+1}\right) \diamond \nu\left(x_n, x_{n+1}, z, \frac{t}{2(p-1)+1}\right)$$
$$\diamond \dots \diamond \nu\left(x_{n+p-1}, x_{n+p}, z, \frac{t}{2(p-1)+1}\right) \diamond \nu\left(x_{n+p-1}, x_{n+p}, z, \frac{t}{2(p-1)+1}\right)$$

3 Fixed Point Results

Definition 3.1. [3] Let $(X, \mu, \nu, *, \diamond)$ be an if2ms and T be a self-mapping on $(X, \mu, \nu, *, \diamond)$. If there is $k \in (0, 1)$ such that

$$\mu(Tx, Ty, z, kt) \geq \mu(x, y, z, t) \text{ and } \nu(Tx, Ty, z, kt) \leq \nu(x, y, z, t)$$

for all $x, y, z \in X, t > 0$, then T is called a contractive mapping. (k is called the fuzzy contractive constant of T.)

Theorem 3.1. [3] Let $(X, \mu, \nu, *, \diamond)$ be a complete if2ms and T be a self-mapping on $(X, \mu, \nu, *, \diamond)$. If T is a contractive mapping, then T has a unique fixed point in X.

Definition 3.2 Let $(X, \mu, \nu, *, \diamond)$ be an if2ms and T be a self-mapping on $(X, \mu, \nu, *, \diamond)$.

(i) T is said to be continuous at $x_0 \in X$ if for all $x_n \to x_0$ implies $Tx_n \to Tx_0$.

(ii) Let $\delta > 0$ and $0 < \lambda < 1$. T is called (δ, λ)-uniform locally contractive if

$$\mu(x, y, z, t) > 1 - \delta \Rightarrow \mu(Tx, Ty, z, t) \geq \mu\left(x, y, z, \frac{t}{\lambda}\right)$$
$$\nu(x, y, z, t) < \delta \Rightarrow \nu(Tx, Ty, z, t) \leq \nu\left(x, y, z, \frac{t}{\lambda}\right)$$

for all $x, y, z \in X$ and $t > 0$. Clearly, every (δ, λ)-uniform locally contractive mapping T is continuous.

Example 3.3. Let $(X, \mu, \nu, *, \diamond)$ be an if2ms and $T: (X, \mu, \nu, *, \diamond) \to (X, \mu, \nu, *, \diamond)$ be given by $Tx = c$ for all $x \in X$ (c is a constant). Let $\delta > 0$ and $0 < \lambda < 1$. Assume that $\mu(x, y, z, t) > 1 - \delta$ and $\nu(x, y, z, t) < \delta$ for all $x, y, z \in X$ and $t > 0$. Therefore, we have that $1 = \mu(Tx, Ty, z, t) \geq \mu\left(x, y, z, \frac{t}{\lambda}\right)$ and $0 = \nu(Tx, Ty, z, t) \leq \nu\left(x, y, z, \frac{t}{\lambda}\right)$ for all $x, y, z \in X$ and $t > 0$. Hence, T is a (δ, λ)-uniform locally contractive mapping.

Remark 3.4. In an if2ms $(X, \mu, \nu, *, \diamond)$, a contractive mapping can be considered as a $(1, \lambda)$-uniform locally contractive mapping.

Definition 3.5. An if2ms $(X, \mu, \nu, *, \diamond)$ is said to be metrically convex if for each $x, y, z \in X$, there is a $u \neq x, y, z$ for which $\mu(x, y, z, t) = \mu(x, u, z, t_0) * \mu(u, y, z, t_1)$ and $\nu(x, y, z, t) = \nu(x, u, z, t_0) \diamond \nu(u, y, z, t_1)$ where $t = t_0 + t_1$ for all $t_0, t_1 > 0$.

Theorem 3.2. Let $(X, \mu, \nu, *, \diamond)$ be a metrically convex if2ms. If a self-mapping T on $(X, \mu, \nu, *, \diamond)$ is (δ, λ)-uniform locally contractive, then T is a contractive mapping with the fuzzy contractive constant λ.

Proof. Let $x, y, z \in X$. Since $(X, \mu, \nu, *, \diamond)$ is metrically convex, there are points $x = x_0, x_1, x_2, \dots, x_{n-1}, x_n = y$ and $t_0, t_1, \dots, t_n > 0$ such that

$\mu(x, y, z, t) = \mu(x_0, x_1, z, t_0) * \mu(x_1, x_2, z, t_1) * \dots * \mu(x_{n-1}, x_n, z, t_n)$ and

$\nu(x,y,z,t) = \nu(x_0,x_1,z,t_0) \diamond \nu(x_1,x_2,z,t_1) \diamond \ldots \diamond \nu(x_{n-1},x_n,z,t_n)$
where $t = t_0 + t_1 + \cdots + t_n$, $\mu(x_{i-1},x_i,z,t) > 1-\delta$ and $\nu(x_{i-1},x_i,z,t) < \delta$ for $i = 1,2,\ldots,n$. Also,
$\mu(Tx,Ty,z,t) = \mu(Tx_0,Tx_1,z,t_0) * \mu(Tx_1,Tx_2,z,t_1) * \ldots * \mu(Tx_{n-1},Tx_n,z,t_n)$ and
$\nu(Tx,Ty,z,t) = \nu(Tx_0,Tx_1,z,t_0) \diamond \nu(Tx_1,Tx_2,z,t_1) \diamond \ldots \diamond \nu(Tx_{n-1},Tx_n,z,t_n)$.
As T is (δ,λ)-uniform locally contractive, we have $\mu(Tx_{i-1},Tx_i,z,t_{i-1}) \geq \mu\left(x_{i-1},x_i,z,\frac{t_{i-1}}{\lambda}\right)$ and $\nu(Tx_{i-1},Tx_i,z,t_{i-1}) \leq \nu\left(x_{i-1},x_i,z,\frac{t_{i-1}}{\lambda}\right)$ for $i = 1,2,\ldots,n$. Hence, we have
$\mu(Tx,Ty,z,t) = \mu(Tx_0,Tx_1,z,t_0) * \mu(Tx_1,Tx_2,z,t_1) * \ldots * \mu(Tx_{n-1},Tx_n,z,t_n)$
$\geq \mu\left(x_0,x_1,z,\frac{t_0}{\lambda}\right) * \mu\left(x_1,x_2,z,\frac{t_1}{\lambda}\right) * \ldots * \mu\left(x_{n-1},x_n,z,\frac{t_n}{\lambda}\right) = \mu\left(x,y,z,\frac{t}{\lambda}\right)$ and
$\nu(Tx,Ty,z,t) = \nu(Tx_0,Tx_1,z,t_0) \diamond \nu(Tx_1,Tx_2,z,t_1) \diamond \ldots \diamond \nu(Tx_{n-1},Tx_n,z,t_n)$
$\leq \nu\left(x_0,x_1,z,\frac{t_0}{\lambda}\right) \diamond \nu\left(x_1,x_2,z,\frac{t_1}{\lambda}\right) \diamond \ldots \diamond \nu\left(x_{n-1},x_n,z,\frac{t_n}{\lambda}\right) = \nu\left(x,y,z,\frac{t}{\lambda}\right)$.
So T is a contractive mapping.

Definition 3.6. Let $(X,\mu,\nu,*,\diamond)$ be an if2ms and $\delta > 0$. A finite sequence $x = x_0,x_1,\ldots,x_{n-1},x_n = y$ is called a δ–chain from x to y if $\mu(x_{i-1},x_i,z,t) > 1-\delta$ and $\nu(x_{i-1},x_i,z,t) < \delta$ for all $z \in X$, $t > 0$ and $i = 1,2,\ldots,n$. An if2ms $(X,\mu,\nu,*,\diamond)$ is called δ-chainable if for every $x,y \in X$ there is a δ-chain from x to y.

Theorem 3.3. Let $(X,\mu,\nu,*,\diamond)$ be a complete and δ-chainable if2ms. If a self mapping T on $(X,\mu,\nu,*,\diamond)$ is a (δ,λ)-uniform locally contractive, then T has a unique fixed point in X.

Proof Let $x \in X$ and $Tx \neq x$ (otherwise x is a fixed point of T). Since $(X,\mu,\nu,*,\diamond)$ is δ-chainable, there is a δ–chain $x = x_0,x_1,\ldots,x_{n-1},x_n = Tx$ from x to Tx. From here, we have $\mu(x_{i-1},x_i,z,t) > 1-\delta$ and $\nu(x_{i-1},x_i,z,t) < \delta$ for all $z \in X$, $t > 0$ and $i = 1,2,\ldots,n$. Applying induction method, we obtain

$$\mu(T^m x_{i-1},T^m x_i,z,t) \geq \mu\left(x_{i-1},x_i,z,\frac{t}{\lambda^m}\right) \quad (3.1)$$

$$\nu(T^m x_{i-1},T^m x_i,z,t) \leq \nu\left(x_{i-1},x_i,z,\frac{t}{\lambda^m}\right) \quad (3.2)$$

for all $t > 0, z \in X, m \in \mathbb{N}$, $i = 1,2,\ldots,n$. Now, we have

$$1 \geq \mu(T^m x,T^{m+1}x,z,t) = \mu(T^m x_0,T^m x_n,z,t)$$
$$0 \leq \nu(T^m x,T^{m+1}x,z,t) = \nu(T^m x_0,T^m x_n,z,t)$$

for all $t > 0, z \in X, m \in \mathbb{N}, i = 1,2,\ldots,n$. Here, using the Lemma 2.1., we get that $\{T^m x\}$ is a Cauchy sequence in X. As $(X,\mu,\nu,*,\diamond)$ is complete, there is point $y \in X$ such that $\lim_{n\to\infty} T^m x = y$. Since T is continuous, we have $\lim_{n\to\infty} T^{m+1}x = Ty$. Hence $Ty = y$ and y is a fixed point of T. To show uniqueness, assume $Tw = w$ for some $w \in X$ $(w \neq y)$. Since $(X,\mu,\nu,*,\diamond)$ is δ-chainable, there is a δ–chain $y = w_0,w_1,\ldots,w_{n-1},w_k = w$ from y to w. Now, for any $h \in \mathbb{N}$, we have

$$1 \geq \mu(y,w,z,t) = \mu(T^h y,T^h w,z,t) = \mu(T^h w_0,T^h w_k,z,t)$$
$$0 \leq \nu(y,w,z,t) = \nu(T^h y,T^h w,z,t) = \nu(T^h w_0,T^h w_k,z,t)$$

for all $t > 0$ $z \in X$, $i = 1,2,\ldots,k$. Using Lemma 2.1., equations (3.1) and (3.2), we have $\mu(y,w,z,t) = 1$ and $\nu(y,w,z,t) = 0$ for all $t > 0, z \in X$. So, we obtain that $y = w$.

Definition 3.7. [13] Let $(X,\mu,\nu,*,\diamond)$ be an if2ms and A,B be two self-mappings on $(X,\mu,\nu,*,\diamond)$. A pair (A,B) is said to be weak compatible if $Ax = Bx$ for some $x \in X$ implies $ABx = BAx$.

Theorem 3.4. Let $(X,\mu,\nu,*,\diamond)$ be a complete, δ-chainable if2ms and the self-mappings A,B,S,T on $(X,\mu,\nu,*,\diamond)$ satisfy the following conditions:
(i) $AX \subseteq TX$ and $BX \subseteq SX$.
(ii) $\exists k \in (0,1)$ such that
$\mu(Ax,By,z,kt) \geq \mu(Sx,Ty,z,t) * \mu(Ax,Sx,z,t) * \mu(By,Ty,z,t) * \mu(Ax,Ty,z,t)$ and
$\nu(Ax,By,z,kt) \leq \nu(Sx,Ty,z,t) \diamond \nu(Ax,Sx,z,t) \diamond \nu(By,Ty,z,t) \diamond \nu(Ax,Ty,z,t)$

for all $x, y, z \in X$ and $t > 0$.
(iii) The pairs (A, S) and (B, T) are weak compatible.
(iv) A and S are continuous.
Then A, B, S and T have a unique common fixed point in X.
Corollary 3.1. Let $(X, \mu, \nu, *, \diamond)$ be a complete, δ-chainable if2ms and the self-mappings A, B, S, T on $(X, \mu, \nu, *, \diamond)$ satisfy the following conditions:
(i) $AX \subseteq TX$ and $BX \subseteq SX$.
(ii) $\exists k \in (0,1)$ such that $\mu(Ax, By, z, kt) \geq \mu(Sx, Ty, z, t)$ and $\nu(Ax, By, z, kt) \leq \nu(Sx, Ty, z, t)$ for all $x, y, z \in X$ and $t > 0$.
Then A, B, S and T have a unique common fixed point in X.
If we take $S, T = I$ in the above corollary, we obtain the following result:
Corollary 3.2. Let $(X, \mu, \nu, *, \diamond)$ be a complete, δ-chainable if2ms and the self-mappings A, B on $(X, \mu, \nu, *, \diamond)$ satisfy the following conditions: $\exists k \in (0,1)$ such that

$$\mu(Ax, By, z, kt) \geq \mu(x, y, z, t)$$
$$\nu(Ax, By, z, kt) \leq \nu(x, y, z, t)$$

for all $x, y, z \in X$ and $t > 0$. Then A and B have a unique common fixed point in X.
If we take $A = B$ in the above corollary, we get the following result:
Corollary 3.3. Let $(X, \mu, \nu, *, \diamond)$ be a complete, δ-chainable if2ms and the self-mapping A on $(X, \mu, \nu, *, \diamond)$ satisfies the following conditions: $\exists k \in (0,1)$ such that

$$\mu(Ax, Ay, z, kt) \geq \mu(x, y, z, t)$$
$$\nu(Ax, Ay, z, kt) \leq \nu(x, y, z, t)$$

for all $x, y, z \in X$ and $t > 0$. Then A has a unique fixed point in X. Hence, we obtain the result stated in Theorem 3.1.

4 Conclusion

In this paper, we established some theorems related to the fixed point theory using some contractive mappings in a complete δ-chainable intuitionistic fuzzy 2-metric space. We obtained some generalizations of previous fixed point theorems in this spaces[4] and stated the relations of these theorems. Further, we investigated some theorems by using weak compatible mappings. In the future scope, we hope that some applications of these results can be founded for the real life problems.

Acknowledgement

We would like to thank the referee for his/her valuable suggestions.

References

1. M. S Bakry, Genaral Mathematics Notes, **27(2)**, 69-84 (2015)
2. S. Gähler, Math. Nachr. **26**, 115-118 (1963/64)
3. E. Güner, V. Çetkin, H. Aygün, submitted.
4. M. S. Khan, Publications de L'Institut Mathematique **27(41)**, 107-112 (1980)
5. O. Kramosil, J. Michalek, Kybernetica **11**, 326-334 (1975)
6. M. Mursaleen, Q. M. D. Lohani, S. A. Mohiuddine,Chaos,Sol. & Fractals **42**,(2009)
7. M. Mursaleen, Q. M. D. Lohani, Chaos, Sol. and Fractals **42**, 2254-2259(2009)
8. J. H. Park, Chaos, Solitions and Fractals **22**, 1039-1046 (2004)
9. S. V. R. Naidu, J. R. Prasad, Indian J. of Pure and App. Math., **17(8)**, 974-993(1986)
10. B. Schweizer, A. Sklar, Pacific J. Math. **10**, 314-334 (1960)
11. S. Sharma, Southeast Asian Bulletin of Mathematics **26**, 133-145 (2002)
12. R. Shrivastava, V. Gupta, N. Vijaywargi, Math. Theory and Mod., **4**, 17-28 (2014)
13. L. A. Zadeh, Inform Control **8** 338-353 (1968)

Inspection of artificially built mechanical failures through innovative condition monitoring techniques

S.Seçkin EROL[1], *Cemal MERAN*[2,*]

[1]Kilis 7 Aralık University, Faculty of Engineering-Architecture, Mechanical Engineering Dept., Kilis, 79000 Turkey
[2,*]Pamukkale University, Faculty of Engineering, Mechanical Engineering Dept., Denizli, 20020, Turkey, Corresponding author: cmeran@pau.edu.tr

Abstract. In this study, the factors influence on root causes of failure initiation are examined with the original test implementation and condition monitoring techniques are emphasized. In working toward this goal, the laboratory test setups and tests which have been created by international research bodies are examined and in order to practise a new research work, a unique test setup system and a test plan is built. In this target of the root cause failure detection, vibration data at radial direction and electrical consumption data are collected through the analysis by comparing two different condition monitoring techniques. In this research, fault detection in modeling fault conditions and vibration, electrical consumption measurement have been let us examination in depth. During the tests, data are collected simultaneously in vibration by four-channel Data Acquisition Card (DAC) and electrical consumption by Motor Condition Monitoring (MCM) system which are integrated with an computer system. Respect to the study results; vibration analysis in detection of defects has been judged to be more successful in comparison with electric consumption analysis under the test conditions in perspective of condition based predictive maintenanceIn the study, it is mentioned that detection of failure initiations at sensitive levels and importance of consequent results.

1 Introduction

In order to reduce the cost of unit products, various optimization methods, logistics, spare parts, work studies as well as six sigma, lean manufacturing techniques and many other methods are applied. In addition to these methods, innovative maintenance techniques can provide significant gains in unit product costs with the reduction of maintenance costs which is one of the main expenses of the enterprises

Diagnostic and prognostic definitions have been transferred from the medical field to the field of mechanical engineering, and these descriptions have become the symbols of ongoing maintenance technologies. Historically, the concept of Prognostic and Health Management (PHM) has been used for the first time in medicine. Medical prognostication has defined the course of the disease as the future, both as a predictor of natural course, and as a predictor of post-treatment course [1]. Prognostication, by the simplest definition, can be described as the ability to consistently make accurate predictions with the detection and visualization of the initial signs of component deterioration [1]. The future development of prognostic and health management will be mutually influenced and even more widespread across different fields outside of engineering [2].

The condition monitoring technique is based on the reading of the equipment sensor measurements, which are indicative of the processes of deterioration, and the identification of certain features and their understanding of the condition monitoring technique and equipment health [3].

Orhan et al. [4] investigated two looseness faults with vibration analysis on two outer rings on the bearings of the fan motors and pumps. When fault detection is performed, it is stated that the general vibration values may be misleading, and that the change in fundamental frequency and harmonics should be followed. In addition, vibration analysis of uncomplicated machines has revealed that unbalance, mechanical looseness and misalignment errors can be detected with precision.

Karacay and Akturk [5] investigated a system with angular contact bearings in operation first by testing the vibration phenomena of the inner ring, outer ring and rolling elements at different rotation speeds with vibration technique, while the bearings were healthy first. He mentioned that even if there is no flaw in the bearings, it gives a vibration response and that in the spectrum graph, the ball passes through the harmonic frequencies. Elbhbah and Jyoti [6] reported that imbalances and imperfections were successfully detected by vibration analysis.

This study is similar in terms of studies in the literature, types of defects and selected detection methods; the sensitivity levels in the imperfections, the number of cycles tested, the electrical frequencies and the approach in the detection methods.

2 Material and method

The test system used in this study was inspired by the climatic fan in a cement plant. Test setup consists of AC induction motor, double suction radial fan, five-legged elastic coupling and frequency converter. The test setup is placed on a stable table made of steel material. The system as a whole is associated with a computer system in which the data acquisition card and the motor status monitoring system are integrated. Figure 1 shows schematic view of the test fixture.

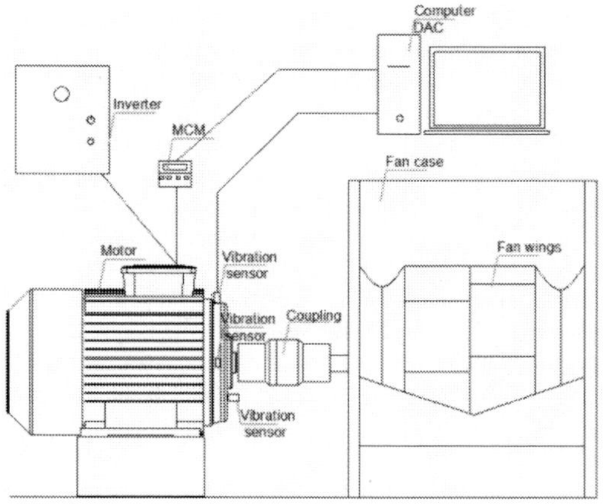

Fig. 1. Test rig

2.1 Frequency harmonics and fault frequencies

The approximate values of the theoretical (T) and actual (A) frequencies used in the tests are given in Table 1. While theoretical frequencies indicate the value digitally set on the frequency converter, the actual frequencies varying numerically by measuring some loss according to the data obtained during the application are measured.

Table 1. Harmonics in theory and actual; (f:frequency (Hz) ; h: harmonics (Hz))

(h)	1x		2x		3x		4x		5x	
(f)	T	A	T	A	T	A	T	A	T	A
50	50	47.61	100	95.22	150	142.83	200	190.44	250	238.05

During the measurements in the tests, the rotational speed was measured as 2880 min^{-1} at the electricity frequency of 50 Hz. Bearing and fan-induced vibrations are included in the calculations while calculating possible failure frequencies and harmonics of the test set.

The basic failure frequencies of the ORS 6202 bearing are calculated and presented in Table 2.

Table 2. Main fault frequencies

f (Hz)	ω_s (Hz)	ω_{bpf} (Hz)	ω_c (Hz)	ω_{bpfo} (Hz)	ω_{bpfi} (Hz)	ω_{bsf} (Hz)
50	47.61	476.1	19	152	248	98.16

The electrical measurements carried out with the motor condition monitoring system that allows detection of defects by following current and voltage changes. Changes can be tracked on trend charts. The types of defects can be selected in trend graph and the changes can be displayed according to standard deviation and time.

2.2 Test parameters

In order to be able to determine the efficiency of the measurement techniques and to increase the sensitivity of the changes in the data obtained from the measurements, an electric motor with a low power of 0.37 kW is preferred to minimize the vibration and electrical consumption values.

Axis eruptive effect was investigated in the range of 0.1 mm – 0.5 mm with a sensitivity of 0.1 mm. The effect of the change in rotation on these defects was tested at 2880 min^{-1} revolutions per minute at the electricity frequency of 50 Hz.

3 Test findings

The test findings were examined in two main titles, vibration data and electricity consumption data, in accordance with the aim of the study.

3.1 Vibration Data and Analysis

Five spectral signals with the highest amplitude in the spectrum charts and RMS values are presented for comparison. For this study, natural frequency data were collected from the motor, motor shaft, coupling, fan, fan shaft, sheet metal, test setup stand and floor where test setup is positioned, using the impact test with a plastic tipped hammer while the system was in motion. Natural frequencies are expressed as f_n.

* The radial direction after this part of operation is expressed only for the vertical direction.

3.1.1 Tests of axial misalignment

Tests are implemented at 2880 min^{-1} revolutions per minute (rpm). As a result of the tests; as the level of axis misalignment increases, there is a higher amplitude increase in the axis misalignment indication signals compared to other signals. Waveform and spectrum findings are given for 0.2 mm and 0.5 mm levels; waterfall spectral data are given broader covering the range of 0.1 mm – 0.5 mm.

a) Test of axial misalignment at 0,2 mm

When the radial direction vibration data are examined, the superharmonic resonance condition at the natural frequency of 96.44 Hz, which is caused by the 95.21 Hz (2x) vibration signal is observed and detected as the most dominant signal. In Fig. 2, the radial direction have been found corresponding to the 2x misalignment harmonics and the natural frequency signals of the $51x\omega_c$ belongs to cage harmonic.

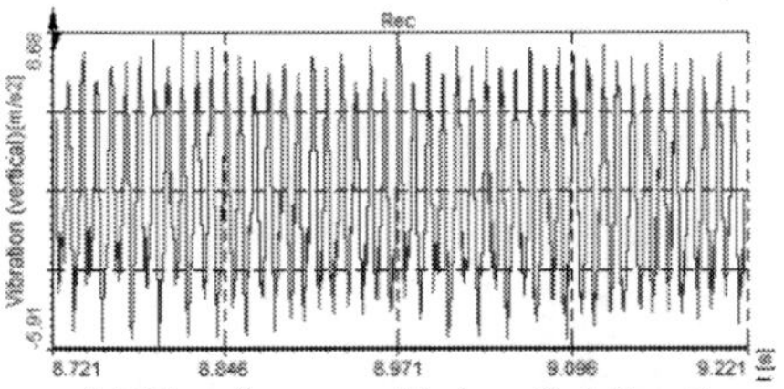

(a) Waveform graphic in radial direction

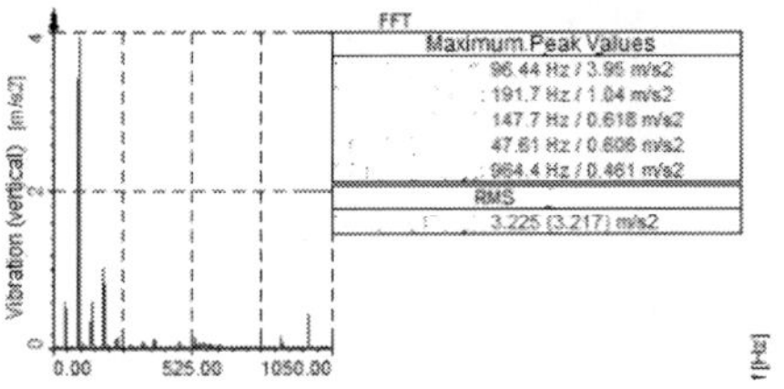

(b) Spectrum graphic in radial direction

Fig. 2. Axial misalignment data of 0.2 mm

In the radial direction, the other vibration signals according to the dominance order are 191.7 Hz (f_n), 147.7 Hz (f_n), 47.61 Hz (1x), 964.4 Hz (f_n). It was observed 1x more dominant as the unbalance harmonic. Only fluctuation effect is observed in waveform graphs. In the radial direction, the signal at natural frequency of 96.44 Hz with the frequency of 95.21 Hz (2x), the natural frequency of 191.7 Hz with the frequency of 190.44 Hz (4x), the frequency of 142.83 Hz (3x) with the natural frequency of 147.7 Hz is coincided and the superharmonic resonance condition has occurred.

b) Test of axial misalignment at 0,5 mm

In Figure 3, the superharmonic resonance effect at the natural frequency of 96.44 Hz, which is caused by the vibration signal at the frequency of 95.21 Hz (2x), is detected as the most dominant signal at the radial direction.

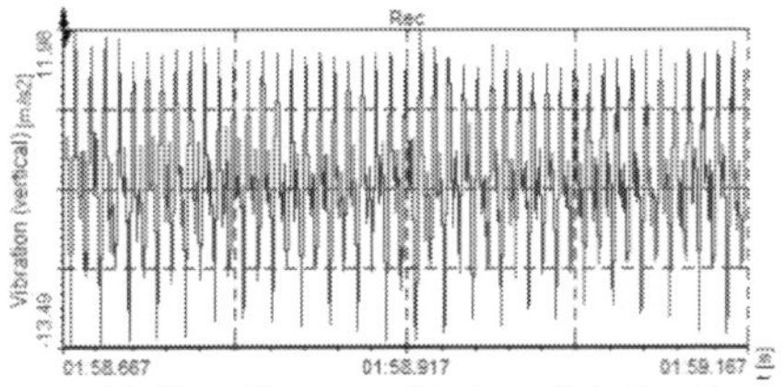

(a) Waveform graphic in radial direction

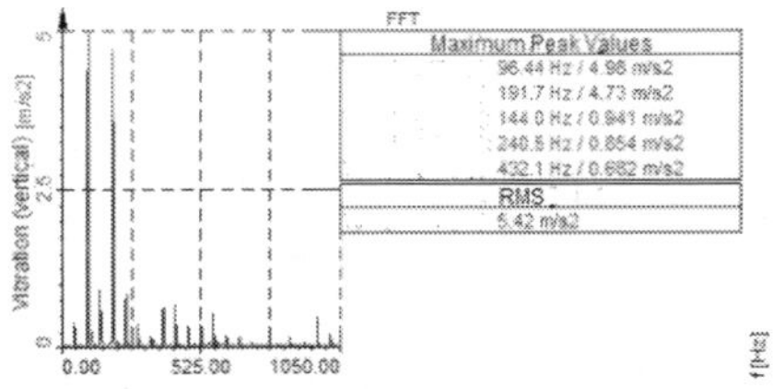

(b) Spectrum graphic in radial direction

Fig. 3. Axial misalignment data of 0.5 mm

At the radial direction, the vibration signal at the frequency of 190.44 Hz (4x) with the natural frequency of 191.7 Hz, the vibration signal at the frequency of 240.5 Hz (5x) with the natural frequency of 240.5 Hz and the vibration signal at the frequency of 428.49 Hz with the natural frequency of 432,1 Hz has caused superharmonic resonance by stimulating its natural frequency. Mechanical looseness harmonics have been observed over 3x and higher orders, including fundamental frequencies that stimulate natural frequencies, more pronounced in the radial direction.

In Table 3, radial harmonics are presented for comparison purposes.

Table 3. Axial misalignment signals at 2880 min^{-1}

MEASUREMENT	1.S	2.S	3.S	4.S	5.S	RMS (m/s^2)
M/2880 min^{-1} / 0,2 mm/(R)	f_n (2x)	f_n (4x)	f_n (3x)	1x	f_n ($51x\omega_c$)	3,217
M/2880 min^{-1} / 0,5mm/(R)	f_n (2x)	f_n (4x)	f_n (3x)	f_n (5x)	f_n (9x)	5,42

M: Misalignment, R: Radial direction, S: Signal

According to Table 3, the natural frequency of the 2x misalignment harmonics is determined as the most dominant signal (1.S) in misalignment tests at 2880 min^{-1} revolutions whereas the natural frequency of the 4x mechanical looseness harmonics. The RMS values increased in parallel with the increase in the axial misalignment. In the radial direction, the two dominant first two signals 3x and 6x harmonics leave their place to the natural frequencies excited by the 2x and 4x harmonics. The misalignment spectrum changes at 2880 min^{-1} revolution are given in Fig 4. In graphs, there is a possibility of confusion with the mechanical looseness indication at the radial direction.

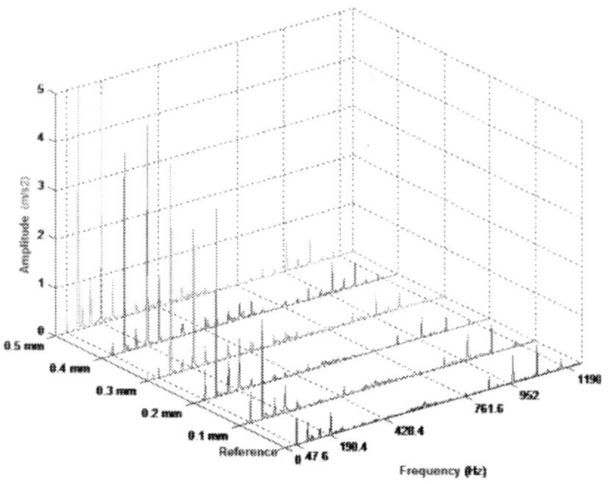

Fig. 4. Waterfall spectrum diagrams of axial misalignment

In Fig. 4, radial and axial direction vibration data at 2880 min^{-1} rpm are given by waveform and spectral imaging techniques. 2x harmonics began to be evident when the axial misalignment level was in the range of 0.1 mm – 0.2 mm, and in the range of 0.3 mm – 0.5 mm, the signal amplitude became visible at the 4x harmonics as well as the 2x harmonic. In the axial direction, only the signal amplitude increase at 2x harmonics becomes evident. Signals have begun to appear in the element transition region, being more pronounced in the radial direction.

3.2 Electrical consumption data and analysis

All the tests performed were evaluated by collecting data on the simultaneous consumption of electricity (current, voltage). The trend data is similar to the time-domain data, and the data can be analyzed as standard deviation respect to the selected parameters.

3.2.1 Tests of axial misalignment with trend data

Since the results are similar, no trend graph is given for all axis misalignment levels; given only for 0.2 mm and 0.5 mm levels. Variations in axial misalignment of 0.2 mm and 0.5 mm are given in Figure 5.

Under the conditions of 0.2 mm misalignment, no indication of any failure was found in the tested speed.

Under the condition of 0.5 mm axis misalignment, there are some malfunctions in the test speed count. The number of revolutions at 2880 min^{-1} (f: 50 Hz) speed level started to increase in loose ground / component, axial misalignment and bearing parameter values.

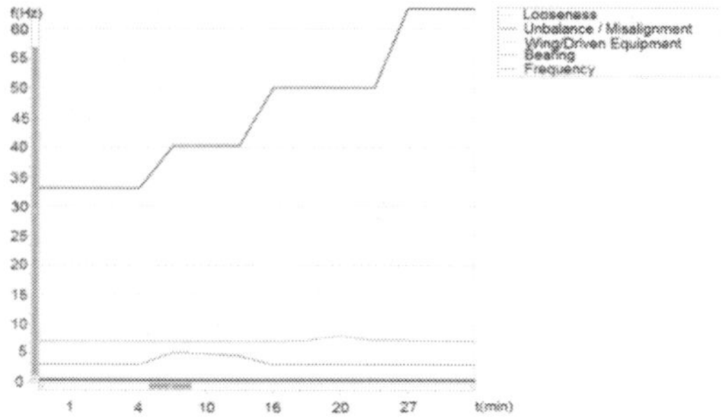

a) Trend data of axial misalignment at level of 0,2 mm

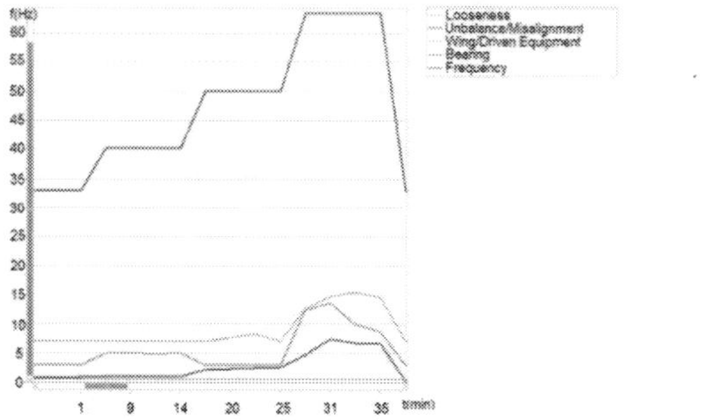

b) Trend data of axial misalignment at level of 0.5 mm

Fig. 5. Trend data of axial misalignment at 0.2 mm and 0.5 mm

4 Conclusions

In this study, two condition monitoring techniques based on vibration and electrical (current, voltage) consumption signals are studied. The results of the analyzes carried out on the basis of the obtained data are valid under the technical conditions under test. In these two condition monitoring techniques; vibration analysis results and trend analysis results of electricity consumption signals are not parallel. On the other hand, the vibration analysis technique was found to be more successful at the end of the study because expected results could not be obtained in the course of electrical trend analysis which is the main defect detection tool of Motor Condition Monitoring (MCM). Especially where vibration data collection is not possible, electrical consumption data can be evaluated; and also vibration analysis evaluations can be supported in the detection of electrical faults.

Acknowledgment

This study was supported by Pamukkale University Scientific Research Projects Coordination Unit with the project number 2012FBE010.

References

1. Hess, A., Prognostics, From the Need to Reality— From the fleet users and PHM System Designer/Developers Perspectives. Proceedings of the IEEE Aerospace Conference, BigSky, MT, USA., 6, 2791–2797, 2002.

2. Moubray, J., Reliability Centred Maintenance, 2nd edition, Butterworth-Heinemann Ltd., Great Britain, 2005.

3. Sudhar, G.N.D.S., Sekhar, A.S., Identification of unbalance in rotor bearing system, Journal of Sound and Vibration, 330, 2299-2313, 2011.

4. Orhan, S., Aktürk, N., ve Çelik, V., Vibration monitoring for defect diagnosis of rolling element bearings as a predictive maintenance tool: comprehensive case studies, NDT&E International, 39, 293-298, 2006.

5. Karacay, T., Akturk, N., Experimental diagnostics of ball bearings using statistical and spectral methods, Tribology International, 42, 836-843, 2009.

6. Elbhbah, K., Jyoti, K.S., 2013: Vibration based condition monitoring of rotating machines using a machine composite spectrum, Journal of Sound and Vibration, 332, 2831-2845.

Assesment of the Image Distortion in Using Various Bit Lengths of Steganographic LSB

Yucel Inan[1],

[1]Near east University, Computer Engineering Department, Nicosia, TRNC, Mersin 10 Turkey

Abstract. Several methods developed and applied for protecting the information. One of these is the stenography. Steganographic techniques are used to transmit the information in the image to the receiver in a secure manner. There are two main principles in the steganographic process. The first one is to hide the message in the image and the second one is to reduce the distortion on the image caused by information hiding. By making changes on digital images, a lot of information can be placed in the image. Nevertheless, changes in the image should not be noticed. In this paper, the effect of using various bit length of the steganographic LSB method on the image distortion is studied. The PSNR, SNR and MSE were used to assess the distortion rates of the images. Histograms were drawn to visualize the differences between original and encoded, "stego-images".

Keywords: Cryptography, Image Steganography, LSB Method, Cover Image, Hidden Secret Image, Stego-Image PSNR, SNR, MSE, Histogram.

1 Introduction

Steganography is an important method of information hiding, it is derived from a work by Johannes Trithemus (1462-1516) entitled "Steganographia" and comes from the Greek (στεγανό-ς, γραφ-ειν) defined as "covered writing"[1]. Steganography is the art and science of communicating information by hiding it from unwanted people [2]. Information can be hidden in the desired "cover-data", which can be in the form of image, text, audio, or video. The resulting encoded data can is called "stego-object" [3-4]. Stego-object does not give a clue about the information that is hidden when compared with the cover-image [5]. The goal in steganography is to hide the existence of a message and to create a covert channel. So, steganography can be seen as part of the cryptology of storing the content of the intended message. But they both have different structures. Cryptography is the method by which the receiver can only decrypt the message

1 Corresponding author: lyinan@hotmail.com

after it has been subjected to the encryption method of the message to be transmitted. And the great advantage of steganography over cryptography is that someone who sees Information never knows whether it is a secret message within the carrier. That is, when considering today's advanced technologies, there is a possibility of broken in the encryption method, even if the password is strong enough but in steganography can not be distinguished because the existence of the secret message is unknown and therefore unnoticeable. While it saying that briefly, steganography takes its power from secrecy, cryptography does this by using encryption algorithms[6]. Steganography classified into two types, Linguistics Steganography and Technical Steganography. Figure 1 displays, four main categories of the formats technical steganography that can be used for steganography [7].

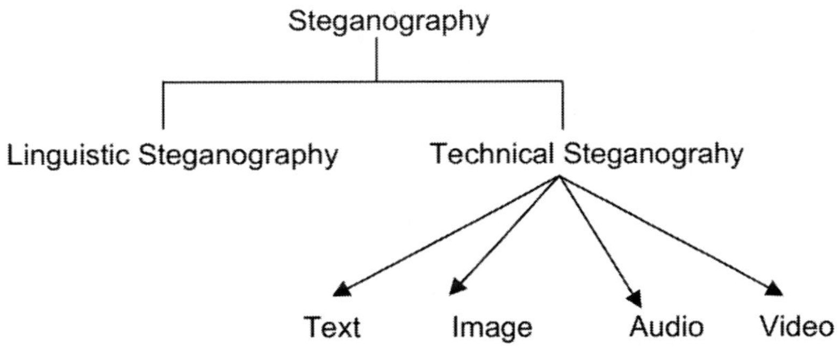

Figure 1: Categories of Technical Steganography

Linguistic steganography is uses text as the cover carrier to hide the secret message. Technique Steganography is invisible ink, hidden places, microdot and computer based, these are methods of hiding data using text, image, audio files[5]. Formally, a steganography system consists of a tripler algorithm that generates a key, encodes a message, and decodes a message.
This is shown in figure 2 below[8].

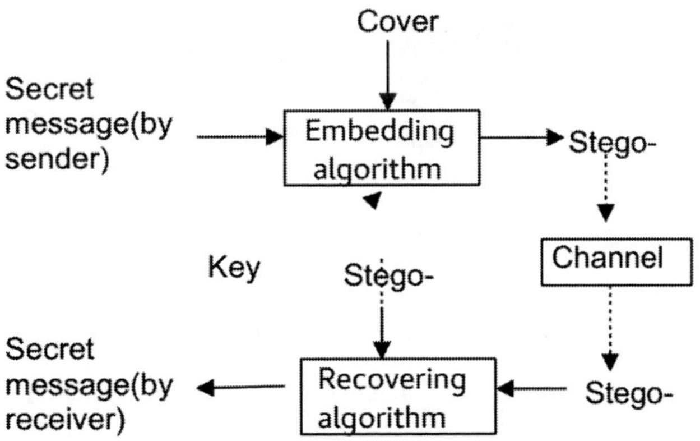

Figure 2: The Mechanism of Steganography

In this study, Cover image Kirlian Photograph[2] was used with the picture. The result is obtained by taking a contains secret message and cover image that both in the same size a 454x459 (jpg and png) in different format. Shows the LSB method experimental results of a pixel gray scale image appearing in the bit-plane in which the image is placed in five separate low order bit planes. This bit grayscale image keeps all this information inside without giving any clue about them and showing no visible decline in image quality. For this we have calculated PSNR and SNR, MSE. At the end of the calculation, PSNR produces a single value. The high value of this value means that the quality is high (the operation on the image is less effect on perceivability). The obtained PSNR, SNR and MSE values are shown in Table 2 and bit-planes image distortion of LSB images are shown in Table 1.

2 Structure of Steganography System

Image Steganography can be seen in the stages for embedding the secret message in image files, by using LSB method.

Figure 3 describes Stegangraphy of system. Firstly the cover image is converted to gray scale image. Then the cover image and the image that contains secret message are both scaled to the same dimension. After that separate LSB applications have been implemented every bit of the data to be hidden is written to the last bit of a byte of the image data. In here, The sender creates a steganogram using a concealment function. The concealment function has two parameters, the carrier channel in which the secret information is to be embedded and the secret information to be hidden. And also, a key password is present in this embedding and extract process to increase the privacy or security of the confidential information. If the recipient receives the image and he or she knows the key code to extract the confidential message, confidential information can be obtained.

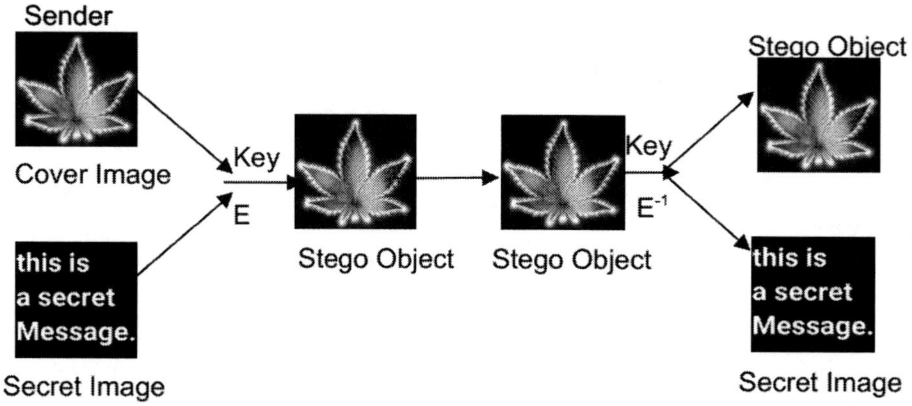

Figure 3: Block diagram of the Steganography System

2http://www.ottophoto.com/kirlian/kirlia_1/album1.html

3 Bit Plane Image Distortion of LSB Method

The substitution systems try to encode secret information by substituting unimportant parts of the cover by secret message bits. The receiver can discover the secret information if he/she has knowledge of the positions where secret information has been embedded. Since only minor changes have been made to the embedding process, the secret message may not be noticeable [9].
This concept is shown in Table 1, in each bit-plane is shown as a separate image to find the contribution of each bit-plane in the generation of the general view. This shows that the Bit-Plane Insert LSB method seriously reduces the cover image [10]. The effect of altering the LSBs up to the 8th bit embedding information with bit-planes for cover image shown in the next table. In Table 1 below shows a gray scale image and its bit planes from the 8th MSB bit- plane to the lowest 1.LSB bit-plane below and in lower bit-planes, the images looking like noise.

Table 1. Image distortions from 8th bit plane to the lowest 1.bit plane LSB.

LSB(Least Significant bit image distortion)Bit Plane

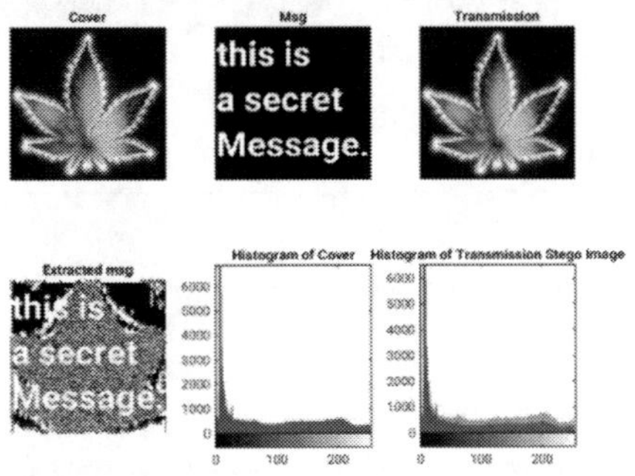 Bit-Plane 1	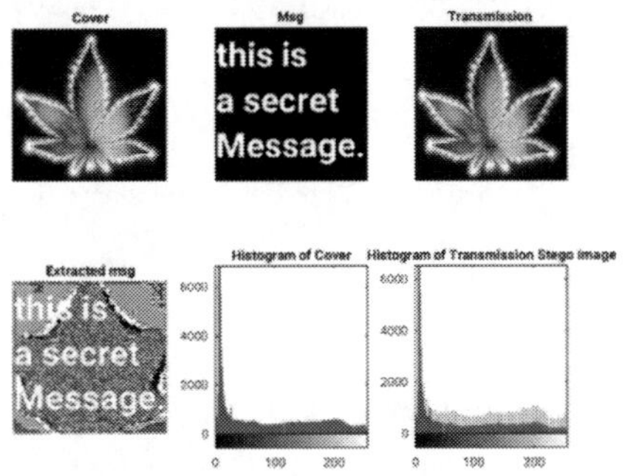 Bit-Plane 2
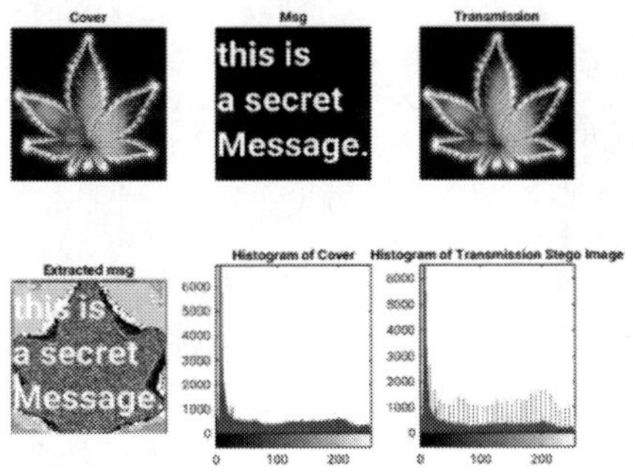 Bit-Plane 3	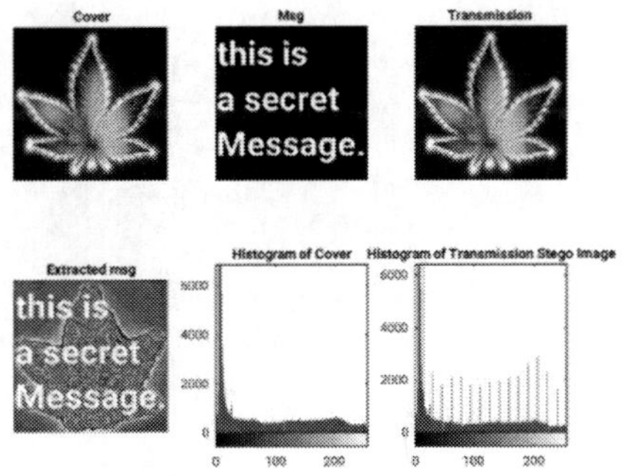 Bit-Plane 4

 https://doi.org/10.1051/itmconf/20182201026

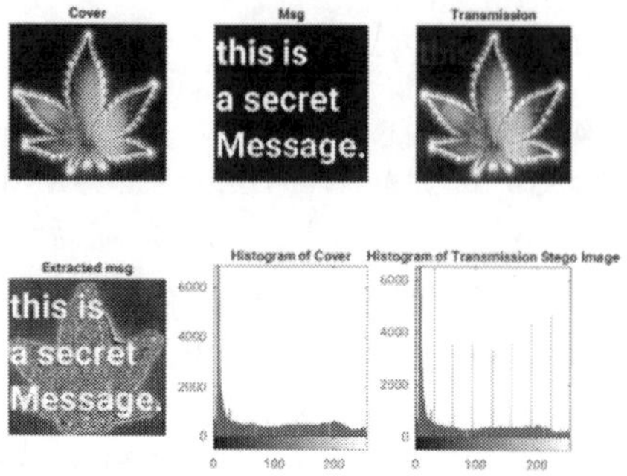 Bit-Plane 5	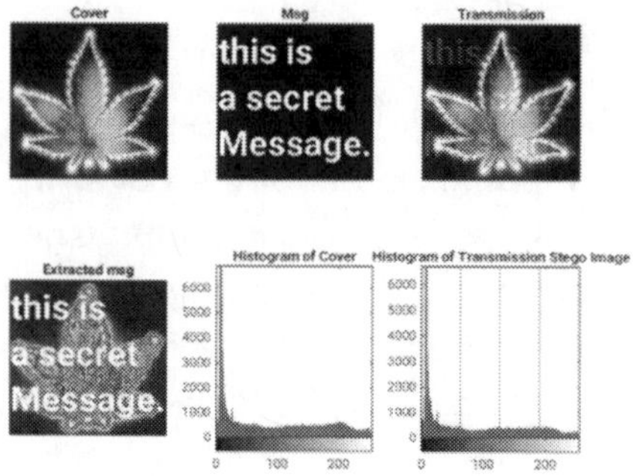 Bit-Plane 6
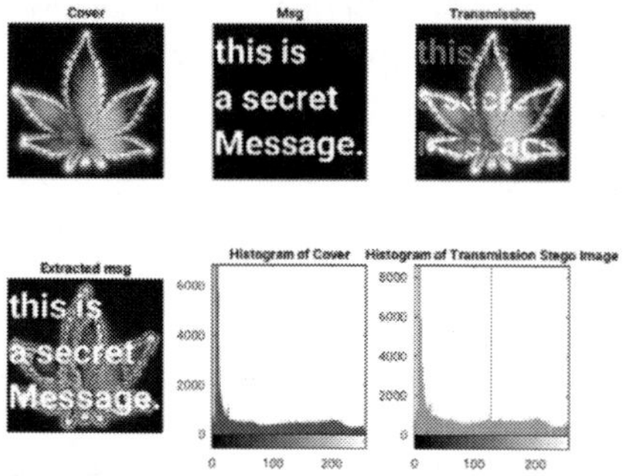 Bit-Plane 7	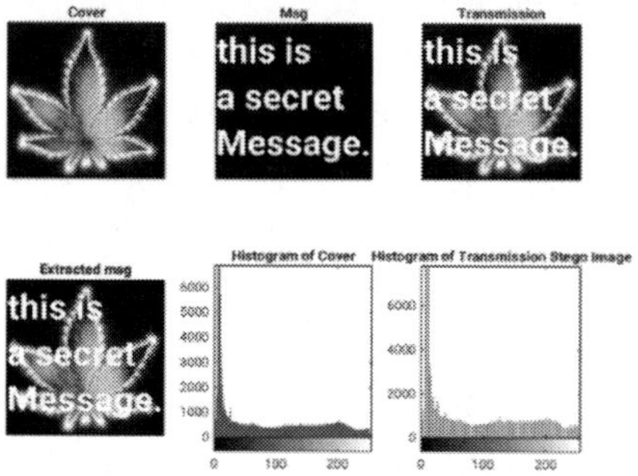 Bit-Plane 8

MSB(Most Significant bit image distortion) Bit plane

Table 2, shows the original RGB cover images and their gray scale images. In table 3 it can be resized cover image and contains secret message images both scaled to the same dimension.

Table 2. Shows the original RGB images and its gray scale images

 RGB Original Cover Image	 Gray scale image

Table 3. The cover image and the secret message images both scaled to the same dimension.

Gray Scale Cover Image

Gray scale Secret image

The steganography of the bit-plane image distortion of LSB method was used to embed hide in the cover image. This approach shows a grayscale image and its bit-planes from the maximum bit-plane at the top to the minimum bit-plane at the bottom. Above the first image is the gray scale original cover image and 8 images were viewed on the screen, the 1st is a 1 bit representation (Least Significant Bit) revealed 5th is 6th bit and so on. The 8th image is revealed the 8 bit (Most Significant Bit) image and the 8th image are obtained after the combination of all 8 bits. This application used for information hiding and extracting was done in the Matlab program environment. PSNR, SNR, MSE, Matlab program used image quality evaluation methods used to see distortions in images. Image quality assessment methods used to see distortions in images have been PSNR, SNR, MSE. In the data hiding applications, it is seen that the LSB method has the least degradation between the cover image and the hidden stego-image.
In the image quality evaluation methods used, the technique with the highest PSNR value is the LSB method. The error bit value with MSE is seen in the LSB method. In situations where the amount of data to be concealed should be small, it is better to choose the LSB method.

4 Image Quality Assessment Method

Evaluation of image quality is used to measure the distortion in the digital image to improve the quality of the resulting image[15]. In this analysis, we have the evaluation cover image and the confidential data to be embedded in this image. The most commonly used image quality measurements are PSNR, SNR, MSE. This study of experimental results, PSNR is used to measure the statistical quality of images containing confidential data. PSNR calculates the quality of similarity between the original image and the image containing the confidential data. image assessment methods that is:

4.1 Peak Signal Noise Ratio(PSNR)

PSNR is used to reveal the similarity between images in digital images. After the information is embedded, the distortions on the stego-object are detected by PSNR. The equation used for this calculation is given below[11]. The SNR measure is an estimate of quality of reconstructed image compared with original image. PSNR is defined as in [13].

$$PSNR = 10\log 10\left(\frac{MAX_i^2}{MSE}\right) \quad (1)$$

4.1 Mean Square Error(MSE)

It is used to compare the original image with the processed image after different types of operations have been performed on an image. The quadratic mean error estimate used to compare differences that may occur in the image. The mathematical function used to estimate the quadratic error is given below[12].

$$MSE = \frac{1}{MN}\sum_{y=1}^{M}\sum_{x=1}^{N}\left[I(x,y) - I'(x,y)\right] \quad (2)$$

In the data hiding applications, it is seen that the bit-plane Image Distortion of LSB method is the most 5th lower position degraded between the cover image and the hidden stego-object. In Table 4. The values of PSNR, SNR and MSE embed values obtained the amount of the application are shown below.

Table 4. Image degradation with different number of bits used for LSB.

Number of bits(LSB)	PSNR(db)	SNR(db)	Mse
1	58.706	51.332	0.088
2	51.170	43.796	0.497
3	44.367	36.993	2.379
4	37.357	29.983	11.951
5	30.591	23.218	56.748
6	23.956	16.582	261.535

7	17.858	10.485	1064.728
8	11.874	4.501	42233.437

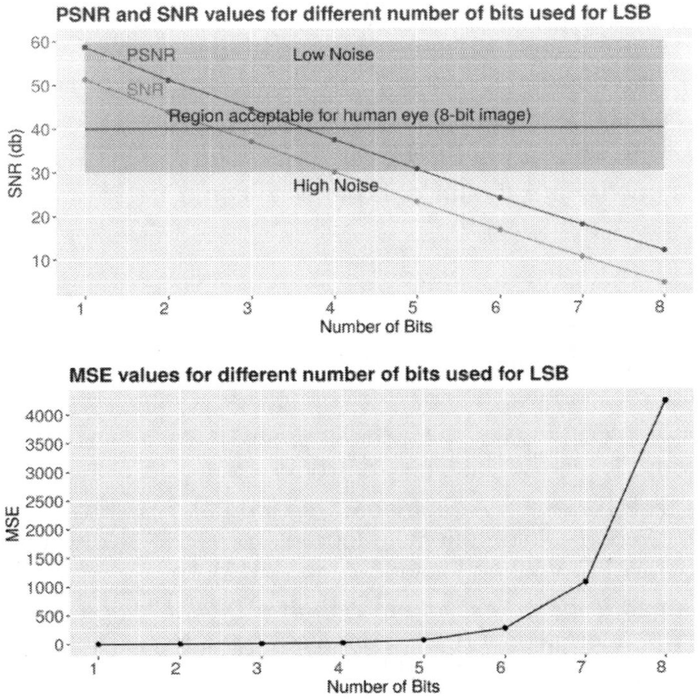

Figure 4. Shows PSNR and SNR, MSE values for different number of bits used LSB.

6 Histogram Analysis

In Steganography it is important that the embedded secret information is not understood when it gets in the hands of undesirable people. Therefore, several methods have been developed to test the robustness of the algorithms developed. The principle containing these methods is called steganaliz.. There are several methods used to detect confidential data. Histogram analysis method was tested from one of these methods. A histogram is a measure of the frequency of each element in a data sequence. It is a method that is used firmly in the field of digital image security. Table 1. Shows a cover image and historgrams of the stego-Image distortions from 8th bit-plane MSB to the lowest 1.bit-plane LSB.

6 Conclusion

In this paper we have studied bit-planes Image Distortion of LSB method for hiding information embedded in image file in gray scale image. We can slicing image into eight bit-planes to extract each bit-planes of original image. We performed the test with different value ranges and selected the opitimum range intensity level range. This range is the optimum range and causes minimal distortion on the entire image. And also MSE(Mean Square Error), SNR(Signal Noise Ratio), PSNR (Peak Signal Noise) and histogram analyszes of stego image is calculated. The histogram is visually distinguishable from the orignal cover image of the cover image and also shows that the PSNR values and the MSE values and the stego quality is a little good. Here, the factor is the distortion.

References

1. L. Y. POR, B. Delina, *Information Hiding: A new Approach in Text Steganography, 7th Wseas Int. Conf. (On Applied Computer & Applied Computational Science (Acacos '08), Hangzhou, China, April 6-8, 2008)*
2. Domenico Bloisi, Luca Iocchi , *Image Based Steganography and Cryptography*, Sapienza University of Rome, Italy
3. D.Seetha, Dr.P.Eswaran, *A Study on Steganography to Hide Secret Message inside an image,* International Journal of P2P Network Trends and Technology(IJPTT)-Volume3 Issue5-June (2013)
4. Nasser Hamad, *Hiding Text Information in a Digital Image Based on Entropy Function,* The International Arab Journal of Information Technology, Vol. 7,No. 2, April (2010)
5. Monika Agarwal , *Text Steganographic Approaches: A Comparison,* International Journal of Network Security & Its Application(IJNSA), Vol.5, No.1.January (2013)
6. Salony Pandey, Prof. Amit .M. Lathigara, *Hiding Text Behinde Image for Secure Communication, Salony Pandey, Prof.Amit .M. Lathigara /International Journal of Engineering Reserach and Applications(IJERA) ISSN:2248-9622 Vol.3,Issue3, May-Jun (2013), pp.1295-1297*
7. Neha Rani, Jyoti Chaudhary, *Text Steganography Techniques: A Review,* International Journal of Engineering Trends and Technology(IJETT)- Volume 4 Issue 7-july (2013)
8. L. Y. POR , B. Delina, Information Hiding: *A New Approach in Text Steganography*, 7th Wseas Int. Conf. On Applied Computer & Applied Computational Science (Acacos '08), Hangzhou, China, April 6-8, (2008)
9. Ali K.Hmood, B.B. Zaidan, A.A. Zaidan and Hamid A. Jalab, *An Overview on Hiding Information Technique in Images*, Journal of Applied Sciences,10: 2094-2100.
10. M.Naseem, Ibrahim M. Hussein, M.Kamran Khan, Aisha Ajmal, *An Optimum Modified Bit Plane Splicing LSB Algorithm for Secret Data Hiding,* International Journal of Computer Applications(0975-8887) Volume 29-No.12,September (2011)

11. Khan Muhammed, Jamal Ahmad, Haleem Farman and Muhammed Zubair, *A novel Image Steganographic Approach for Hiding Text in Color Images HSI Color Model.* Middle-East Journal of Scientific Research 22(5):647-654,(2014)
12. Rengarajan Amirtharajan and John Bosco Balaguru Rayappan,*Pixel Authorized by Pixel to Trace with SFC on Image to Sabotage Data Mugger: A Comparative Study on PI Stego.* Research Journal of Information Technology, 4: 124-139,June 27,(2012)
13. Yusra A. Y.Al-Najjar, Dr.Der Chen Soong, *Comparison of Image Quality Assessment: PSNR, HVS, SSIM, UIQI.* International Journal of Scientific & Engineering Research, Volume 3, Issue 8, August-(2012) 1 ISSN 2229-5518

Reproducing kernel functions for linear tenth-order boundary value problems

Ali Akgül[1,*], *Esra* Karatas Akgül[2,**], *Baris* Orcan[3,***], and *Mustafa* Inc[4,****]

[1]Siirt University, Art and Science Faculty, Department of Mathematics, 56100, Siirt/TURKEY
[2]Siirt University, Faculty of Education, Department of Mathematics, 56100, Siirt/TURKEY
[3]Siirt University, Art and Science Faculty, Department of Mathematics, 56100, Siirt/TURKEY
[4]Firat University, Faculty of Science , Department of Mathematics, 23119, Elazig/TURKEY

Abstract. Higher order differential equations have always been an onerous problem to investigate for the mathematicians and engineers. Different numerical methods were applied to get numerical approximations of such problems. This paper gives some reproducing kernel functions to find approximate solutions of the tenth-order boundary value problems (BVPs). These reproducing kernel functions are very important in the reproducing kernel Hilbert space method.

1 Introduction

A numerical approximation of tenth-order BVPs is rarely presented in the literature [1]. When heating an endless flat film of fluid from below, under the supposition that fluid is subject to the action of rotation and uniform magnetic field across the fluid is applied in the same direction as gravity, unsteadiness starts. When unsteadiness begins in as usual convection, then it can be modeled by tenth-order BVP [2]. Pervaiz et al. [3] introduced the numerical approximations of twelfth-order BVPs by applying non-polynomial cubic spline method. Omotayo et al. [4] presented non-polynomial spline technique for the fourth order BVPs. Usmani [5] introduced his work to approximate the fourth order BVPs by using the quartic spline method. Twizell and Boutayeb [6] enhanced and showed the numerical approximations for higher order value problems. The approximation of second order BVPs was introduced by Alberg and Ito [7]. Siraj-ul-Islam et al [8] introduced a non-polynomial spline method to approximate the sixth-order BVPs. Papamichael and Worsey [9] studied the cubic spline algorithm for solving linear fourth-order BVPs. Siddiqi and Twizell [10, 11] enhanced numerical approximations of tenth and twelfth-order BVPs.

*e-mail: aliakgul00727@gmail.com
**e-mail: esrakaratas@siirt.edu.tr
***e-mail: barisorcan@gmail.com
****e-mail: minc@firat.edu.tr

2 MATERIALS AND METHODS

The main aim of this study is to find some new functions for numerical approximations of linear tenth-order BVPs. We consider

$$\begin{aligned} h^{(x)}(t) + a_1(t)h^{(ix)}(t) + a_2(t)h^{(viii)}(t) + a_3(t)h^{(vii)}(t) + a_4(t)h^{(vi)}(t) \\ +a_5(t)h^{(v)}(t) + a_6(t)h^{(iv)}(t) + a_7(t)h'''(t) + a_8(t)h''(t) \\ +a_9(t)h'(t) + a_{10}(t)h(t) = M(t), \quad t \in [a,b], \end{aligned}$$

with boundary conditions:

$$\begin{aligned} &h(a) = \alpha_0, \quad h(b) = \beta_0, \quad h''(a) = \alpha_1, \quad h''(b) = \beta_1, \\ &h^{(iv)}(a) = \alpha_2, \quad h^{(iv)}(b) = \beta_2, \quad h^{(vi)}(a) = \alpha_3, \quad h^{(vi)}(b) = \beta_3, \\ &h^{(viii)}(a) = \alpha_4, \quad h^{(viii)}(b) = \beta_4, \end{aligned}$$

where, $\alpha_j, \beta_j, j = 0, 1, 2, 3, 4$ are arbitrary fixed real constants, $a_j(t), j = 1, 2, ..., 10$ and $M(t)$ are continuous functions described on $[a, b]$.

We obtain our reproducing kernel functions for following test problem.

Test Problem

We consider the following tenth-order equation as a second experiment:

$$h^{(10)}(t) = -(80 + 19t + t^2)\exp(t), \quad 0 \le t \le 1,$$

with boundary conditions:

$$\begin{aligned} h(0) &= 0, & h(1) &= 0, \\ h^2(0) &= 0, & h^2(1) &= -4\exp(1), \\ h^4(0) &= -8, & h^4(1) &= -16\exp(1), \\ h^6(0) &= -24, & h^6(1) &= -36\exp(1), \\ h^8(0) &= -48, & h^8(1) &= -64\exp(1). \end{aligned}$$

The exact solution to the above BVP is given by [1]:

$$h(t) = t(1-t)\exp(t).$$

3 Reproducing kernel functions

We define $V_2^{11}[0, 1]$ as

$$V_2^{11}[0,1] = \left\{ \begin{array}{l} u, u', u'', u''', u^{(4)}, u^{(5)}, u^{(6)}, u^{(7)}, u^{(8)}, u^{(9)}, u^{(10)} \text{ are absolutely continuous functions,} \\ u^{(11)} \in L^2[0,1], \quad u(0) = u''(0) = u^{(4)}(0) = u^{(6)}(0) = u^{(8)}(0) = 0 \\ u(1) = u''(1) = u^{(4)}(1) = u^{(6)}(1) = u^{(8)}(1) = 0. \end{array} \right\}$$

We give

$$\begin{aligned} \langle u, G_z\rangle_{V_2^{11}[0,1]} &= u(0)G_z(0) + u'(0)G_z'(0) + u''(0)G_z''(0) + u'''(0)G_z'''(0) + u^{(4)}(0)G_z^{(4)}(0) \\ &+ u^{(5)}(0)G_z^{(5)}(0) + u^{(6)}(0)G_z^{(6)}(0) + u^{(7)}(0)G_z^{(7)}(0) + u^{(8)}(0)G_z^{(8)}(0) \\ &+ u^{(9)}(0)G_z^{(9)}(0) + u^{(10)}(0)G_z^{(10)}(0) + \int_0^1 u^{(11)}(x)G_z^{(11)}(x)d(x). \end{aligned}$$

We have

$$G_z(0) = G_z''(0) = G_z^{(4)}(0) = G_z^{(6)}(0) = G_z^{(8)}(0) = 0$$

$$G_z(1) = G_z''(1) = G_z^{(4)}(1) = G_z^{(6)}(1) = G_z^{(8)}(1) = 0,$$

by boundary conditions. Therefore, we get

$$\begin{aligned}
\langle u, G_z \rangle_{V_2^{11}[0,1]} &= u'(0)G_z'(0) + u'''(0)G_z'''(0) \\
&+ u^{(5)}(0)G_z^{(5)}(0) + u^{(7)}(0)G_z^{(7)}(0) \\
&+ u^{(9)}(0)G_z^{(9)}(0) + u^{(10)}(0)G_z^{(10)}(0) + \int_0^1 u^{(11)}(x)G_z^{(11)}(x)d(x).
\end{aligned}$$

We obtain

$$\begin{aligned}
\langle u, G_z \rangle_{V_2^{11}[0,1]} &= u'(0)G_z'(0) + u'''(0)G_z'''(0) + u^{(5)}(0)G_z^{(5)}(0) + u^{(7)}(0)G_z^{(7)}(0) \\
&+ u^{(9)}(0)G_z^{(9)}(0) + u^{(10)}(0)G_z^{(10)}(0) + u^{(10)}(1)G_z^{(11)}(1) - u^{(10)}(0)G_z^{(11)}(0) \\
&- u^{(9)}(1)G_z^{(12)}(1) + u^{(9)}(0)G_z^{(12)}(0) + u^{(8)}(1)G_z^{(13)}(1) - u^{(8)}(0)G_z^{(13)}(0) \\
&- u^{(7)}(1)G_z^{(14)}(1) + u^{(7)}(0)G_z^{(14)}(0) + u^{(6)}(1)G_z^{(15)}(1) - u^{(6)}(0)G_z^{(15)}(0) \\
&- u^{(5)}(1)G_z^{(16)}(1) + u^{(5)}(0)G_z^{(16)}(0) + u^{(4)}(1)G_z^{(17)}(1) - u^{(4)}(0)G_z^{(17)}(0) \\
&- u'''(1)G_z^{(18)}(1) + u'''(0)G_z^{(18)}(0) + u''(1)G_z^{(19)}(1) - u''(0)G_z^{(19)}(0) \\
&- u'(1)G_z^{(20)}(1) + u'(0)G_z^{(20)}(0) + u(1)G_z^{(21)}(1) - u(0)G_z^{(21)}(0) \\
&- \int_0^1 u(x)G_z^{(22)}(x)d(x)
\end{aligned}$$

by integration by parts. Then, we have

$$\begin{aligned}
\langle u, G_z\rangle_{V_2^{11}[0,1]} &= u'(0)G_z'(0) + u'''(0)G_z'''(0) + u^{(5)}(0)G_z^{(5)}(0) + u^{(7)}(0)G_z^{(7)}(0) \\
&+ u^{(9)}(0)G_z^{(9)}(0) + u^{(10)}(0)G_z^{(10)}(0) + u^{(10)}(1)G_z^{(11)}(1) - u^{(10)}(0)G_z^{(11)}(0) \\
&- u^{(9)}(1)G_z^{(12)}(1) + u^{(9)}(0)G_z^{(12)}(0) - u^{(7)}(1)G_z^{(14)}(1) + u^{(7)}(0)G_z^{(14)}(0) \\
&- u^{(5)}(1)G_z^{(16)}(1) + u^{(5)}(0)G_z^{(16)}(0) - u'''(1)G_z^{(18)}(1) + u'''(0)G_z^{(18)}(0) \\
&- u'(1)G_z^{(20)}(1) + u'(0)G_z^{(20)}(0) - \int_0^1 u(x)G_z^{(22)}(x)d(x).
\end{aligned}$$

We have

$$G_z'(0) + G_z^{(20)}(0) = 0, \quad G_z'''(0) + G_z^{18}(0) = 0$$

$$G_z^{(5)}(0) + G_z^{16}(0) = 0, \quad G_z^{(7)}(0) + G_z^{14}(0) = 0$$

$$G_z^{(9)}(0) + G_z^{12}(0) = 0, \quad G_z^{10}(0) - G_z^{11}(0) = 0$$

$$G_z^{(11)}(1) = 0, \quad G_z^{(12)}(1) = 0, \quad G_z^{(14)}(1) = 0$$

$$G_z^{(16)}(1) = 0, \quad G_z^{(18)}(1) = 0, \quad G_z^{(20)}(1) = 0$$

Thus,we acquire

$$\langle u, G_z\rangle_{V_2^{11}[0,1]} = -\int_0^1 u(x)G_z^{(22)}(x)d(x) = u(z).$$

Therefore, we obtain

$$G_z(x) = \begin{cases} \sum_{i=1}^{11} c_i x^{i-1} & , \quad x \le z \\ \sum_{i=1}^{11} d_i x^{i-1} & , \quad x > z \end{cases}$$

by

$$G_z^{(22)}(x) = -\delta(x - z).$$

4 Conclusion

We obtained some reproducing kernel functions to find approximate solutions of the tenth-order boundary value problems in this paper. We need these reproducing kernel functions for applications of the reproducing kernel Hilbert space method.

References

[1] M. J. Iqbal , S. Rehman, A. Pervaiz, and A. Hakeem, Approximations for Linear Tenth-order Boundary Value Problems through Polynomial and Non-polynomial Cubic Spline Techniques, Proceedings of the Pakistan Academy of Sciences 52 (4): 389–396 (2015).

[2] Chandrasekhar. *Hydrodynamic and Hydromagnetic Stability.* Clarendon Press, Oxford, England (1961).

[3] Pervaiz, A., Zafar, Z. & M.O. Ahmad. A non-polynomial spline method for solving linear twelfth order boundary-value problems. *Pakistan Academy of Sciences* 51(2): 157-165 (2014).

[4] Omotayo, A.T. & O.M. Ogunlaran. A non-polynomial spline method for solving linear fourth order boundary-value problems. *International Journal of the Physical Sciences* 6: 3246-3254 (2011).

[5] Usmani, R.A. The use of quartic splines in the numerical solution of the fourth order boundary value problem. *Journal of Computational and Applied Mathematics* 44:187-199 (1992).

[6] Twizell, E.H. & A. Boutayeb. Numerical methods for eight, tenth and twelfth order eigenvalue problems arising in thermal instability. *Advances in Computational Mathematics* 2:407-436 (1994).

[7] Alberg, J.H. & T. Ito. A collocation method for two-point boundary value problems. *Mathematics of Computation* 29: 761-776 (1975).

[8] Siraj-ul-Islam, I.A. Tirmizi, Fazal-i-Haq & M.A. Khan. A non-polynomial spline approach to the solution of sixthorder BVPs. *Applied Mathematics and Computation* 195: 277-284 (2008).

[9] Papamichael, N. & A.J. Worsey. A cubic spline method for the solution of linear fourth-order boundary value problems. *Journal of Computational and Applied Mathematics* 7: 187-189 (1981).

[10] Siddiqi, S.S. & E.H. Twizell. Spline solution of linear twelfth-order boundary value problems. *Journal of Computational and Applied Mathematics* 78: 371-390 (1997).

[11] Siddiqi, S.S. & E.H. Twizell. Spline solution of linear tenth-order boundary value problems. *Journal of Computational and Applied Mathematics* 68: 345-362 (1998).

[12] Siddiqi, S.S. & Ghazala Akram. Spline solutions of tenth-order boundary value problems. *Journal of Computational and Applied Mathematics* 185: 115-127 (2007).

Reproducing kernel functions for the generalized Kuramoto-Sivashinsky equation

Ali Akgül[1,*], *Esra* Karatas Akgül[2,**], *Sahin* Korhan[3,***], and *Mustafa* Inc[4,****]

[1]Siirt University, Art and Science Faculty, Department of Mathematics, 56100, Siirt/TURKEY
[2]Siirt University, Faculty of Education, Department of Mathematics, 56100, Siirt/TURKEY
[3]Siirt University, Art and Science Faculty, Department of Mathematics, 56100, Siirt/TURKEY
[4]Firat University, Faculty of Science , Department of Mathematics, 23119, Elazig/TURKEY

Abstract. Reproducing kernel functions are obtained for the solution of generalized Kuramoto–Sivashinsky (GKS) equation in this paper. These reproducing kernel functions are valuable in the reproducing kernel Hilbert space method. They will be useful for interested researchers.

1 Introduction

We consider the following problem[1]:

$$\frac{\partial v}{\partial t} + v\frac{\partial v}{\partial x} + \alpha\frac{\partial^2 v}{\partial x^2} + \beta\frac{\partial^3 v}{\partial x^3} + \gamma\frac{\partial^4 v}{\partial x^4} = 0, \tag{1}$$

where α , β and γ are nonzero [2]. This equation has been obtained in the context of plasma instabilities, flame front propagation, and phase turbulence in reaction–diffusion system [3]. We obtain reproducing kernel functions to get approximate solutions of the following problem:

Example In this example, we consider the Kuramoto–Sivashinsky equation [5]

$$v_t + vv_x + v_{xx} + v_{xxxx} = 0,$$

with the initial condition [5]

$$v(x,0) = exp(-x^2),$$

and the boundary conditions

$$v(a,t) = 0, \quad v(b,t) = 0, \quad v_x(a,t) = 0, \quad v_x(b,t) = 0.$$

*e-mail: aliakgul00727@gmail.com
**e-mail: esrakaratas@siirt.edu.tr
***e-mail: sahin_3192@hotmail.com
****e-mail: minc@firat.edu.tr

 https://doi.org/10.1051/itmconf/20182201028

2 Reproducing kernel functions

We define $V_2^5[0,1]$ as

$$V_2^5[0,1] = \left\{ \begin{array}{ll} v, v', v'', v''', v^{(4)} & \text{are absolutely continuous functions,} \\ v^{(5)} \in L^2[0,1], & v(a) = v'(a) = v(b) = v'(b) = 0. \end{array} \right\}$$

We give

$$\begin{aligned} \langle v, S_z \rangle_{V_2^5[a,b]} &= v(a)S_z(a) + v'(a)S_z'(a) + v''(a)S_z''(a) + v'''(a)S_z'''(a) \\ &+ v^{(4)}(a)S_z^{(4)}(a) + \int_a^b v^{(5)}(t)S_z^{(5)}(t)d(t). \end{aligned}$$

We obtain

$$\begin{aligned} \langle v, S_z \rangle_{V_2^5[a,b]} &= v(a)S_z(a) + v'(a)S_z'(a) + v''(a)S_z''(a) + u'''(a)S_z'''(a) \\ &+ v^{(4)}(a)S_z^{(4)}(a) + v^{(4)}(b)S_z^{(5)}(b) - v^{(4)}(a)S_z^{(5)}(a) \\ &- v^{(3)}(b)S_z^{(6)}(b) + v^{(3)}(a)S_z^{(6)}(a) \\ &+ v''(b)S_z^{(7)}(b) - v''(a)S_z^{(7)}(a) - v'(b)S_z^{(8)}(b) \\ &+ v'(a)S_z^{(8)}(a) + v(b)S_z^{(9)}(b) - v(a)S_z^{(9)}(a) \\ &- \int_a^b v(t)S_z^{(10)}(t)d(t). \end{aligned}$$

We have

$$\begin{aligned} S_z(a) &= 0 \\ S_z(b) &= 0 \\ S_z'(a) &= 0 \\ S_z'(b) &= 0 \end{aligned}$$

by boundary conditions. Therefore, we obtain

$$\begin{aligned} \langle v, S_z \rangle_{V_2^5[a,b]} &= v''(a)S_z''(a) + v'''(a)S_z'''(a) + v^{(4)}(a)S_z^{(4)}(a) \\ &+ v^{(4)}(b)S_z^{(5)}(b) - v^{(4)}(a)S_z^{(5)}(a) - v^{(3)}(b)S_z^{(6)}(b) + v^{(3)}(a)S_z^{(6)}(a) \\ &+ v''(b)S_z^{(7)}(b) - v''(a)S_z^{(7)}(a) - \int_a^b v(t)S_z^{(10)}(t)d(t). \end{aligned}$$

We have

$$
\begin{aligned}
S_z''(a) - S_z^{(7)}(a) &= 0 \\
S_z'''(a) + S_z^{(6)}(a) &= 0 \\
S_z^{(4)}(a) - S_z^{(5)}(a) &= 0 \\
S_z^{(5)}(b) &= 0 \\
S_z^{(6)}(b) &= 0 \\
S_z^{(7)}(b) &= 0
\end{aligned}
$$

Therefore, we obtain

$$\langle v, S_z \rangle_{V_2^5[a,b]} = -\int_a^b v(t) S_z^{(10)}(t) d(t) = v(z).$$

Thus, we get

$$S_z^{(10)}(t) = -\delta(t - z).$$

When $t \neq z$

$$S_z^{(10)}(z) = 0.$$

Then, we acquire

$$
S_z(t) = \begin{cases} \sum\limits_{i=1}^{10} c_i(z) t^{i-1} &, \quad t \leq z, \\ \sum\limits_{i=1}^{10} d_i(z) t^{i-1} &, \quad t > z. \end{cases}
$$

3 Conclusion

In this paper, we obtained very useful reproducing kernel function for solving the generalized Kuramoto–Sivashinsky equation. This reproducing kernel function is very useful to apply reproducing kernel Hilbert space method.

References

[1] A.H. Khater, R.S. Temsah, Numerical solutions of the generalized Kuramoto–Sivashinsky equation by Chebyshev spectral collocation methods, Comput. Math. Appl. 56 1465-1472 (2008).

[2] Y. Kuramoto, T. Tsuzuki, Persistent propagation of concentration waves in dissipative media far from thermal equilibrium, Prog. Theor. Phys. 55 356-369 (1976).

[3] J. Rademacher J, R. Wattenberg, Viscous shocks in the destabilized Kuramoto–Sivashinsky, J. Comput. Nonlinear Dynam. 1 336-347 (2006).

[4] R. Conte R, Exact solutions of nonlinear partial differential equations by singularity analysis, in: Lecture Notes in Physics, Springer, Berlin, 2003, pp. 1-83.

[5] M. Lakestani, M. Dehghan, Numerical solutions of the generalized Kuramoto-Sivashinsky equation using B-spline functions, Applied Mathematical Modelling 36 605-617 (2012).

[6] F. Genga, F. Shenb, Solving a Volterra integral equation with weakly singular kernel in the reproducing kernel space, Mathematical Sciences, Vol. 4, No. 2 159-170 (2010).

ITM Web of Conferences **22**, 01029 (2018) https://doi.org/10.1051/itmconf/20182201029
CMES-2018

Example of A Kinetic Mathematical Modeling in Food Engineering

Özlem ERTEKİN[1,*]

[1] Department of Food Engineering, Munzur University, Tunceli, Turkey

Abstract. Mathematical modeling of biochemical, chemical reaction processes facilitates understanding. The kinetics of these reaction processes can be analyzed mathematically and kinetics are presented as systems of differential equations. Mathematical model of a reaction kinetic is studied in this study. Bernoulli-Sub equation function method is used in this study. This example can be new model for food engineering applications.

1 Introduction

Reaction kinetic is a science that examines the fast and mechanisms of reactions. Mechanism of reaction is defined as the detailed description of the reaction based on the behavior of atoms, molecules and ions. Many reactions occur with mechanisms involving multiple steps. Various chemical or biochemical reactions occur during the processing and storage of food. Food quality can be lost with these reactions. The aim is to obtain the good quality product by ensuring these reactions at the lowest level. Knowledge the rate and mechanism of reactions allows for the preparation of the good condution in performing any food process and establishment of storage conditions [1].

About a mathematical reference on chemical kinetics see, for instance, [2] and for food reaction technology reference see, for example, [3]. About references on mathematical models, see [4, 5, 6].

According to [7] mathematical modeling is used in cases where experimental studies are limited or impossible. Mathematical modeling has widely potential for application in food processing research and food industry in Turkey. According to another explanation, the expression of models inactivation of microorganisms in food is important because these models allow later estimation in similar cases without experiment [8].

Mathematical modeling can be used in areas different food engineering such as food chemistry, food microbiology, food processing. Recently, some scientists have focused on food engineering in mathematical modeling. For example, quality properties of traditional fermented sausage including total mesophilic aerobic bacteria, lactic acid bacteria, total *Enterobacteriaceae* and *Staphylococcus*/*Micrococcus* counts and other quality features were studied during storage duration, and kinetic model estimation for the changes in these quality characteristics were applied [9].

Moreover, nowadays mathematical modeling is essential to determine the growth kinetics of microorganisms and it is significant that can determine the temperature the relationship in perishable foods.

There are many different models used to symbolize from various real world problems to specific case studies [10,11,15].

* Corresponding author: oertekin@munzur.edu.tr

This report is arranged as follows: In section 2, we describe the Bernoulli Sub-Equation function method (BSEFM), In section 3, we consider the following kinetic model defined as

$$\frac{d[B]}{dt} = k_+[A][B] - k_-[B]^2, \tag{1a}$$

where $[A]$ is concentration and $[B]$ is concentration, k is the rate constant, k_+ is forward side, k_- is back side, t is time [12].

Moreover, for simplicity, we can rewrite Eq.(1a) as following

$$\frac{dU}{dt} = CU - PU^2, \tag{1b}$$

where $U = [B]$, is dependent variable, $C = k_+[A]$ and $P = k_-$ are real constants with non- zero.

We apply the tool to the model which ordinary differential equation. Finally, we introduce a general conclusion about the results obtained by BSEFM in section 4.

2 Fundamental Properties of Bernoulli-Sub-Equation Function Method

An approach to the mathematical models including partial differential equations will be exhibited in this sub- section of study. The steps of the this technique can be taken as below [13,14].

Step 1: We theorize the partial differential equation in two variables such as x, t and a dependent variable u;

$$P(u_x, u_t, u_{xx}, u_{xt} \dots) = 0, \tag{2}$$

and take the wave transformation

$$u(x,t) = U(\eta),\ \eta = x - ct, \tag{3}$$

where $c \neq 0$. Substituting Eq.(3) into Eq.(2), it gives us the following nonlinear ordinary differential equation ;

$$N(U, U', U'', U''', \dots) = 0, \tag{4}$$

where $U = U(\eta)$, $U' = \frac{dU}{d\eta}$, $U'' = \frac{d^2U}{d\eta^2}$, ...

Step 2: Taking the trial solution for Eq. (4) as:

$$U(\eta) = \sum_{i=0}^{n} a_i F^i(\eta) = a_0 + a_1 F(\eta) + a_2 F^2(\eta) + \dots\dots + a_n F^n(\eta), \tag{5}$$

where

$$F' = \alpha F(\eta) + \beta F^M(\eta), \tag{6}$$

where $\alpha \neq 0$, $\beta \neq 0$, $M \in R\backslash\{0,1,2\}$, and $F(\eta)$ is Bernoulli differential polynomial. Substituting above relations into Eq.(4), we acquire an equation of polynomial $\Omega(F(\eta))$ of $F(\eta)$;

$$\Omega(F(\eta)) = \rho_s F^s(\eta) + \dots + \rho_1 F(\eta) + \rho_0 = 0. \tag{7}$$

According to the balance principle, we can get values of and M and n.

Step 3: Let us assume the coefficients of $\Omega(F(\eta))$ all be zero, we will obtain an algebraic equations system:

$$\rho_i=0,\ i=0,\ldots,s. \tag{8}$$

Solving this system, we will determine the values of $a_0,\ldots a_n$.

Step 4: When we solve nonlinear Bernoulli differential equation that is Eq. 6, we obtain the following to situations according to α and β;

$$F(\eta)=[\frac{-\beta}{\alpha}+\frac{E}{e^{\alpha(M-1)\eta}}]^{\frac{1}{1-M}},\ \alpha\neq\beta, \tag{9}$$

$$F(\eta)=[\frac{(E-1)+(E+1)\tanh(\frac{\alpha(1-M)\eta}{2})}{1-\tanh(\frac{\alpha(1-M)\eta}{2})}]^{\frac{1}{1-M}},\ \alpha=\beta,\ E\in R. \tag{10}$$

Using a complete discrimination system for polynomial, we obtain the solutions to Eq.(4) via some computer programming and classify the exact solutions to Eq.(4). For a better comprehension of findings found in this way, we can plot two and three imensional surfaces of solutions by taking into consideration favorable parameter values.

3 Application of BSEFM

In this section, it has succesfully applied the BSEFM to the equation for getting some new complex travelling wave solutions. If we consider Eq.(1b), as a simplicity, $[B]=U$, we co-write following.

$$U'-CU+PU^2=0. \tag{11}$$

Applying the homogeneous blancing technique between the highest derivative U' and the highest nonlinear term U^2, yields the following relationship among n and M;

$$M=n+1,\quad n,M\in Z^+. \tag{12}$$

If we take as n=3 and M=4 in Eq. (5), we can write following equation;

$$U=a_0+a_1F+a_2F^2+a_3F^3. \tag{13}$$

For the first derivation of U, we can write following equation;

$$U'=a_1\alpha F+a_1\beta F^4+2a_2\alpha F^2+2a_2\beta F^5+3a_3\alpha F^3+3a_3\beta F^6, \tag{14}$$

where $a_3\neq 0$, $\alpha\neq 0$, $\beta\neq 0$. When we put Eq.(13,14) in Eq.(11), we obtain an algebraic equation including various power degree of F. When we consider to zero all coefficients of F, we can find an algebraic system. Solving this system with the help of computer programming, we can detect many values of parameters, If we choose the suitable values of such parameters, we detect the following solutions;

Case 1: For α≠β, it can be considered the following coefficients;

$$a_1 = 0, a_2 = 0, \alpha = -\frac{Pa_0}{3}, \beta = -\frac{Pa_3}{3}, \ C = Pa_0$$

Substituting these coefficients in Eq. (13) along with Eq.(9), we obtain the following exponential function solution for Eq.(1b);

$$U(t) = a_\circ + \frac{a_3}{e^{pta_0}EE - \frac{a_3}{a_0}}, \tag{15}$$

where EE, a_0, a_3, P are real constants with non-zero.

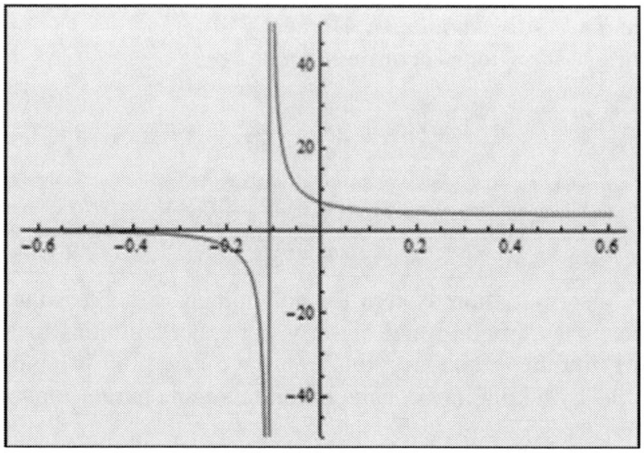

Fig.1. The 2D surface of Eq.(15) under the values of $P = 2$, $EE = 3$, $a_0 = 4$, $a_3 = 5$, -0.6<t<0.6

4 Conclusions

In this paper, the BSEFM is applied to the nonlinear kinetic model equation. We successfully obtained exponential function solution. We presented the 2D graphs of the obtained solution with the help of some powerful software program.

It has been observed that this solution can be converted into the exist result introduced to the literature when we use exponential function properties.

This result comes from only Case-1. Moreover, if we chose as M=5 and n=4.

$$U = a_\circ + a_1F + a_2F^2 + a_3F^3 + a_4F^4.$$

Where

$$F' = \alpha F + \beta F^5.$$

This will gives more different solution. Finally, if we consider M=6, n=5, then, we can write another different solution form as following;

$$U = a_\circ + a_1F + a_2F^2 + a_3F^3 + a_4F^4 + a_5F^5.$$

With

$$F' = \alpha F + \beta F^6.$$

The values of M and n is infinitive. This property is belong to the BSEFM. We can obtain some other new traveling wave solutions to the nonlinear kinetic model equation. As far as we know, the application of BSEFM to the model considered in this paper has not been submitted to the literature in advance. Therefore, this method is a powerful and efficient mathematical tool that can be used to handle various nonlinear mathematical models.

References

1. N. Aktaş, Gıda Teknolojisinde Reaksiyon Kinetiği. Atatürk Üniversitesi, Ziraat Fakültesi Ders Yayınları (2008).
2. D.A. Beard, H. Qian, Chemical biophysics: quantitative analysis of cellular systems. Cambridge University Press, Cambridge (2008).
3. M. Earle, R. Earle, Fundamentals of Food Reaction Technology. Leatherhead Food International. Leatherhead, England (2003).
4. J.D. Murray,Mathematical biology, 3rd edn,Springer, Berlin (2003).
5. R. Plonsey, R.C. Barr, Bioelectricity: a quantitative approach, 2nd edn. Kluwer, New York (2000).
6. J. Keener, J. Sneyd, Mathematical physiology, 2nd edn, Springer, New York (2009).
7. Y.O. Devres, M. Pala, Gıda Sanayiinde Matematiksel Modellemenin Önemi ve Uygulama Alanları. GIDA, 18(3) 173-181 (1993).
8. N. Çoksöyler, Gıdalarda Mikroorganizmaların İnaktivasyonunun Modellenmesi. Türkiye 9. Gıda Kongresi, Bolu, 24-26 Mayıs (2006).
9. S.G. Özkal, H. Ercoşkun, Kinetic Modeling of Quality Aspects of Fermented Sausage (Sucuk) During Storage, Kafkas Üniversitesi Veteriner Fakültesi Dergisi, 23(2),195-200 (2017).
10. F. Düşünceli, M. Şan, B. Bulur, New Exponential and Complex Traveling Wave Solutions to the Konopelchenko-Dubrovsky Model, 3rd International Conference on Computational Mathematics and Engineering Sciences- (CMES2018), Girne/Cyprus, May 04-06 (2018).
11. T.A. Sulaiman, A. Yokus, N. Gulluoglu, H.M. Baskonus, Regarding the Numerical and Stability Analysis of the Sharma-Tosso-Olver Equation, 3rd International Conference on Computational Mathematics and Engineering Sciences- (CMES2018), Girne/Cyprus, May 04-06(2018).
12. J.W. Cain, Chemical Reaction Kinetics: Mathematical Underpinnings. Molecular Life Sciences,Springer, NewYork (2014).DOI 10.1007/978-1-4614-6436-5_564-1.
13. H. Bulut, G. Yel, H.M. Baskonus, An Application of Improved Bernoulli Sub-Equation Function Method to The Nonlinear Time-Fractional Burgers Equation. Turk. J. Math. Comput. Sci., 5,1-7 (2016)
14. H. Bulut, S.S. Atas, H.M. Baskonus, Some Novel Exponential and Complex Structural Properties of the Fisher Equation Arising in Mathematical Bioscience. CMES2017,ITM Web of Conferences, 13,01017 (2017).
15. E.Yuce, A. Ö. Tarakçioglu, Attitudes of Computer Engineering Department Students Towards ESP Courses Integrated to Foreign Language Courses: Tunceli University Case, Bilim ve Gençlik Dergisi, 1(1), 47-57(2013).

Some New Characterizations Of Parallel Translation Surface According To Bishop Frame With Timelike M_1 In Minkowski 3-Space

Gülden Altay Suroğlu[1]

[1]Department of Mathematics, University of Fırat, Elazığ, Turkey

Abstract. In this paper we consider parallel translation surfaces, which are generated by spacelike curves, according to Bishop frame with timelike M_1 in Minkowski 3- space. Then, we obtain some characterizations of these surface.

1 Introduction

Creation of paralel surfaces is useful in design and manufacture. A surface M^r is parallel to M if points of M^r are at a constant distance along the normal from the surface M. So, there are infinite numbers of parallel surfaces, [8].

Kızıltuğ and Yaylı studied timelike curves on timelike parallel surfaces in Minkowski 3-spaces $E_1{}^3$. Then, they gave some characterization for its image curve in, [3].

Unluturk and Ekici obtain paralel surfaces of timelike ruled surfaces which are developable are timelike ruled Weingarten surfaces. Then, some properties of that kind paralel surfaces are obtained in Minkowski 3-space in [7].

Safiulina gave a complete classification of the existence and geometry of such two-dimensional Riemannian submanifolds (surfaces). Moreover, it is shown that in E_s^n with s>0 do exist not totally geodesic minimal semiparallel space-like surfaces, [6].

In [1], Calvaruso and Van der Veken completely classified surfaces with parallel second fundamental form in all non-symmetric homogeneous Lorentzian three manifolds. In [2], Fukui and Hasegawa investigate singularities of all parallel surfaces to a given regular surface.

2 Fundamental Properties of the Method

Let $\alpha: I \to E_1{}^3$ be a non geodesic spacelike curve on the $E_1{}^3$ parameterized by arc length. Let {T,N,B} be the Frenet frame. Then, for spacelike curves we have the following Frenet formulas:

Corresponding Author : guldenaltay23@hotmail.com

$$\frac{d}{ds}\begin{bmatrix}\mathbf{T}(s)\\ \mathbf{N}(s)\\ \mathbf{B}(s)\end{bmatrix}=\begin{bmatrix}0 & \kappa(s) & 0\\ -\varepsilon\kappa(s) & 0 & \tau(s)\\ 0 & \tau(s) & 0\end{bmatrix}\begin{bmatrix}\mathbf{T}(s)\\ \mathbf{N}(s)\\ \mathbf{B}(s)\end{bmatrix}, \tag{1}$$

where $\kappa \neq 0$ and $\tau \neq 0$ and

$$g(T,T) = 1, g(N,N)=\varepsilon=\pm 1, g(B,B)=1.$$

The Bishop frame is an option when the spacelike curve has a vanishing second derivative. The Bishop frame is

$$\frac{d}{ds}\begin{bmatrix}\mathbf{T}(s)\\ \mathbf{M}_1(s)\\ \mathbf{M}_2(s)\end{bmatrix}=\begin{bmatrix}0 & k_1(s) & -k_2(s)\\ k_1(s) & 0 & 0\\ k_2(s) & 0 & 0\end{bmatrix}\begin{bmatrix}\mathbf{T}(s)\\ \mathbf{M}_1(s)\\ \mathbf{M}_2(s)\end{bmatrix}, \tag{2}$$

where

$$g(T,T) = -g(M_1, M_1)=g(M_2, M_2)=1.$$

For timelike curves, we have the following Frenet formulas:

$$\frac{d}{ds}\begin{bmatrix}\mathbf{T}(s)\\ \mathbf{N}(s)\\ \mathbf{B}(s)\end{bmatrix}=\begin{bmatrix}0 & \kappa(s) & 0\\ \kappa(s) & 0 & \tau(s)\\ 0 & -\tau(s) & 0\end{bmatrix}\begin{bmatrix}\mathbf{T}(s)\\ \mathbf{N}(s)\\ \mathbf{B}(s)\end{bmatrix} \tag{3}$$

Then, the Bishop frame is expressed as

$$\frac{d}{ds}\begin{bmatrix}\mathbf{T}(s)\\ \mathbf{M}_1(s)\\ \mathbf{M}_2(s)\end{bmatrix}=\begin{bmatrix}0 & k_1(s) & k_2(s)\\ k_1(s) & 0 & 0\\ k_2(s) & 0 & 0\end{bmatrix}\begin{bmatrix}\mathbf{T}(s)\\ \mathbf{M}_1(s)\\ \mathbf{M}_2(s)\end{bmatrix} \tag{4}$$

Here, the set $\{T, M_1, M_2\}$ is called Bishop trihedra, k_1 and k_2 as Bishop curvatures, $\tau(s)=\theta'(s)$ and $\kappa(s)=\sqrt{(|k_2^2-k_1^2|)}$.

3 Parallel Translation Surfaces in E_1^3

Let $\alpha(u)$ and $\beta(v)$ be spacelike curves and $\theta(u,v)$ be a translation surface which is parametrized as

$$\theta(u,v)=\alpha(u)+\beta(v). \tag{5}$$

Then, parallel surfaces of the translation surface is

$$\phi(u,v)=\theta(u,v)+aU(u,v) \tag{6}$$

where a is a constant and $U(u,v)$ is unit normal vector field of the translation surface $\theta(u,v)$. So, it parameterized as

$$\phi(u,v)=\alpha(u)+\beta(v)+\frac{a}{\sin\alpha}(T_\alpha\times T_\beta) \tag{7}$$

where α is the constant angle between unit tangent vector field of curves T_α and T_β.

Theorem 1. Let $\alpha(u)$ and $\beta(v)$ be spacelike curves where M_1^α and M_1^β are timelike vector fields in Minkowski three space. Then if $\phi(u,v)$ is minimal,

$$E(-a^2(k_1^{\beta 2}-k_2^{\beta 2})^2+k_1^{\beta}<M_1^{\beta},U>-k_2^{\beta}<M_2^{\beta},U>+\frac{ak_1^{\beta\,\prime}}{\sin\alpha}<W,U>-\frac{ak_2^{\beta\,\prime}}{\sin\alpha}<X,U>)$$
$$-2F(\frac{ak_1^{\beta}}{\sin\alpha}(k_1^{\beta}<Q,U>-k_2^{\beta}<P,U>)+\frac{ak_2^{\alpha}}{\sin\alpha}(k_1^{\beta}<R,U>-k_2^{\beta}<S,U>) \quad (8)$$
$$+G(-a^2(k_1^{\alpha 2}-k_2^{\alpha 2})^2+k_1^{\alpha}<M_1^{\alpha},U>-k_2^{\alpha}<M_2^{\alpha},U>$$
$$+\frac{ak_1^{\alpha\,\prime}}{\sin\alpha}<W,U>-\frac{ak_2^{\alpha\,\prime}}{\sin\alpha}<X,U>)=0.$$

Proof. If we take derivatives of the parallel surface of the translation surface according to u and v, then from bishop equations, we have

$$\phi_u(u,v)=T_\alpha+\frac{ak_1^{\alpha}}{\sin\alpha}(M_1^{\alpha}\times T_\beta)-\frac{ak_2^{\alpha}}{\sin\alpha}(M_2^{\alpha}\times T_\beta)\,. \quad (9)$$

Similarly,

$$\phi_v(u,v)=T_\beta+\frac{ak_1^{\beta}}{\sin\alpha}(T_\alpha\times M_1^{\beta})-\frac{ak_2^{\beta}}{\sin\alpha}(T_\alpha\times M_2^{\beta}) \quad (10)$$

The compenents of the first fundamental form of $\phi(u,v)$ are

$$E=1-ak_1^{\alpha}<U,M_1^{\alpha}>+ak_2^{\alpha}<U,M_2^{\alpha}>+\frac{1}{\sin^2\alpha}(ak_1^{\alpha 2}(\|M_1^{\alpha}\|^2-<M_1^{\alpha},T_\beta>)$$
$$+ak_2^{\alpha 2}(\|M_2^{\alpha}\|^2-<M_2^{\alpha},T_\beta>)), \quad (11)$$

$$F=\cos\alpha+\frac{a^2k_1^{\alpha}k_1^{\beta}}{\sin^2\alpha}(<M_1^{\alpha},T_\alpha><M_1^{\beta},T_\beta>-\sin\alpha<M_1^{\alpha},M_1^{\beta}>)$$
$$-\frac{a^2k_1^{\alpha}k_2^{\beta}}{\sin^2\alpha}(<M_1^{\alpha},T_\alpha><M_2^{\beta},T_\beta>-\sin\alpha<M_1^{\alpha},M_2^{\beta}>)$$
$$-\frac{a^2k_2^{\alpha}k_1^{\beta}}{\sin^2\alpha}(<M_2^{\alpha},T_\alpha><M_1^{\beta},T_\beta>-\sin\alpha<M_2^{\alpha},M_1^{\beta}>) \quad (12)$$
$$-\frac{a^2k_2^{\alpha}k_2^{\beta}}{\sin^2\alpha}(<M_2^{\alpha},T_\alpha><M_2^{\beta},T_\beta>-\sin\alpha<M_2^{\alpha},M_2^{\beta}>)$$

$$G=1-ak_1^{\beta}<U,M_1^{\beta}>+ak_2^{\beta}<U,M_2^{\beta}>+\frac{1}{\sin^2\alpha}(ak_1^{\beta 2}(\|M_1^{\beta}\|^2-<M_1^{\beta},T_\alpha>) \quad (13)$$
$$+ak_2^{\beta 2}(\|M_2^{\beta}\|^2-<M_2^{\beta},T_\alpha>)).$$

The unit vector field of the parallel translation surface is

$$U=\frac{1}{\sqrt{EG}\sin\gamma}(\phi_u\times\phi_v) \quad (14)$$

where γ is the angle between vector fields ϕ_u and ϕ_v.

Then, components of the second fundamental form of the $\phi(u,v)$ are

$$h_{11} = -a^2(k_1^{\alpha 2} - k_2^{\alpha 2})^2 + k_1^{\alpha} < U, M_1^{\alpha} > -k_2^{\alpha} < U, M_2^{\alpha} >$$
$$+\frac{ak_1^{\alpha}{}'}{\sin\alpha} < U, W > -\frac{ak_2^{\alpha}{}'}{\sin\alpha} < U, X >, \tag{15}$$

$$h_{12} = \frac{ak_1^{\alpha}}{\sin\alpha}(k_1^{\beta} < Q, U > -k_2^{\beta} < P, U >) + \frac{ak_2^{\alpha}}{\sin\alpha}(k_1^{\beta} < R, U > -k_2^{\beta} < S, U >), \tag{16}$$

$$h_{22} = -a^2(k_1^{\beta 2} - k_2^{\beta 2})^2 + k_1^{\beta} < U, M_1^{\beta} > -k_2^{\beta} < U, M_2^{\beta} >$$
$$+\frac{ak_1^{\beta}{}'}{\sin\alpha} < U, W > -\frac{ak_2^{\beta}{}'}{\sin\alpha} < U, X >, \tag{17}$$

Where

$$M_1^{\alpha} \times T_{\beta} = W,\ M_2^{\alpha} \times T_{\beta} = X,\ T_{\alpha} \times M_1^{\beta} = Y,\ T_{\alpha} \times M_2^{\beta} = Z,\ M_1^{\alpha} \times M_1^{\beta} = Q,$$
$$M_1^{\alpha} \times M_2^{\beta} = P,\ M_2^{\alpha} \times M_1^{\beta} = R,\ M_2^{\alpha} \times M_2^{\beta} = S. \tag{18}$$

Then the proof is complete.

Corollary 1. Let α(u) be timelike and β(v) be spacelike curves where M_1^{β} is timelike vector field . Then if parallel surfaces of the translation surface $\tilde{\phi}(u,v)$ is maximal,

$$\tilde{E}(a^2(k_1^{\beta 2} - k_2^{\beta 2})^2 + k_1^{\beta} < M_1^{\beta}, \tilde{U} > -k_2^{\beta} < M_2^{\beta}, \tilde{U} > + \frac{ak_1^{\beta}{}'}{\cosh\alpha} < W, \tilde{U} > - \frac{ak_2^{\beta}{}'}{\cosh\alpha} < X, \tilde{U} >)$$
$$-2\tilde{F}(\frac{ak_1^{\alpha}}{\sin\alpha}(k_1^{\beta} < Q, \tilde{U} > -k_2^{\beta} < P, \tilde{U} >) + \frac{ak_2^{\alpha}}{\cosh\alpha}(k_1^{\beta} < R, \tilde{U} > -k_2^{\beta} < S, \tilde{U} >) \tag{19}$$
$$+\tilde{G}(a^2(k_1^{\alpha 2} - k_2^{\alpha 2})^2 + k_1^{\alpha} < M_1^{\alpha}, \tilde{U} > -k_2^{\alpha} < M_2^{\alpha}, \tilde{U} >$$
$$+\frac{ak_1^{\alpha}{}'}{\cosh\alpha} < W, \tilde{U} > - \frac{ak_2^{\alpha}{}'}{\cosh\alpha} < X, \ddot{U} >) = 0$$

where $\tilde{E}$, $\tilde{F}$ and $\tilde{G}$ are components of the first fundamnetal form.

References

1. Giovanni Calvaruso and Joeri Van der Veken, Parallel Surfaces in Three Dimensional Lie Groups, Taiwanese Journal of Mathematics, **14**, 223-250 (2010)
2. Toshizumi Fukui, Masaru Hasegawa, Singularities of Parallel Surfaces, Tohoku Mathematical Journal, **64**, 387-408 (2012)
3. Sezai Kızıltuğ, Yusuf Yaylı, Timelike Curves on Timelike Parallel Surfaces in Minkowski 3-space $E_1{}^3$, Mathematica Aeterna, **2**, 689 - 700 (2012)
4. Talat Körpınar, Essin Turhan, Spacelike Biharmonic New Type B-Slant Helices According to Bishop Frame in the Lorentzian Heisenberg Group H^3, General Mathematics Notes, **10**,36–42 (2012)
5. Talat Körpınar, Essin Turhan, Parallel Surfaces to Normal Ruled Surfaces of General Helices in the Sol Space Sol^3, Boletim Sociedade Paranaense de. Matematica, **31**, 245-253 (2013)

6. Elena Safiulina, Parallel and Semi- parallel Space-like Surfaces in Pseudo-Euclidean Spaces,Proc.Estonian Acad. Sci. Phys.Math., **50**, 16-33, (2001)
7. Yasin Unluturk, Cumali Ekici, On Parallel Surfaces of Timelike Ruled Weingarten Surfaces, Balkan Journal of Mathematics, **1**, 72-91 (2003)
8. Dae Woon Yoon, Some properties of parallel surfaces in Euclidean 3-spaces, Honam Mathematical Journal, **30**, 637-644 (2008)
9. Süha Yılmaz, Bishop spherical images of a spacelike curve in Minkowski 3-space, International Journal of the Physical Sciences, **5**, 898-905 (2010)

The Reverse Operation Of Knot Digraph Notation

Tamer UĞUR[1] **Ceren Sultan ELMALI**[2*], **Ferit YALAZ**[3]

[2]*Department of Mathematics, Faculty of Science and Arts, Atatürk University, 25240 Erzurum TURKEY*

[2]*Department of Mathematics, Faculty of Science and Arts, Erzurum Technical University, 25100 Erzurum-TURKEY*

[3]*Graduate School of Natural and Applied Science, Erzurum Technical University, 25100 Erzurum-TURKEY*

Abstract. It is well known that bitopologies associated with knot digraphs is finded by using knot digraph notation. In this work, we have developed a method that we called reverse of knot digraph notation to find out which knot belongs to when a bitopology associated with the knot is given.

1.Introduction

The notion of bitopological space (X, τ_1, τ_2), that is, a set X equipped with two arbitrary topologies τ_1, τ_2 was first formulated by J.C.Kelley [4] Girija and Plakkat showed that bitopology is get from a digraph which has orientation by using quasi-pseudo metric entitled with 'Bitopological space associated with digraphs' [3] in 2013. A quasi pseudo metric on nonempty set X is a nonnegative real valued function ρ on $X\times X$ such that $\rho(x,x) = 0, \forall x \in X$ and $\rho(x,z) \leq \rho(x,y) + \rho(y,z), \forall x,y,z \in X$associated with ρ there is another quasi pseudometric q such that $\rho(x,y) = q(y,x)$.In this study the authors is called digraph if the vertex of graph is reachable from edge of it. Let be V the vertex set of a digraph. The function $p: V\times V \to \mathbb{R}$ and $q: V\times V \to \mathbb{R}$ defined by

$$p(x,y) = \begin{cases} 0, & If\ x\ is\ reachable\ from\ y \\ 1, & otherwise \end{cases}, \quad q(x,y) = \begin{cases} 0, & If\ y\ is\ reachable\ from\ x \\ 1, & otherwise \end{cases}$$

respectively. These functions are quasi-pseudo metrics on V and induces a unique bitopology on V with

$S_p(x,\varepsilon) = \{y: p(x,y) < \varepsilon\}$ and $S_q(x,\varepsilon) = \{y: q(x,y) < \varepsilon\}$ as a basis. The collection of all open ρ-balls forms a basis for a topology on X.Let this topology be denoted by τ_1. Similarly the collection of all open q-balls forms a basis for a topology on X.Let this topology be denoted by τ_2.Thus we get the bitopology (X, τ_1, τ_2). A bitopological space (X, τ_1, τ_2) is a set X with two topologies, τ_1 and τ_2, on it.

We know that this method can be applied for the oriented knot graph which is obtained form regular diagram of a knot. This process is called knot digraph notation. We know that a bitopology matched with any knot by using knot digraph notation. In this work, we have developed an algorithm. The knots that correspond to bitopological spaces will be found with the aid of this algorithm. We called this algorithm as 'reverse operation of knot digraph notation'.

2.Knot and Knot Graphs

A *knot* is a simple closed curve obtained by embedding of the circle S^1 into R^3 (or S^3) [1]. A regular projection of knots or links on a 2- dimensional plane divides the plane into several regions, each of which is homeomorphic to an open disk. It is a frequently used method to separate these regions (domains) into two classes, white-black or shaded-unshaded or shadow-unshadow regions, in the study of the knot theory.

A regular projections of knot K divides plane into $n + 2$ regions where n is the number of double point A (A_1, A_2, ... ,A_n). The Figure 1 show that the number of double point of Trefoil knot is 3. Therefore The number of regions is 5.

* Corresponding Author: ceren.elmali@erzurum.edu.tr

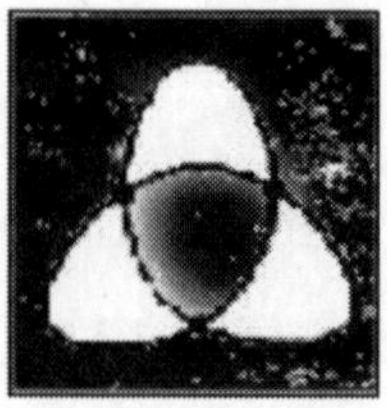

Figure 1. The number of double point of Trefoil knot

Starting with the outer most region, we can color the regions either black or white where neighboring regions are never this same color. We are labeled white and black regions, respectively, such that $\alpha = \{\alpha_1,\alpha_2,\ldots,\alpha_k\}$ and $\beta = \{\beta_1,\beta_2,\ldots,\beta_l\}$. The point at infinity is contained in a region of black. Take points $P_i \in \alpha_i$ $(i = 1,2,\ldots,k)$ and connect this points by n- non-intersection arcs $d_1,d_2,\ldots,d_n$ in such a way each d_t corresponds to A_t $(t = 1,2,\ldots,n)$ and P_i and P_j connected by d_t, if and only if α_i, α_j have a common double point A_t on their boundaries. In this way, we obtain from regular projection of the knot a plane graph G. From same consideration about class β, we get another graph G^d. We call this the dual graph of G. This method is called as Tait's method. Figure 2(a) and 2(b) show that graph and dual graph of Trefoil knot, respectively,

(a) (b)

Figure 2. Graph and dual graph of Trefoil knot

The regions of class α in all can be considered as a projection of a surface spanning K which is twisted 180° at each double point of projection. From this consideration, we define the sign of d_t in such a way that it is + or – according as the surface is twisted at A_t as the right screw or the left one, respectively (Figure 3).

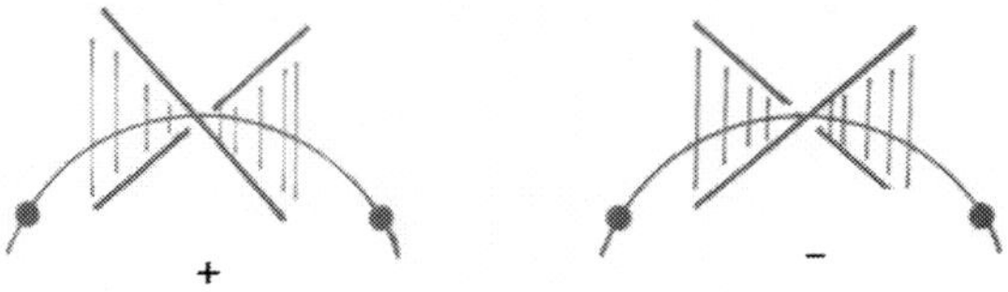

Figure 3. + or – according as the surface is twisted at A_t

It is well known that every deformation of a knot is equivalent to a combination of the fundamental deformation (Reidemeister Movies) Ω_1, Ω_2 and Ω_3 in [7].

Figure 4. Reidemeister Movies

These fundamental deformations can be translated into the operations on graphs as follows [8].

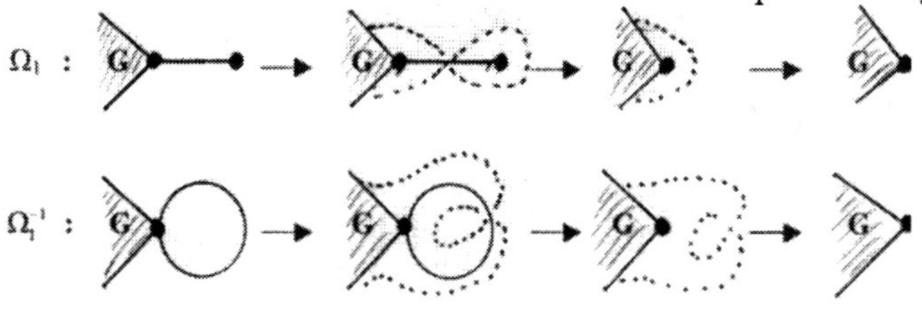

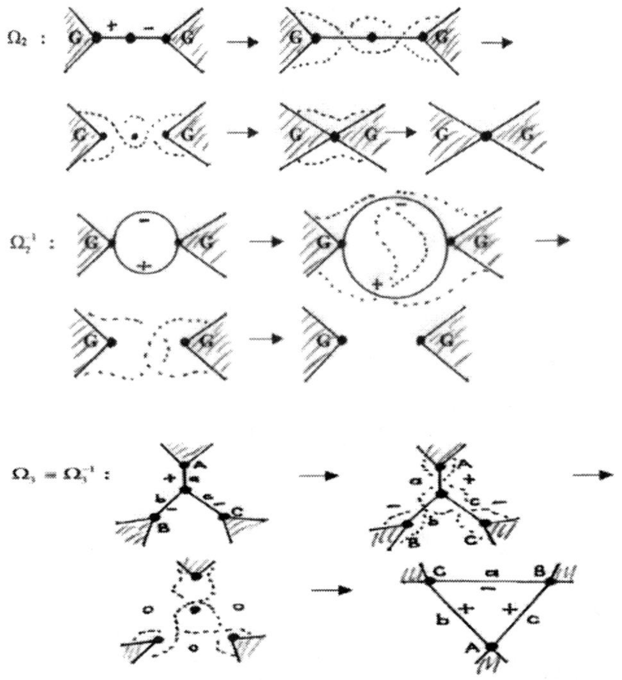

Figure 5.Graph Deformations

We recommend [2,5,6] for details about this section.

3. Knot digraph Notation

Firstly, signed graph is obtained from regular diagram of a knot by using Tait methods explained in Section 2.Secondly, orientation is given depending on the signed of the signed knot graph as Figure 6.If the signed of edge is +, orientation of this edge is given from a to b. If the signed of edge is -, orientation of this edge is given from b to a. So we can get two directed graphs, when we consider graph and dual graph of a knot. We called this directed graph as knotdigraph.

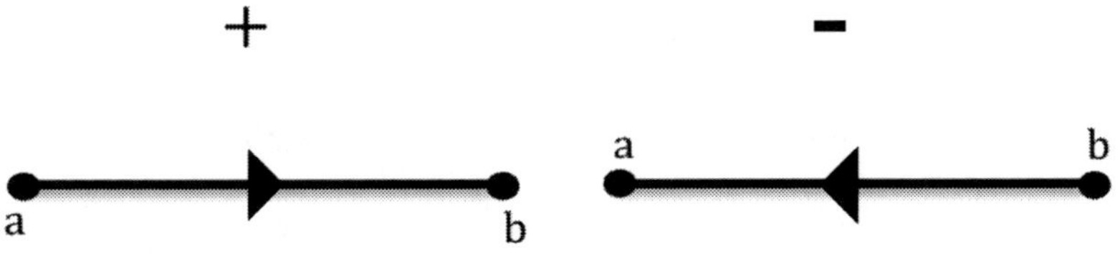

Figure 6. Orientation of knot graph depend on the signed of the signed knot graph

After that, every vertices of knot digraph are labelled by lowercases. Finally, the base is obtained from knot digraph by using quasi-pseudo metric for bitopology. Thus, every knot is matched with bitopology obtained by using this method.We called this method as Knot digraph Notation

4. The Knots Obtaining with the Help of Bitopology

Definition 4.1: Let (X, τ_1, τ_2) is a bitopological space. The set τ_1 has the largest number of elements, except the X set, is called the strong set of τ_1 and is denoted by $\overline{\tau_1}$.Similiarly, the set which is the largest number of elements, except X is called the strong set of τ_2 and is denoted by $\overline{\tau_2}$.

Definition 4.2: Let (X, τ_1, τ_2) is a bitopological space. The set of τ_1 with at least one element excluding the null set is called the weak set of τ_1 and is denoted by $\underline{\tau_1}$. Similiarly, the set is called the weak set of τ_2 with at least one element excluding the null set, denoted by $\underline{\tau_2}$.

The number of elements of $\overline{\tau_1}$,the strong set of τ_1 topology, is the corner number (node) of the knot digraph corresponding to the bitopological space of (X, τ_1, τ_2). The number of elements of $\overline{\tau_1}$ is denoted by $s(\overline{\tau_1}) = V$. The number of elements of the set $\underline{\tau_2}$, the weak set of $\tau_2$topology, shows how many impairment are made between digits in the knot digraph corresponding to the bitopology (X, τ_1, τ_2). We denote the number of elements of $\underline{\tau_2}$ by $s\left(\underline{\tau_2}\right) = B$

The regular diagrams of $m_n$knots corresponding to the bitopologies using loop distortion with $m = 3,4 \dots 10$ and $n = 1,2,3$ will be found by the following six steps.

Step 1.A node corresponding to the bitopological space (X, τ_1, τ_2) is marked by $s(\overline{\tau_1}) = V$. All points that are made are numbered 1, 2, ..., V as shown in Figure 7 in clockwise direction.

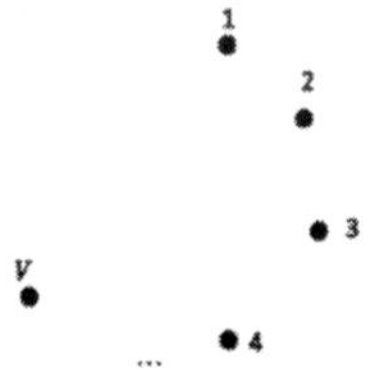

Figure 7. Corner number of the knot given the bitopology

Step 2. The number of element $\varepsilon_k^{ij}$of $\underline{\tau_2}$ shows how many corruptions are made between i and j. For example, while the ε_1^{13}element shows an impairment made between 1 and 3 corners, the ε_2^{13}element shows two impairments made between 1 and 3 corners. This impairment is as following Figure 8.

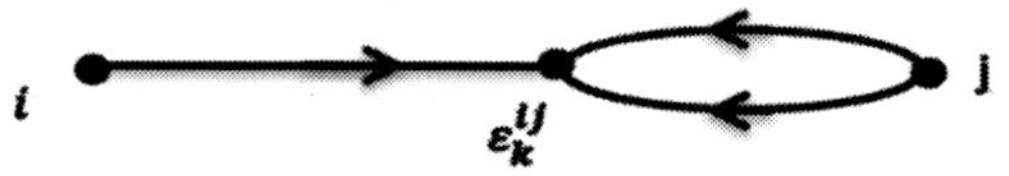

Figure 8. Cycle degradation

Step 3. $V + 2B - 1$ formula gives the number of edges of knot digarph. Namely it gives crossing number of given knot

Step 4. After process of the impairment $V - 1$ edges are added clockwise from the numbered edge by 1.

Step 5. Graph movements transform digraph into each other. Especially, the movement shown in the following Figure 9 is often used.

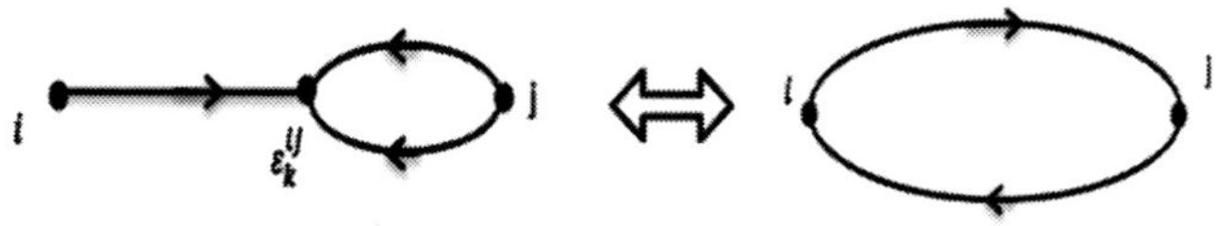

Figure 9. Cycle disturbance and cycle

Step 6. In the first five steps a knot digraph corresponding to the given bitopology is found. We can get knot form this knot digraph by using Tait's method.

Now,the knots that correspond to the following bitopological spaces will be found with the aid of the given algorithm.

Example: Let (X, τ_1, τ_2) is a bitopological space such that

$$\tau_1 = \{X, \emptyset, \{1\}, \{1,2\}, \{1,2,3\}\}$$
$$\tau_2 = \{X, \emptyset, \{2,3, \varepsilon_1^{13}\}, \{3, \varepsilon_1^{13}\}, \{\varepsilon_1^{13}\}\}$$

which knot corresponds to this bitopology?

1. $\overline{\tau_1} = \{1,2,3\}$, $s(\overline{\tau_1}) = V = 3$. So let's place 3 points of flattening and give this number in clockwise direction as in Figure 10.

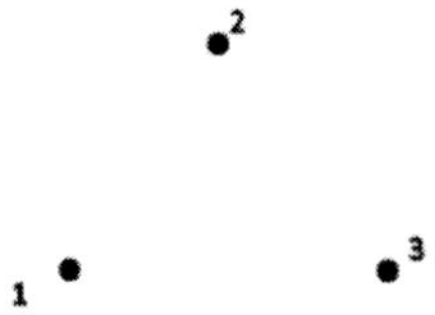

Figure 10.Placement of the points

2. $\underline{\tau_2}$=$\{\varepsilon_1^{13}\}$, $s\left(\underline{\tau_2}\right) = B = 1$. Then an impairment is placed between points 1 and 3 as in Figure 11.

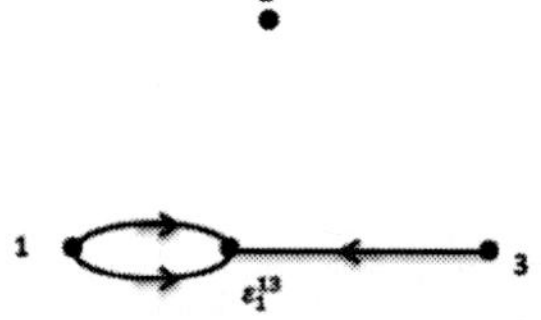

Figure 11. Placement of the impairment

3. $V + 2B - 1 = 3 + 2 - 1 = 4$,this shown that drawn digraph has got 4 edges .

4. $V - 1 = 3 - 1 = 2$, this shown that 2 edges will be added clockwise from 1 as in Figure12.

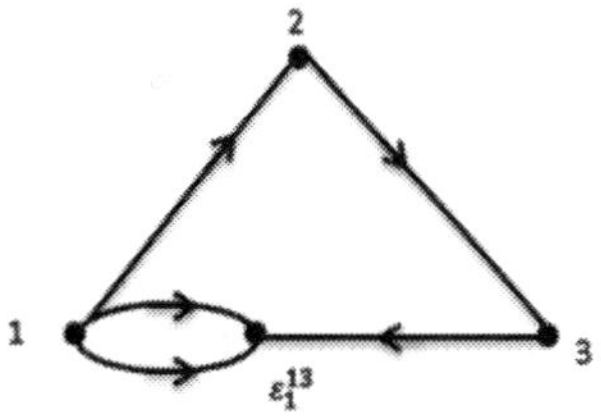

Figure 12. Placement of the edges

5. The impairment between points 1 and 3 is transformed by using graph movies as in Figure13.

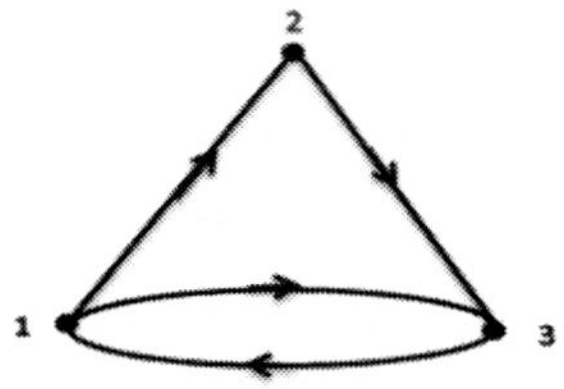

Figure 13. Transformed digraph

6. Using the Tait method, $\mathbf{4_1}$ knot are obtained.

Figure 14. Regular diagram of $\mathbf{4_1}$ knot

References

1. Alexander, J.W., G.B., *On types of knotted curve.* Ann. Math., 28, 562-586, 1956.
2. Bankwitz, C., *Uber die Torsionszahlen der alternierden Knoten.* Math. Ann., 103, 145-161, 1930.
3. Girija B., Pilakkat R., *Bitopological spaces associated with digraphs*, South Asian Journal of Mathematics, Vol.3 (1):56-65, 2013.
4. Kelley J.C., *Bitopological Spaces*, Proc. London Math.(3), 13, 71-89, 1963.
5. Murasugi, K., *On invariants of graphs with applications to knot theory.* Trans. Amer. Math. Soc., 314, 1-49, 1989.
6. Murasugi, K., *Knot Theory and Its Application.*Boston,1993.
7. K. Reidemeister: *Knotentheorie*, Julius Springer, 1932.
8. Yajima T., Kinoshita S., *On the graphs of knots,* Osaka Math. J., 9, 155-163, 1957.

Complex Acoustic Gravity Wave Behaviors to a Mathematical Model Arising in Nonlinear Mathematical Physics

Tolga Akturk[1 *], *Tukur Abdulkadir Sulaiman*[2,,3], *Haci Mehmet Baskonus*[4] *and Hasan Bulut*[2,5]

[1]Department of Mathematics, Ordu University, Ordu, Turkey

[2]Department of Mathematics, Firat University, Elazig, Turkey

[3]Department of Mathematics, Federal University Dutse, Jigawa, Nigeria

[4]Department of Computer of Engineering, Munzur University, Tunceli, Nigeria

[5]Department of Mathematics Education, Final University, Girne, Cyprus

Abstract. In this article, we utilize the powerful sine-Gordon expansion method (SGEM) in constructing some new solutions to the (2 + 1)-dimensional Boiti-Leon-Pempinelli equation by using the Mathematica software. We successfully obtain some new travelling solutions bearing some new structures such as trigonometric function, exponential function and hyperbolic function structures. We claim that some of our results are complex in structure. All the solutions obtained verified the the (2 + 1)-dimensional Boiti-Leon-Pempinelli equation. To illustrate our results, present the numerical simulation of all the obtained solutions in this study by selecting appropriate values of the parameters. Furthermore, we give the physical interpretation of all the graphics. We also give the physical meaning to some of the obtained results in this study.

1 Introduction

Many aspects in nonlinear science such as mathematical physics, biological science, physics and chemistry and many more nonlinear phenomena can be explicited in the style of nonlinear partial differential equations (NPDEs). Of late years, different analytical techniques have been formulated and employed to search for some new solitary wave solutions like this models for instance the modified expansion function method [1], the generalized Kudryashov method [2], the extended homoclinic test function method [3], the simple equation method [4], the extended hyperbolic function method [5], the $\tan(\frac{F(\xi)}{2})$-expansion method [6], the generalized tanh-function type method [7], the modified tanh-coth method [8], the generalized Bernoulli sub-equation function method [9] and many more powerful analytical methods [10-21].

* Corresponding author: tolgaakturkk@gmail.com

However, in this study, we used the powerful SGEM [22, 23] in studying the solutions behaviors of the (2 + 1)-dimensional BoitiLeon-Pempinelli (BLP) equation given by [24]

$$u_{ty} = (u^2 - u_x)_{xy} + 2v_{xxx}$$
$$v_t = v_{xx} + 2uv_x \tag{1.1}$$

The (2+1)-dimensional BLP equation which arises in mathematical physics, describe the interaction of the two waves with different distributional relationships. [25].

2 Application

In this section, we employ the SGEM [10, 11] to Eq. (1.1). Performing the travelling wave transformation,
$u = u(x, y, t) = U(\xi)$, $v = u(x, y, t) = V(\xi)$, $\xi = \mu(\text{x}+\text{y}-\text{ct})$ on Eq. (1.1) :

$$\mu^2 U'' - 2U^3 - 3U^2 - c^2 U = 0. \tag{2.1}$$

By SGEM, Eq. (3.1) is assumed to have the following solutions [10, 11]:

$$\text{U}(\xi) = \sum_{i=1}^{m} \tanh^{i-1}(\xi)[\text{B}_i \text{sech}(\xi) + \text{A}_i \tanh(\xi)] + A_0, \tag{2.2}$$

$$\text{U}(w) = \sum_{i=1}^{m} \cos^{i-1}(w)[\text{B}_i \sin(w) + \text{A}_i \cos(w)] + A_0. \tag{2.3}$$

we get $m = 1$ by homogenous balance technique. Using Eqs. (2.2) and (2.3) with $m = 1$, we obtain the following :

$$\text{U}(\xi) = B_1 \,\text{sech}(\xi) + \text{A}_1 \tanh(\xi) + A_0 \tag{2.4}$$

and

$$\text{U}(w) = B_1 \sin(w) + \text{A}_1 \cos(w) + A_0, \tag{2.5}$$

respectively.
Putting Eq. (2.5) and its second derivative into Eq. (2.1), produces an equation in trigonometric functions. Equating all summation of the coefficients of the trigonometric functions of the same power to zero, we get the a set of algebraic equations. Reducing the set of algebraic equations gives the values of the coefficients inclusive. Substituting the values of the coefficients into Eq. (2.4), gives the solutions to Eq. (1.1).

Case-1 :

$$A_0 = -\frac{1}{2}, A_1 = \frac{1}{2}, B_1 = 0, \mu = \frac{1}{2}, c = 1,$$

which gives :

$$u_1(x,y,t)=\frac{1}{2}\left(-1+\tanh\left[\frac{1}{2}\left(x+y-t\right)\right]\right), \tag{2.6}$$

$$v_1(x,y,t)=\frac{1}{2}\tanh\left[\frac{1}{2}\left(x+y-t\right)\right]. \tag{2.7}$$

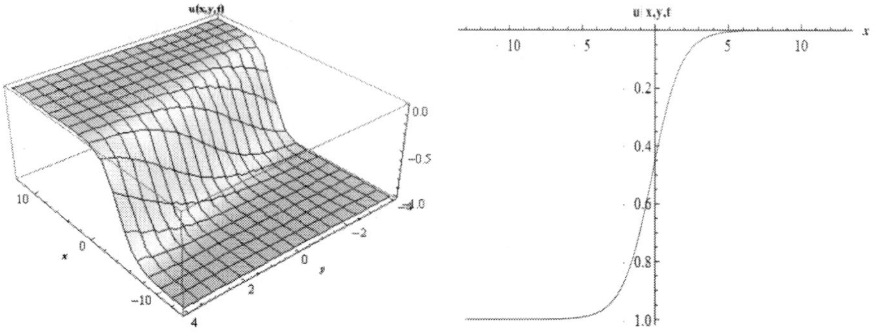

Fig.1. : The kink-type solution shape of Eq. (2.6) under the values $t = 0.5$, -13 < x < 13 , -2 < y < 2 and $y = 0.7$ for the 2D graphic.

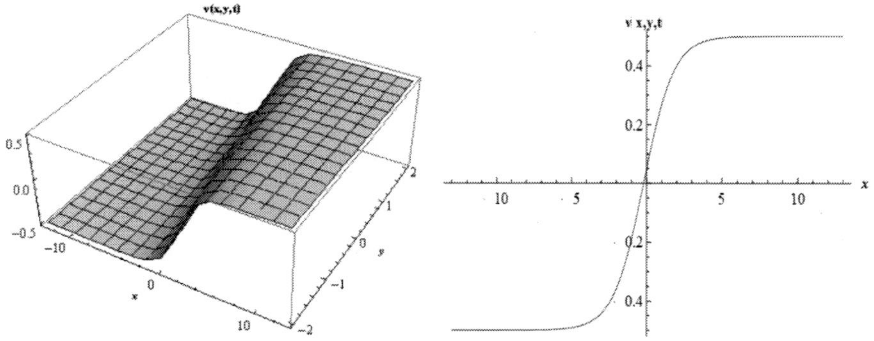

Fig. 2. : The kink-type solution shape of Eq. (2.7) under the values $t = 0.5$, -13 < x < 13 , -2 < y < 2 and $y = 0.7$ for the 2D graphic.

Case-2 :

$$A_0=-\frac{1}{2},\ A_1=\frac{1}{2}, B_1=-\frac{i}{2}, \mu=1, c=1,$$

$$u_2(x,y,t)=v_2(x,y,t)=\frac{1}{-1+ie^{x+y-t}}. \tag{2.8}$$

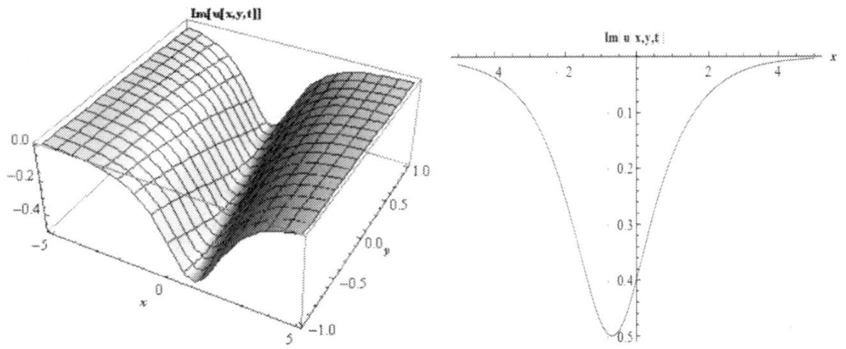

Fig. 3. : The singular solution shape of imaginary Eq. (2.8) under the values $t = 0.5$, -13 < x < 13 , -1 < y < 1 and $y = 0.7$ for the 2D graphic.

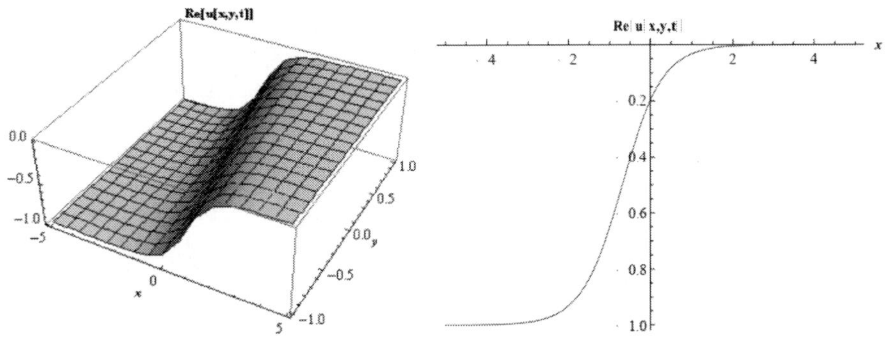

Fig. 4. : The kink-type shape of real part Eq. (2.8) under the values $t = 0.5$, -13 < x < 13 , -1 < y < 1 and $y = 0.7$ for the 2D graphic.

3 Conclusion

This study is assigned in constructing some new solutions to the (2 + 1)-dimensional Boiti-Leon-Pempinelli equation that is used to describe the interactions of two waves with different dispersion relations in mathematical physics. We successfully found new solutions with some novel structures such as complex hyperbolic function, trigonometric function and exponential function structures. We give the physical meaning of some obtained solutions in this study. We also present the 3D and 2D surfaces of all the obtained solutions together with their physical interpretations. We compare our results with some substantial results in the literature and we sighted that our results is only just constructed results with some good structures that have important physical meaning. With these results, we can say that SGEM is indeed powerful and efficient mathematical tool that can provide you with the varieties of solution structures when employed to various nonlinear evolution equations.

References

1. F. Ozpinar, H.M. Baskonus. H. Bulut, Entropy, 17(12), 8267-8277 (2015)
2. E.M.E. Zayed, G.M. Moatimid, A.G. Al-Nowehy, Scientific Journal of Mathematics Research, 5(2) , 19-39 (2015)
3. C. Wang, Nonlinear Dynamics, 85(2), 1119-1126 (2016)
4. T.A. Nofal, Journal of the Egyptian Mathematical Society, 24, 204-209 (2016)
5. Y. Shang, Chaos, Solitons and Fractals, 36(3), 762-771 (2008)
6. J. Manafian, Optik, 127, 4222-4245 (2016)
7. W. Zhang, Applied Mathematics, 4, 11-16 (2013)
8. L. Wazzan, Communications in Nonlinear Science and Numerical Simulation, 14(6), 2642-2652 (2009)
9. B. Zheng, WSEAS Transactions on Mathematics, 7(11), 618-626 (2012)
10. H. Bulut, H.A. Isik, T.A. Sulaiman, ITM Web of conferences, 13, 01019 (2017)
11. S. Duran, M. Askin, T.A. Sulaiman, IJOCTA, 7(3), 240-247 (2017)
12. H.M. Baskonus, H. Bulut, A. Atangana, Smart Materials and Structures, 25(3), 035022 (2016)
13. H. Bulut, T.A. Sulaiman, H.M. Baskonus, Eur. Phys. J. Plus, 132, 459 (2017)
14. H.M. Baskonus, H. Bulut, T.A. Sulaiman, Eur. Phys. J. Plus, 132, 482 (2017)
15. H. Bulut, T.A. Sulaiman, B. Demirdag, Nonlinear Dyn, 91, 1985-1991 (2018)
16. H.M. Baskonus, T.A. Sulaiman, H. Bulut, Opt Quant Electron, 50, 14 (2018)
17. H. Bulut, T.A. Sulaiman, H.M. Baskonus, T. Akturk, Opt Quant Electron, 50, 19 (2018)
18. T.A. Sulaiman, T. Akturk, H. Bulut, H.M. Baskonus, Journal of Electromagnetic Waves and Applications, 32(9) 1093-1105 (2017)
19. A. Yokus, T.A. Sulaiman, H. Bulut, Opt Quant Electron, 50 31 (2018)
20. S.Ş. Şener, Y Saraç, M. Subaşı, Applied Mathematical Modelling, 37, 2623-2629 (2013)

21. M. Subaşı, S.Ş. Şener, Boundary Value Problems, 1-16 (2015)

22. C. Yan, Physics Letters A, 22(4), 77-84 (1996)

23. H. Bulut, T.A. Sulaiman, F. Erdogan, H.M. Baskonus, Eur. Phys. J. Plus, 132, 350 (2017)

24. H.M. Baskonus, H. Bulut, Waves in Random and Complex Media, 26(2), 201-208 (2016)

25. M. Kumar, R. Kumar, A. Kumar, Computers and Mathematics with Applications, 70, 212-221 (2015)

On the wave solutions to the TRLW equation

Tukur Abdulkadir Sulaiman[1, 2], *Canan Unlu*[3]*, and *Hasan Bulut*[1,4]

[1]Department of Mathematics, Firat University, Elazig, Turkey

[2]Department of Mathematics, Federal University Dutse, Jigawa, Nigeria

[3]Department of Mathematics, Istanbul University, Istanbul, Turkey

[4]Department of Mathematics Education, Final University, Girne, Cyprus

Abstract. In this study, a nonlinear model is investigated, namely; the time regularized long wave equation. Various solitary wave solutions are constructed such as the non-topological, compound topological-non-topological bell-type, singular and compound singular soliton solutions. Under the choice of suitable parameters values, the 2D and 3D graphs to all the obtained solutions are plotted. The reported results in this study may be helpful in explaining the physical meanings of some important nonlinear models arising in the field of nonlinear science.

1Introduction

For the past two decades, the investigation of the solutions to the various kind of nonlinear evolution equations (NLEEs) has attracted the attention of many researchers. Nonlinear evolution equations are often used to model various nonlinear complex physical aspects that arise in the various fields on nonlinear physical sciences, especially in physics plasma, biology and fluid mechanics and so on. Various analytical techniques have been used to explore search of several NLEEs [1–11].

However, in the present study, the extended sinh-Gordon equation expansion method [12-15] (ShGEEM) will be used to investigate the wave solutions to the time regularized long wave (TRLW) equation [16]. The time regularized long wave is given by

$$u_t + u_x + \beta u u_x + u_{xtt} = 0, \tag{1.1}$$

where$u(x,t)$ is the unknown function of x; the spatial coordinate and the time t, respectively, and β is a nonzero constant.

3Application

In this section, the application of the extended sinh-Gordon equation expansion method to the TRLW is presented.

Consider the travelling wave transformation

----

*Corresponding Author: unlucanan@hotmail.com

$$u = \Psi(\eta),\ \eta = \mu(x - vt). \tag{2.1}$$

Inserting Eq. (2.1) into Eq. (1.1), yields

$$2\mu^2 v^2 \Psi'' + 2(1 - v)\Psi + \beta\Psi^2 = 0. \tag{2.2}$$

By the extended ShGEEM [12], the solutions of any given nonlinear partial differential equation are assumed to be of the forms

$$\Psi(w) = \sum_{j=1}^{m}\left[B_j \sinh(\theta) + A_j \cosh(w)\right]\cosh^{j-1}(w) + A_0, \tag{2.3}$$

$$\Psi(\eta) = \sum_{j=1}^{m}\left[\pm iB_j \operatorname{sech}(\eta) \pm A_j \tanh(\eta)\right]\tanh^{j-1}(\eta) + A_0, \tag{2.4}$$

$$\Psi(\eta) = \sum_{j=1}^{m}\left[\pm B_j cs\operatorname{ch}(\eta) \pm A_j \coth(\eta)\right]\coth^{j-1}(\eta) + A_0, \tag{2.5}$$

where $i = \sqrt{-1}$ and $w' = \sinh(w)$ [5]

Balancing the terms Ψ'' and Ψ^2, yields $m = 2$.

With $m = 2$, Eqs. (2.3), (2.4) and (2.5) take the forms

$$\Psi(w) = B_1 \sinh(w) + A_1 \cosh(w) + B_2 \cosh(w)\sinh(w) + \mathrm{A}_2 \cosh^2(w) + A_0 \tag{2.6}$$

Inserting Eq. (2.6) and its second derivative into Eq. (2.2), gives a polynomial in powers of hyperbolic functions. Summing each coefficients of the hyperbolic functions of the same degree and equating each summation to zero, gives a group of algebraic equations. The group of algebraic equations is then simplified to secure the values of the parameters involved into different cases. For each case, inserting the obtained values of the parameters into each of Eqs. (2.4) and (2.5) with $m = 2$, yields the solutions to Eq. (1.1).

Case-1: When

$$A_0 = \frac{6(-1+v)}{\beta}, A_1 = 0, B_1 = 0, A_2 = \frac{6-6v}{\beta}, B_2 = -\frac{6\sqrt{(-1+v)^2}}{\beta}, \mu = -\frac{\sqrt{-1+v}}{v},$$

we get

$$u_1(x,t) = \frac{1}{\beta}\left(6(-1+\nu)\,\text{sech}\left[\frac{\sqrt{-1+\nu}(x-t\nu)}{\nu}\right]^2 + 6i(\nu-1)\text{sech}\left[\frac{\sqrt{-1+\nu}(x-t\nu)}{\nu}\right]\tanh\left[\frac{\sqrt{-1+\nu}(x-t\nu)}{\nu}\right]\right) \quad (2.7)$$

$$u_2(x,t) = -\frac{6\sqrt{(-1+\nu)^2}\coth\left[\frac{\sqrt{-1+\nu}(x-t\nu)}{\nu}\right]\text{csch}\left[\frac{\sqrt{-1+\nu}(x-t\nu)}{\nu}\right] + 6(-1+\nu)\,\text{csch}\left[\frac{\sqrt{-1+\nu}(x-t\nu)}{\nu}\right]^2}{\beta} \quad (2.8)$$

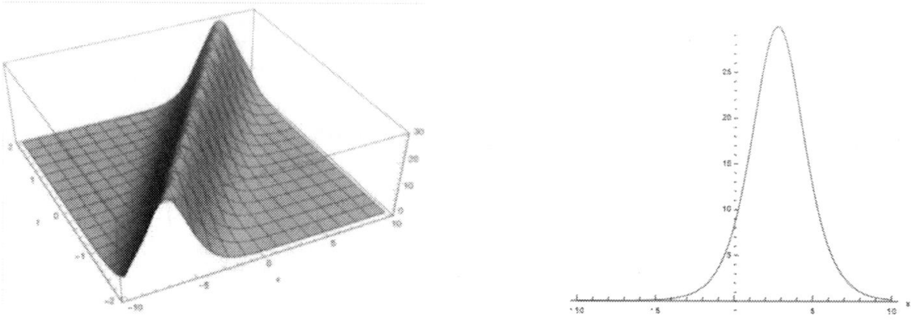

(a) (b)

Figure 1: The 3D and 2D graphs of the real part of Eq. (2.7).

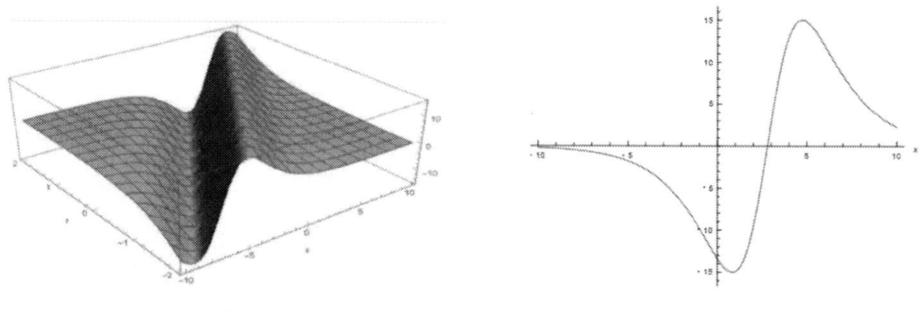

(a) (b)

Figure 2: The 3D and 2D graphs of the imaginary part of Eq. (2.7).

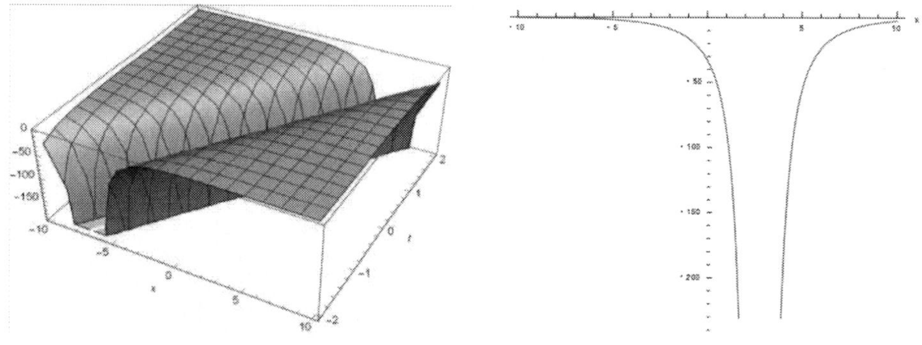

(a) (b)

Figure 3: The 3D and 2D graphs of Eq. (2.8).

Case-2: When

$$A_0 = \frac{1-\nu}{\beta}, A_1 = 0, B_1 = 0, A_2 = \frac{3(-1+\nu)}{\beta}, B_2 = 0, \mu = \frac{\sqrt{-(\nu-1)}}{2\nu},$$

we get

$$u_3(x,t) = \frac{(-1+\nu)(-2+\cosh[\frac{\sqrt{1-\nu}(x-t\nu)}{\nu}])sech^2[\frac{\sqrt{1-\nu}(x-t\nu)}{\nu}]}{\beta} \tag{2.9}$$

And

$$u_4(x,t) = \frac{(-1+\nu)(2+\text{Cosh}[\frac{\sqrt{1-\nu}(x-t\nu)}{\nu}])csch^2[\frac{\sqrt{1-\nu}(x-t\nu)}{\nu}]}{\beta} \tag{2.10}$$

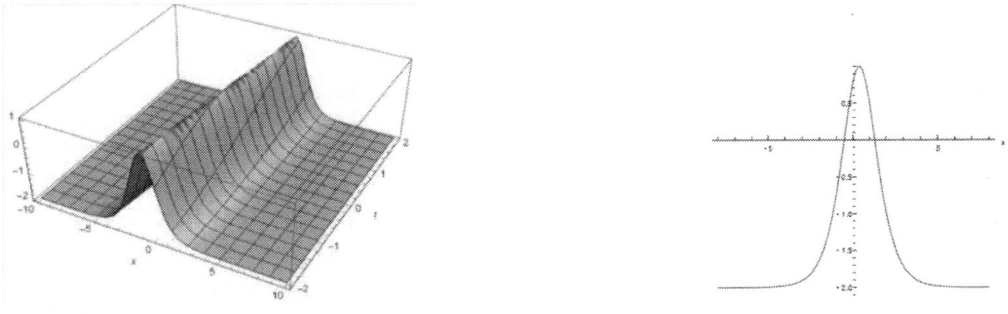

(a) (b)

Figure 3: The 3D and 2D graphs of Eq. (2.9).

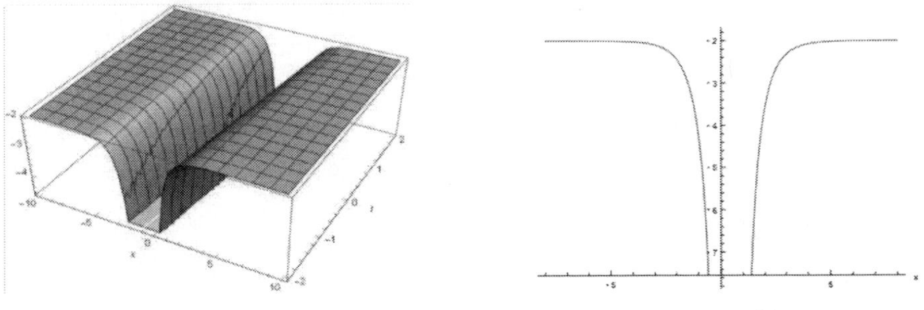

(a) (b)

Figure 4: The 3D and 2D graphs of Eq. (2.10).

3 Conclusion

In this study, the time regularized long wave equation is investigated by using the extended sinh-Gordon equation expansion method with the aid of Wolfram Mathematica 11. Various soliton solutions such as the non-topological, compound topological, non-topological bell-type, singular and compound singular solitons are successfully extracted. Using suitable values of parameters, the 2D and 3D graphics to all the reported solutions are plotted.

References

1. Z. Yan, MM Research Preprints, **22** 363-375 (2003)
2. H. Bulut, T.A. Sulaiman and H.M. Baskonus, Opt Quant Electron, **48** 564 (2016)
3. H. Bulut, T.A. Sulaiman, H.M. Baskonus and A.A. Sandulyak, Optik, **135** 327-336 (2017)
4. Md. Nur Alam, F.M. Belgacem and M.A. Akbar, Journal of Applied Mathematics and Physics, **3** 1571-1579 (2015)
5. H. Bulut, H.A. Isik, T.A. Sulaiman, ITM Web of conferences, 13, 01019 (2017)
6. S. Duran, M. Askin, T.A. Sulaiman, IJOCTA, 7(3), 240-247 (2017)
7. H. Bulut, T.A. Sulaiman, H.M. Baskonus, Eur. Phys. J. Plus, 132, 459 (2017)
8. H.M. Baskonus, H. Bulut, T.A. Sulaiman, Eur. Phys. J. Plus, 132, 482 (2017)
9. H. Bulut, T.A. Sulaiman, B. Demirdag, Nonlinear Dyn, 91, 1985-1991 (2018)
10. T.A. Sulaiman, T. Akturk, H. Bulut, H.M. Baskonus, Journal of Electromagnetic Waves and Applications, 32(9) 1093-1105 (2017)
11. A. Yokus, T.A. Sulaiman, H. Bulut, Opt Quant Electron, 50 31 (2018)
12. H. Bulut, T.A. Sulaiman and H.M. Baskonus, Optik, **163** 1-7 (2018)
13. H.M. Baskonus, T.A. Sulaiman, H. Bulut, Opt Quant Electron, 50, 14 (2018)
14. C. Cattani, T.A. Sulaiman, H.M. Baskonus andH. Bulut, Eur. Phys. J. Plus **133**, 228 (2018)
15. C. Cattani, T.A. Sulaiman, H.M. Baskonus and H.Bulut, **50**, Opt Quant Electron 138 (2018)
16. K.S. Hossain, M.A. Akbar and Md.A.K. Azad, Cogent Physics, **4** 396948 (2017)

Investigation of Earthquake Behaviour of Construction System and Materials in Traditional Turkish Architecture

Seyhan Yardımlı[1,*], *Murat* Dal[2] , and *Esma* Mıhlayanlar [3]

[1]Faculty of Architecture and Design, İstanbul Aydın University, İstanbul, Turkey

[2]Faculty of Engineering, Munzur University, Tunceli, Turkey

[3]Faculty of Architecture, Trakya University, Edirne, Turkey

Abstract. In this study, it is aimed to present a point of view regarding the behaviour of construction systems implemented in traditional Turkish architecture against earthquakes. In the scope of the study, examples of civil architecture were considered and their structures were evaluated as building elements such as foundation, wall and flooring. Traditional Turkish architecture construction systems can be evaluated in two parts. One of them is the wooden carcass system and the other is the unreinforced masonry system. In the wooden carcass system, the carrier is the load bearing elements used in horizontal and vertical directions. Intermediate parts (strut, diagonal etc.) are placed between these elements to form triangles. The triangles (strut, diagonal) used in the wooden skeleton system comprise highly resistant forms against earthquakes. Moreover, due to the internal structure and physical properties of the wood, which is the skeleton material, the flexibility that it maintains can meet the lateral loads of earthquakes. The second construction system which is the system addressed in this application, is the unreinforced masonry system. In this system, the loadbearing system itself is the walls, which are not resistant to lateral loads. In order to provide this flexibility beams (hatıllar) are installed at certain intervals. After the wall is built to a certain height, a different material is laid allowing a plane of movement on the wall. Thus, when the wall is exposed to a lateral load, it escapes from the planes where the beams (hatıllar) are present, and is protected against large damages by absorbing the earthquake load. In order to establish that the foundation of the structure can withstand earthquakes by movement, wood is placed at the lower part of the foundation above a layer of sand ensuring lateral movement and flexibility of the building. In traditional buildings the slabs are connected to the building walls with beams (hatıllar). Through the agency of the beams (hatıllar) formed at the connection points, the slabs can act as a mass so that they can meet the earthquake load. Allowing the structure to move makes it resistant to earthquakes, seismic isolators are used for this purpose by absorbing the earthquake load and moving the foundation of the structure.

* Corresponding author: seyhanyardimli@gmail.com

1 Introduction

One of Turkey's most active earthquake belts is situated along the Alpine Himalayan earthquake fault [1]. From time to time, much destruction occurs in this geography. Structures that are built using traditional methods as well as those with reinforced concrete have been reduced to rubble in these natural disasters, leading to discussions surrounding the importance of both construction systems as well as building quality. In some cases, local people use durable building construction techniques and these techniques have not yet been tested by modern engineers [2].
From the standpoint of carrier characteristics of buildings that incorporate the traditional system in Turkey, these can generally be taken up in two categories, the wooden framed and masonry construction systems. In considering these two systems from the aspect of earthquake resistance, the wooden frame system appears to be more resistant. In touching briefly upon the frame system in this study, we'll be actually dealing with the masonry system, and take up construction elements that protect the structure against lateral loads and boost the system's resistance against earthquakes in today's structures.

2 Behavior of a Wooden-Frame Structure During Earthquakes

The frame system is comprised from horizontal and vertical bearings positioned on a prepared foundation. Composed of horizontal and vertical elements, these main carriers form rectangular or square forms. The durability of these forms is boosted by placing cross elements between them. Because a wooden framed system is comprised of a form made from triangles, it is extremely rigid (Table 1.a). While square or rectangular forms are more susceptible to deformation due to the lateral effect, triangular forms are much less susceptible to deformation via lateral load (Table 1.b). While it was observed wooden framed structures weren't damaged, numerous reinforced structures collapsed [3] during the 1999 Marmara-region earthquakes centered in Gölcük and Düzce.

Table 1. The utilization of the triangle form in the wooden frame system

a. The formation of the triangle shape in the wooden frame system	b. Lateral effect behavior of the rectangle or square vs. the triangle shape

3 Behavior of a Masonry System Structure in an Earthquake

The behavior of the foundation, walls and the roof of a building constructed in the masonry system will be taken up in this section.

3.1. The Impact of Earthquake on Masonry Building Foundations

In taking a look at foundations of historical structures in which traditional construction system have been applied, it is observed a system was utilized in these elements that dampen the effect of an earthquake. A layer of 15-20 cm thick of sand or pebbles was spread, or else a wooden grid that facilitates the possibility of movement in this system was first laid beneath the foundation (Table 2) [4]. Moreover, the foundations of historical structures were protected from the effects of water and ventilated with peripheral wells or galleries [5].

Table 2. The positioning of the foundation in a traditional construction system to counter earthquake loads

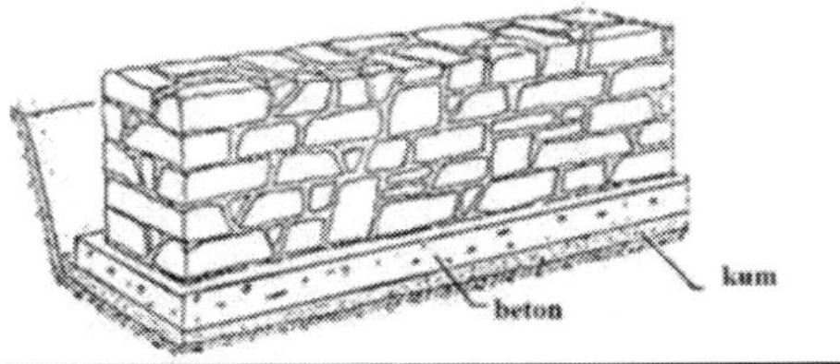	
Spreading sand beneath foundations of traditional structures [6]	Use of wood in a masonry building foundation at Sirkeci [4]

In recent years, interest has intensified on systems that absorb the effect of earthquakes within the structure [7]. The objective is to ensure that the structure behaves rigidly on the vertical side and flexibly on the horizontal side in countering the effect of an earthquake. Fort his, floor insulators were positioned in the foundation [8]. Examples of shock absorbers are seen in the foundation in Table 3.

Table 3. Examples of floor insulators and absorbers which are imbedded in the foundation in contemporary structures [7]

	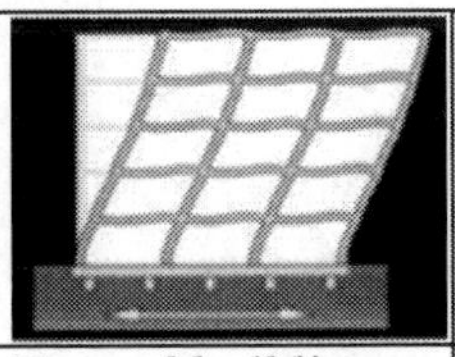	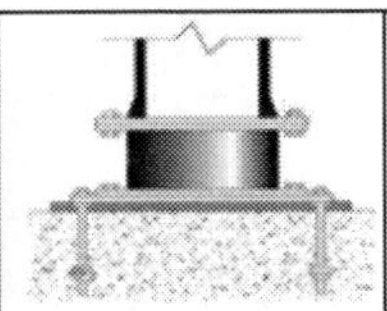	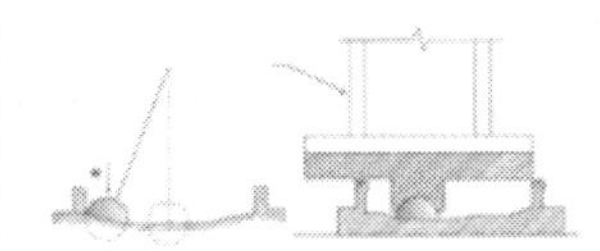
With seismic insulation	Normal building	Rubber insulator	Insulators based on rubbing

3.2. Earthquake Effects in Wall Elements

In comparing masonry structures to reinforced concrete framed structures, the wall in masonry systems are simple and continual, serving as a curtain door [9]. In masonry structure walls, the tensile strength of wall blocks is very low. For this reason, damage occurs in the face in for this reason, damage is seen in the direction of the plane of the wall or against the load that stands perpendicular to the wall. The beam element is utilized in traditional structures in order for earthquake loads to be absorbed within the wall in a masonry system. The Hatil facilitates the hatil arrangement sliding plane that allows the wall to move. This sliding plane ensures that the structure wall remains stable by absorbing the tension the earthquake causes in the horizontal direction. Beams begin to be imbedded in the wall from a height of 80 – 100 cm above the floor and are spaced at intervals of 100 –

150 cm. These dampening planes reduce the earthquake load by pulling it from the height of the wall down to the height of the beam. This proves is seen in Table 4.

Table 4. Earthquake behavior in masonry structure walls

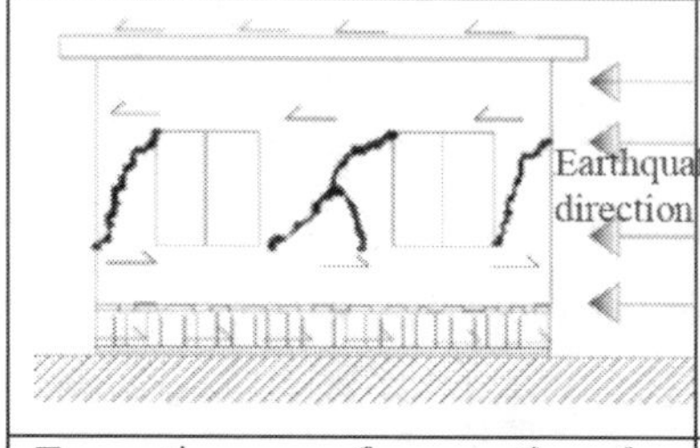		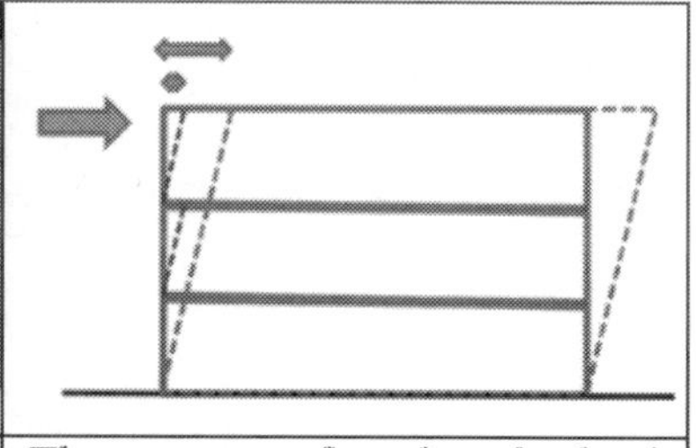
Formation of earthquake damage in a masonry wall [6]	The use of beams in masonry buildings	The amount of earthquake load absorption with beams

'In some regions of the world, timber or brick bands are used to enhance the wall stability in both uncoursed random rubble and semi-dressed masonry. This is a traditional practice in some parts of Nepal, İndia, Pakistan, Turkey and Greece' [10]. Recommended use of timber and reinforced concrete bands (hatils) in Turkish stone masonry are shown in the figure 1.

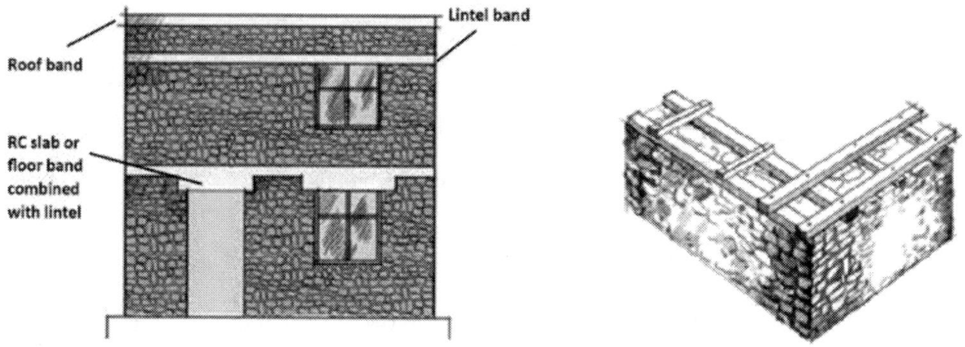

Fig. 1. Reinforced concrete and timber bands [10]

The wooden reinforced stone wall technique known as Bhatar is used for post-earthquake housing needs in northern parts of Pakistani (Fig 2) [2].

Fig. 2. New house in Tarand, Battagram [2]

To increase the strength of the buildings, T. Schacher recommends plinth beam, sill band, lintel band and bond beam (Figure 3) [11].

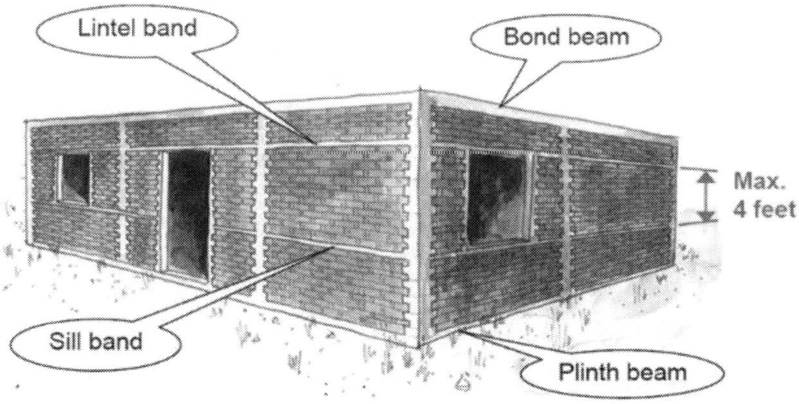

Fig.3. Increasing the stability of a building against earthquake [11]

It is crucial for the walls of masonry structure to be of the required quality and standards. In reality, it is observed that wall formation regulations were complied with inadequately in numerous collapsed structures (Table 5).

Table 5. 2017 Causes of damage in collapsed walls in the Çanakkale Ayvacık earthquake

İnsufficient corner connections	The non-formation of cross-members	Positioning the joints along the same alignment

Beams that are installed in traditional masonry structures are made from materials such as wood, concrete and brick (Table 6).

Table 6. Samples of beam material types

Wooden beam sample [12]	Brick Beam sample	Concrete Beam sample

3.3. Earthquake Impact on the Roof Element as a Building Cover

Only roof coverings applied in housing were considered within the study of civil architecture examples. As the roof covering for masonry construction systems, the general application is either a flat roof covered with soil, or else a wooden room covered with adobe tiles. The roof must be of the quality to shift the load to the upper portion of the walls and ensure rigidity in itself. Concurrently, the walls must be able to bear the load coming from the roof itself [13]. The roof must be properly positioned over trailing edge elements at the top of the wall. This trailing edge element can either be wooden floor joist or a reinforced concrete beam. During an earthquake, the separation of the joists from the wall, and subsequent roof collapse is due to the little distance of the roof beams positioned over the wall. Excessive loads of soil-covered rooftops also increases the rate of impact with an earthquake load. As a consequence of the collapse of the walls, the roofs on top of them collapse as their ability to bear the load disappears.
In determinations made in the Ayvacık earthquake zone, it was observed excessive damage occurred due to the failure to apply proper construction techniques (Table 7).

Table 7. Roof damage

Positioning the roof purlin to the Wall without the use of floor joists	Positioning the joists to the wall with a short distance	Roof collapse as a result of wall collapse.

4 Conclusion

Earthquakes have occured throughout history in this geography which is in an earthquake belt. In this regard, several examples can be provided. In reviewing one or two of these with the objective of assessing the construction systems; 14,156 buildings were damaged in the 1995 Dinar Earthquake, whereas wooden framed structures sustained little or no damage. The amount of damage sustained in masonry structures in the 1992 Erzincan and 1995 Dinar earthquakes was found to be of a lower rate than other structures [9]. Langenbach stated that masonry structures remained more intact than reinforced concrete buildings as a result of the earthquake that struck Armenia on 7 December, 1988 [14]. Engineers examine mainly reinforced concrete structures in research conducted in earthquake regions, whereas buildings constructed with traditional systems are not paid the necessary attention and care.

In 2005 Pakistani earthquake showed that not only traditional masonry buildings but also reinforced concrete frame structures were collapsed down. On the other hand, traditional timber-reinforced buildings resisted well enough [2].

In conducting inspections, it is seen that most buildings damaged in this quake did not comply with required construction techniques and that the structures were erected haphazardly. In reality, the necessary precautions for rendering masonry structures resistant to earthquakes are known. The problem often arises from the failure to implement these measures.

In the traditional construction system, the planning of the building foundation so that it absorbs earthquake shocks is being used by adapting today's conditions as well as in the technology that is available at present. For this reason, it appears that technique applied in traditional masonry structures is correct.

The use of beams in the wall system also creates earthquake load absorption surfaces. Thus, the wall is not affected by the earthquake load as a large mass, but rather as a short distance between two beams. The load on the roof is also reduced as a result of earthquake load dampening by the beams imbedded in the wall.

The preservation of traditional systems is required from the aspect of building culture continuity. In terms of building physics, these buildings are also more advantageous in regards to providing thermal comfort. In terms of production and utilization of materials that go into the production process in their production systems are also more eco-friendly.

Therefore, rather than pretending structures produced with the traditional system don't exist, it is more beneficial to approach this type of home structure, which is still inhabited to a major extent in rural areas, in a constructive and solution-minded manner.

References

1. http://www.icisleriafad.gov.tr/ulkemizin-deprem-riski-haritası, Access:19.04.2018
2. T. Schacher, Good Engineering without Appropriate Communication doesn't lead to Seismic Risk Reduction: some thoughts about appropriate knowledge transfer tools, The 14th World Conference on Earthquake Engineering, Beijing, China, (2008)
3. Z. Ahunbay, Geleneksel Yapılar ve R. Langenbach'la Söyleşi, Yapı Mimarlık Kültür ve Sanat Dergisi, Vol: 228, 101, (2000)
4. G. Arun, TMMOB İnşaat Mühendisleri Odası Gaziantep Eğitim Semineri, Slide:19, Slide:10, (2016)
5. A. Bayraktar, 'Tarihi Yığma Yapıların Depreme Karşı Güçlendirilmesi', YDGA 2005-Yığma Yapıların Deprem Güvenliğinin Arttırılması Çalıştayı, 4, (2005)
6. G. Arun, Yığma Kagir Yapı Davranışı, YDGA 2005-Yığma Yapıların Deprem Güvenliğinin Arttırılması Çalıştayı, ODTÜ, Ankara, 6,1, (2005)
7. T.C. M E B, İnşaat Teknolojisi Deprem İzolatör Sistemleri 440FB000, 5-28, (2011)
8. S. Oymael, To provide the earthquake resistance in the structures and the sustainability of neighborhood relations, X. INTERNATIONAL SINAN SYMPOSIUM, 319-326, (2017)
9. V. Koç, 'Depreme Maruz Kalmış Yığma ve Kırsal Yapı Davranışlarının İncelenerek Yığma Yapı Yapımında Dikkat Edilmesi Gereken Kuralların Derlenmesi', Çanakkale Onsekiz Mart Üniversitesi Fen Bilimleri Enstitüsü Dergisi, 1-57, (2016)
10. J. Bothara, S. Brzev, A Tutorial: Improving the Seismic Performance of Stone Masonry Buildings, WHE-2011-01, Oakland, California, 12, 35-36, (2011) http://www.world-housing.net/wp-content/uploads/2011/06/Stone_Masonry_English.pdf

11. T. Schacher, Confined Masonry For one and two story buildings in low-tech environments, A guidebook for technicians and artisans, nicee National Information Centre of Earthquake Engineering, 13, (2009) https://www.pseau.org/outils/ouvrages/nicee_confined_masonry_1_2_storey_buildings_schacher.pdf Access: 22.10.2017
12. http://sivrihisar.web.tr/sivrihisar-zaimaga-konagi.html, Access: 10.11. (2016)
13. A. Budak, H. Uysal, A. C. Aydın, Kırsal Yapıların Deprem Karşısındaki Davranışı, Atatürk Ziraat Fak. J. 35 (3-4), 209-219, (2004)
14. R. Langenbach. "Intuition from the Past: What we can learn from Traditional Construction", KEYNOTE ADDRESS AT THE ISTANBUL CONFERENCE, (2000) http://www.conservationtech.com/IstanCon/keynote.htm. Access: 22.10.2017

New Wave Solutions for Nonlinear Differential Equations using an Extended Bernoulli Equation as a New Expansion Method

Serbay DURAN[1,*] and *Doğan KAYA*[2]

[1]Adiyaman University, Faculty of Education, 020040 Adiyaman, Turkey
[2]Istanbul Commerce University, Department of Mathematics, 34762 Istanbul, Turkey

Abstract. In this paper, we presented a new expansion method constructed by taking inspiration for the Kudryashov method. Bernoulli equation is chosen in the form of $F' = BF^n - AF$ and some expansions are made on the auxiliary Bernoulli equation which is used in this method. In this auxiliary Bernoulli equation some wave solutions are obtained from the shallow water wave equation system in the general form of "n-order". The obtained new results are simulated by graphically in 3D and 2D. To sum up, it is considered that this method can be applied to the several of nonlinear evolution equations in mathematics physics.

1 Introduction

Mathematical studies from nonlinear phenomena to particular case works play a significant role in applied mathematics, physics, engineering, education and so on [1]. Where at the laws of the natural and physical world are generally modeled by partial differential equations. Searching the solutions of partial differential equations is become one of the main areas of scientific works [2]. Many analytic methods have been introduced in literature. Some of them are: the Jacobi elliptic-function method [3], the sine-Gordon expansion method [4], the improved Bernoulli sub-equation function method [5] and so on.

The aim of this work, to investigate solutions of the system of the Shallow Water Wave equation with the help of the new expansion method which is a useful and popular tool for constructing exact solutions.

2 Analysis of the Method

2. 1 The Kudryashov Method

This section is cover the Kudryashov method for finding wave solutions of nonlinear partial differential equations [6-7]. Accordingly, a general form of nonlinear PDE is defined as follows

* Corresponding author: sduran@adiyaman.edu.tr

$$Q_1(\tilde{u}_t, \tilde{u}_x, \tilde{u}_{xx}, \tilde{u}_{xxx}, \ldots, x, t) = 0\,, \tag{2.1}$$

$$\tilde{u}(x,t) = y(\xi), \quad \xi = kx - wt\,, \tag{2.2}$$

where k and w are respectively the number of waves and the wave velocity. After (2.2) transformation is applied to the Eq. (2.1), we obtain a nonlinear ordinary differential equation for $y(\xi)$

$$Q_2(-wy_\xi, ky_\xi, k^2 y_{\xi\xi}, k^3 y_{\xi\xi\xi}, \ldots) = 0. \tag{2.3}$$

It is assumed that the solution $y(\xi)$ of the nonlinear equation given in Eq. (2.3) can be written in the following form:

$$\tilde{u}(x,t) = y(\xi) = \sum_{i=0}^{M} a_i F^i, \tag{2.4}$$

where a_i, $1 \le i \le M$ are constants and M is positive integers. M will be determine with homogeneous balance principle. $F(\xi)$ is the following function

$$F(\xi) = \frac{1}{1+e^{\xi}}. \tag{2.5}$$

This function satisfies to the first order ordinary differential equation

$$F_\xi = F^2 - F. \tag{2.6}$$

Eq. (2.6) is necessary to calculate the derivatives of function $y(\xi)$. Differentiating the expression (2.6) with respect to ξ and taking into account (2.6) we have

$$y_\xi = \sum_{i=1}^{M} a_i i(F-1)F^i. \tag{2.7}$$

We substitute expressions (2.7) in Eq. (2.1). After it we take $y(\xi)$ from (2.4) into account. Thus Eq. (2.1) takes the form $P[F(\xi)] = 0$ where $P[F(\xi)]$ is a polynomial of function $F(\xi)$. Then we collect all terms with the same powers of function $F(\xi)$ and equate this expression equal to zero. As a result we obtain system of algebraic equations. Solving this system we get the values of unknown parameters.

2. 2 New Expansion Method

Instead of the auxiliary equation given in the Kudryashov method, we have specifically defined a new equation as follows.

$$F_\xi = A_1 F^r - A_2 F \text{ (where } F_\xi = \frac{dF}{d\xi}\text{, } A_1 \text{ and } A_2 \text{ are constants).} \tag{2.8}$$

Similarly, applying steps described in the Kudryashov method (2.1), wave solutions in general form were obtained. The fundamental step of this new technique is to take full advantage of the Bernoulli equation. Auxiliary Eq. (2.8) which obtained at last step is an extended form of Eq. (2.6) in the Kudryashov method. In this new expansion method, general forms of wave solutions were obtained for special A_1 and A_2 values different from Kudryashov method. According to case of A_1 and A_2 generalized solutions of Bernoulli equation given by Eq. (2.8) are written as following:

$$F(\xi) = \left(1 + cCosh[(r-1)\xi] + cSinh[(r-1)\xi]\right)^{\frac{1}{1-r}}, \tag{2.9}$$

where $A_1 = 1$, $A_2 = 1$; c is an arbitrary constant from Eq. (2.8).

3 Exact Solitary Wave Solutions of the Nonlinear Differential Equations

3.1 Exact solitary wave solutions of the system of the shallow water wave equation

In this example we consider the system of the shallow water wave equation [8]

$$\begin{aligned} &\tilde{u}_t+\tilde{u}_x\tilde{v}+\tilde{v}_x\tilde{u}+\tilde{v}_{xxx}=0,\\ &\tilde{v}_t+\tilde{u}_x+\tilde{v}\tilde{v}_x=0. \end{aligned} \tag{3.1}$$

$$\tilde{u}(x,t)=\tilde{u}(\xi) \text{ and } \tilde{v}(x,t)=\tilde{v}(\xi), \quad \xi=kx+wt. \tag{3.2}$$

Using Eq. (3.2) on Eq. (3.1) and integrating with respect to ξ yields, we yield following equation,

$$\begin{aligned} &w\tilde{u}+k\tilde{u}\tilde{v}+k^3\tilde{v}''=0,\\ &w\tilde{v}+k\tilde{u}+\frac{k}{2}\tilde{v}^2=0. \end{aligned} \tag{3.3}$$

The solutions of the systems, we are looking for in stated in the form

$$\tilde{u}(x,t)=\tilde{u}(\xi)=\sum_{i=0}^{M_1}a_iF^i, \qquad \tilde{v}(x,t)=\tilde{v}(\xi)=\sum_{i=0}^{M_2}b_iF^i. \tag{3.4}$$

Balancing the highest linear terms with the highest nonlinear terms Eq. (3.3) we can found $M_1=2r-2$, $M_2=r-1$ and

$$\tilde{u}=a_0+a_1F+a_2F^2+\ldots+a_{2r-2}F^{2r-2}, \quad \tilde{v}=b_0+b_1F+\ldots+b_{r-1}F^{r-1}. \tag{3.5}$$

Substituting (3.5) into Eq. (3.3) yields a set of algebraic equations for $k,w,a_0,a_1,\ldots,a_{2r-2},b_0,b_1,\ldots,b_{r-1}$. With the aid of Mathematica from the solutions of the system, we can found

Case I:

$$\begin{aligned} &k\neq0,\quad w\neq0,\quad m\in(0,r-1),\quad l\in(r-1,2r-2),\quad where\quad m,l\in IN^+,\\ &a_{r-1}=(2r-2)A_2w,\quad a_{2r-2}=-\frac{(2r-2)A_2^2w}{A_1},\quad b_0=\pm2\sqrt{(r-1)A_1w},\\ &b_{n-1}=\mp\frac{2A_2\sqrt{(r-1)A_1w}}{A_1},\quad k=\mp\frac{\sqrt{w}}{\sqrt{(r-1)A_1}},\\ &a_0=a_m=a_l=b_m=0,\quad A_1\neq0,\quad \alpha\neq0,\quad r>2. \end{aligned} \tag{3.6}$$

Case II:

$$\begin{aligned} &k\neq0,\quad w\neq0,\quad m\in(0,r-1),\quad l\in(r-1,2r-2),\quad where\quad m,l\in IN^+,\\ &a_{r-1}=-(2r-2)A_2w,\quad a_{2r-2}=\frac{(2r-2)A_2^2w}{A_1},\quad b_0=\pm2i\sqrt{(r-1)A_1w},\\ &b_{r-1}=\mp\frac{2iA_2\sqrt{(r-1)w}}{\sqrt{A_1}},\quad k=\pm\frac{i\sqrt{w}}{\sqrt{(r-1)A_1}},\\ &a_0=a_m=a_l=b_m=0,\quad A_1\neq0,\quad \alpha\neq0,\quad i^2=-1,\quad r>2. \end{aligned} \tag{3.7}$$

Solution 1:

We obtain the solution of Eq. (3.1) for case 1 as following

$$\tilde{u}(x,t)=(2r-2)A_2w\left(\left(\frac{A_2}{A_1}+cCosh\left[A_1(r-1)\xi\right]+cSinh\left[A_1(r-1)\xi\right]\right)^{\frac{1}{1-r}}\right)^{r-1}-$$

$$=-\frac{(2r-2)A_2^2w}{A_1}\left(\left(\frac{A_2}{A_1}+cCosh\left[A_1(r-1)\xi\right]+cSinh\left[A_1(r-1)\xi\right]\right)^{\frac{1}{1-r}}\right)^{2r-2},\tag{3.8}$$

$$\tilde{v}(x,t)=\pm 2\sqrt{(r-1)A_1w}\mp\frac{2A_2\sqrt{(r-1)w}}{\sqrt{A_1}}\left(\left(\frac{A_2}{A_1}+cCosh\left[A_1(r-1)\xi\right]+cSinh\left[A_1(r-1)\xi\right]\right)^{\frac{1}{1-r}}\right)^{r-1},$$

$$\xi=\pm\frac{\sqrt{w}}{\sqrt{(r-1)A_1}}x+wt.$$

Solution 2:

We obtain the solution of Eq. (3.1) for case 2 as following

$$\tilde{u}(x,t)=-(2r-2)A_2w\left(\left(\frac{A_2}{A_1}+cCosh\left[A_1(r-1)\xi\right]+cSinh\left[A_1(r-1)\xi\right]\right)^{\frac{1}{1-r}}\right)^{r-1}+$$

$$=+\frac{(2r-2)A_2^2w}{A_1}\left(\left(\frac{A_2}{A_1}+cCosh\left[A_1(r-1)\xi\right]+cSinh\left[A_1(r-1)\xi\right]\right)^{\frac{1}{1-r}}\right)^{2r-2},\tag{3.9}$$

$$\tilde{v}(x,t)=\pm 2i\sqrt{(r-1)A_1w}\mp\frac{2iA_2\sqrt{(r-1)w}}{\sqrt{A_1}}\left(\left(\frac{A_2}{A_1}+cCosh\left[A_1(r-1)\xi\right]+cSinh\left[A_1(r-1)\xi\right]\right)^{\frac{1}{1-r}}\right)^{r-1},$$

$$\xi=\pm\frac{i\sqrt{w}}{\sqrt{(r-1)A_1}}x+wt.$$

Specifically, in the solution (3.8) we obtained solution below for $n=2$, $A=1$ and $B=1$.

$$\tilde{u}(x,t)=\mp\frac{2cw\left(Cosh\left[tw\pm\sqrt{wx}\right]+Sinh\left[tw\pm\sqrt{wx}\right]\right)}{\left(1+cCosh\left[tw\pm\sqrt{wx}\right]+cSinh\left[tw\pm\sqrt{wx}\right]\right)^2},$$

$$\tilde{v}(x,t)=\mp 2\sqrt{w}\left(1-\frac{1}{1+cCosh\left[tw\pm\sqrt{wx}\right]+cSinh\left[tw\pm\sqrt{wx}\right]}\right).\tag{3.10}$$

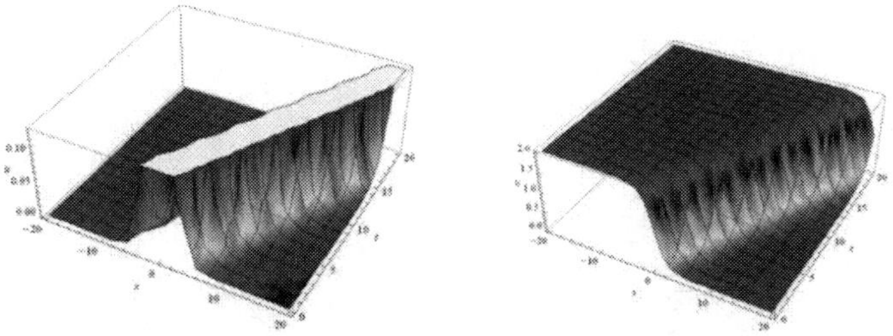

Fig. 1. $\tilde{u}(x,t)$ and $\tilde{v}(x,t)$ three dimensional wave view, by substituting the values $\alpha=1$, $w=1$, $c=1$ respectively for solution (3.10) of the Eq. (3.1).

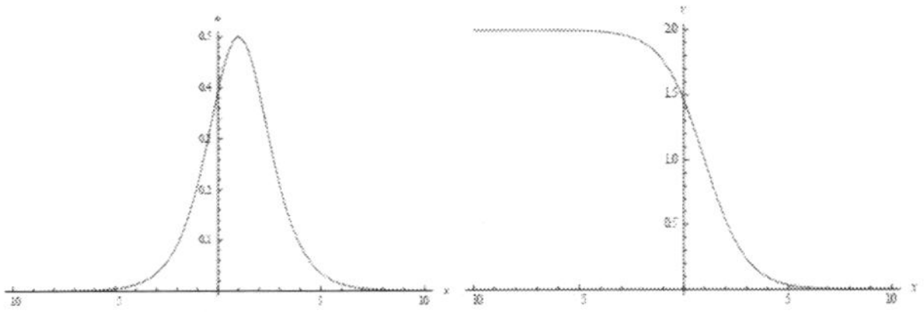

Fig. 2. $\tilde{u}(x,t)$ and $\tilde{v}(x,t)$ two dimensional wave view, by substituting the values $\alpha = 1$, $w = 1$, $c = 1$ and $t = 1$ respectively for solution (3.10) of the Eq. (3.1).

4 Conclusions

In this study, the new expansion method is given by using auxiliary equation. Solutions different from literature and more general solutions have been achieved by applying this method, these solutions have been illustrated by 2 and 3 dimensional graphics. The expansion method we have devised may be used to create solutions to other nonlinear equations and nonlinear system of equations.

References

1. E. Yuce, A. Ö. Tarakçioglu, Attitudes of Computer Engineering Department Students towards ESP Courses Integrated to Foreign Language Courses: Tunceli University Case, Bilim ve Gençlik Dergisi, 1(1), 47-57, (2013).
2. A. Yokus, H. M. Baskonus, T. A. Sulaiman & H. Bulut, Numerical simulation and solutions of the two-component second order KdV evolutionary system. Numerical Methods for Partial Differential Equations, 34(1), 211-227, (2018).
3. L. Shikuo, Jacobi elliptic function expansion method and periodic wave solutions of nonlinear wave equations, Physics Letters A 289.1-2, 69-74, (2001).
4. A. Yokus, T. A. Sulaiman, H. Bulut, On the analytical and numerical solutions of the Benjamin–Bona–Mahony equation. Opt. Quant. Electron. 50, 31, (2018).
5. H. M. Baskonus and H. Bulut, Exponential Prototype Structure for (2+1)-Dimensional Boiti-Leon-Pempinelli systems in Mathematical Physics, Waves in Random and Complex Media, 26(2), 189-196, (2016).
6. N. A. Kudryashov, On Types of Nonlinear Nonintegrable Equations with Exact Solutions, Physics Letters A, 155, 269-275, (1991).
7. N. A. Kudryashov, One method for finding exact solutions of nonlinear differential equations, Communications in Nonlinear Science and Numerical Simulation 17.6, 2248-2253, (2012).
8. S. Duran, and D. Kaya, Applications of a New Expansion Method for Finding Wave Solutions of Nonlinear Differential Equations, World Applied Sciences Journal 18.11, 1582-1592, (2012).

Regarding the numerical solutions of the Sharma-Tasso-Olver equation

Tukur Abdulkadir Sulaiman[1,2], *Asif* Yokus[3], *Nesrin* Gulluoglu[4,*], and *Haci Mehmet* Baskonus[5]

[1]Department of Mathematics, Firat University, Elazig, Turkey
[2]Department of Mathematics, Federal University, Dutse, Jigawa, Nigeria
[3]Department of Actuary, Firat University, Elazig, Turkey
[4]Department of Mathematics, University of Harran, Sanliurfa, Turkey
[5]Department of Computer Engineering, Munzur University, Tunceli, Turkey

Abstract. With aid of the Wolfram Mathematica package, this study investigates the solutions of a nonlinear model with strong nonlinearity, namely; the Sharma-Tasso-Olver equation. We use the improved Bernoulli sub-equation function method in acquiring the analytical solution to this equation, we successfully obtain one-singular soliton solution with exponential function structure. Through the obtained analytical solution, the finite forward difference method is used in approximating the exact and numerical solutions to this equation. We check the stability of the finite forward difference method with this equation using the Fourier-Von Neumann stability analysis. We find the L_2 and L_∞ norm error to the numerical approximation. We present the interesting 3D and 2D figures of the obtained singular soliton solution. We also plot the graphics of the numerical error, exact and numerical approximations data obtained in this study by using the MATLAB package.

1 Introduction

In the recent decades there has been a rapid development in the search of solutions to the nonlinear evolution equations (NEEs) that arise in the various fields of nonlinear sciences such as physics, biological sciences, chemistry, mathematical physics and engineering. Several scientists have developed various technique for finding solutions to such type of problems [1–19].

However, in this study we investigate the solutions of the Sharma-Tasso-Olver equation [20] by using the improved Bernoulli sub-equation function method (IBSEFM) [21] and the finite forward difference method (FDM) [22]. The Sharma-Tasso-Olver equation:

$$u_t + 3au_x^2 + 3au^2u_x + 3auu_{xx} + au_{xxx} = 0, \tag{1}$$

*Corresponding Author: ngulluoglu@harran.edu.tr

where α is a real constant, u is the unknown function which depends on the temporal variable t and the spatial variable x, $u(x,t)$ describes the wave motion. This equation contains both linear dispersive term and the double nonlinear terms.

Taking $a = 1$ [20], we have

$$u_t + 3u_x^2 + 3u^2u_x + 3uu_{xx} + u_{xxx} = 0. \tag{2}$$

2 Analysis of IBSEFM

The IBSEFM is developed by extending the Bernoulli sub-equation function method [21]. In the section we present the steps to follow in using the method.

Step 1. Let us consider the following nonlinear partial differential equation;

$$P(u,\ u_xu^2\ u_x,\ u_t,\ \ldots), \tag{3}$$

performing the wave transformation $u(x,\ y,\ z,\ t) = U(\xi)$, $\xi = x + y + z - ct$, reduces Eq. (3) to the following nonlinear ordinary differential equation (NODE);

$$Q(U,\ U'\ U'',\ U^3,\ \ldots), \tag{4}$$

Step 2. Consider the following as a trial solution to Eq. (3);

$$U(\xi) = \frac{\sum_{i=0}^{n} a_i G^i}{\sum_{j=0}^{m} b_j G^j} = \frac{a_0 + a_1G + a_2G^2 + \ldots + a_nG^n}{b_0 + b_1G + b_2G^2 + \ldots + b_mG^m}, \tag{5}$$

according to the Bernoulli theory,

$$G' = wG + dG^M,\ w \neq 0,\ d \neq 0,\ M \in \mathbb{R} \setminus \{0, 1, 2\}, \tag{6}$$

where $F = F(\xi)$ is a Bernoulli differential polynomial. Substituting Eq. (6) into (4) produces an equation of polynomial $\Psi(G)$ function of G:

$$\Psi(G) = \rho_sG^s + \ldots + \rho_1G + \rho_0. \tag{7}$$

We obtain the values of n, m and M by using the balancing technique.

Step 3. We obtain a system of algebraic equations by letting the summations of every coefficients of $\Psi(G)$ with the same power to zero. We obtain the values of the coefficients $a_0, \ldots, a_n$ and $b_0, \ldots, b_n$ by solving the system of equations.

Step 4. Solving the nonlinear Bernoulli differential equation (Eq. (6)), gives the following two cases based on w and d:

$$G(\xi) = \left[\frac{-d}{w} + \frac{\epsilon}{e^{w(M-1)\xi}}\right]^{\frac{1}{1-M}},\quad w \neq d \tag{8}$$

$$G(\xi) = \left[\frac{(\epsilon - 1) + (\epsilon + 1)tanh\left(\frac{w(1-M)\xi}{2}\right)}{1 - tanh\left(\frac{w(1-M)\xi}{2}\right)}\right]^{\frac{1}{1-M}},\quad w = d, \epsilon \in \mathbb{R}. \tag{9}$$

Using the complete discrimination of the polynomial of G, we solve the algebraic system of equations with the help of Wolfram Mathematica 9 and obtain the various cases for the results of the coefficients. We obtain the exact solutions to Eq. (3) by substituting the obtained values of the coefficients into the trial solution that is Eq. (5).

3 The Analysis of FDM

Presenting the analysis of the finite forward difference method needs the following notations;

1. Δx which is the spatial step
2. Δt which is the time step
3. $x_i = a + i\Delta x,\ i = 0, 1, 2, ..., N$ stands for the coordinates points of mesh and $N = \frac{b-a}{\Delta x}$, $t_j = j\Delta x,\ j = 0, 1, 2, ..., M$ and $M = \frac{T}{\Delta t}$.
4. the function $u(x,t)$ stands for the values of these solutions at the grid points which is $u(x_i, t_j) \approx u_{i,j}$.
5. the $u_{i,j}$ represent the numerical approximatios of the exact values of $u(x,t)$ at the point (x_i, t_j)

We then collect the following finite forward difference operators:

$$H_t u_{i,j} = u_{i,j+1} - u_{i,j}, \tag{10}$$

$$H_x u_{i,j} = u_{i+1,j} - u_{i,j}, \tag{11}$$

$$H_{xx} u_{i,j} = u_{i+1,j} - 2u_{i,j} + u_{i-1,j}. \tag{12}$$

$$H_{xxx} u_{i,j} = u_{i+2,j} - 2u_{i+1,j} + 2u_{i-1,j} - u_{i-2,j}. \tag{13}$$

We the give the approximate forms of the partial derivatives for the difference operators as:

$$\left.\frac{\partial u}{\partial t}\right|_{i,j} = \frac{H_t u_{i,j}}{\Delta t} + O((\Delta t)), \tag{14}$$

$$\left.\frac{\partial u}{\partial x}\right|_{i,j} = \frac{H_x u_{i,j}}{\Delta x} + O((\Delta x)), \tag{15}$$

$$\left.\frac{\partial^2 u}{\partial x^2}\right|_{i,j} = \frac{H_{xx} u_{i,j}}{(\Delta x)^2} + O((\Delta x)^2), \tag{16}$$

$$\left.\frac{\partial^3 u}{\partial x^3}\right|_{i,j} = \frac{H_{xxx} u_{i,j}}{(\Delta x)^3} + O((\Delta x)^2), \tag{17}$$

Eq. (2) may be written in the finite difference operator form as:

$$\frac{H_t u_{i,j}}{\Delta t} + 3\Big(\frac{H_x u_{i,j}}{\Delta x}\Big)^2 + 3u_{i,j}^2 \frac{H_x u_{i,j}}{\Delta x} + 3u_{i,j}\frac{H_{xx} u_{i,j}}{(\Delta x)^2} + \frac{H_{xxx} u_{i,j}}{(\Delta x)^3} = 0. \tag{18}$$

Inserting Eqs. (10), (11), (12) and (13) into Eq. (18), yields the following indexed form:

$$u_{i+1,j} = \frac{1}{6}\Bigg(\frac{1}{(\Delta x)} + 6u_{i,j} - 3(\Delta x)u_{i,j}^2 - \frac{1}{\sqrt{(\Delta t)(\Delta x)}}\Big((\Delta t)$$
$$+6(\Delta t)(\Delta x)u_{i-2,j} - 12(\Delta t)(\Delta x)u_{i-1,j} + 12(\Delta t)(\Delta x)u_{i,j}$$
$$+12(\Delta x)^4 u_{i,j} - 36(\Delta t)(\Delta x)^2 u_{i-1,j}u_{i,j} + u_{i,j} + 66(\Delta t)(\Delta x)^2 u_{i,j}^2$$
$$+9(\Delta t)(\Delta x)^4 u_{i,j}^4 - 12(\Delta x)^4 u_{i,j+1} - 36(\Delta t)(\Delta x)^2 u_{i,j}u_{i,j+1}$$
$$-6(\Delta t)(\Delta x)u_{i+2,j}\Big)^{\frac{1}{2}}\Bigg). \tag{19}$$

Remark-1: We consider $u_{i,0} = u_0(x_i)$ as initial value.

4 Application of IBSEFM

In this section, we employ the IBSEFM in searching for the solutions of the Sharma-Tasso-Olver equation given in Eq. (2).

Using the wave transformation $u = U(\xi)$, $\xi = x - kt$ on Eq. (2), gives the following NODE;

$$U^3 - kU + 3UU^{'} + U^{''} = 0. \tag{20}$$

We apply the balancing technique to Eq. (20) there by using the highest power nonlinear term U^3 and the highest derivative $U^{''}$, we therefore obtain this relation $n+1 = M+m$. Choosing $M = 3$ and $m = 1$, we get $n = 3$.

We may therefore write the following equation by using Eq. (5) along with $m = 1$ and $n = 3$;

$$U(\xi) = \frac{a_0 + a_1 G(\xi) + a_2 G^2(\xi) + a_3 G^3(\xi)}{b_0 + b_1 G(\xi)} = \frac{\Phi}{\Omega}, \tag{21}$$

differentiating Eq. (21) for the first and the second time, we have the following equations;

$$U^{'}(\xi) = \frac{\Phi^{'}(\xi)\Omega(\xi) - \Phi(\xi)\Omega^{'}(\xi)}{\Omega^2(\xi)}, \tag{22}$$

$$U^{''}(\xi) = \frac{\Phi^{''}(\xi)\Omega(\xi) - \Phi(\xi)\Omega^{'}(\xi)}{\Omega^2(\xi)} - \frac{[\Phi(\xi)\Omega^{'}(\xi)]^{'}\Omega^2(\xi) - 2\Phi(\xi)[\Omega^{'}(\xi)]^2\Omega(\xi)}{\Omega^4(\xi)}, \tag{23}$$

where $G^{'} = wG + dG^3$, $w \neq 0$, $d \neq 0$. Substituting Eq. (22) and (23) into Eq. (20) and simplifying, we obtain an equation that involves the polynomials of G. We therefore collect the system of equations there by equating the summation of the coefficients of G that are having the same power to zero. We solve the system of equations with the help of Wolfram Mathematica 9 and obtain the following case;

Case-1: For $w \neq d$, we have

Case-1.1:

$$a_0 = -2wb_0, a_1 = -2wb_1, a_2 = -4db_0, a_3 = -4db_1, k = 4w^2$$

,

substituting these coefficients into Eq. (21), we obtain the following singular soliton solution to Eq. (2)

$$u(x,\ t) = 2w\left(1 + \frac{2\epsilon w}{de^{2w(x-4w^2t)} - \epsilon w}\right). \tag{24}$$

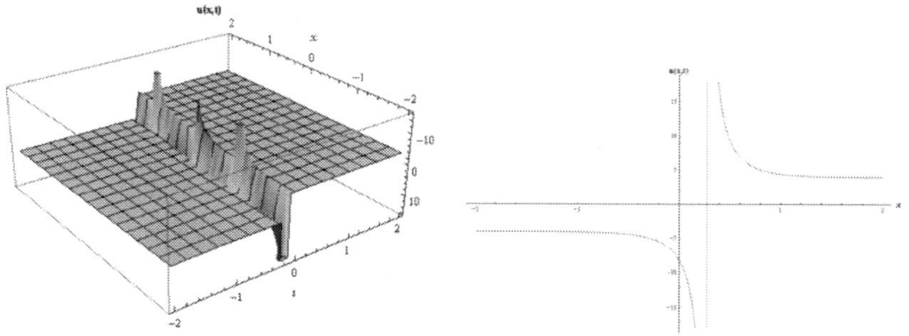

Figure 1. The singular soliton surface of Eq. (24) by substituting the values $w = 2$, $d = 0.2$, $\epsilon = 0.3$, $-13 < x < 13$, $-2 < t < 2$ and $t = 0.0000001$ for the 2D graphic.

5 Von-Neumann Stability Analysis

In this section, we utilize the Von-Neumann stability analysis in investigating the stability of the the finite forward difference method with Eq. (2). We consider ξ^n to be the amplication factor. The growth factor of a typical Fourier mode is therefore defined as:

$$u_m^n = \xi^n e^{im\Phi}, \tag{25}$$

where $i = \sqrt{-1}$.

The Sharma-Tasso-Olver equation is an equation with strong nonlinearity. To check the stability of the numerical scheme, the nonlinear terms u_x^2, u^2u_x and uu_x in Eq. (2) are linearized by making the quantity $\widehat{u_1} = 3u_x$, $\widehat{u_2} = 3u^2$ and $\widehat{u_3} = 3u$ the local constants. Thus, the nonlinear terms $3u_x^2$, $3u^2u_x$ and $3uu_{xx}$ in Eq. (2) become $\widehat{u_1}u_x$, $\widehat{u_2}u_x$ and $\widehat{u_3}u_{xx}$ respectively. In this case Eq. (2) becomes

$$u_t + \widehat{u_1}u_x + \widehat{u_2}u_x + \widehat{u_3}u_{xx} + u_{xxx} = 0. \tag{26}$$

Putting Eq. (14), (15), (16), (17) and Eq. (25) into Eq. (26), hence we get:

$$\xi = X + iY, \tag{27}$$

where

$$X = \frac{1}{((\Delta x)^2 + (\Delta t)\widehat{u_3})}\Big((\Delta x)^2 + (\Delta x)(\Delta t)\widehat{u_1} + (\Delta x)(\Delta t)\widehat{u_2} + 2(\Delta t)\widehat{u_3} - (\Delta x)(\Delta t)\widehat{u_1}cos(\Phi) - (\Delta x)(\Delta t)\widehat{u_2}cos(\Phi) - (\Delta t)\widehat{u_3}cos(\Phi)\Big) \tag{28}$$

and

$$Y = -\frac{(\Delta t)sin(\Phi)}{(\Delta x)((\Delta x)^2 + (\Delta t)\widehat{u_3})}\Big(-2 + (\Delta x)^2\widehat{u_1} + (\Delta x)^2\widehat{u_2} - (\Delta x)\widehat{u_3} + 2cos(\Phi)\Big), \tag{29}$$

which imply that

$$|\xi|^2 = X^2 + Y^2. \tag{30}$$

If this implies $|\xi| \leq 1$ conditional, the method is unconditionally stable [23].

Remark-2: The numerical approximation and stability of the numerical scheme with Eq. (2) depend on how small (Δx) and (Δt) are chosen.

6 L_2 and L_∞ Error Norms

We investigate the numerical and exact approximation of the Sharma-Tasso-Olver equation in this study. We show that how numerical and exact approximations are close to each other by using the L_2 and L_∞ error norms. The L_2 and L_∞ error norms can be defined as [22]:

$$L_2 = \left\|u^{exact} - u^{numeric}\right\|_2 = \sqrt{h\sum_{j=0}^{N}\left|u_j^{exact} - u_j^{numeric}\right|^2},$$

$$L_\infty = \|u^{exact} - u^{numeric}\|_\infty = \underset{j}{Max}|u_j^{exact} - u_j^{numeric}|.$$

7 Numerical Example

This section presents the numerical and exact approximations obtained from Eq. (2) under the following data $w = 2$, $d = 0.2$, $\epsilon = 0.3$. Inserting these values into Eq. (2), yields the following special exact solution:

$$u(x,t) = 4\left(1 + \frac{1.2}{-0.6 + 0.2e^{4(x-16t)}}\right). \tag{31}$$

A $t = 0$, we get the following equations:

$$u_0(x) = 4\left(1 + \frac{1.2}{-0.6 + 0.2e^{4x}}\right). \tag{32}$$

According to the finite forward difference method, inserting the values $(\Delta x) = (\Delta t) = 0.0000001$ into Eq. (19), yields the following:

$$u_{i+1,j} = 1.0666 \times 10^{-13}\Big(1.5625 \times 10^{19} + 9.375 \times 10^{12}u_{i,j} - 468750u_{i,j}^2 + \Big(2.44141 \times 10^{38} + 1.46484 \times 10^{32}u_{i-2,j} - 2.92969 \times 10^{32}u_{i-1,j} + 2.9296 \times 10^{32}u_{i,j} - 8.78906 \times 10^{26}u_{i,j}^2 + 2.19727 \times 10^{11}u_{i,j}^4 - 2.92964 \times 10^{18}u_{i,j+1} - 8.78906 \times 10^{25}u_{i,j}u_{i,j+1} - 1.46484 \times 10^{32}u_{i+2,j}\Big)^{\frac{1}{2}}\Big). \quad (33)$$

We present in Tables 1 and 2 below, the exact and numerical values and the L_2, L_∞ error norm obtained with the finite forward difference method respectively.

Table 1.

x_i	t_j	Numerical solution	Exact Solution	Eror
0	0.01	-8.00024	-7.99616	4.07756×10^{-5}
0.01	0.01	-8.00048	-7.99640	4.07788×10^{-5}
0.02	0.01	-8.00072	-7.99664	4.07821×10^{-5}
0.03	0.01	-8.00096	-7.99688	4.07853×10^{-5}
0.04	0.01	-8.00120	-7.99712	4.07886×10^{-5}
0.05	0.01	-8.00144	-7.99736	4.07919×10^{-5}
0.06	0.01	-8.00168	-7.99760	4.07951×10^{-5}

Table 1: Numerical and exact solutions of Eq. (2) and absolute errors at $\Delta(x) = \Delta(t) = 0.0000001$ and $0 \le x \le 1$ by considering Eq. (31).

Table 2.

$x_i = t_j$	L_2	L_∞
0.001	1.99556×10^{-1}	9.60626×10^{-1}
0.0001	4.20020×10^{-3}	4.39084×10^{-2}
0.00001	1.28802×10^{-4}	4.10969×10^{-3}
0.0000001	1.28378×10^{-7}	4.08043×10^{-5}

Table 2: L_2 and L_∞ error norm when $0 \le h \le 1$ and $0 \le x \le 1$

Figure 2. Numerical and exact solution of Eq. (3) by considering Eq. (8).

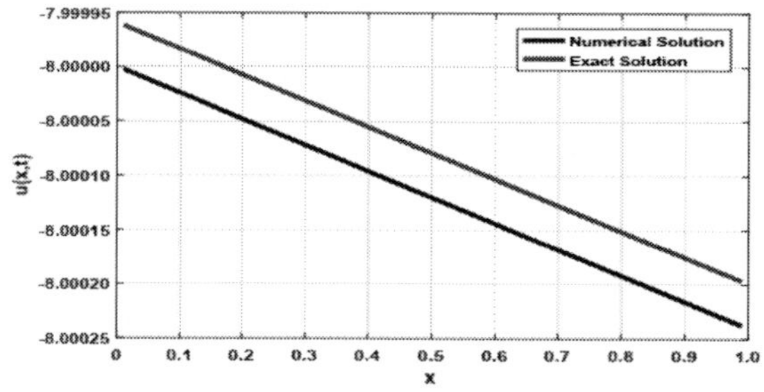

Figure 3. Absolute error graphic at $\Delta(x) = \Delta(t) = 0.0000001$.

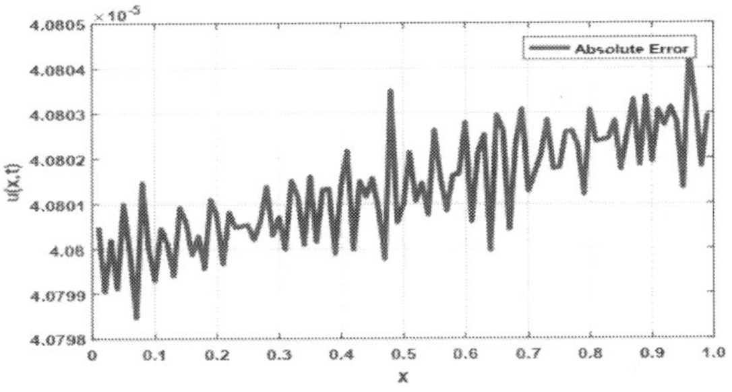

8 Conclusion

I this study, with the aid of the Wolfram Mathematica package, we implemented the improved Bernoulli sub-equation function method in obtaining the analytical solution to the Sharma-Tasso-Olver equation. We successfully obtained one-singular soliton solution in exponential function form. The obtained analytical solution indeed satisfy the considered model. We presented the 2D and 3D graphics of the obtained analytical solution in this study. On the other hand, the finite forward difference scheme is used in approximating the numerical and exact solutions of the considered model. The comparison between exact and numerical approximation is presented in tables 1 and 2 and supported by figure 2 and 3. We observed from figure 2, as Δx approaching 0 the truncation errors are very small, which shows that numerical and exact approximations are coming closer to each other. In summary, the improved Bernoulli sub-equation function method is a powerful and efficient mathematical approach which can be applied to various nonlinear evolution equations and gives a very good results. Furthermore, the analytical solutions obtained by using the improved Bernoulli sub-equation function method can be used to approximate the exact and numerical solutions of various nonlinear evolution equations. To the best of our knowledge this approach is not submitted to literature beforehand.

References

[1] X. Li, International Journal of Differential Equations **2012**, 596762 (2012)
[2] R.I. Nuruddeena, L. Muhammad, A.M. Nass and T.A. Sulaiman, Palestine Journal of Mathematics **7(1)** 262–280 (2018)
[3] T.A. Sulaiman, T. Akturk, H. Bulut and H.M. Baskonus, Journal of Electromagnetic Waves and Applications **32(9)**, 1093–1105 (2017)
[4] H. Bulut, T.A. Sulaiman and B. Demirdag, Nonlinear Dyn. **91** 1985–1991 (2018)
[5] A. Yokus, H.M. Baskonus, T.A. Sulaiman and H. Bulut, Numer Methods Partial Differential Eq. **34**, 211–227 (2018)
[6] Y. Zhao, Journal of Applied Mathematics **2013**, 895760 (2013)
[7] O.A. Ilhan, T.A. Sulaiman, H. Bulut and H.M. Baskonus, Eur. Phys. J. Plus **133**, 27 (2018)
[8] Z. Jin-Ming and Z. Yao-Ming, Chin. Phys. B **21(1)**, 010205 (2011)
[9] H.M. Baskonus, T.A. Sulaiman and H. Bulut, Optik **131**, 1036-1043 (2017)

[10] H. Bulut, T.A. Sulaiman and H.M. Baskonus, Opt. Quant. Electron **48**, 564 (2016)

[11] H. Bulut, T.A. Sulaiman, H.M. Baskonus and A.A. Sandulyak, Optik **135** 327-336 (2017)

[12] S. Momani and S. Abuasad, Chaos, Solitons and Fractals, **27(5)** 1119-1123 (2006)

[13] M. Dehghan and F. Shakeri, Journal of Computational and Applied Mathematics **214**, 435-446 (2008)

[14] O.H. El-Kalaawy, International Journal of Basic and Applied Sciences **3(4)**, 466-478 (2014)

[15] S. Duran, M. Askin and T.A. Sulaiman, IJOCTA **7(3)**, 240–247 (2017)

[16] R. Pourgholi and A. Saeedi, Numer Methods for Partial Differential Equations **33**, 88-104 (2017)

[17] B. Karaagac and A. Esen, Numer Methods for Partial Differential Equations 34(5), 1637-1644 (2018)

[18] O. Omer, A. Esenand F. Bulut, Engineering with Computers 1-12 (2018)

[19] A. Esen, B. Karaagac and O. Tasbozan, Appl. Math. Inf. Sci. Lett 4(1), 1-4 (2016)

[20] I.E. Inan and D. Kaya, Physica A **381** 104-115 (2007)

[21] H. Bulut, H.A. Isik and T.A. Sulaiman, ITM Web of Conferences **13**, 01019 (2017)

[22] A. Yokus, T.A. Sulaiman and H. Bulut, Opt Quant Electron **50**, 31 (2017)

[23] A. Golbabai and E. Mohebianfar, Journal of Nonlinear Science and Applications **49(2)**, 271-288 (2016)

Al-Khwârizmî's Place and Importance in the History of Mathematics

Serbay DURAN[1], *Hüseyin SAMANCI*[1,*]

[1]Department of Mathematics Education, University of Adiyaman, Adiyaman, Turkey

Abstract. The aim of this study is to introduce Muḥammad ibn Mûsâ al-Khwârizmî and his works in terms of history of mathematics and mathematics education. Muḥammad ibn Musa al-Khwârizmî an Iraqi Muslim scholar and it is the first of the Muslim mathematicians who have contributed to this field by taking an important role in the progress of mathematics in his own period. He found the concept of Algorithm in mathematics. In some circles, he was given the nickname Abu Ilmi'l-Hâsûb (the father of the account). He carried out important studies in algebra, triangle, astronomy, geography and map drawing. Algebra has carried out systematic and logical studies on the solution of inequalities at second level in the development of the algebra. He with all these studies have contributed to mathematical science and today was a guide to the works done in the field of mathematics.

Al-Khwârizmî's life

Muḥammad ibn Mûsâ al-Khwârizmî is an Iraqi Muslim scholar who lived in ninth century. Although the exact date of birth was unknown, he was born around 780s CE and died in 847 CE. He was called al-Khwârizmî and Abû Ja'far. The origin of the al-Khwârizmî today is from "Khuwa" city, locates in south of the Aral Sea in Asia (Uzbekistan) His family moved to live in Baghdad, the center of knowledge for his period. He lived in Baghdad during the period of Abbasid Caliphate Ma'mûn [1-2].

He was always attached to Caliphate Ma'mun who was very generous to himself. Caliphate, worked in the House of Wisdom, where he founded for the scholars. At that time, House of Wisdom became famous in the fields of research, translation and publication. The scientists working there translated to Arabic, especially in scientific and philosophical works in Greek [1-2].

Al-Khwârizmî, in the library of Ma'mûn, benefited from many books and interested in mathematics, geography, astronomy, and history. He was also interested in scientific studies of Greek and Indian civilizations [2].

His reputation spread after he came out in the field of astronomy and mathematics and he was the first of the Muslim mathematicians who have contributed to this field by taking an important role in the progress of mathematics in his own period [2].

* Corresponding author: hsamanci@adiyaman.edu.tr

He learned about account transactions, found the concept of Algorithm in mathematics. In some circles, he was given the nickname Abû Ilm al-Hâsûb (the Father of the Accounting) [3].

Al-Khwârizmî in the field of mathematics

Al-Khwârizmî important works in the field of mathematics were a result of his special research. He discussed the mathematics works on Greek and Indian civilizations before he developed the knowledge he gained on this issue. He showed a special interest in combining this knowledge with logic [2-3].

Al-Khwârizmî, carried out important studies in algebra, triangle, astronomy, geography and map drawing and in 830 CE he wrote his work Kitâb al-Jabr wa'l Muqâbala. This term was translated as algebra in many languages with its English name. His work has followed a methodological and logical path to the solution of in the second degree inequalities [2].

Al-Khwârizmî's found the concept of al-Khwârizmîyya in mathematics and calculus that allowed him to be given the nickname of Father of Accounting. Algorithm and algebra terms belong to al-Khwârizmî's famous works in the field of mathematics. These two names are derived from Arabic terms الخوارزمي (al-Khwârizmî-algorithm) and الجبر (al-Jabr-algebra) [3].

The concept of al-Khwârizmî's is derived from his name and is referred to English as
He made a fundamental change in the concept of numbers and he used Arabic numbers. It brought the concept of zero as a number [3].

The works of al-Khwârizmî contributed to the transfer of Arabic numbers and the concept of algebra in mathematics to Europe and in the development of algebra science, systematic and logical studies have carried out on the solution of the second degree inequalities [3].

Al-Khwârizmî's great contribution to the calculations process, as well as the astronomy province, was revealed. He has done new research on triangles. He made the astronomy Zîj tables.These so-called Zîj tables have had a major impact on the tables that the Arabs created in the field of astronomy later [3].

Kitâb al-Jabr wa'l Muqâbala dealt with inheritance, wills, property sharing, commercial affairs and decision-making in relation among people, making calculations of two sites and dealing with engineering work [3].

One of the most important contributors of al-Khwârizmî's scientific work has been the addition of Ptolemy's geographical studies as both text and map [3-4].

Al-Khwârizmî made some adjustments by conducting special research on the geographical map created by Greek Ptolemy [3-4]. He also took part in geographical studies in order to reveal the first world map at that time and he wrote the most important work in the field of geography Sûrat al-Ard. This book explains the positions of the countries of his time [4].

Al-Khwârizmî's Kitâb al-Jabr wa'l Muqâbala

Kitâb al-Jabr wa'l Muqâbala is the most important work that al-Khwârizmî wrote as a result of his important researches in mathematics [1-2]. Kitab al-Jabr was the first of the books on the systematic solution of linear and quadratic equations. He benefited from Greek and Indian sources in the past to gather and develop the knowledge he possesses. He showed a special interest in evaluating the knowledge obtained logically and combining that with his own considerations [3].

Al-Khwârizmî is obviously the father of algebra and shares the same narrative with Greek mathematician Diophantus. His theorem went into the areas of accountancy with his Latin translations [3].

Al-Khwârizmî has made algebra a science which to be benefited with full independence through its own method and rules [4]. He brought new terms to understand calculations and mathematical operations with it. Algebra, derived from the al-jabr term, is one of the two operations used to solve linear equations.

Al-Khwârizmî showed a great ability to understand and to grasp wide possibilities of algebra. He was able to solve engineering problems by algebraic methods. He alerted to the situation where it is impossible to find real value for the unknown and called it "impossible problems." This term remained in circulation until the late 18th century [5-6].

Al-Khwârizmî completely changed his understanding of numbers and he used the world of Arabic numbers. He brought the concept of zero. In the 12th century, he transferred the system of decimals from Indian numbers to the west. Then nine Arabic numbers have spread all over Europe [5]. When the West transferred their numbers from the Arabs, They moved with it their way of reading numbers from right to left [6].

Al-Khwârizmî's efforts were not limited to educating the West in writing numbers and calculations. It has gone beyond that complex math problems. His calculation of the rules even today still carry his name like a flag [7].

He is the first person to set the base for the calculation of the logarithm and this science is called by this name. The logarithm term derives from the Persian term "al-Khwârizm." The name is so known in Latin and algorithm is a division between calculus and algebra [7].

Al-Khwârizmî developed an applied method to find the area of the surfaces, the area of the circle, the area of a part of the circle and the area of the triangle. He arrived to calculate the size of the triple pyramid and the size of the quadratic pyramid and he put the method of multiplication of roots [8].

Al-Khwârizmî brought the terms of the equations of the first and second degrees and found the cues for them. After al-Khwârizmî, Muslim mathematicians came. They also worked on the development and dissemination of equations [2-3]. He is the first to replace the signs (- or +) when moved from one side of the equation to the other [3].

He wrote a booklet other than Kitâb al-Jabr wa'l Muqâbala that related to accounting science. He explained the usage system of Indian figures in this booklet and addition, subtraction, division, multiplication and calculation of fractions. This booklet was translated into Latin in the 12th century. The manuscript of the book is in the Vienna Balat Library. The second version of the book was found in Deir Salem and is now preserved in Heidelberg in Germany [5-6].

When he dealt with his book location of zero such as thirty-eight minus twenty-eight equals ten in the addition and subtraction operations, he said that In the subtraction operations If there was no more, we put zero. We do not leave the empty place until there is no confusion between the units digit and the dozens digit.

And he adds "the zero should be on the right of the number." Because the zero is on the left of the one. İt does not change its value and does not make it ten [6].

We see that Western translators of Arabic sources have translated this system literally into Latin, and transferred from the system of writing and reading from the Arabs. In other words from right to left [7-8].

Al-Khwârizmî also corrected the researches of the Greek scientist Ptolemy in geography. He also oversaw the work of 70 geographies to complete the first map of the world known at the time. Of the most famous books in the geography book of the Earth image [8].

Al-Khwârizmî's works

1. **Kitâb al mukhtasar fi Hesab al jabr wa'l muqâbala**
2. **Kitâb fi Jourafia Sharh fihi Ârâi Ptolemy**
3. **Kitâb jadâvil li'n-Nujûm ve Harakâtiha min Mujelladayn**
4. **Kitâb fit-Tariqâti Ma'rifat al-Waqt bi Wasâtat al-Shams**
5. **Kitâb jamî fihi bayna'l-Hisâb wa'l-Hendese wa'l-Mûsîgâ wa'l-Falak**
6. **Kitâb al-Amâl bi'l-Ousturlab**
7. **Kitab Sun'i'l-Ousturlab**
8. **Kitâb Wazhi fihi Tariqât al-Jam wa't-Tarkh**
9. **Kitâb Sûrat al-Ard ve Jourafiatihâ**
10. **Kitâb Surat al-Ard**
11. **Kitâb al-Mârifa**
12. **Kitâb Zîj al- al-Khwârizmî al-Avval**
13. **Kitâb Zîj al-Khwârizmî al-Sânî**
14. **Risâlat al-Nisbat at-Takrîbiyya ve Qıymatih ar-Riyâziyya**
15. **Risâlat Wazhi fihâ Ma'nâ al-Wahdat al-Musta'mal fi'l-Masâhâti wa al-nujûm**
16. **Risâlet Zikri fihâ Burhânan Âhira li Nazariyya Pisagor Mustakhdiman Musallas Qâim al-Zâviyat wa Mustevâ al-Sâqayn**
17. **Kitâb Rasm al-Ma'mûr**
18. **Kitâb al-Târīkh**
19. **Kitâb al-jam'i wa al-Tafrîq**
20. **Kitâb Hay'at al-Ard**
21. **Kitâb al-Rukhâma**
22. **Kitâb al-Mu'âmalât**

References

1. H. Az-Ziriklî, Al-a'lâm, Dâru'l-'İlmi li'-l-Melâyîn, Beyrût, (2002).
2. K. Tarek, The Universal Effect for Arabic İslamic Civilization, İskenderiyye, (2009).
3. K. A. Salah, Ulemâu'l-Muslimîn fi'l-Ulûm ve'l-Teknolojia, (2010).
4. A. A. Al-Daffa, The Muslim Contribution to Mathematics, Atlantic Highlands, NJ: Humanities Press, (1977).
5. J. Hoyrup, Al-Khwârizmî, Ibn Turk, and the Liber Mensurationum, on the Origins of Islamic Algebra, Erdem 10.5, (1986).
6. S. Fuat, Al-Kitâb al-mukhtasar fi hisâb al-jabr wa-l-muqâbala by Muhammad ibn Mûsâ al-Khwârizmî, Western Translations and Adaptations, (2006).
7. Van der Waerden, L. A. Bartel, history of algebra, from al-Khwārizmī to Emmy Noether, Springer Science & Business Media, (2013).
8. R. Rashed, al-Khwârizmî le commencement de l'algèbre, Quadrature 23, (2009).

Pre-Service Lower Secondary School Mathematics Teachers' Ability of Drawing the Symmetry of a Figure according to a Line

Ebru Mutlu[1,*] and *Asuman* Duatepe-Paksu[1]

[1]Department of Mathematics and Science Education, Pamukkale University, Turkey

Abstract. The aim of the present study is to investigate pre-service lower secondary school mathematics teachers' ability of drawing a given figure's symmetry according to the lines in various positions. The sampling of the study comprised eight pre-service mathematics teachers who were attending a mathematics education department and willing to participate in the study. The criterion sampling methods was used. Six open-ended questions were asked to determine the ability of the prospective teachers to take the symmetry of a given figure according to the lines in different situations. Furthermore, interviews were conducted with the participants on their definitions of symmetry and symmetry according to a line. The data of the study were the prospective teachers' drawing papers and the video recordings of the interviews including the definitions of the concepts of symmetry and symmetry according to a line. Content analysis was used to analyze the data. The findings indicated that the prospective teachers were successful in drawing the symmetry of a given figure according to the vertical, horizontal and oblique lines, and they used an informal language to define the concepts of symmetry and symmetry according to a line.

1 Introduction

One of the aims of the mathematics curriculum at primary education is to train individuals who can understand mathematics and use it in daily life. This fact has been emphasized for teaching mathematics to use the new curriculum which has been prepared by focusing on the national and international research studies, mathematics curriculum of developed countries and the experiences of mathematics education in our country (MoNE, 2018).

It is evident in the related literature that many different terms are used to express the difficulties in mathematics learning. The terms "difficulty" and "mistakes" are the terms most often used by the students to express the difficulties they have experienced in learning mathematics (Aksoy and Bayazit, 2010). Students, teacher candidates and teachers are faced with many difficulties regarding many concepts specific to mathematics. Some of these studies focus on the concept of 'symmetry' (Köse, 2012, Aktaş, 2015). Since the concept of symmetry is interwoven with the real life, it is one of the fundamental issues for the students to get prepared for the life. In mathematics, the concept of symmetry is

* Corresponding author: emutlu@pau.edu.tr

intertwined with nature (Dreyfus & Eisenberg, 1989). Studies in the literature on symmetry are generally conducted on students, and they aim to determine their current knowledge and misconceptions (Köse 2012, Karadeniz et. al 2015). There is a limited research on what pre-service mathematics teachers pay attention when taking the symmetry according to a line, and on determining their existing misconceptions. The purpose of this study is to reveal the existing misconceptions of pre-service teachers regarding the concept of reflection symmetry.

2 Aim of the Study

The aim of the current study is to find out how pre-service primary school mathematics teachers define symmetry and their knowledge about drawing the symmetry of a figure according to a given symmetry line. In response to this general objective, answers to the following questions are investigated.
Sub-questions:
1-) How is the ability of pre-service secondary school mathematics teachers in taking the symmetry of a given shape according to the horizontal, vertical and oblique lines when the shape cuts the symmetry line or not?
2-) How do the pre-service secondary school mathematics teachers define the concepts of 'symmetry' and 'according to straight line'?

3 Method

In this study, which aims at determining the pre-service mathematics teachers' definitions of the 'symmetry' and 'symmetry according to the straight line' concepts. A case study design of qualitative research methods was used in collecting, analyzing and interpreting the data obtained from the participants.

3.1 Participants of the Study

The sampling of the study consists of eight senior pre-service mathematics teachers (5 females and 3 males), who are attending a mathematics education department in the study. The aim of selecting final year prospective teachers is that they are considered to have the attainments regarding the concepts of symmetry and reflection symmetry in geometry lessons. The average geometry course grade of the participants' is 85 and above. The criterion sampling method, which is non-probabilistic method was used to choose the participants.

3.2 Data Collection

Six open-ended questions were used to determine the ability of prospective teachers to take the symmetry of a figure according to the vertical, horizontal and oblique lines for two cases whether the figure intersect or does not intersects with the line (Figure 1). Subsequently, interviews were held on the candidates' definitions of the concepts of 'symmetry' and 'symmetry according to a line". The data of the study were collected through one-to-one interview method. Each interview with the participants took approximately 30 minutes and they were recorded. In this way, the meanings that the prospective teachers attributed to the idea of taking the symmetry of an object are elaborated. A semi-structured interview method inspired from Köse (2012) was created. Expert opinions were also collected to form the interview questions.

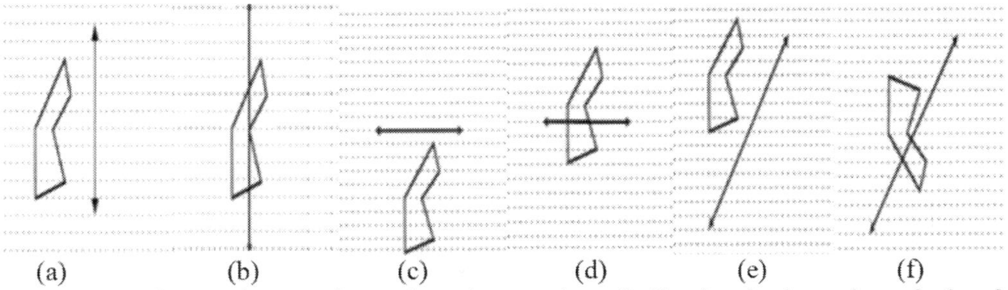

Fig.1. A figure's intersecting and not intersecting of a line in a horizontal, vertical and oblique position

3.3 Data Analysis

In the analysis of the data, the following four categories adapted by Lima (2006; as cited in Köse, 2012) were utilized to take the symmetry of a given shape according to a line: 1. Completely correct drawing, 2. Partially correct drawing, 3. Wrong drawing, 4. Blank or incomplete drawing.

4 Findings

Some examples of pre-service mathematics teachers drawing of the symmetry of the figures according to lines in vertical, horizontal and oblique positions are shown in Fig 2-4. Fig 2 shows the completely correct drawing of symmetry according to vertical line drawn by pre-service teachers 2 (PT2). Fig 3 shows the partially correct drawing of symmetry according to horizontal line drawn by PT6. Fig 4 shows the wrong drawing of symmetry according to oblique line drawn by PT4.

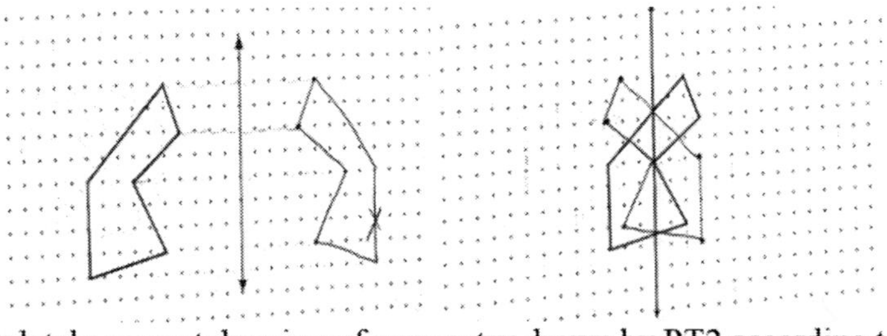

Fig.2. Completely correct drawing of symmetry drawn by PT2 according to vertical line

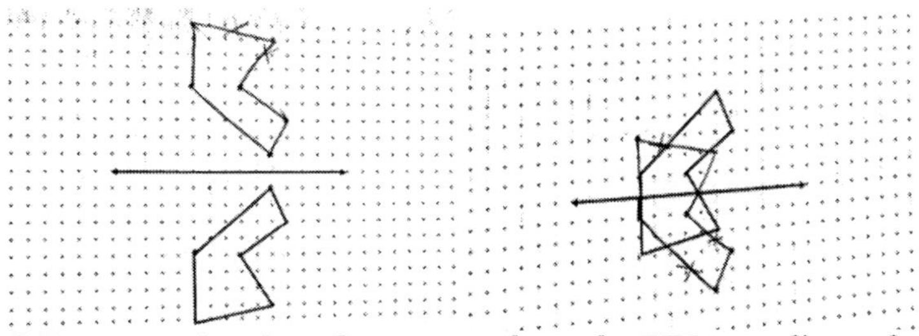

Fig.3. Partially correct drawing of symmetry drawn by PT6 according to horizontal line

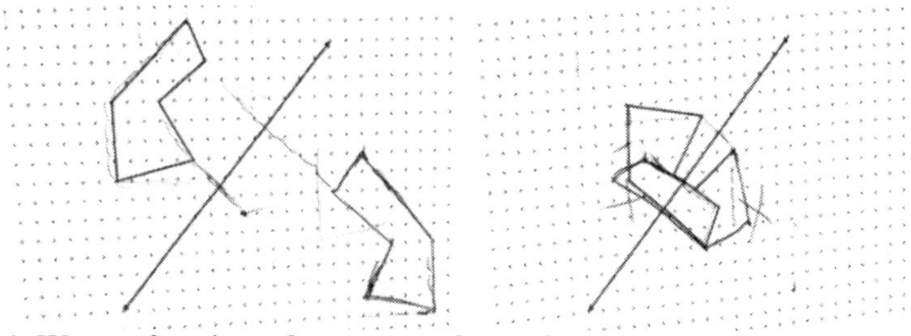

Fig.4. Wrong drawing of symmetry drawn by PT4 according to oblique line

Table 1 shows the answers of pre-service mathematics teachers to six open-ended questions regarding figures given in Figure 1.

Table 1. The frequency and percentage of responses of teacher candidates to the question of taking the symmetry of a figure according to a line in vertical, horizontal and oblique positions [f (%)]

Figures	Completely correct	Partially correct	Wrong	Blank or incomplete
Figure 1(a)	7 (87.5)	1 (12.5)	0 (0)	0 (0)
Figure 1(b)	7 (87.5)	1 (12.5)	0 (0)	0 (0)
Figure 1(c)	5(62.5)	3 (37.5)	0 (0)	0 (0)
Figure 1(d)	2 (25)	6 (75)	0 (0)	0 (0)
Figure 1(e)	4 (50)	2 (25)	2 (25)	0 (0)
Figure 1(f)	5 (62.5)	3 (37.5)	0 (0)	0 (0)

The definitions of the pre-service mathematics teachers regarding the symmetry concept are given in Table 2. For instance, pre-service teacher 5 (PT5) defined the symmetry concept as "*By taking according to that line, it is crossing over a certain point or any object.*"

Table 2. Participants' definitions regarding the symmetry concept

Definitions	Frequency	Pre-service Teacher (PT)
Mirror image	1	PT1
Reflection	2	PT2, PT7
Equidistance	1	PT4
Mirror image and reflection	1	PT5
Equidistance and reflection	1	PT8
Equidistance and mirror image	2	PT3, PT6

Definitions of the pre-service teachers regarding the concept of symmetry according to a line are given in Table 3. For instance, pre-service teacher 2 (PT2) defined the symmetry according to a line as "*Symmetry according to the line is the co-image of a line on the other side of the line, the shortest distance of a point to the point on that line.*"

Table 3 Participants' definitions regarding the concept of symmetry according to a line

Definitions	Frequency	Pre-service Teacher (PT)
Reflection	4	PT3, PT4, PT6, PT7
Co-image	1	PT2
Equal distance image	2	PT1, PT8
Crossing over	1	PT5

Based on the findings of the study, there are some variations on the ability of fourth grade pre-service secondary school mathematics teachers to take the symmetry of a given figure according to a line in different positions. The findings of the present study are in line with those of a study conducted by Karadeniz et. al. (2015) on prospective teachers. When the answers given by the prospective teachers were examined, it is clear that they were the most successful at the question about taking the symmetry of a given figure according to a vertical line. On the other hand, they had the lowest success at the question including a figure's intersecting and not intersecting of an oblique line.

It was found that the teacher candidates used one or more of the terms of reflection, mirror, equidistance, co-image in their expressions to explain the concept of symmetry and symmetry according to a straight line. When the definitions were examined, it was seen that the teacher candidates used an informal language. Teacher candidates' use of that informal language to express the symmetry concept similar with the findings of the study of Kose (2012) on the 8th grade students. It is noteworthy that the data obtained from the current research sampling are similar with those of the 8th grade students. Only one teacher candidate tried to explain the concepts of symmetry and symmetry according to a line by using correct mathematical expressions. However, he could not make an appropriate definition. In this context, it can be also considered that the participating teacher candidates may not have efficient language skills.

5 Conclusion and Implication

Based on the analysis of the research data, it can be said that prospective teachers are generally successful in their skills to take the symmetry of a given figure according to the different cases of a line in contrast to the findings in the literature (Köse, 2012). This is thought to be due to the sampling of the study which consists of 4th grade pre-service mathematics teachers. It can be said that they are likely to be more successful when they consider the basic components while taking symmetry according to a straight line. According to the findings of the Aktas's study (2015), the achievements of pre-service teachers can be further increased by using dynamic geometry software to take the symmetry of a figure according to a line in each case. Considering that teacher candidates have taken geometry and geometry teaching courses, it can be anticipated that these courses may have affected their ability of taking symmetry.

Moreover, given the definitions of the pre-service teachers regarding the concepts of symmetry and the symmetry according to a line, it is concluded that they cannot use mathematical expressions efficiently when making the definitions. For the development of teacher candidates' skills of definition, topics including activities to enhance that skill can be involved in the course contents taken during undergraduate education.

Studies can be conducted with a broader sample group for undergraduate students and in technology-supported settings. It is especially considered that the studies to measure the pre-service teachers' skills of definition or to develop these skills will also contribute to the existing literature. It is thought that the pre-service teachers who will graduate from the education faculties should be supported with activities including verbal skills along with mathematical skills.

References

1. Aksoy, Y. ve Bayazit, İ. (2010). Simetri kavramının öğrenim ve öğretiminde karşılaşılan zorlukların analitik bir yaklaşımla incelenmesi. E. Bingölbali and M. F. Özmantar (Eds.), İlköğretimde karşılaşılan matematiksel zorluklar ve çözüm önerileri içinde (2. Baskı., s. 187-215). Ankara: Pegem Akademi.
2. Aktaş, M. (2015). 7.Sınıf matematik dersinde bilgisayar animasyonları ve aktiviteleri ile simetri öğretiminin akademik başarıya etkisi. GEFAD / GUJGEF 35(1): 49-62
3. Dreyfus, T., & Eisenberg, T. (1989). Symmetry in mathematics learning. zentralblatt fur didaktik der mathematik. 90(2), 53-59.
4. Karadeniz, M.H., Baran, T., Bozkuş, F. & Gündüz, N. (2015). İlköğretim Matematik Öğretmeni Adaylarının Yansıma Simetrisi ile İlgili Yaşadıkları Zorluklar. Turkish Journal of Computer and Mathematics Education. Vol.6 No.1, 117-138.
5. Köse, N.Y. (2012).İlköğretim öğrencilerinin doğruya göre simetri bilgileri. Hacettepe University Journal of Education Faculty, 42: 274-286.
6. Ministry of National Education [MoNE]. (2018). 6-8th grade primary school mathematics course curriculum. Board of Education and Discipline.
7. National Council of Teachers of Mathematics [NCTM]. (2000). Principles and Standards for School Mathematics. Reston, VA.

ITM Web of Conferences **22**, 01039 (2018) https://doi.org/10.1051/itmconf/20182201039
CMES-2018

A Note on a Binary Relation Corresponding to a Bipartite Graph

Hatice Kübra Sarı[1*], *Abdullah Kopuzlu*[1] ,

[1]Department of Mathematics, University of Atatürk, Erzurum, Turkey

Abstract. In this paper, we firstly define a binary relation corresponding to the bipartite graph and study its properties. We also establish a relationship between the independent sets of the bipartite graph and the definable sets of binary relations corresponding to the bipartite graph.

1 Introduction

Rough set theory was initially defined in [9] for overcome the uncertainty in information systems. This theory enabled approximately characterization an subset of an universe by using operators called as lower approximation and upper approximation operator. Rough set theory has many application in fields, e.g. machine learning, knowledge discovery, decision analysis.

In rough set theory, the space that Pawlak mentions is the approximation space created by the equivalence relation R . But, such an equivalence relation is inadequate to using in some applications [5,15,19,20]. Because of this, many extensions of equivalence relations such as arbitrary binary relations [10,17,16], tolerance relations [7,8,11,13], similarity relations [1,12,14] and pre-order relations [4,6] have been proposed.

It is shown that a relation R can be represented by a graph [4]. In [3], authors say that an undirected simple graph can be represented a corresponding binary relation. Moreover, they show that the binary relation induced by a graph has irreflexive property. A graph is named as a bipartite graph if set of vertices of the graph can be divided into two independent sets. In this paper, we study the binary relation corresponding to the bipartite graph. We prove that this relation has irreflexive and nontransitive properties.

2 Preliminaries

We present in section, preliminary definitions that we shall use in our work related generalized approximation spaces and graph theory.

2.1. Approximation Spaces and Approximation Operators

Definition 1. [3] Consider a finite non-empty set U.

i) An irreflexive relation is the relation R on U satisfies condition "For all $x \in U, (x,x) \notin R$ "

* Corresponding author: hatice.sari@atauni.edu.tr

ii) A serial relation is the relation R on U satisfies condition "For all $x \in U$, there exists $y \in U$ such that $(x, y) \in R$."

Definition 2. [18] Consider a finite non-empty set U and a binary relation R on U. The pair (U,R) is a generalized approximation space. Also the space is marked as GA-space. When R is an equivalence relation on U, GA-space (U,R) coincide with Pawlak approximation space [10,17].

Definition 3. [15] Given a (U,R) GA-space. For any $X \subseteq U, \underline{R}: P(U) \to P(U), \overline{R}: P(U) \to P(U)$ the lower and the upper approximation operators of approximation space (U,R) respectively are defined as:

$\underline{R}X = \{x \in U : R_s(x) \subseteq X\}, \quad \overline{R}X = \{x \in U : R_s(x) \cap X \neq \varnothing\}$.

Here $R_s(x) = \{y \in U : (x,y) \in R\}$.

The subset X is named definable set iff $\underline{R}X = \overline{R}X$. Or else, X is named a rough set.

2.2. Bipartite Graphs

Because bipartite graphs frequently used in data modeling in computer science have an important place in graph theory, characterization of bipartite graphs is important [2].

Definition 4. [2] Consider a finite non-empty set U. The pair $G = (U,E)$ whose sets of vertices and edges repectively is U and $E \subseteq U \times U$ is a graph. A vertice is called isolated vertice if it is not an end of any edge. Arbitrary two vertices of G are called as adjacent if there is at least one edge linking them. The graph G whose every edge links different two vertices is called as a simple graph. A set of vertices whose elements are pairwise non-adjacent is called an independent set [2].

Definition 5. [2] A bipartite graph is a graph G whose vertex set U can be divided into non-empty subsets X and Y such that every one edge has one vertice in X and one vertice in Y. That is, a graph G is called bipartite iff the vertex set U of G can be divided into two independent sets.

Binary relations on U can be represented by graphs whose vertices are element of U, where there is an edge that vertex x is related vertex y iff $(x,y) \in R$ and $(y,x) \in R$. The graph induced by R is called a graph of R [4]. An undirected simple graph can also be represented a corresponding binary relation [3]. When such the graph $G = (U,E)$ is given corresponding approximation space can be defined (U,R) by: $(x,y) \in E$ iff $(x,y) \in R$ and $(y,x) \in R$. The relation R is called relation induced by G.

Now we give properties of the relation induced by G.

Definition 6. [3] Let R be a serial, irreflexive and symmetric relation on U. Then R is a serial preclusivity relation (SP-relation) on U. The approximation space (U,R) is a serial preclusivity approximation space (SPA-space).

Clearly, the relation R induced by the undirected simple graph G is SP-relation. SP-relations and undirected simple graphs can be represented by each other [3].

3 Main Result

We firstly shall give following theorem that allows to characterize the independent sets of a simple graph by the way of the upper approximation of R binary relation induced by this graph.

Theorem 1. [3] Given a simple graph $G=(U,E)$ and the relation R induced by G. For $X \subseteq U$, X is an independent set of G iff $\overline{R}X \cap X = \varnothing$.

Definition 7. Let U be a finite non-empty set and R be a relation on U. If there are not elements that provide transitive property with regard to R on U, then R is called nontransitive relation on U.

Definition 8. Let U be a finite nonempty set and R be a serial, irreflexive, symmetric relation. If R is nontransitive, then it is called R serial-nontransitive relation on U. (U,R) is called serial-nontransitive space.

Now we shall prove the following theorem.

Theorem 2. Let $G=(U,E)$ be a simple graph and R be the relation induced by

G. If G is bipartite then R is nontransitive relation on U.

Proof. We suppose that G is a bipartite graph. Then there are A and B independent sets such that $\overline{R}A=B,\ \overline{R}B=A,\ A\cap B=\varnothing,\ A\cup B=U.$ We assume that there exist $x,y,z\in U$ which provide the transitive property with regard to R on U. Let $A=\{a_1,a_2,...,a_m\}, B=\{b_1,b_2,...,b_n\}$. Since $A\cup B=U$ and $A\cap B=\varnothing$, $x=a_i\in A$ or $x=b_j\in B.$ Let $x=a_i\in A$. Since A is independent set, $(a_i,y)\in R$ iff $y\in B.$ Thus, we have $y=b_j\in B.$ Since B is independent set, $(b_j,z)\in R$ iff $z\in A.$ Thus, we have $z=a_k\in A.$ Since there exist $a_i,b_j,a_k\in U$ such that $(a_i,b_j)\in R$ and $(b_j,a_k)\in R$, we obtain $(a_i,a_k)\in R.$ This contradicts with A being an independent set. Similarly, if we assume that $x=b_j\in B$ then we obtain contradict with B being independent set. Therefore, for any $x,y,z\in U$ such that $(x,y)\in R$ and $(y,z)\in R, (x,z)\notin R.$ Hence R is nontransitive relation on $U.$

Example 1. Bipartite graph $G=(U,E)$ is given by Fig. 1. Here $U=\{x,y,z,t,u,v\}$, $E\subseteq U\times U$. The relation R induced by the bipartite graph G is

$R = \{(x,y),(y,x),(x,z),(z,x),(x,u),(u,x),(y,t),(t,y),(y,v),(v,y),(z,t),(t,z),(t,u),(u,t)\}$.
R is serial-nontransitive relation.

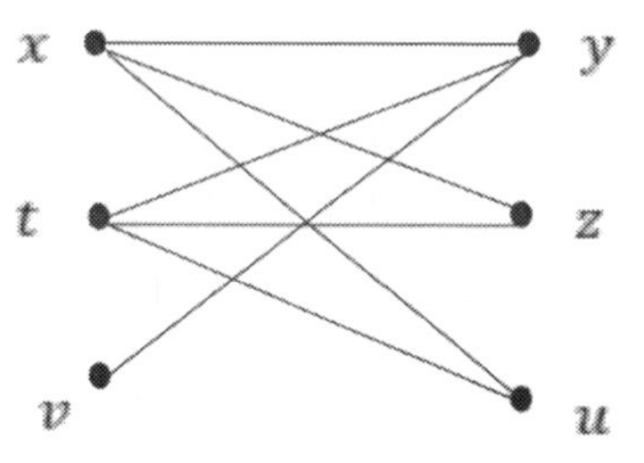

Fig. 1. Bipartite Graph G

Example 2. Let $U = \{x,y,z,t,u\}$, $R = \{(x,y),(y,x),(x,u),(u,x),(z,t),(t,z),(t,u),(u,t)\}$.
R is serial-nontransitive relation. The graph of R is shown as follows:

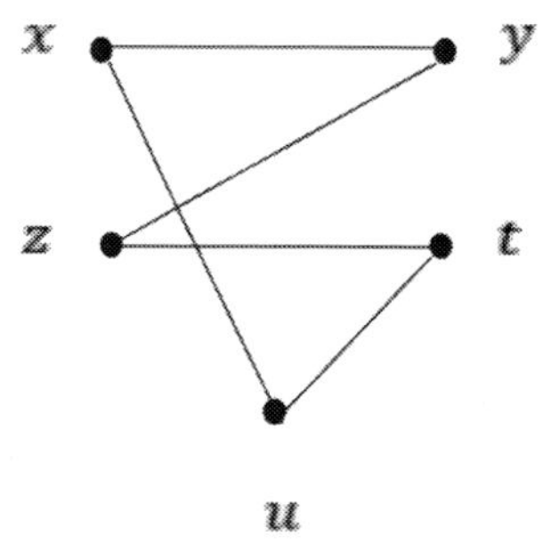

Fig. 2. Graph of Relation R

As seen above, the graph of serial-nontransitive relation R on U is not bipartite graph.

Example 3. Let's search definable sets of the above space (U,R). Since $\underline{R}\{x,t,v\} = \overline{R}\{x,t,v\} = \{y,z,u\}$ and $\underline{R}\{y,z,u\} = \overline{R}\{y,z,u\} = \{x,t,v\}$, the sets $X = \{x,t,v\}$ and $Y = \{y,z,u\}$ are definable sets of serial-nontransitive space (U,R).

It is seen that X and Y being disjoint independent sets such that $X \cup Y = U$ in a bipartite graph are definable sets with regard to binary relation induced by this bipartite graph in examples in above.
Now we give relationship between independent sets of a bipartite graph and definable sets of serial-nontransitive binary relation corresponding to this bipartite graph in the following theorem.

Theorem 3. Given a bipartite graph $G=(U,E)$ and the relation R induced by graph G. Then the disjoint independent sets X and Y such that $X \cup Y = U$ are definable sets of GA-approximation space (U,R).

Proof. Since subsets X and Y are independent sets, $\overline{R}X \cap X = \emptyset$ and $\overline{R}Y \cap Y = \emptyset$. We must show that $\overline{R}X = \underline{R}X$ and $\overline{R}Y = \underline{R}Y$ to show that sets X and Y are definable sets. Since $R_s(x) \subseteq X \Rightarrow R_s(x) \cap X \neq \emptyset$, then $\underline{R}X \subseteq \overline{R}X$. So we must show that $\overline{R}X \subseteq \underline{R}X$. Let $x \in \overline{R}X$. Since $\overline{R}X \cap X = \emptyset$, $X \cap Y = \emptyset$ and $X \cup Y = U$, then $x \in Y$. If $x \in \overline{R}X \Rightarrow R_s(x) \cap X \neq \emptyset$, there is $a \in U$ such that $a \in R_s(x)$ and $a \in X$, from which $(x,a) \in R$ and $a \in X$. Suppose that there is $b \in Y$ such that $(x,b) \in R$. $b \in R_s(x)$ and $b \in Y \Rightarrow R_s(x) \cap Y \neq \emptyset \Rightarrow x \in \overline{R}Y \Rightarrow x \notin Y$ (since $\overline{R}Y \cap Y = \emptyset$) It is a contradiction. Hence we have $R_s(x) \cap Y = \emptyset$ which gives $R_s(x) \subseteq X$, that is, $x \in \underline{R}X$. Thus we obtain $\overline{R}X \subseteq \underline{R}X$. Similarly we obtain $\overline{R}Y = \underline{R}Y$. Then X and Y are definable sets of GA-approximation space (U,R).

Acknowledgment

The first author would like to thank TUBITAK (The Scientific and Technological Research Council of Turkey) for their financial supports during her PhD studies.

References

1. E. A. Abo-Tabl, Inform. Sci., **81,** 2587-2596 (2011)
2. J.A. Bondy, U.S.R. Murty, Elsevier Science Publishing Co., Inc.,(1976)
3. J. Chen, J. Li,, Inform. Sci., **201**, 114-127 (2012)
4. J. Järvinen, Transactions on Rough Sets, VI, LNCS **4374**, 400-498 (2007)
5. J. Jarvinen, Fund. Inform., **53**, 135-153 (2002)
6. J. Kortelainen, Fuzzy Sets and Systems, , **61**, 91-95 (1994)
7. J. Nieminen,, Fund. Inform., **11(3)**, 289-296 (1998)
8. Y. Ouyang, Z. D. Wang, H. P. Zhang, Inform. Sci., **180,** 532-542 (2010)
9. Z. Pawlak, Internat. J. Comput. Inform. Sci., **11(5)**, 341-356 (1982)
10. Z. Pawlak A. Skowron, Inform. Sci., **177**, 3-27 (2007)
11. J.A. Pomykala, Lecture Notes in Artificial Intelligence, **2475**, 175-182 (2002)
12. Z. H. Shi, Z. T. Gong, Inform. Sci., **180**, 3745-3763 (2010)
13. A. Skowron, J. Stepaniuk, Fund. Inform., **27**, 245-253 (1996)
14. R. Slowinski, D. Vanderpooten, IEEE Trans. Knowl. Data Eng., **12**,331-336 (2000)
15. Y. Y. Yao, Inform. Sci., **109**, 21-47 (1998)
16. Y. Y. Yao, Physica-Verlag, Heidelberg, 286-318 (1998)
17. Y. Y. Yao, Rough sets and current trends in computing, LNAI 1424, 21-47 (1998)
18. Y.Y. Yao, Internat. J. Approx. Reason., **15 (4)**,291-317 (1996)
19. W. Zhu, Inform. Sci., **177**, 4997-5011 (2007)
20. W. Zhu, Inform. Sci., **179**, 210-225 (2009)

Brief Note on System Signature and Long Term Method

Mehmet Gurcan[1,] *Ayse T. Bugatekin*[2*]

[1]Department of Statistics, University of Fırat, Elazığ, Turkey,

Abstract. In this paper, the system signature of the system is creating by using the long-term method and then the transition matrices of each step are calculated. In this way, it is possible to have information about the reliability of the system after each unit becomes failure. Also, in this study, three systems are presented as examples.

1 Introduction

One of the materials used in the calculation of system reliability is the system signature, in recent years. There have been many studies on reliability of technical systems and system signature [1-7].System signature is obtained by calculating the probabilities of order statistics that disrupt the system from within the possible sequences, regardless of the distributions of the units that make up the system. The system signature is a probability vector as obtained from possible states. However, while the components of this probability vector are show the system corruption probability of order statistics up to last term, them last term is the form adapted to be 1 sum of probabilities. In fact, no matter what system, the system corruption probability of maximum order statistics is 1 always. In the study, if the *i th* order statistic caused the disruption of the system, Bernoulli variable $\xi_i = 0$ and otherwise $\xi_i = 1$. Also, a step transition matrix $P(i)$, which represent a step transition between the corrupt and robust states $\{0,1\}$ of the system, has been established. Since the designed Markov chain is not homogeneous, the transition matrix at each step is different. So the system signature has been created for a given system and then the transition matrix for each step can be calculated. Thus, it is possible to have information about the reliability of the system after the failure of each unit.

Corresponding Author: ayseturan23@hotmail.com

The long term method used in the study is designed with this principle. In this study, three systems are presented as examples.

2 Construction of the Used Method

The operating principle of a technical system is generally defined by some conditions placed on the units forming the system. In a series connected technical systems, if any of the units is corrupted the technical systems is corrupted. In parallel connected technical systems, if all units fail, the technical system is fail. Also, in k out of n technical system, if and only if at most k machine are fail, the technical system is fail. In similar ways, these principles can be changed to define different technical systems and operating principles. It is evident from this that the order statistics of the distributions describing the working times of the units that constitute a technical system are very important in calculating the reliability of the technical system. If the system fails when the r *th* unit fails in a technical system, the life-time of r *th* order statistic describes the reliability of the system. Let Y_r be r *th* order statistic. The reliability of the system is calculated as follows,

$$R = \Pr\{Y_r > t\} \tag{1}$$

Accordingly, when the r *th* disruption is occurs, the disruption of the system is expressed by $\xi_r = 0$ and otherwise $\xi_r = 1$. The transition probabilities between the states of the system are defined as follows,

$$\alpha_r(i,j) = \Pr\{\xi_{r+1} = j | \xi_r = i\},\ i,j = 0,1 \tag{2}$$

In this case, the $P(r) = [\alpha_r(i,j)]$ matrix $\{\xi_r : r \geq 1\}$ that defined by the above probabilities, will be a step transition matrix of the non homogeneous stochastic process. However, let $1-\alpha_r = \Pr\{\xi_r = 1\}$ be possibility of system operation when r *th* disruption occurs and α_r possibility of system corruption when r *th* disruption occurs. If a technical system consists of n units, then the n-component vector consisting of α_r components is denoted by $\eta = (\alpha_r : 1 \leq r \leq n)$. If $\alpha_n = 1-(\alpha_1 + \ldots + \alpha_{n-1})$ is taken, the vector η will be transformed into the system signature. At the same time, the following relationship exists between α_r possibilities,

$$(\alpha_r, 1-\alpha_r)P(r) = (\alpha_{r+1}, 1-\alpha_{r+1}) \tag{3}$$

It constitutes the idea that the basis of this designed analysis can be defined by a technical system (α_r) sequence. If the number of terms in this sequence is not limited by the number of units that make up the technical system, the sequence (α_r) is one that is convergent. This situation can only be applied to repairable systems. In literature reviews of repairable technical systems, it is assumed that the units have first performance after repair. However, this is unlikely in reality. In reality, even it can be repaired; each unit has a finite working cycle.

3 Illustrative Examples

1. The following table shows that the possible order of units of a technical system with a system signature $(6/24, 14/2, 4/24, 0)$ and the order of disruption that causes the system to deteriorating system according to the specified order,

Table 1.

Possible order/ **(Order statistics that disrupt the system)**																			
1	2	3	4	**(1)**	2	1	3	4	**(2)**	2	3	1	4	**(3)**	2	3	4	1	**(3)**
1	3	2	4	**(1)**	3	1	2	4	**(2)**	3	2	1	4	**(3)**	3	2	4	1	**(3)**
1	2	4	3	**(1)**	2	1	4	3	**(2)**	2	4	1	3	**(2)**	2	4	3	1	**(2)**
1	3	4	2	**(1)**	3	1	4	2	**(2)**	3	4	1	2	**(2)**	3	4	2	1	**(2)**
1	4	2	3	**(1)**	4	1	2	3	**(2)**	4	2	1	3	**(2)**	4	2	3	1	**(2)**
1	4	3	2	**(1)**	4	1	3	2	**(2)**	4	3	1	2	**(2)**	4	3	2	1	**(2)**

Thus, the η vector $\eta = (6/24, 14/2, 1, 0)$ $\eta = (6/24, 14/24, 1, 0)$and the elements of the transition matrices, $\alpha_1(1{,}0) = 4/9$ $\alpha_2(1{,}0) = 1$.

2. In the second example, let's define the technical system $\alpha_r = 1 - \exp((1-r)/l), l > 0$ $r = 1{,}2{,}...$. In this case, each $\alpha_r(1{,}0)$ is obtained as $1 - \exp(-1/l)$. Here, l is the adaptation constant of the long term method.

3. In this example, let's examine a repairable technical system consisting of two units connected in parallel. Let the distribution of the operating times of both units be F_1 and F_2 and the distribution of the repair time be $G(u)$. After the system has started running, the first distorted unit will be repaired. When the second disruption takes place within the repair period, the system is corrupted. According to this, the operating time of the working unit for the deterioration of the system must be smaller than the repairing time of the units which are completed after the disruption in single row. In this case, the system signature consists of the following components,

$$\alpha_{2r-1} = 0, \alpha_{2r} = \Pr\{Y_{2r} < U\}, r = 1,2,... \qquad (4)$$

Here, the components of the transition matrix are obtained as $\alpha_{2r}(1,0) = \Pr\{Y_{2r} < U\}$.

4 Conclusion

The operating principles of the technical systems described in the literature are increasing day by day according to the possibilities of technology and application areas. The identification of the technical system is equivalent to the identification of reliability in a way. In this study, the system signature is briefly discussed and a different method of using the constructed system signature has been put forward. Since the system signature is defined independently of the distribution and the operating time of the system, a time-dependent dynamic analysis is not possible. However, this study can be done when the stochastic process that is introduced to describe the system in the study can be obtained with continuous parameters. In this regard, a technical system can be defined by rate matrix of a continuous parametric process. This definition will also lead to the introduction of many new systems.

References

1. P. K. Andersen, Q. Borgan, R. Grill and N. Keiding, Springer. New York (1992)
2. T. Avent, H. Haukaas, Reliab. Eng. and Syst. Safety, **58**, 43-53 (1997)
3. R. Barlow, F. Proschan, Wiley. New York (1965)
4. A. Birolini, Springer. New York (1994)
5. B.V. Gnedenko, I. A. Ushakov, Wiley. New York (1995)
6. V.V. Kalashnikov, Syst. Anal. Model. Simul. **6**, 833-851(1989)
7. J. Keilson, Springer. Berlin (2015)

Power Side Channel Analysis and Anomaly Detection of Modular Exponentiation Method in Digital Signature Algorithm Based Fpga

Burcu Sönmez[1]*Ahmet Bedri* Özer[2]

[1] Department of Computer Engineering, University of Agri Ibrahim Cecen, Agri, Turkey

[2]Department of Computer Engineering, University of Firat, Elazig, Turkey

Abstract. In this study, digital signature application was performed on FPGA with classical RSA and Chinese Remainder Theorem (CRT). The power consumption of the system was observed when the digital signature process was performed on the FPGA. In order to distinguish the modular exponentiation methods as the classical RSA and the Chinese Remainder Theorem (CRT), the anomaly detection method was applied to the digital signature application using the power side channel analysis of the system. According to the obtained result, it is proved that information about the structure of the algorithm executing in the system can be obtained by using the power information consumed by a cryptographic device.

1 Introduction

RSA-based digital signature applications are one of the most used signature applicationsin today. In a public key cryptosystem such as RSA [1] and ElGamal [2], the legitimate user can use the recipient's public key to encrypt a document. A particular recipient uses the secret key to decrypt the encrypted document. On the other hand, a user can sign a document with his private key and any recipient verify the signature using the public key of signer.Therefore, the digital signature can be used to prove the identity of the sender, to protect the privacy of the transmitted document, and to verify the integrity of the received document. The RSA algorithm is the most popular algorithm to secure the public key cryptosystem. It was proposed by Riyest, Shamir and Adleman in 1977 [1]. RSA algorithm is used as a public key cryptosystem in implemented digital signature application.In the classical RSA algorithm, modular exponentiation, left-to-right squared and multiplication algorithm are used. Modular exponentiation is a costly step in creating digital signatures. Especially when very large prime numbers are used in terms of security, modular exponentiation and modular multiplication operations are very costly. There are many commonly used rapid exponentiation and fast multiplication methods to increase efficiency. These methods are referred to in the literature as the Montgomery Multiplication Method and the Chinese Remainder Theorem [3,4]. The method used for the fast modular exponentiation is the Chinese Remainder Theorem. With this theorem, modular exponentiation process is performed as two separate modules. Then, the modules which are divided into two in the signature creation phase are combined [4]. In this case, the classical RSA-based digital signature application is performed in two separate modules instead of the one-step exponentiation in the Chinese Remainder Theorem and faster results are obtained in terms of performance. In this study, Digital Signature was implemented on FPGA using RSA open key cryptosystem with China Remainder Theorem. The power consumption of the digital signature application using a device is measured. The power consumption of the Chinese Remainder Theory and the classical RSA-based digital signature applications are compared.

2 Rsa Digital Signature Algorithm

To sign an M message with the RSA, an N value is obtained, where p and q are two large prime numbers, and the M message is transformed using a hash function representing the $m \epsilon Z_N$ and a padding operation.Then the $m^d \ modN$ is calculated using the secret key of the signer. Thus, a digital signature is obtained for the M message. Here d represents a secret key [5,6]. RSA is based on the modular exponentiation process which can be performed by algorithms of base, square, multiplication and modular exponentiation of digital signature generation calculations. The implemented classical RSA based digital signature algorithm is given in Algorithm 1.In algorithm 1, two prime numbers are chosen as p and q for the calculation of the N module to be used to generate the public (e) and private (d) keys. In step 4, if gcd (e, (n)) = 1 is true, it is certain that e is an inverse of the N module. Thus the secret key d is always present. In the given algorithm, public and private keys are calculated in steps 4 and 5.

Corresponding Author: bsonmez@agri.edu.tr

The power differences in the Chinese Theorem-based and classical RSA-based digital signature application arise in the 5th step, ie the modular exponential phase. As a result of detecting the obtained abnormality, modular exponentiation method can be easily distinguished for both methods.

Algorithm 1:

1. Two prime numbers p and q close to each other are selected.
2. N=p*q and ϕ (N) =(p-1)(q-1) are calculated.
3. An integer e is selected between 1 <e <(N).
4. An integer d is selected such that e*d = 1 (mod (N)).
5. $S = M^{d}$ modNis computed. (S: digital signature)

3 Chinese Remainder Theorem (CRT)

RSA application with China Remainder Theorem (CRT) is a widely used algorithm that requires high performance of the RSA digital signature algorithm [7].Firstly, the signer first computes each prime factor p and q in the signing modüle, separately. Then, signature S modN is calculated using China Remainder Theorem. Since the size of p and q is approximately half that of N, exponentiation with CRT is four times faster than direct exponentiation. The implemented CRT based RSA digital signature algorithm is given in Algorithm 2.Important examples of side-channel attacks on CRT-based RSA digital signature implementation are given in [8] and [9].

Algorithm 2:

1. Two prime numbers p and q close to each other are selected.
2. N=p*q and ϕ (N) =(p-1)(q-1) are calculated.
3. An integer e is selected between 1 <e <(N).
4. An integer d is selected such that e*d = 1 (mod (N)).
5. $m_{p=}$ m (modp) and m_q= m (modq) are calculated.
6. d_p= d(mod (p-1)) and d_q=d(mod (q-1)) are calculated.
7. $xp = m_p^{dp}$ modpvexq = m_q^{dq} modq are calculated separately.

4 Power Side Channel Analysis and Anomaly Detection of ModularExponentiation

Power analysis attacks use dependency between the power consumption on a cryptographic device and the data being processed or performed [10]. With power analysis, the power consumption of the device is monitored by taking advantage of the system's current differences. In this way, there is a relationship between the power consumption of the device and the confidential information. Firstly, the power consumption of the device must be measured. For this purpose, a small-valued resistor is placed on the line between the circuit and the source, and current information is obtained by taking advantage of the voltage differences at both ends of the resistor [11]. The measurement setup is shown in Fig 1.

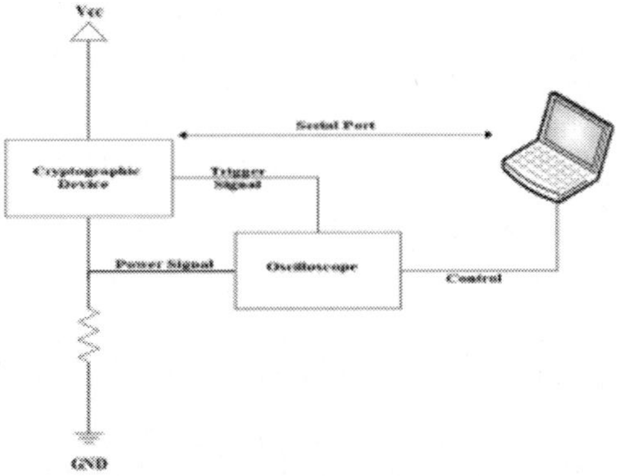

Fig 1 Measurement system for monitoring power consumption

In order to determine the modular exponentiation method of the digital signatures, the power side channel of the system is observed using the system in Fig 1.In [12], the necessary hardware information for power attacks is given in detail.One of the methods developed in recent years to prevent side channel attacks is the detection of the attack process using anomaly detection [13,14]. Anomaly detection is a method of machine learning techniques based on the problem of finding unexpected behaviors in data. These unsuitable templates are generally expressed in terms of anomalies, outliers, anomalous observations, exceptions, deviations [15].

In the performed power analysis attack, the power consumption of the digital signatures using the classical RSA algorithm and Chinese Remainder Theorem was used as an attribute. Information about the modular exponentiation method of the system according to the power side channel analysis has been obtained. The results obtained are given in the next section.

5 Experimental Results

In this study, Digital Signature was implemented on FPGA using RSA open key cryptosystem with China Remainder Theory. The Chinese Remainder Theorem compares the power consumption of classical RSA based digital signatures. As a result of the implemented digital signature application, it is observed that the Chinese Remainder Theorem consumes less power. In the implemented application, the message to be signed is processed as two separate sections as Mp and Mq. The digital signature is obtained by combining the separately performed modular exponentiation operations. Altera EP4CE115F29C7 set was used in the application. The ZXCT1021 low offset high-side current monitor integrated is used to detect the power consumed. A 0.1 ohm resistor is used in the application circuit for precisely measurement. Fig 2 shows the application circuit.

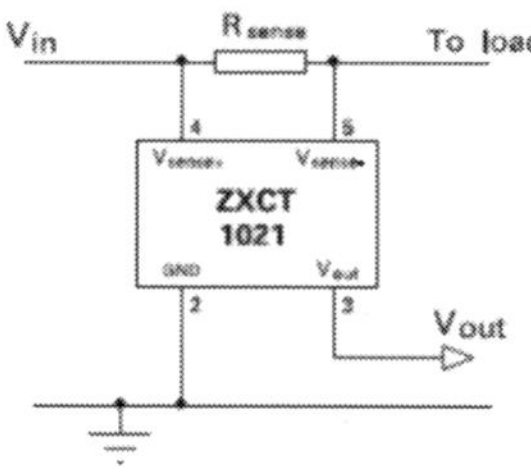

Fig 2 ZXCT1021 integrated circuit

Fig 3 shows the power consumption signal measured from the Vout pin of the ZXCT1021 integrated circuit shown in fig 3 as a result of the implementation of digital signature application with the Chinese Remainder Theory and the classical RSA on the experimental set. As a result of the obtained tension differences, it is observed that the digital signing application of Chinese Remainder Theorem in the modular exponential phase reduces the workload.

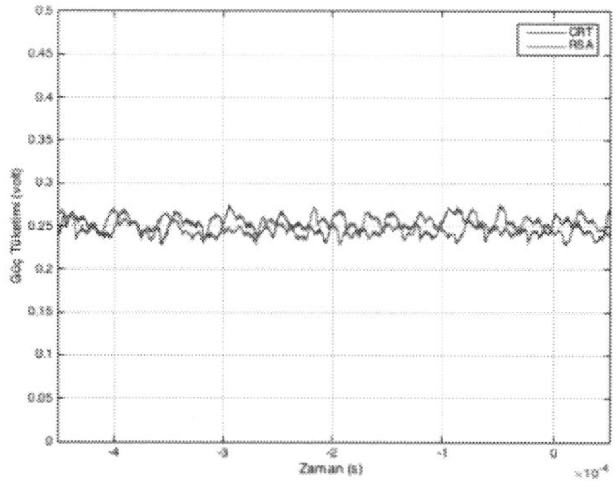

Fig 3 Chinese Remainder Theorem and RSA power traces

Especially when very large prime numbers are used, Chinese Remainder Theorem will be more advantageous than the classical RSA-based digital signature implementation of the Theorem.In order to distinguish the method used in a digital signature application from the values in this chart, an anomaly detection method is applied by using power consumption as information leak of the system. In this method, the k-average clustering algorithm is applied to the power consumption values of the system. The values outside the cluster were determined as anomaly values belonging to the Chinese Remainder Theorem, as shown in Fig 4.

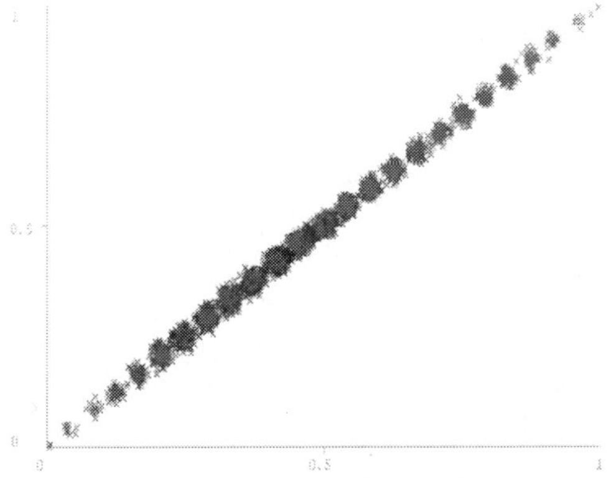

Fig 4 Anomaly detection results

The x and y axis in Fig 4 represent power consumption values. Blue marked values belong to the classical RSA algorithm, red marked value and Chinese Remainder Theorem (CRT) values. As can be seen in this figure, values greater than 0.75 have been found to be anomaly for the Chinese Remainder Theorem. According to the obtained result, it has been proved that information about the structure of the algorithm executing in the system can be obtained by using machine learning techniques by using the power information consumed by a cryptographic device. As a result of this study, it is observed that if the side channel measures are not taken for the systems, confidential information such as transactions, ciphers, keys can be obtained by using side channel information leaking from a system. This study attention to side channel analysis attacks. With the evolving technology, cryptographic devices need to be designed taking into account side channel attacks.

6 Conclusion

As a result of this study, it has been observed that if the side channel measures are not taken for the systems, information such as the processes and methods performed by the system can be obtained by using side channel information leaked from a system. With the evolving technology, cryptographic devices need to be designed to resist side channel attacks. In recent years, however, the use of cryptography in combination with artificial intelligence has become widespread as a support for information security. The study shows that artificial intelligence can be used in security measures such as intrusion detection methods.As a result of the information obtained, it has been proved that machine learning techniques can be used to detect attacks and obtain information about applied methods.

Preferences

1. R.*Rivest,* A.*Shamir,* L.*Adleman,* A method for obtaining digital signatures and public-key cryptosystems, Communications of the ACM, 120-126, 1978.
2. T.Elgamal, A public-key cryptosystem and a signature scheme based on discrete logarithms, IEEE Transactions on Information Theory, pp. 469–472, 1985.
3. P.Lawrence Montgomery, Modular Multiplication Without Trial Division, Mathematics of Computation, 4A 170, 519-521, 1985.
4. P.A.Fouque, N.Guillermin, D.Leresteux, M.*Tibouchi,* J.C.Za- palowicz, Attacking RSA-CRT Signatures with Faults on Montgomery Multiplication, Cryptographic Hardware and Embeded Sysytems CHES 2012.
5. C.Paar, J.Pelzl, Understanding Cryptography, Springer Heidelberg Dordrecht London New York, 2009.
6. C.D. Walter, Ç.K.Koç, Christof Paar, Cryptographic Hardware and Embedded Systems-CHES 2003, 5th International Workshop Cologne, 254-269, 2003.
7. J.J.Quisquater, C. Couvreur, Fast decipherment algorithm for RSA public-key cryptosystem, Electronics letters, 18(21), 905-907, 1982.
8. Kim, C., Ha, J., Kim, S. H., Kim, S., Yen, S. M., & Moon, S. A secure and practical CRT-based RSA to resist side channel attacks. In *International Conference on Computational Science and Its Applications*, pp. 150-158, Springer, Berlin, Heidelberg, May 2004.

9. Fournaris, A. P.,& Koufopavlou, O. CRT RSA Hardware Architecture with Fault and Simple Power Attack Countermeasures. In *Digital System Design (DSD), 2012 15th Euromicro Conference on* ,pp. 661-667, IEEE, September 2012.
10. Le T.H., Canovas C., Clédière J.,An overview of side channel analysis attacks, Proceedings of the 2008 ACM symposium on Information, computer and communications security Pages 33-43, Tokyo, March 18 – 20.
11. J.J.Quisquater, M.Rizk, Side Channel Attacks, Access Date: 24 March 2018,https://www.ipa.go.jp/security/enc/CRYPTREC/fy15/doc/1047_Side_Channel_report.pdf, October 2002.
12. Gamaarachchi, H.,& Ganegoda, H. Power Analysis Based Side Channel Attack. *arXiv preprint arXiv:1801.00932*, 2018.
13. M.Chiappetta, REAL TIME DETECTION OF CACHE-BASED SIDE-CHANNEL ATTACKS USING HARDWARE PERFORMANCE COUNTERS, Computer Science and Engineering, Master's Thesis, 2016.
14. M.Chiappetta, E.Savas, C.Yilmaz, Real time detection of cache-based side-channel attacks using hardware performance counters, Applied Soft Computing, Volume 49, 1162-1174, December 2016.
15. Chandola, Banerjee and Kumar, Anomaly Detection: A Survey, A modified version of this technical report will appear in ACM Computing Surveys, September 2009.

STUDY OF THREE DIMENSIONAL LAMINAR FORCED CONVECTION OF NANO FLUID FLOW THROUGH ECCENTRIC RECTANGULAR ANNULAR DUCT

Amr G. Eltorky [a*], *Mohamed Elhelw* [b], *Mohamed Fayed* [c], *Abdelhamid Attia* [d]
[a]*Faculty of Engineering, Mechanical Department, Pharos University In Alexandria, Egypt*
[b, c, d] *Faculty of Engineering, Mechanical Department, Alexandria University, Egypt*
*Corresponding author: amr.gaber@pua.edu.eg

Abstract: Heat transfer through the horizontal eccentric annulus between eccentric rectangular ducts with different eccentricities has been calculated with various aspect ratio of inner duct. Boundary conditions are used with constant hot temperature on inner duct and constant cold temperature on the outer duct. Space between two ducts included TiO_2-water Nano-fluid with different solid Volume fraction (φ = 0, 2, 5, 10 %). The eccentricity was changed with different values (E = 0.025, 0.05, 0.075 m) in left direction and aspect ratio was changed with different value (AR= 0.25, 0.375, 0.5). Results show that with an increase of aspect ratios, the average Nusselt number increases. Alsowith an increase in eccentricity value,the average Nusselt numberremains constant, and then an increase begins to occur. And the average Nusselt number increase due to an increase in nanoparticle concentration.

1 Introduction

Science and engineering include many applications such as a generator, solar heating system, ventilation ducts, atomic reactors, electrical motor, heating and cooling of electric cables. Convection heat transfer inside eccentric annular duct may be used in above applications. Lately, implementing the impact of eccentricity and aspect ratio on heat transfer has been an interesting topic and using Nano-fluid to enhance the thermal conductivity of the fluid and other fluid properties. However, much exploring related to this subject has been managed by two dimensional circular annular duct or square annular duct and relatively little three dimensional rectangular annular duct studies have been carried out.Hussein Togun et. al [1] surveyed studies on natural, mixed and forced heat transfer to Nano-fluid and fluid flow in an annulus passage configurations, and the result show that Nusselt number values for the boundary condition of constant wall temperature are lower than those at uniform heat flux. Batra and Eissa [2] studied heat transfer by laminar and forced convection of a model fluid in an eccentric annulus. For same values of the cross-section parameters, a Sutterby model fluid has been found to yield a lower average Nusselt number as compared to that for a Newtonian fluid. Manglik and Fang [3] make a Numerical solutions for fully developed, laminar and heat transfer by forced convective in eccentric annulus are given with an insulated on an outer surface, result show Increasing the eccentricity and reducing the radius ratio of the annulus geometry results in large non-uniformities in the axial velocity and temperature fields.Matin [4] studied non-Newtonian power-law liquid (fluid) by natural convection of two dimensional in steady state assumption between two eccentric square horizontal ducts at a constant temperature. The results prove that with increasing n from 0.6 to 1.4, observe that the mean Nusselt number decreases and decrease of heat transfer rate. Arefmanesh et. al [5] studied analysis of heat transfer between two horizontal square ducts and passage full of (water-TiO_2) Nano-fluid and observed from the results that, an increase in average Nusselt number due to an increase in volume fraction of particles. El-Maghlany et. al [6] studied natural convection through an open horizontal eccentric passage between two eccentric square ducts. Results mention that with an increase in eccentricity in upper and lower direction, a decrease occurs in convection heat transfer coefficient.Ciampi et. al [7] studied heat transfer by mixed convection flow through concentric and horizontal tube annular passage and observed that the Nusselt number at the bottom of the cylinder lower than at the top. Ingham and Patel [8] studied mixed convection in a vertical eccentric annular passage for the fully developed flow of a Bingham plastic. Khandekar and Moharana [9] using doubly and singly connected tube for different cross sections, applying two-dimensional technique, laminar flow, fully developed and estimating pressure drop through this passage. Macêdo et. al [10] studied Generalized Integral Transform Technique by using a compound of numerical–analytical solution and obtained for non-Newtonian fluids and laminar flow into doubly connected ducts.

Jeevan et. al [11] are focused on discussing the heat transfer attitude of a square spongy annulus subjected to outer cold temperature and inner hot temperature and it is noticed that the Nusselt number at the bottom is higher than at the top hot wall of the annular passage. Mahmoodi and Mazrouei [12] studied free convection flow and heat rate of composition (Cu– water) Nano-fluid through an annular passage between two square bodies and the results show that for every Rayleigh numbers without $Ra = 10^4$ the Nusselt number rises with an increase in the solid volume fraction of the particles.Gavara and Kanna [13] studied Laminar free convection between coaxial two vertical rectangular pipes and observed that The average Nusselt number of lower face is further than 2.4 times of Nusselt number averaged on other faces for the Grashof of this study.Xing et. al [14] studied laminar flow through the rectangular duct and how it is affected by the change in aspect ratio and result show that the Poiseuille number rise as the aspect ratio reduces. Shah and London [15] pointed out some works that dealt with laminar flow inside annular passages and various other types of doubly connected geometries.Moreover, as well governing equations for finite volume process are taken from book for Versteeg and Malalasekera [16] has documentary the book of the preamble to computational fluid dynamics. There are some rules for each program; ANSYS-FLUENT 16.0 is used in this research, which required guides are theory guides, user guides and tutorial [17]. Wibulswas [18] studied laminar flow heat transfer for rectangular ducts and equilateral triangular ducts of various aspect ratios with the thermal boundary conditions of constant wall temperature and constant heat input per unit length of the duct. Bartlett and Viskanta [19] studied the enhancement of forced convection in a heated duct full of with high thermal conductivity media.Heat transfer enhancement from a heated surface using porous structures has a potential application in electronics cooling.

Nomenclature

B	Width of outer rectangle, (m)
b	Width of inner rectangle, (m)
Cp	Specific heat at constant pressure, (kJ/kg.K)
D_h	hydraulic diameter
g	Acceleration of gravity, (m/s^2)
H	Height of outer rectangle, (m)
h	Height of inner rectangle, (m)
h_{avg}	Average heat transfer coefficient, (W/m^2K)
L	Length of duct, (m)
Nu_{avg}	Average Nusselt number,
P	Pressure, (N/m^2)

T	Temperature, (K)
t	Time, (s)
u,v,w	Velocity components in the x, y and z
ΔT	Temperature difference, (K)
q_{avg}	Average heat flux (W/m2)

Subscripts

avg	Average
c	Cold
f	Fluid
h	Hot
nf	Nano fluid
p	Particle

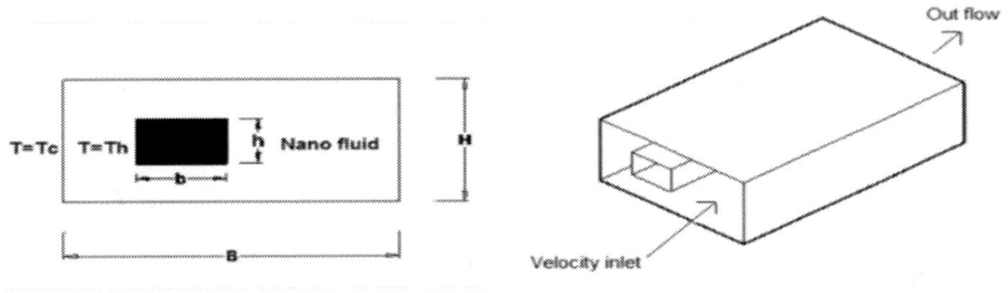

Fig. 1. Schematic of the eccentric rectangle with boundary conditions (current work).

2 Scope and objectives

Heat transfer calculations of forced convection in TiO_2-water Nano-fluid from the surface of a heated horizontal inner duct to a parallel outer rectangular duct due to a few work in this area, are offered for different eccentricities,the solid volume fraction of particles and aspect ratio. The scope and objective of the current study are concentrated on finding the effects of eccentricities and aspect ratio on 3D annular rectangular ductassociated with forced convection through annulusas shown in Fig.1, resulting from the temperature difference between inner and outer duct.

3 Problem formulation

Consider the two eccentric ducts,which are shown in Fig.1.The height and the width of the outer duct are denoted by H and B respectively. Moreover, the height and the width of the inner duct are denoted by h and b respectively. The eccentricity (E) is defined as a distance between the centers of duct, eccentricity ratio ($E_R = E/D_h$). The aspect ratio of the annulus is known as AR = h/H = b/B. The length of the outer and the inner ducts orthogonal to the plane of the paper = L. The outer and inner rectangular duct are keeping at different constant temperatures of T_c and T_h, respectively, with $T_h > T_c$. The annular (the volume between the two rectangular duct) is filled with a composed of TiO_2 spherical nanoparticles and water. The Nano-fluid is an incompressible flow, and the nanoparticles are supposed to be in thermal equilibrium together with the water. The eccentricity was changed with different values (E = 0.025, 0.05, 0.075 m) in left direction and aspect ratio was changed with different value (AR= 0.25, 0.375, 0.5). Moreover, there is no slip between the particles and water. Table 1 presented the physical properties of water and the nanoparticles. Thermal properties of Nano-fluid are a blend of TiO_2 spherical nanoparticles and water can be calculated by equations included volume fraction of nanoparticles. The properties of the Nano-fluid (TiO_2 - water) are supposed to be constant.

Table.1. physical properties of the base fluid and nanoparticles [5]

Physical properties	water	Nanoparticles(TiO_2)
C_p(J/kg.K)	4179	686.2
ρ(kg/m^3)	997.1	4250
k (W/m.K)	0.613	8.9538
$\beta \times 10^5$ (K^{-1})	21	0.9
μ(kg/m-s)	0.001003	–

3.1 Thermophysical properties of the Nano-fluid

The following equations proposed by A. Arefmanesh [5] included the density(ρ_{nf}), the thermal expansion coefficient (β_{nf}) and the specific heat capacity, (Cp_{nf}) of the Nano-fluid respectively.

$$\rho_{nf} = (1-\phi)\rho_f + \phi\rho_p \tag{1}$$

$$\beta_{nf} = \frac{(1-\phi)(\rho\beta)_f + \phi(\rho\beta)_p}{(1-\phi)\rho_f + \phi\rho_p} \tag{2}$$

$$C_{p_{nf}} = \frac{(1-\phi)\rho_f C_{p_f} + \phi\rho_p C_{p_p}}{(1-\phi)\rho_f + \phi\rho_p} \tag{3}$$

The Nano-fluid viscosity (μ_{nf}) and thermal conductivity (k_{nf}) of Nano-fluid are denoted by:

$$\mu_{nf} = \mu_f(1 + 4.62\phi + 199.21\phi^2) \tag{4}$$

$$k_{nf} = k_f(1 + 4.82\phi + 125.62\phi^2) \tag{5}$$

The thermal diffusivity equation of the Nano-fluid α_{nf}.

$$\alpha_{nf} = \frac{k_{nf}}{(\rho C_p)_{nf}} \tag{6}$$

The average Nusselt number on the outer surface based on hydraulic diameter of the annulus is evaluated from the following relation:

$$Nu_{avg} = \frac{q_{avg} D_h}{k_{nf}\Delta T} \tag{7}$$

3.2 Basic Governing Equations

The components, x, y and z of Continuity, momentum, and energy equations for the three-dimensional steady and laminar forced convection for Nano-fluid flow are given respectively.

The Three Dimensional continuity equation for fluid flow is given in Cartesian coordinates:

$$\frac{\partial \rho u}{\partial x}+\frac{\partial \rho v}{\partial y}+\frac{\partial \rho w}{\partial z}=0 \quad (8)$$

Three Dimensional Navier Stoke equations and energy equation for fluid flow are given in Cartesian coordinates:

$$-\frac{\partial P}{\partial x}+\mu\left[\frac{\partial^2 u}{\partial x^2}+\frac{\partial^2 u}{\partial y^2}+\frac{\partial^2 u}{\partial z^2}\right]=\rho\left[u\frac{\partial u}{\partial x}+v\frac{\partial u}{\partial y}+w\frac{\partial u}{\partial z}\right] \quad (9)$$

$$-\frac{\partial P}{\partial y}+\mu\left[\frac{\partial^2 v}{\partial x^2}+\frac{\partial^2 v}{\partial y^2}+\frac{\partial^2 v}{\partial z^2}\right]=\rho\left[u\frac{\partial v}{\partial x}+v\frac{\partial v}{\partial y}+w\frac{\partial v}{\partial z}\right] \quad (10)$$

$$\rho g_z-\frac{\partial P}{\partial z}+\mu\left[\frac{\partial^2 w}{\partial x^2}+\frac{\partial^2 w}{\partial y^2}+\frac{\partial^2 w}{\partial z^2}\right]=\rho\left[u\frac{\partial w}{\partial x}+v\frac{\partial w}{\partial y}+w\frac{\partial w}{\partial z}\right] \quad (11)$$

$$\rho C_p\left[u\frac{\partial T}{\partial x}+v\frac{\partial T}{\partial y}+w\frac{\partial T}{\partial z}\right]=k\left[\frac{\partial^2 T}{\partial x^2}+\frac{\partial^2 T}{\partial y^2}+\frac{\partial^2 T}{\partial z^2}\right] \quad (12)$$

Few assumptions are made such as, constant fluid properties, incompressible flow, radiation is neglected, laminar flow, and steady state conditions are assumed.

$$\mathrm{Gr}=\frac{g\beta_{nf}(\Delta T)D_h^3}{\nu_{nf}^2} \quad (13)$$

$$\mathrm{Re}=\frac{\rho_{nf}V_{in}D_h}{\mu_{nf}} \qquad \mathrm{Pr}=\frac{\nu_{nf}}{\alpha_{nf}}\mathrm{Ri}=\frac{\mathrm{Gr}}{\mathrm{Re}^2} \quad (14)$$

Where the Grashof Number (Gr), the Reynolds number (Re), the Prandtl number (Pr), inlet velocity (V_{in}) and Richardson number (Ri) to calculate if the flow is forced convection, natural convection or mixed convection. Richardson number has been lower than 1 to simulate forced convection dominant flow and in this case Ri = 0.006.

3.3 Computer Program Validation

To ensure that the FLUENT extractresults are in good agreement with previous work acomparison with a previous experimental result are made.

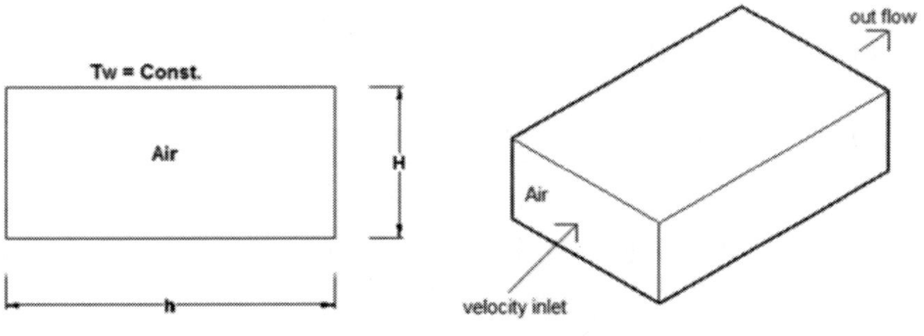

Fig.2. Schematic of rectangle duct by Wibulswas [18]

A comparison is used in the case of constant wall temperature in rectangle duct and the test section was made of copper which is a good heat conductor. Its dimensions were originally 2 ft. in length with aspect ratio 2, 2 x 1 in^2 in cross section and a wall of 1/8 in. in thickness were chosen and space filled by air as shown in Fig.2. From this comparison we observed that the various between the FLUENT result and experimental results [18] are between (0.5 to 7.5 %)as shown in Fig.3. It can be noticed that the average Nusselt number of the present study is in good agreement with the Wibulswas [18]results.

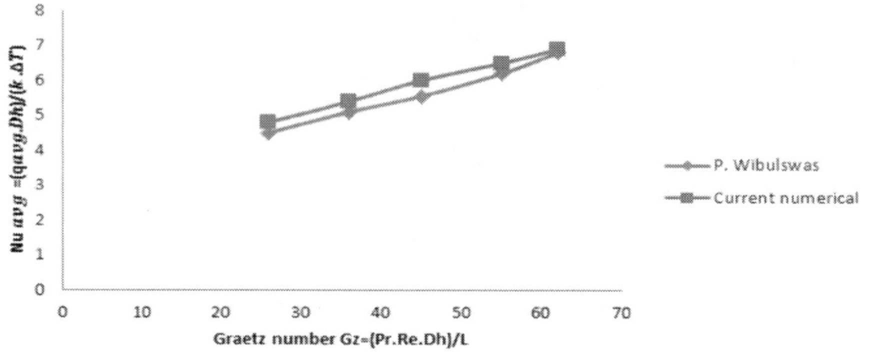

Fig.3. Average Nusselt number for laminar flow through a rectangle duct as a function of non-dimensional Graetz number (Gz) and validation of the current numerical with Wibulswas [18]

3.4 Grid independence study

In order to set a suitable grid for the present numerical solution, a grid independence study is managed for the forced convection heat transfer in the annulus between eccentric rectangular ducts shown in Fig.1. The aspect ratio is AR = 1/2, eccentricity ratio E_R= 0.375 and the computation are completed for TiO_2–water Nano-fluid with φ = 2% at ΔT=65°c, Re=10^3 and D_h/L= 0.011. Five various grid system sizes, namely, (9×18, 19×38, 29 × 58, 39×78 and 49×98) are utilized for the numerical simulations. As shown in Fig.4 that the average Nusselt number on outer duct for these grids. It is observed from Fig.4 that the appropriate results at less time a domain grid of 29 × 58 was adopted. Therefore, due to these results, a 29×58 grid system is utilized in all the subsequent numerical calculations.

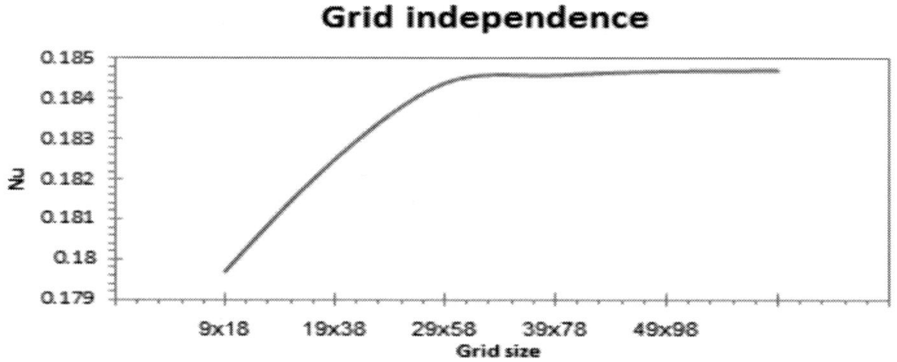

Fig.4. The average Nusselt number on the outer surface with different grid size

4 Results and discussions

Investigation of the effect of eccentricity and aspect ratio on heat transfer has become a subject of interest. The suggested code is used to study the laminar forced convection of Nano-fluid flow through the annular duct and heat transfer in the space between two eccentric rectangular ducts as shown in Fig.1, full of TiO_2–water Nano-fluid. The results obtained are for a range of aspect ratios between 0.25 to 0.5 and a range of eccentricity from 0.025 to 0.075 m, and solid volume fractions φ = 2

% for aspect ratio effect case and eccentricity effect case, $\Delta T=65°c$, $Re=10^3$, L=12 m, H=0.2 m, B=0.4 mwith several hydraulic diameter according to various aspect ratio. Fig.5 to 7&Fig. 9 to 11 show the streamlines and the isotherms in the annulus with different aspect ratios and different eccentricities. The fluid is heated by the sides of the inner rectangular duct and extend as it shifts upward and the fluid is cooled by the sides of the outer rectangular duct as it shifts downward.

4.1 Effect of aspect ratio (AR)

The average Nusselt number was calculated on the outer surface of the duct at various aspect ratios of the duct from FLUENT numerical solution, Fig. 5, 6, 7 shows the fluid flow through the different aspect ratios at the outlet port, the aspect ratio varying from 0.25 to 0.5 and eccentricity = 0.05 m, in streamline figures two counter-rotating vortexes, a clockwise and a counterclockwise, are decided in the right and the left of the annulus, respectively. The heat convection taking space from the portion of the hot surface of the inner rectangle forms a streamline plume according to the flow of the fluid in the counter-rotating vortexes in this space as shown in Fig.5. By comparing streamlines at outlet cross section port of Fig. 5, 6, 7 with increasing aspect ratio with the same eccentricity, noted that the streamlines become more crowded adjacent to sides of the outer and the inner rectangle. Also, as Fig.7 shows, rotary vortex eyes will disappear in the left side of the space with increasing aspect ratio for the same eccentricity, rotary vortex eyes are converted from circular to ellipse cross section in the right side of the space with increasing aspect ratio. As far as the isotherms are involved, in some ways, isotherms distributed within the entire annulus with some special thermal boundary layers are created around the inner rectangle as shown in Fig.7. Furthermore, the thermal boundary layer thickness decrease and lead to an increasein heat transfer by increasing the aspect ratio. From Fig.8, it is observed that the convection heat transfer coefficient increasesby increasing the aspect ratio. This phenomenon occurs because with increasing the aspect ratio the gap between the ductsbecome small, then the fluid traps in the annulus left side and can't move freely. However, in annulus right side the fluid velocity increase and lead to an increase in average velocity and also average heat transfer coefficient.So,the average Nusselt number increases with increasing the aspect ratio due to an increase in temperatureprofile and also an increase in average temperature through the annulus. These observations mean that an increase in aspect ratio is associated with an increase in heat transfer rate.

A) Streamline **B) Isotherms**

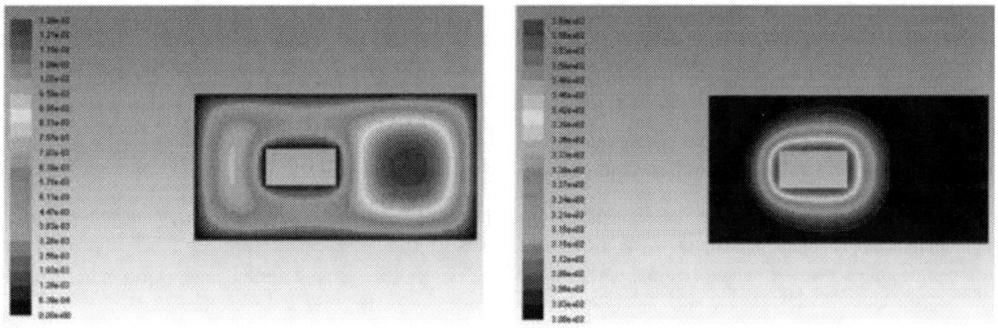

Fig.5. Streamline and Isotherms for AR=0.25 &E_R=0.25

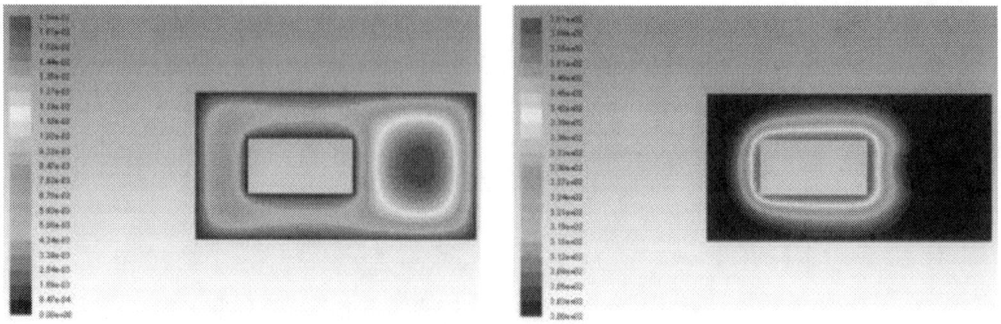

Fig.6. Streamline and Isotherms for AR=0.375 & E_R=0.3

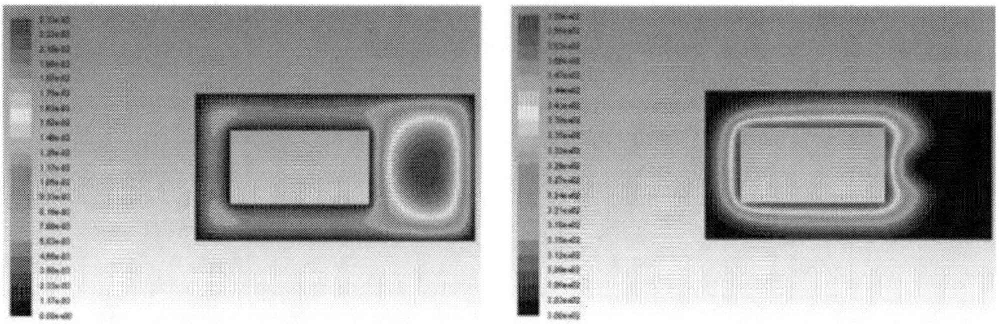

Fig.7. Streamline and Isotherms for AR=0.5 & E_R=0.375

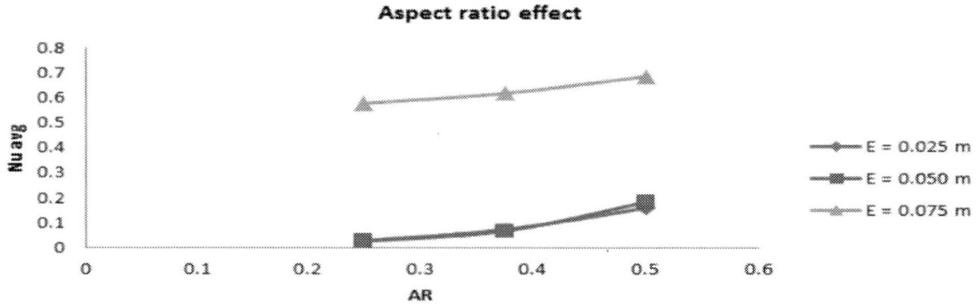

Fig.8. Effect of aspect ratio on the average nusselt number

4.2 Effect of Eccentricity (E)

Forced convection in an open-ended horizontal eccentric annulus between two eccentric horizontal rectangular ducts with different eccentricities has been studied with constant wall temperature. Fig. 9, 10, 11 show streamlines and isotherms at constant aspect ratio and different eccentricity at outlet cross section port of the annulus. The boundary conditions on inner and outer ducts are as follows: the constant hot temperature on the inner duct, the constant cold temperature on the outer duct. The eccentricity ratio was changed within a range between (E_R = 0.1 to 0.6) in left direction for same aspect ratio. Fig.11 shows the flow it takes the right side path of the annulus and disappears from left side due to increasing eccentricity of the inner rectangle and lead to decreasing heat transfer on the left side but increasing on other sides. As far as the isotherms are involved, in some ways, isotherms distributed within the entire annulus and Fig.11 shows how heat transfer increases on the sides by increasing eccentricity due to thermal boundary layer thickness. The results of the average Nusselt number on the outer rectangle are presented in Fig.12 and indicate that with an increase in the value of eccentricity ratio, the convection heat transfer coefficient remains constant within a range of eccentricity ratio between zero and 0.35. Eccentricity effect on heat transfer coefficientin this range not significantly different from those of concentric annuli, and then an increase begins to occur. However, the average convection heat transfer coefficient is very sensitive to eccentricity ratio for a value greater than 0.35. The highest values of average Nusselt number are accomplished at the higher level of eccentricity. Because the narrow left section flow blockage results in increased velocities in the wider annular gap and also increase in heat transfer coefficient. As a result, the average Nusselt number remains constant before an increase begins to occur. These observations mean that an increase in eccentricity is associated with an increase in heat transfer rate.

A) Streamline **B) Isotherms**

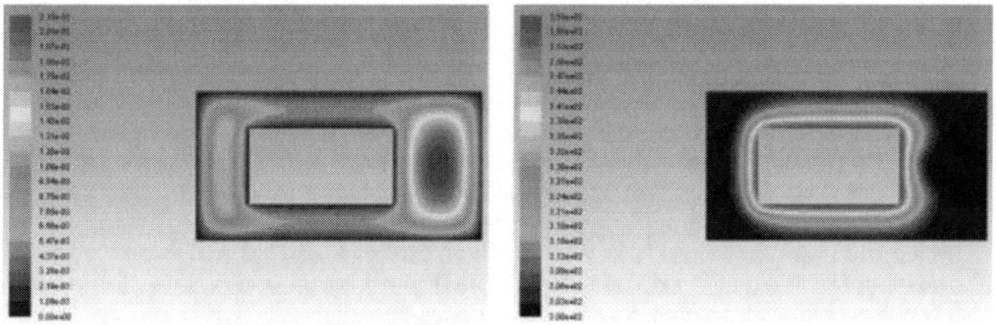

Fig.9. Streamline and Isotherms for AR=0.5 & E_R=0.187

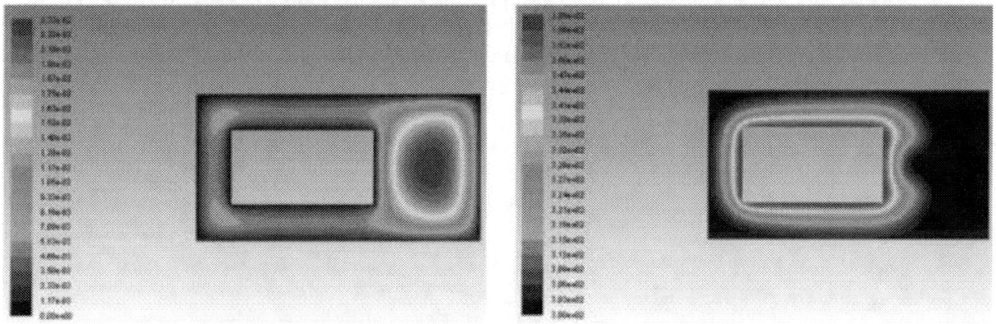

Fig.10. Streamline and Isotherms for AR=0.5 & E_R=0.375

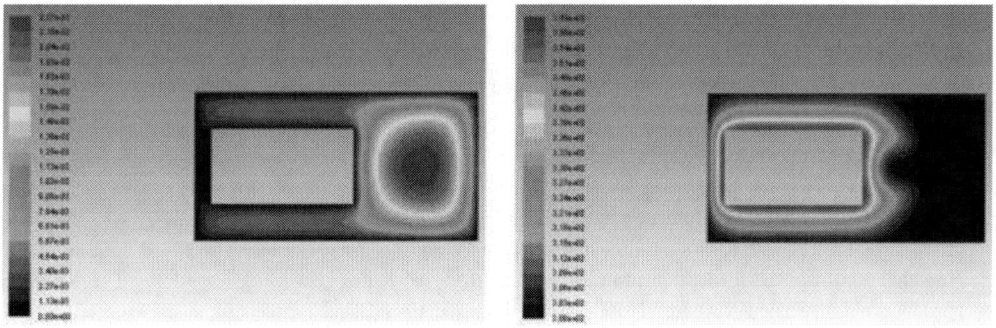

Fig.11. Streamline and Isotherms for AR=0.5 & E_R=0.562

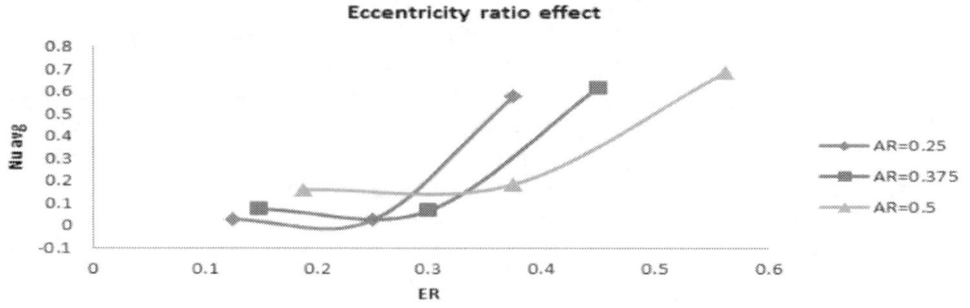

Fig.12. Effect of eccentricity on average nusselt number

4.3 Effect of Particle Volume fraction (φ)

Nano-fluids are comprised of Nano-sized solid particles that may change in size/shape in a liquid. The results of thermal conductivity presented show that an increase in temperature and volume concentrations is due to an increase in thermal conductivity of Nano-fluid. The aspect ratio is AR = 1/2, eccentricity ratio is equal to 0.375, and the computations are completed for TiO_2–water Nano-fluid with solid Volume fraction (φ) range between 0 % and 10 % and constant wall temperature on inner and outer duct at ΔT=65°c and D_h/L= 0.011. The calculated data also showed that the viscosity of Nano-fluids increaseby increasing particle volume concentration.Finally, pure water had the lowest viscosity. The influence of nanoparticle concentrations on average Nusselt number is shown in Fig.13. Results indicate, that the Nusselt number increases due to an increase in nanoparticle concentration given that the aspect ratio, eccentricity remain constant. The highest values of average Nusselt number are accomplished at higher level of concentration of solid nanoparticles which is related to an increase in thermal boundary layer profile.

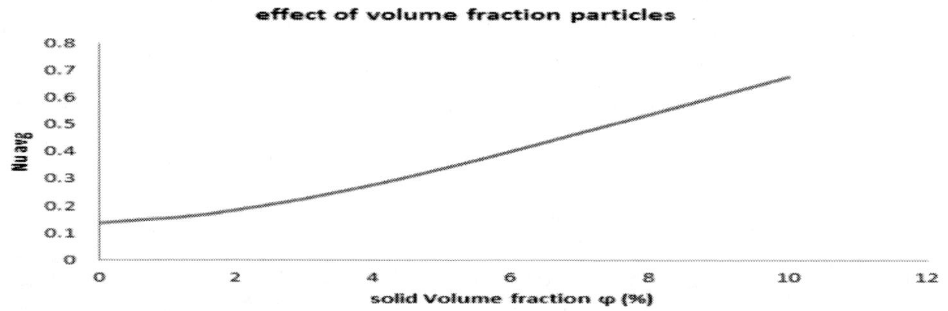

Fig.13. Effect of Solid Volume Fraction on Average Nusselt number

5 Conclusions

The heat transfer and fluid flow through two rectangular duct annuli filled with TiO_2–water Nano-fluid flow are presented to determine the effect of a number of variables. The finite volume method together with the SIMPLE algorithm is proceeded to solve the governing equations. The numerical procedure is verified via simulations, forced convection heat transfer in a differentially-heated rectangular cavity filled with TiO_2-water.A variables study involving the effects of the annulus aspect ratio, eccentricity of the inner duct on horizontal axis to left side and heat transfer are conducted. The results show:

1- The average Nusselt number on the outer duct is increase with increasing the aspect ratio. These observations mean that an increase in aspect ratio is associated with an increase in heat transfer rate.
2- Also, an increase in the value of eccentricity,the average Nusselt number remains constant and then an increase begins to occur. These observations mean that an increase in eccentricity is associated with an increase in heat transfer rate.
3- Nusselt number increase by increasing solid Volume fraction.

References

[1] Togun H, Abdulrazzaq T, Kazi SN, Badarudin A, Kadhum AAH, Sadeghinezhad E. A review of studies on forced, natural and mixed heat transfer to fluid and nano fluid flow in an annular passage 2014;39:835–56.

[2] Tobergte DR, Curtis S. Summary for Policymakers. In: Intergovernmental Panel on Climate Change, editor. Clim. Chang. 2013 - Phys. Sci. Basis, vol. 53, Cambridge: Cambridge University Press; 2013, p. 1–30.

[3] Manglik RM, Fang PP. Effect of eccentricity and thermal boundary conditions on laminar fully developed flow in annular ducts. Int J Heat Fluid Flow 1995;16:298–306. doi:10.1016/0142-727X(95)00030-T.

[4] Matin MH, Pop I, Khanchezar S. Journal of Non-Newtonian Fluid Mechanics Natural convection of power-law fluid between two-square eccentric duct annuli. J Nonnewton-Fluid Mech 2013; 197:11–23.

[5] Arefmanesh A, Amini M, Mahmoodi M, Najafi M. European Journal of Mechanics B / Fluids Buoyancy-driven heat transfer analysis in two-square duct annuli filled with a nanofluid.Eur JMech B/Fluids2012;33:95–104.

[6] El-maghlany WM, Sorour MM, Hozien O. Experimental study of natural convection in an annulus between two eccentric horizontal square ducts 2015;65:65–72.

[7] Ciampi M, Faggiani S, Grassi W, Tuoni G, Incropera FP. Mixed convection heat transfer in horizontal, concentric annuli for transitional flow conditions. Int J Heat Mass Transf 1987;30:833–41.

[8] Studies AM. M i x e d c o n v e c t i o n f l o w of a Bi plastic in an e c c e n t r i c annulus 1994;15:132–41.

[9] Appl M, Engrg M. Generalized formulation for estimating pressure drop in fully-developed laminar flow in singly and doubly connected channels of non-circular. Comput Methods Appl Mech Eng 2013;259:64–76.

[10] Monteiro ER, Macêdo EN, Quaresma JNN, Cotta RM. International Journal of Heat and Mass Transfer Laminar flow and convective heat transfer of non-Newtonian fluids in doubly connected ducts. Int J Heat Mass Transf 2010;53:2434–48.

[11] Anjum I, Al-rashed AAAA, Ahmed NJS, Kamangar S, Jeevan K. International Journal of Heat and Mass Transfer Natural convection in a square porous annulus. Int J Heat Mass Transf 2012;55:7175–87.

[12] Mahmoodi M, Mazrouei S. Superlattices and Microstructures Natural convection in a square cavity containing a nanofluid and an adiabatic square block at the center. Superlattices Microstruct 2012;52:261–75.

[13] Gavara M, Kanna PR. Study of conjugate natural convection between vertical coaxial rectangular cylinders . Int Commun Heat Mass Transf 2012;39:904–12.

[14] Xing D, Yan C, Wang C, Sun L. Progress in Nuclear Energy A theoretical analysis about the effect of aspect ratio on single-phase laminar fl ow in rectangular ducts. Prog Nucl Energy 2013;65:1–7.

[15] Irvine TF, Hartnett JP. Edited by FORCED CONVECTION. n.d. shah and london.

[16] H. K. Versteeg and W. Malalasekera, "Introduction to computational fluid dynamics - Finite Volume Method" 1995.

[17] ANSYS, Inc. ," Fluent Theory Guide".

[18] Wibulswas, P. Laminar-flow heat-transfer in non-circular ducts (Doctoral dissertation, University of London).

[19] R. F. Bartlett, Raymond Viskanta. Enhancement of Forced Convection in an Asymmetrically Heated Duct Filled with High Thermal Conductivity Porous Media

Some Spectral Properties of p-Laplacian Diffusion Boundary Value Problem on Time Scales

Tuba Gulsen[1,*], *Emrah Yilmaz*[1], and *Meltem Kayalı*[1]

[1]Department of Mathematics, Firat University, Elazig, Turkey

Abstract. In this paper, we consider p – Laplacian type Diffusion boundary value problem on an arbitrary time scale. We generalize some spectral properties of p – Laplacian Diffusion problem to an arbitrary time scale. Moreover, we prove Picone's identity for this problem.

1 Introduction

The theory of time scale calculus was introduced for the first time by Stefan Hilger in his Ph.D. thesis in 1988 [1]. A time scale $\mathbb{T}$ is an arbitrary, non-empty, closed subset of $\mathbb{R}$. For standard notions and notations connected to time scale calculus, we refer to [2].

The significance of working boundary value problems on time scales was understood and some important results were obtained. In 2002, Agarwal et al. [3] studied linear dynamic equations and initial value problems on time scales. Guseinov and Kaymakçalan [4] considered second order linear dynamic equations and gave some sufficient conditions for nonoscillation on time scales in 2002. Erbe et al. [5] gave several comparison theorems for second order linear dynamic equation on time scales. In 2003, Guseinov [6] investigated Riemann and Lebesgue integration on time scales and gave the fundamental theorems of calculus. In 2007, Rynne [7] defined space of square-integrable functions and Sobolev type spaces on time scales and gave practices of the functional analytic results to Sturm-Liouville type boundary value operator. Grace et al. [8] studied second order half-linear dynamic equations on time scales and investigated some oscillation criteria in 2009.

In recent years, several authors have obtained many important results about p – Laplacian boundary value problem on time scales (see [9-12]). Anderson et al. [13] proved that there is at least one positive solution for a one dimensional p – Laplacian delta-nabla dynamic equation on time scales in 2004. He [14] proved that there exists at least double positive solutions with three-point boundary conditions for p – Laplacian dynamic equation on time scales in 2005. Sun and Li [15] considered a one dimensional p – Laplacian boundary value problem and showed existence of this problem on time scales in 2007. Binding et al. [16] considered p – Laplacian nonlinear eigenvalue problem on a half line in 2010. Su and

* Corresponding author: tubagulsen87@hotmail.com

Feng [17] studied a second-order $p-$Laplacian dynamic boundary value problem on a periodic time scale and showed existence of solutions of this problem in 2017.

In this study, we want to prove some basic spectral properties of $p-$Laplacian Diffusion eigenvalue problem of the form

$$Ly^{(p-1)} = -(y^{\Delta(p-1)})^{\Delta}(t) + (p-1)\{q(t) + 2\lambda r(t)\} y^{\sigma(p-1)}(t) = \lambda^2 (p-1) y^{\sigma(p-1)}(t), \quad (1)$$

with the separated boundary conditions

$$y^{\Delta(p-1)}(\rho(a)) = 0, \quad (2)$$

$$\alpha y^{(p-1)}(b) + \beta y^{\Delta(p-1)}(b) = 0, \quad (3)$$

where λ is a spectral parameter and $t \in [\rho(a), b] \cap \mathbb{T}$. During this study, we suppose that $q, r \in L_2^{\Delta}[\rho(a), b]$ are continuous functions ; $p > 1$; $a, b \in \mathbb{T}$ with $y^{\sigma(p-1)} = y^{(p-1)}(\sigma)$, $a < b$, $(\alpha^2 + \beta^2) \neq 0$. $y^{(p-1)} \in C_{rd}^2([\rho(a), b], \mathbb{R})$ is eigenfunction of the problem (1)-(3). By setting $\mathbb{T} = \mathbb{R}$ in (1), we get below classical $p-$Laplacian Diffusion equation

$$-(y'^{(p-1)})'(t) = (p-1)(\lambda^2 - q(t) - 2\lambda r(t)) y^{(p-1)}(t). \quad (4)$$

For $p = 2$ in (4), we have

$$-y''(t) + (q(t) + 2\lambda r(t)) y(t) = \lambda^2 y(t), t \in [a, b]. \quad (5)$$

(5) is known as classical Diffusion equation (or quadratic pencil of differential pencil) in literature. There are numerous studies in spectral theory related to this equation. The studies of $p-$Laplacian type Diffusion equation is as follows. In [18], Koyunbakan considered the inverse problem for $p-$Laplacian Schrödinger equation with energy-dependent potential function and found asymptotic formulas of eigenvalues, nodal points and nodal lenghts by using Prüfer substitution in 2013. In [19], Gulsen and Yilmaz obtained asymptotic formulas of eigenvalues and nodal parameters and reconstructed the potential function $p-$Laplacian Diffusion problem in 2016. In [20], Gulsen et al. took into consideration Diffusion and impulsive Diffusion eigenvalue problems, respectively and showed some spectral properties of these problems on time scales. This study is organized as follows: In Section 2, we prove some basic theorems and Picone's identity for the $p-$Laplacian Diffusion equation on $\mathbb{T}$. In Section 3, we give a conclusion to summarize our study.

2 Main Results

In this section, we give some valuable results for $p-$Laplacian Diffusion problem on a time scale $\mathbb{T}$ whose all points are right dense. It is well known that the problem (1)-(3) has only real, simple eigenvalues and eigenfunctions which are orthogonal to each other for $\mathbb{T} = \mathbb{R}$ [21]. The following results will generalize these basic results.

Theorem 1. The eigenvalues of the problem (1)-(3) are all simple.

Theorem 2. Let $y_1^{(p-1)}, y_2^{(p-1)} \in C_{rd}^2([\rho(a), b], \mathbb{R})$ be the eigenfunctions of the problem (1)-(3). Then, below equalities hold.

a) $(Ly_1^{(p-1)}) y_2^{\sigma(p-1)} - (Ly_2^{(p-1)}) y_1^{\sigma(p-1)} = W^{\Delta}(y_1^{(p-1)}, y_2^{(p-1)})$, on $[\rho(a), b] \cap \mathbb{T}$.

b) $< Ly_1^{(p-1)}, y_2^{\sigma(p-1)} > - < Ly_2^{(p-1)}, y_1^{\sigma(p-1)} > = W(y_1^{(p-1)}, y_2^{(p-1)})(b) - W(y_1^{(p-1)}, y_2^{(p-1)})(\rho(a))$,

where $W(y_1^{(p-1)}, y_2^{(p-1)}) = y_1^{(p-1)} y_2^{\Delta(p-1)} - y_2^{(p-1)} y_1^{\Delta(p-1)}$.

Above theorems can be easily proved.

***Theorem** 3*. The eigenfunctions $y_1^{(p-1)}(t,\lambda_1), y_2^{(p-1)}(t,\lambda_2)$ of the problem (1)-(3) corresponding to distinct eigenvalues λ_1 and λ_2, are orthogonal, i.e.

$$(p-1)(\lambda_1+\lambda_2)\int_{\rho(a)}^{b} y_1^{\sigma(p-1)}(t,\lambda_1)y_2^{\sigma(p-1)}(t,\lambda_2)\Delta t-2(p-1)\int_{\rho(a)}^{b} r(t)y_1^{\sigma(p-1)}(t,\lambda_1)y_2^{\sigma(p-1)}(t,\lambda_2)\Delta t=0.$$

Proof. We have

$$-(y_1^{\Delta(p-1)})^{\Delta}(t)+(p-1)\left\{q(t)+2\lambda_1 r(t)\right\}y_1^{\sigma(p-1)}(t)=\lambda_1^2(p-1)y_1^{\sigma(p-1)}(t),$$

$$-(y_2^{\Delta(p-1)})^{\Delta}(t)+(p-1)\left\{q(t)+2\lambda_2 r(t)\right\}y_2^{\sigma(p-1)}(t)=\lambda_2^2(p-1)y_2^{\sigma(p-1)}(t).$$

Multiplying above equations by $y_2^{\sigma(p-1)}, y_1^{\sigma(p-1)}$, respectively, we have

$$\begin{aligned}&-(y_1^{\Delta(p-1)})^{\Delta}(t)y_2^{\sigma(p-1)}(t)+(p-1)\left\{q(t)+2\lambda_1 r(t)\right\}y_1^{\sigma(p-1)}(t)y_2^{\sigma(p-1)}(t)\\&=\lambda_1^2(p-1)y_1^{\sigma(p-1)}(t)y_2^{\sigma(p-1)}(t),\end{aligned}\tag{6}$$

$$\begin{aligned}&-(y_2^{\Delta(p-1)})^{\Delta}(t)y_1^{\sigma(p-1)}(t)+(p-1)\left\{q(t)+2\lambda_2 r(t)\right\}y_2^{\sigma(p-1)}(t)y_1^{\sigma(p-1)}(t)\\&=\lambda_2^2(p-1)y_2^{\sigma(p-1)}(t)y_1^{\sigma(p-1)}(t).\end{aligned}\tag{7}$$

Subtracting (6) by (7) leads to

$$\begin{aligned}&\left\{y_2^{(p-1)}(t,\lambda_2)y_1^{\Delta(p-1)}(t,\lambda_1)-y_1^{(p-1)}(t,\lambda_1)y_2^{\Delta(p-1)}(t,\lambda_2)\right\}^{\Delta}\\&=(p-1)\left\{(\lambda_2^2-\lambda_1^2)-2(\lambda_2-\lambda_1)r(t)\right\}y_1^{\sigma(p-1)}(t,\lambda_1)y_2^{\sigma(p-1)}(t,\lambda_2).\end{aligned}\tag{8}$$

Applying $\Delta-$integral to the last equality from $\rho(a)$ to b, we get

$$\begin{aligned}&\int_{\rho(a)}^{b}\left\{y_2^{(p-1)}(t,\lambda_2)y_1^{\Delta(p-1)}(t,\lambda_1)-y_1^{(p-1)}(t,\lambda_1)y_2^{\Delta(p-1)}(t,\lambda_2)\right\}^{\Delta}\Delta t\\&=(p-1)(\lambda_2^2-\lambda_1^2)\int_{\rho(a)}^{b} y_1^{\sigma(p-1)}(t,\lambda_1)y_2^{\sigma(p-1)}(t,\lambda_2)\Delta t\\&-2(p-1)(\lambda_2-\lambda_1)\int_{\rho(a)}^{b} r(t)y_1^{\sigma(p-1)}(t,\lambda_1)y_2^{\sigma(p-1)}(t,\lambda_2)\Delta t\\&=-y_1^{(p-1)}(b,\lambda_1)y_2^{\Delta(p-1)}(b,\lambda_2)+y_2^{(p-1)}(b,\lambda_2)y_1^{\Delta(p-1)}(b,\lambda_1)\\&\quad+y_1^{(p-1)}(\rho(a),\lambda_1)y_2^{\Delta(p-1)}(\rho(a),\lambda_2)-y_2^{(p-1)}(\rho(a),\lambda_2)y_1^{\Delta(p-1)}(\rho(a),\lambda_1)=0,\end{aligned}$$

and then,

$$(p-1)(\lambda_1+\lambda_2)\int_{\rho(a)}^{b} y_1^{\sigma(p-1)}(t,\lambda_1)y_2^{\sigma(p-1)}(t,\lambda_2)\Delta t-2(p-1)\int_{\rho(a)}^{b} r(t)y_1^{\sigma(p-1)}(t,\lambda_1)y_2^{\sigma(p-1)}(t,\lambda_2)\Delta t=0,$$

for $\lambda_1\neq\lambda_2$. Hence, it shows that the eigenfunctions $y_1^{(p-1)}(t,\lambda_1), y_2^{(p-1)}(t,\lambda_2)$ are always orthogonal.

Theorem 4. The equality

$$y^{(p-1)}(t,\lambda)\frac{\partial}{\partial\lambda}y^{\Delta(p-1)}(t,\lambda)-y^{\Delta(p-1)}(t,\lambda)\frac{\partial}{\partial\lambda}y^{(p-1)}(t,\lambda)$$
$$=-2\lambda(p-1)\int_{\rho(a)}^{t}(y^{\sigma(p-1)}(\tau,\lambda))^2\Delta\tau+2(p-1)\int_{\rho(a)}^{t}r(\tau)(y^{\sigma(p-1)}(\tau,\lambda))^2\Delta\tau,$$

holds where $t\in[\rho(a),b]\cap\mathbb{T}$ and $\lambda\in\mathbb{R}$ for the problem (1)-(3).

Theorem 5. All eigenvalues of the problem (1)-(3) are real.

Proof. Let $\bar{\lambda}_0$ be a complex eigenvalue and $\bar{y}^{(p-1)}(t)$ be an eigenfunction corresponding to the eigenvalue $\bar{\lambda}_0$ of the problem (1)-(3). Then, we obtain

$$\left\{y^{\Delta(p-1)}\bar{y}^{(p-1)}-\bar{y}^{\Delta(p-1)}y^{(p-1)}\right\}^{\Delta}$$
$$=(p-1)(\bar{\lambda}_0^2-\lambda_0^2)\left|y^{\sigma(p-1)}\right|^2-2(p-1)(\bar{\lambda}_0-\lambda_0)\left|y^{\sigma(p-1)}\right|^2.$$

If we use $\Delta-$integral for the last equality from $\rho(a)$ to b, and by considering the boundary conditions (2), (3), we get

$$(p-1)(\bar{\lambda}_0+\lambda_0)\int_{\rho(a)}^{b}\left|y^{\sigma(p-1)}(t)\right|^2\Delta t-2(p-1)\int_{\rho(a)}^{b}r(t)\left|y^{\sigma(p-1)}(t)\right|^2\Delta t=0\Rightarrow y^{\sigma(p-1)}(t)=0,$$

for $\bar{\lambda}_0\neq\lambda_0$ and $(\bar{\lambda}_0+\lambda_0)-2r(t)>0$. This is a contradiction and the proof is completed.

Now, we will prove Picone's identity which is an important formula for the problem (1)-(3) to prove oscillation criteria. There are many studies about the Picone's identity (see [22-26]). In 1998, Allegretto and Xi [27] obtained a Picone's identity for the $p-$Laplace operator and Bal [28] showed a generalized Picone's identity for the $p-$Laplace operator and then proved Sturmian comparison principle and a Liouville type theorem (see also [29-31]). Picone's identity for dynamic equations on time scales was studied in [32-34].

Let denote the kernel of the matrix A by $Ker\,A$, and put $defM=dimKerA$, and $\alpha^{\Diamond}=1/\alpha$ for $\alpha\neq0$ and $0^{\Diamond}=0$. It is given below Picone's identity for the problem (1)-(3).

Theorem 6. Let $Ker\,y^{(p-1)}(t)\subset Ker\,x^{(p-1)}(t)$ and $Ker\,y^{\sigma(p-1)}(t)\subset Ker\,x^{\sigma(p-1)}(t)$, $x^{(p-1)}\in C_{rd}^2([\rho(a),b],\mathbb{R})$ and $y^{(p-1)}$ is a solution of $Ly^{(p-1)}+\lambda^2(1-p)y^{\sigma(p-1)}=0$. Then,

$$\left\{L(x^{(p-1)})-\lambda^2(p-1)x^{\sigma(p-1)}\right\}x^{\sigma(p-1)}$$
$$=[x^{(p-1)}(y^{(p-1)})^{\Diamond}W(x^{(p-1)},y^{(p-1)})]^{\Delta}+(y^{(p-1)}y^{\sigma(p-1)})^{\Diamond}W^2(x^{(p-1)},y^{(p-1)}),$$

where $W(x^{(p-1)},y^{(p-1)})=x^{(p-1)}y^{\Delta(p-1)}-y^{(p-1)}x^{\Delta(p-1)}$.

Proof. Suppose that $(y^{(p-1)}y^{\sigma(p-1)})(t)\neq0$. Then, we get

$$[x^{(p-1)}(y^{(p-1)})^{\Diamond}W(x^{(p-1)},y^{(p-1)})]^{\Delta}=\frac{x^{\sigma(p-1)}}{y^{\sigma(p-1)}}W^{\Delta}(x^{(p-1)},y^{(p-1)})$$

$$+\left(\frac{x^{\Delta(p-1)}y^{(p-1)}-x^{(p-1)}y^{\Delta(p-1)}}{y^{(p-1)}y^{\sigma(p-1)}}\right)W(x^{(p-1)},y^{(p-1)})$$

$$=\frac{x^{\sigma(p-1)}}{y^{\sigma(p-1)}}\left\{-L(y^{(p-1)})x^{\sigma(p-1)}+L(x^{(p-1)})y^{\sigma(p-1)}\right\}-\frac{W^{2}(x^{(p-1)},y^{(p-1)})}{y^{(p-1)}y^{\sigma(p-1)}}$$

$$=\frac{x^{\sigma(p-1)}}{y^{\sigma(p-1)}}\left\{-\lambda^{2}(p-1)y^{\sigma(p-1)}x^{\sigma(p-1)}+L(x^{(p-1)})y^{\sigma(p-1)}\right\}-\frac{W^{2}(x^{(p-1)},y^{(p-1)})}{y^{(p-1)}y^{\sigma(p-1)}}$$

$$=\left\{L(x^{(p-1)})-\lambda^{2}(p-1)x^{\sigma(p-1)}\right\}x^{\sigma(p-1)}-(y^{(p-1)}y^{\sigma(p-1)})^{\Diamond}W^{2}(x^{(p-1)},y^{(p-1)}).$$

If t is right-scattered and $(y^{(p-1)}y^{\sigma(p-1)})(t)=0$, then $(x^{(p-1)}x^{\sigma(p-1)})(t)=0$ and

$$W(x^{(p-1)},y^{(p-1)})=x^{(p-1)}y^{\Delta(p-1)}-y^{(p-1)}x^{\Delta(p-1)}$$

$$=x^{(p-1)}(y^{\sigma(p-1)}-y^{(p-1)})/\mu-y^{(p-1)}(x^{\sigma(p-1)}-x^{(p-1)})/\mu$$

$$=(x^{(p-1)}y^{\sigma(p-1)}-y^{(p-1)}x^{\sigma(p-1)})/\mu=0.$$

If t is left-scattered, $x^{(p-1)}(y^{(p-1)})^{\Diamond}W(x^{(p-1)},y^{(p-1)})$ is continuous at t. If t is left dense, $x^{(p-1)}(y^{(p-1)})^{\Diamond}W(x^{(p-1)},y^{(p-1)})$ is also continuous at t. By using L'Hospital rule, we get

$$\lim_{s\to t^{-}}[x^{(p-1)}(y^{(p-1)})^{\Diamond}W(x^{(p-1)},y^{(p-1)})](s)=\frac{x^{\Delta(p-1)}(t)}{y^{\Delta(p-1)}(t)}W^{\Delta}(x^{(p-1)},y^{(p-1)})]=0$$

$$=[x^{(p-1)}(y^{(p-1)})^{\Diamond}W(x^{(p-1)},y^{(p-1)})](t).$$

$$x^{(p-1)}(y^{(p-1)})^{\Diamond}W(x^{(p-1)},y^{(p-1)})]^{\Delta}$$

$$=\frac{[x^{(p-1)}(y^{(p-1)})^{\Diamond}W(x^{(p-1)},y^{(p-1)})]^{\sigma}-x^{(p-1)}(y^{(p-1)})^{\Diamond}W(x^{(p-1)},y^{(p-1)})}{\mu}$$

$$=\frac{[x^{(p-1)}(y^{(p-1)})^{\Diamond}W(x^{(p-1)},y^{(p-1)})]^{\sigma}}{\mu}=x^{\sigma(p-1)}(y^{\sigma(p-1)})^{\Diamond}W^{\Delta}(x^{(p-1)},y^{(p-1)})$$

$$=x^{\sigma(p-1)}(y^{\sigma(p-1)})^{\Diamond}\left\{-L(y^{(p-1)})x^{\sigma(p-1)}+L(x^{(p-1)})y^{\sigma(p-1)}\right\}$$

$$=\frac{x^{\sigma(p-1)}}{y^{\sigma(p-1)}}\left\{-\lambda^{2}(p-1)y^{\sigma(p-1)}x^{\sigma(p-1)}+L(x^{(p-1)})y^{\sigma(p-1)}\right\}$$

$$=\left\{L(x^{(p-1)})-\lambda^{2}(p-1)x^{\sigma(p-1)}\right\}x^{\sigma(p-1)}-(y^{(p-1)}y^{\sigma(p-1)})^{\Diamond}W^{2}(x^{(p-1)},y^{(p-1)}).$$

If t is right-dense point, the Picone's identity holds.

3 Conclusion

$p-$Laplacian type Diffusion eigenvalue problems arise from many fields of science and have various applications in mathematics and physics. We notice that much attention paid on discussing the solutions for this type problems for different operators on time scales. Hence, we have considered some spectral properties for p – Laplacian Diffusion eigenvalue problem on an arbitrary time scale and generalized basic results. We have obtained some important results to fill the gaps in this area.

References

1. S. Hilger, Ph.D. Thesis, Universitat Wurzburg, (1988)
2. M. Bohner, Boston(MA), Birkhauser, Boston Inc, (2001)
3. R. Agarwal, M. Bohner, D. O'Regan, A Peterson, J. Comput. Appl. Math., **141**, 1-26 (2002)
4. G.Sh. Guseinov, B. Kaymakçalan, J. Comput. Appl. Math., **141**, 187-196 (2002)
5. L. Erbe, A. Peterson, P. Řehák, J. Math. Anal. Appl., **275**, 418-438 (2002)
6. G.Sh. Guseinov, J. Math. Anal. Appl., **285**, 107-127 (2003)
7. B.P. Rynne, J. Math. Anal. Appl., **328**, 1217-1236 (2007)
8. S.R. Grace, M. Bohner, R.P. Agarwal, J. Difference Equ. Appl., **15**, 451-460 (2009)
9. C. Song, P. Weng, Nonlinear Analy. Theory, **68**, 208-215 (2008)
10. W. Han, S. Kang,. Math. Comput. Model., **49**, 527-535 (2009)
11. J. R. Graef, L. Kong, J. Difference Equ. Appl., **17**, 831-839 (2011).
12. A. Dogan, Adv. Difference Equ., **2013**, 238 (2013).
13. D. Anderson, R. Avery, J. Henderson, J. Difference Equ. Appl., **10**, 889-896 (2004)
14. Z. He, J. Comput. Appl. Math., **182**, 304-315 (2005)
15. H.R. Sun, W.T. Li, J. Differential Equations, **240**, 217-248 (2007).
16. P. Binding, P.J. Browne, I.M. Karabash, Proc. Edinb. Math. Soc., **53**, 271-291 (2010)
17. Y.H. Su, Z. Feng, Appl. Anal., 1-19 (2017)
18. H. Koyunbakan, Bound. Value Probl., **2013**, 8 pp. (2013)
19. T. Gulsen, E. Yilmaz, Commun. Fac. Sci. Univ. Ank. Sér. A1 Math. Stat., **65**, 23-36 (2016)
20. T. Gulsen, S.S.M. Sian, E. Yilmaz, H. Koyunbakan, Int. J. Anal. Appl., **16**, 137-148 (2018)
21. M.G. Gasymov, G.Sh. Guseinov, Dokl. Akad. Nauk Az. SSSR, **37**, 19-23 (1981)
22. K. Kreith, , J. Math. Anal. Appl., **31**, 297-308 (1970)
23. C.A. Swanson, Picone's identity, Rend. Mat, **8**, 373-397 (1975)
24. G.J. Etgen, , R.T. Lewis, Proc. Edinb. Math. Soc., **22**, 277-290 (1979)
25. D. Hinton, Birkhäuser Basel. pp. 1-27 (2005)
26. T. Kusano, J. Jaroš, N. Yoshida, Nonlinear Analysis: Theory, Methods and Applications, **40**, 381-395 (2000)
27. W. Allegretto, H.Y. Xi, Nonlinear Analy. Theory, **32**, 819-830 (1998)
28. K. Bal, Electron. J. Differ. Equ., **2013**, 1-6 (2013)
29. X. Yang, Appl. Math. Comput., **136**, 181-193 (2003)
30. A. Tiryaki, Electronic J. Differential Equations, **2016**, 1-7 (2016)
31. O. Došlý, S. Hilger, J. Comput. Appl. Math., **141**, 147-158 (2002)
32. B. Belinskiy, J.R. Graef, S. Petrovic. Int. J. Difference Equ., **2**, 25-35 (2007)
33. D.R., Anderson, J.R. Graef, Appl. Anal. Discrete Math., **4**, 338-346 (2010)
34. R.P. Agarwal, M. Bohner, P. J. Wong, Appl. Math. Comput., **99**, 153-166 (1999)

Approximate Solutions of Two-Dimensional Burgers' and Coupled Burgers' Equations by Residual Power Series Method

Selahattin Gulsen[1,*], *Mustafa Inc*[1] , and *Harivan R. Nabi*[2]

[1]Department of Mathematics, Firat University, Elazig, Turkey
[2]Department of Road Construction, Duhok Polytechnic University, Duhok, Iraq

Abstract. In this study, two-dimensional Burgers' and coupled Burgers' equations are examined by the residual power series method. This method provides series solutions which are rapidly convergent and their components are easily calculable by Mathematica. When the solution is polynomial, the method gives the exact solution using Taylor series expansion. The results display that the method is more efficient, applicable and accuracy and the graphical consequences clearly present the reliability of the method.

1 Introduction

Initial value problems (IVPs) of differential equations (DEs) appear in a significant number of applications such as physical, biological and chemical processes [1-3]. Since solutions of IVPs cannot be obtained with analytical methods, numerical and approximate methods have been always constituted to get approximate solutions by some authors (see [4-6]). The previous studies do not give a methodical way to find solutions of IVPs and also previous studies require more time and more work to accomplish the results.

In this study, we apply the residual power series method (RPSM) to obtain approximate solutions of two-dimensional Burgers' and inhomogeneous coupled Burgers' equations. RPSM was firstly introduced by Arqub [7] to find series solutions of linear and nonlinear fuzzy problems in 2013. Arqub applied RPSM to the first-order and second-order fuzzy DEs and gave series solutions of these equations. Then Al-Smadi [8] carried out RPSM to linear and nonlinear IVPs in 2013. Alquran [9] fulfilled RPSM to nonlinear time-fractional foam drainage equation to achieve approximate solution in 2014. El-Ajou [10] et al. found approximate solution of nonlinear fractional KdV-Burgers equation with time-space-fractional derivatives with RPSM in 2015. Moaddy et al. [11] utilized RPSM to differential algebraic equations system in 2015. Tchier et al. [12] considered nonlinear time fractional reaction-diffusion equation with RPSM in 2016. Momani et al. [13] introduced RPSM for linear and nonlinear systems of IVPs and carried out on the effectiveness of RPSM in 2016. Mahmood and Yousif [14] applied RPSM to the coupled strong system nonlinear Boussinesq-Burgers equations to achieve approximate solutions

* Corresponding author: sgulsen2301@gmail.com

and compared these solutions with other approximate solutions which found in other methods in 2017. Many scientists have applied RPSM to different equations and found approximate solutions of these equations (see [15-18]). RPSM is a semi analytical method which can be directly applied to IVPs to achieve series solution of them without linearization, perturbation and discretization [7]. In polynomial form, an analytical approximate solution is established by this method. RPSM and traditional higher order Taylor series method different from each other. As traditional Taylor series method is very expensive in high order computations, RPSM is very cheap and applicable to high orders. In practise this method can be called an alternative method to acquire analytic Maclaurin series solution of IVPs. Considering error concept for the method, we get a series solution, indeed this solution is a truncated series solution. The RPSM method consists of the following steps; firstly the method construct Maclaurin expansion of the solution, exact solution can be obtained when the solution is polynomial. Furthermore the solutions and all derivatives can be applied for each arbitrary point. Secondly, RPSM needs less time to obtain high precision solution with small computational requirements. In this article, we examine two-dimensional Burgers' and inhomogeneous coupled Burgers' equations. The Burgers equation has been found to explain the appromixate flow theory from a schock wave moving in a viscous fluid [19,20]. Fletcher [21] used the Hopf-Cole transformation to get an analytic solution of two-dimensional Burgers' equations in 1983. Because of the wide applicability of these equations, many scientists have applied different numerical methods to find approximate solutions of these equations (see [22-27]). This study consists of the following steps : In Section 2, we give the fundemental scheme of RPSM on two-dimensional Burgers' equation. In Sections 3 and 4, RPSM is applied to two-dimensional Burgers' equation and inhomogeneous coupled Burgers' equation with appropriate initial conditions and obtained results are shown in graphs. Finally, consequences are given in section 5.

2 Analysis of Residual Power Series Method (RPSM) for Two-Dimensional Burgers' Equation

To demonstrate the fundamental scheme of RPSM, we consider two-dimensional Burgers' equation

$$u_t + uu_x + vu_y - \frac{1}{Re}(u_{xx} + u_{yy}) = 0, \tag{1}$$

$$v_t + uv_x + vv_y - \frac{1}{Re}(v_{xx} + v_{yy}) = 0, \tag{2}$$

with the initial conditions

$$u(x,y,0) = \frac{3}{4} - \frac{1}{4\left(1+\exp\left(\frac{Re(-x+y)}{8}\right)\right)}, \quad v(x,y,0) = \frac{3}{4} + \frac{1}{4\left(1+\exp\left(\frac{Re(-x+y)}{8}\right)\right)}, \tag{3}$$

where Re is the Reynolds number.

The solutions of the equations (1) and (2) depending on the conditions (3) are expressed as a series expansion around an initial point $t = 0$ in the RPSM. Now, suppose that the solutions have the forms as

$$u(x,y,t) = \sum_{m=0}^{\infty} f_m(x,y)t^m, \quad v(x,y,t) = \sum_{m=0}^{\infty} g_m(x,y)t^m, \tag{4}$$

where $m = 0,1,2...$. Let us then define the expressions $u_k(x,y,t)$ and $v_k(x,y,t)$ as the k-th truncated series of $u(x,y,t)$ and $v(x,y,t)$, respectively, i.e.

$$u_k(x,y,t) = \sum_{m=0}^{k} f_m(x,y)t^m, \quad v_k(x,y,t) = \sum_{m=0}^{k} g_m(x,y)t^m, \tag{5}$$

where $k = 1,2,3,...$. Clearly, $u(x,y,t)$ and $v(x,y,t)$ provides the initial conditions (3), therefore the 0-th RPSM approximate solutions of $u(x,y,t)$ and $v(x,y,t)$ are

$$u(x,y,0) = f_0(x,y) = f(x,y), \quad v(x,y,0) = g_0(x,y) = g(x,y). \tag{6}$$

Then, the estimated initial approach the first RPSM approximation solutions of $u(x,y,t)$ and $v(x,y,t)$ using equations (5) should be

$$u_1(x,y,t) = f(x,y) + f_1(x,y)t, \quad v_1(x,y,t) = g(x,y) + g_1(x,y)t. \tag{7}$$

So, expansion of equations (5) are reformulated as

$$u_k(x,y,t) = f(x,y) + f_1(x,y)t + \sum_{m=2}^{k} f_m(x,y)t^m, \quad v_k(x,y,t) = g(x,y) + g_1(x,y)t + \sum_{m=2}^{k} g_m(x,y)t^m, \tag{8}$$

where $k = 2,3,...$. To find the values of coefficients $f_m(x,y)$ and $g_m(x,y)$, $m = 1,2,3,...,k$ in series expansion of the equations (8), one can define the residual functions as

$$Res_u(x,y,t) = u_t + uu_x + vu_y - \frac{1}{Re}(u_{xx} + u_{yy}), \quad Res_v(x,y,t) = v_t + uv_x + vv_y - \frac{1}{Re}(v_{xx} + v_{yy}) \tag{9}$$

and so the $k-$th residual functions can be expressed as

$$Res_{u,k}(x,y,t) = (u_k)_t + u_k(u_k)_x + v_k(u_k)_y - \frac{1}{Re}((u_k)_{xx} + (u_k)_{yy}), \tag{10}$$

$$Res_{v,k}(x,y,t) = (v_k)_t + u_k(v_k)_x + v_k(v_k)_y - \frac{1}{Re}((v_k)_{xx} + (v_k)_{yy}), \tag{11}$$

where $k = 1,2,3,...$ Clearly, $Res(x,y,t) = 0$, $\lim_{k\to\infty} Res_k(x,y,t) = Res(x,y,t)$ for all $x \in I$ and $t \geq 0$. Therefore, when $t = 0$ for each $s = 0,1,...,k$, $\partial^s Res(x,y,t)/\partial t^s = 0$. To obtain the coefficients $f_m(x,y)$ and $g_m(x,y)$, $m = 1,2,3,...,k$, we carry out the following steps: Firstly, to employ the $m-$th truncated series of $u(x,y,t)$ and $v(x,y,t)$, we put on the derivative formula $\partial^s Res(x,y,t)/\partial t^s = 0$ in $Res_k(x,y,t)$, $m = 1,2,3,...,k$. In the obtained formula, we write $t = 0$ and we equate it to zero. Lastly, to achieve the other coefficients, we solve the found algebraic formula

$$(\partial^s Res_k(x,y,t)/\partial t^s) = 0 \ (t = 0), \ s = 1,2,3,...,k. \tag{12}$$

Thus all coefficients of the multiple power series of equations (1)-(4) can be calculated.

3 Application of Residual Power Series Method (RPSM) for Two-Dimensional Burgers' Equation

The exact solutions of two-dimensional Burgers' problem (1)-(4) are

$$u(x,y,t)=\frac{3}{4}-\frac{1}{4\left(1+\exp\left(\dfrac{Re(-t-4x+4y)}{32}\right)\right)},\quad v(x,y,t)=\frac{3}{4}+\frac{1}{4\left(1+\exp\left(\dfrac{Re(-t-4x+4y)}{32}\right)\right)}. \tag{13}$$

With the iteration of RPSM for $n=0,1,\ldots$ it is obtained

$$u_1(x,y,t)=\frac{3}{4}-\frac{1}{4\left(1+\exp\left(\dfrac{Re(-x+y)}{8}\right)\right)}-\frac{Rt\exp\left(\dfrac{Re(-x+y)}{8}\right)}{128\left(1+\exp\left(\dfrac{Re(-x+y)}{8}\right)\right)^2}, \tag{14}$$

$$v_1(x,y,t)=\frac{3}{4}+\frac{1}{4\left(1+\exp\left(\dfrac{Re(-x+y)}{8}\right)\right)}+\frac{Rt\exp\left(\dfrac{Re(-x+y)}{8}\right)}{128\left(1+\exp\left(\dfrac{Re(-x+y)}{8}\right)\right)^2}, \tag{15}$$

and in this way $u_2(x,y,t), u_3(x,y,t),\ldots,$ and $v_2(x,y,t), v_3(x,y,t),\ldots,$ can be obtained and then the solutions are found as following

$$u(x,y,t)=u_1(x,y,t)+u_2(x,y,t)+\ldots.,\quad v(x,y,t)=v_1(x,y,t)+v_2(x,y,t)+\ldots. \tag{16}$$

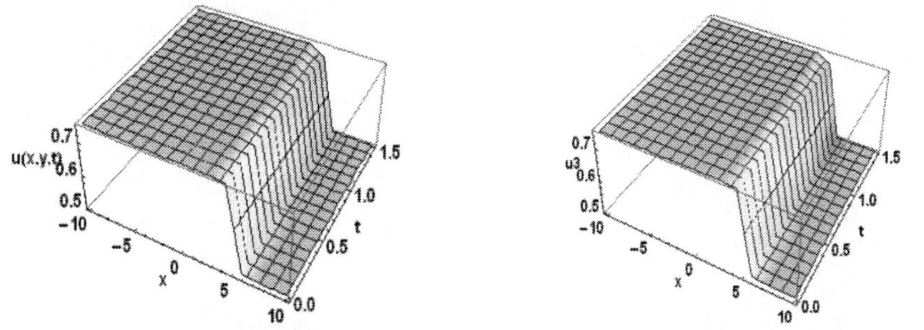

Figure 1. The surface graphs of the exact solution $u(x,y,t)$ and the approximate solution $u_3(x,y,t)$ of two-dimensional Burgers' problem when $y=5,\ R=50$, respectively.

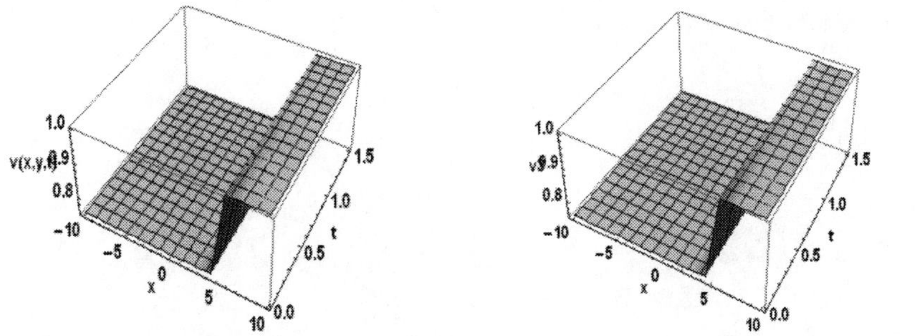

Figure 2. The surface graphs of the exact solution $v(x,y,t)$ and the approximate solution $v_3(x,y,t)$ of two-dimensional Burgers' problem when $y=5,\ R=50$, respectively.

https://doi.org/10.1051/itmconf/20182201044

4 Application of Residual Power Series Method (RPSM) for Inhomogeneous Coupled Burgers' Equation

We consider inhomogeneous coupled Burgers' equation

$$u_t - u_{xx} + 2uu_x + (uv)_x = F(x,t), \tag{17}$$

$$v_t - vv_{xx} + 2vv_x + (uv)_x = G(x,t), \tag{18}$$

with the initial conditions

$$u(x,0) = 0,\ v(x,0) = 0, \tag{19}$$

where

$$F(x,t) = x^2 - 2t + 2x^3t^2 + t^2,\ \ G(x,t) = \frac{1}{x} - \frac{2t}{x^3} - \frac{t^2}{x^3} + t^2. \tag{20}$$

The exact solutions of inhomogeneous coupled Burgers' problem (26)-(28) are

$$u(x,t) = x^2t,\ \ v(x,t) = \frac{t}{x}. \tag{21}$$

With the iteration of RPSM for $n = 0,1,...$ it is found

$$u_1(x,t) = x^2t,\ \ v_1(x,t) = \frac{t}{x}, \tag{22}$$

and similarly

$$u_2(x,t) = x^2t,\ \ v_2(x,t) = \frac{t}{x},\ \ u_3(x,t) = x^2t - \frac{2t^3x^3}{3},\ \ v_3(x,t) = \frac{t^3}{3x^3} + \frac{t}{x}, \tag{23}$$

and in this way $u_4(x,t), u_5(x,t),...,$ and $v_4(x,t), v_5(x,t),...,$ can be obtained and then the solutions are found as following

$$u(x,t) = u_1(x,t) + u_2(x,t) + ...,\qquad v(x,t) = v_1(x,t) + v_2(x,t) + ... \tag{24}$$

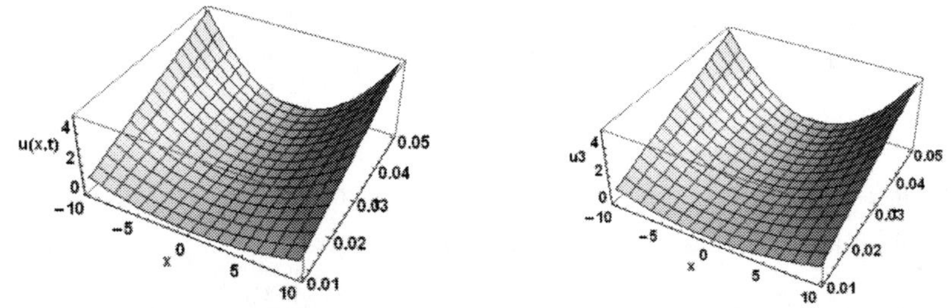

Figure 3. The surface graphs of the exact solution $u(x,t)$ and the approximate solution $u_3(x,t)$ of inhomogeneous coupled Burgers' problem, respectively.

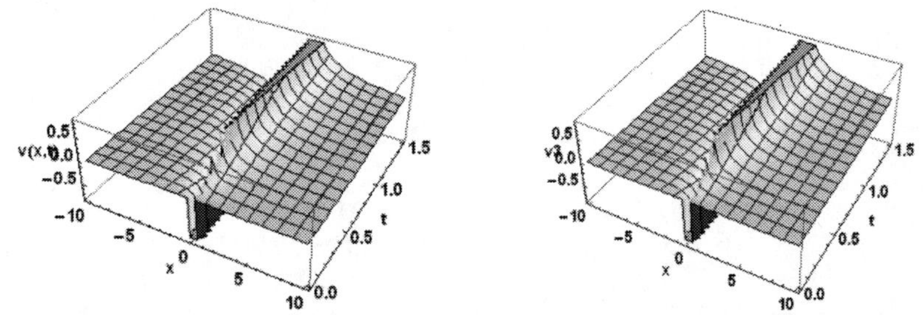

Figure 4. The surface graphs of the exact solution $v(x,t)$ and the approximate solution $v_3(x,t)$ of inhomogeneous coupled Burgers' problem, respectively.

5 Conclusion

In this study, approximate solutions of the two-dimensional Burgers equation and inhomogeneous coupled Burgers equation were calculated by RPSM. It is seen that the results obtained with the RPSM are very close to the exact solutions. The obtained results were compared with the exact solutions and the obtained results are shown in graphs.

References

1. G. B. Whitham, John Wiley & Sons, (1974)
2. M.W. Hirsch, S. Smale, Academic, New York, (1974)
3. L. Collatz, New York: John Wiley & Sons, (1986)
4. J.R. Cash, A.H. Karp, ACM Trans. Math. Software (TOMS), **16**, 201-222 (1990)
5. M.J. Jang, C.L. Chen, Y.C. Liy, Appl. Math. Comput., **115**, 145-160. (2000)
6. I.A.H. Hassan, Chaos, Solitons & Fractals, **36**, 53-65 (2008)
7. O.A. Arqub, J. Adv. Res. Appl. Math., **5**, 31-52 (2013)
8. M. Al-Smadi, Theoret. Math. Appl., **3**, 199-210 (2013)
9. M. Alquran, Math. Sci., **8**, 153-160 (2014)
10. A. El-Ajou, O.A. Arqub, S. Momani, J. Comput. Phys., **293**, 81-95 (2015)
11. K. Moaddy, M. Al-Smadi, I. Hashim, Discrete Dyn. Nat. Soc., **2015**, 12 pp. (2015)
12. F. Tchier, M. Inc, Z.S. Korpinar, D. Baleanu, Adv. Mech. Eng., **8**, 1-10 (2016)
13. S. Momani, O.A. Arqub, M.A. Hammad, A. Alsaedi, Appl. Math. Inf. Sci, **10**, 765-775 (2016)
14. B.A. Mahmood, M.A. Yousif, Cogent Math. Stat., **4**, 11 pp. (2017)
15. I. Komashynska, M. Al-Smadi, A. Ateiwi, S. Al-Obaidy, Appl. Math, **10**, 1-11 (2016)
16. A. Kumar, S. Kumar, M. Singh, Commun. Numer. Anal., **2016**, 1-10 (2016)
17. L. Wang, X. Chen, Entropy, **17**, 6519-6533 (2015)
18. M. Ali, I. Jaradat, M. Alquran, J. Math. Computer Sci., **17**, 106-114 (2017)
19. J.D. Cole, Quart. Appl. Math., **9**, 225-236 (1951)
20. J.M. Burger, Adv. Appl. Mech., **1**, 171-199 (1948)
21. C.A. Fletcher, Int. J. Numer. Methods Fluids, **3**, 213-216 (1983)
22. P.C. Jain, D.N. Holla, Internat. J. Non-Linear Mech., **13**, 213-222 (1978)
23. A.R. Bahadır, Appl. Math. Comput., **137**, 131-137 (2003)
24. S.M. El-Sayed, D. Kaya, Appl. Math. Comput., **158**, 101-109 (2004)
25. M.A. Abdou, A.A. Soliman, J. Comput. Appl. Math., **181**, 245-251 (2005)
26. A.K. Alomari, M.S.M. Noorani, R. Nazar, Appl. Math. Sci., **2**, 1963-1977 (2008)
27. R. Abazari, A. Borhanifar, Comput. Math. Appl., **59**, 2711-2722 (2010)

On the Solutions of Fractional Cauchy Problem Featuring Conformable Derivative

Mehmet Yavuz[1,*], and *Necati* Özdemir[2]

[1] Department of Mathematics-Computer Sciences, Necmettin Erbakan University, Konya, Turkey

[2] Department of Mathematics, Balikesir University, Balikesir, Turkey

Abstract. In this study, we have obtained analytical solutions of fractional Cauchy problem by using q-Homotopy Analysis Method (q-HAM) featuring conformable derivative. We have considered different situations according to the homogeneity and linearity of the fractional Cauchy differential equation. A detailed analysis of the results obtained in the study has been reported. According to the results, we have found out that our obtained solutions approach very speedily to the exact solutions.

1 Introduction

In recent years, mathematical modelling with fractional PDEs in some special areas such like physics, mathematics, engineering, finance, biology and medicine [1-8] have been studied. Analytical solutions for fractional differential equations (FDEs) are not very popular in the literature, but the methods applied to avoid effective engineering analyzes, and therefore, numerical methods are frequently applied [9-13].

Recently, [14-16] recommended a new derivative operator namely "conformable" (CDO) and some applications of this operator in various fields have been improved. In this context, some researchers [5, 7, 17-20] applied the conformable operator to solve the problems in engineering, finance, biology, medicine, physics and applied mathematics. In this study, we have solved the linear/nonlinear fractional Cauchy problem with the proposed q-HAM described by using the conformable operator.

2 Fundamental Properties of the Conformable Operator

Definition 1. Let $u:[0,\infty)\to\mathbb{R}$ be a function. The conformable derivative of u order $\alpha\in(0,1]$ is defined by [14]:

$$CD_t^{\alpha}(u)(t)=\lim_{\varepsilon\to 0}\frac{u\left(t+\varepsilon t^{1-\alpha}\right)-u(t)}{\varepsilon} \tag{1}$$

for all $t>0$.

* Corresponding author: mehmetyavuz@konya.edu.tr

Definition 2. [14, 16] The $\alpha-$ fractional integral of u function starting is defined by

$$I_a^\alpha(u)(t)=I_a^1\left(t^{\alpha-1}u\right)=\int_a^t \frac{u(x)}{x^{1-\alpha}}dx, \quad \alpha\in(0,1). \tag{2}$$

3 q-Homotopy Analysis Method in the Conformable Sense

Let us take the nonlinear FPDE:

$$\mathcal{N}\left(CD_t^\alpha u(x,t)\right)=f(x,t), \tag{3}$$

where CD_t^α shows the conformable derivative of order α. We show the nonlinear term with $\mathcal{N}$, the known function with $f(x,t)$, and the unknown function with $u(x,t)$. We construct the zero-order modified equation which is related with standard homotopy method as [21]:

$$(1-nq)L\left(\omega(x,t;q)-u_0(x,t)\right)=qhH(x,t)\left(\mathcal{N}\left(CD_t^\alpha\omega(x,t;q)\right)-f(x,t)\right), \tag{4}$$

where $n\geq 1$, $q\in[0,1/n]$ represents the embedded parameter, L is a linear operator, $H(x,t)$ is non-zero supportive function. Distinctly, since $q=0$ and $q=1/n$, we can see Eq.(4) as

$$\omega(x,t;0)=u_0(x,t), \qquad \omega\left(x,t;\frac{1}{n}\right)=u(x,t), \tag{5}$$

respectively. Therefore, q increases from zero to $1/n$, the solution $\omega(x,t;q)$ varies from the initial value $u_0(x,t)$ to the solution $u(x,t)$. If the parameters are chosen appropriately, solution of Eq.(5) is available. Now, we have the following expansion of $\omega(x,t;q)$

$$\omega(x,t;q)=u_0(x,t)+\sum_{m=1}^{\infty}u_m(x,t)q^m, \tag{6}$$

where

$$u_m(x,t)=\frac{1}{m!}\frac{\partial^m\omega(x,t;q)}{\partial q^m}\bigg|_{q=0}. \tag{7}$$

After that we get

$$u(x,t)=u_0(x,t)+\sum_{m=1}^{\infty}u_m(x,t)\left(\frac{1}{n}\right)^m. \tag{8}$$

Let the vector

$$\vec{u}_n=\left\{u_0(x,t),u_1(x,t),\ldots,u_n(x,t)\right\}. \tag{9}$$

Then we have mth- order altered equation as [22]

$$L\left[u_m(x,t)-\chi_m^* u_{m-1}(x,t)\right]=hH(x,t)\Re_m\left(\vec{u}_{m-1}(x,t)\right) \tag{10}$$

with initial conditions

$$u_m^{(k)}(x,0)=0, \quad k=0,1,2,\ldots,m-1, \tag{11}$$

where

$$\Re_m\left(\vec{u}_{m-1}(x,t)\right)=\frac{1}{(m-1)!}\frac{\partial^{m-1}\left[\mathcal{N}\left(CD_t^\alpha\omega(x,t;q)\right)-f(x,t)\right]}{\partial q^{m-1}}\Bigg|_{q=0} \tag{12}$$

and

$$\chi_m^* = \begin{cases} 0, & m \le 1 \\ n, & m > 1. \end{cases} \tag{13}$$

Finally, we get the components of solution series as,

$$u_m(x,t) = \chi_m^* u_{m-1}(x,t) - \chi_m^* u_{m-1}(x,0^+) + hH(x,t)\Im_t^\alpha \Re_m(\vec{u}_{m-1}(x,t)). \tag{14}$$

4 Numerical Examples

Example 1. We consider a special case of fractional Cauchy equation [23, 24]

$$CD_t^\alpha u(x,t) + u_x(x,t) = x, \quad x \in R, \quad t > 0, \quad 0 < \alpha \le 1, \tag{15}$$

with initial condition

$$u(x,0) = e^x, \quad x \in R. \tag{16}$$

If we choose $H(x,t)=1$, we can construct the zeroth-order deformation equation as

$$L\left[u_m(x,t) - \chi_m^* u_{m-1}(x,t)\right] = h\Re_m(\vec{u}_{m-1}(x,t)), \tag{17}$$

with initial condition for $m \ge 1$, $u_m(x,0)=0$, χ_m^* is as defined in (13) and

$$\Re_m(\vec{u}_{m-1}(x,t)) = CD_t^\alpha u_{m-1} + (u_{m-1})_x - x. \tag{18}$$

Therefore, the solution of Eq.(15) for $m \ge 1$ becomes

$$u_m(x,t) = \chi_m^* u_{m-1}(x,t) + h\Im_t^\alpha \Re_m(\vec{u}_{m-1}(x,t)). \tag{19}$$

Then using Eq.(19) we can obtain the q-HAM consequence as

$$\begin{aligned}
&u_0(x,t) = e^x, \\
&u_1(x,t) = h(e^x - x)\frac{t^\alpha}{\alpha}, \\
&u_2(x,t) = nh(e^x - x)\frac{t^\alpha}{\alpha} + h^2(e^x - x)\frac{t^\alpha}{\alpha} - hx\frac{t^\alpha}{\alpha} + h^2(e^x - 1)\frac{t^{2\alpha}}{2\alpha^2}, \\
&u_3(x,t) = (n^2h + 2nh^2 + h^3)(e^x - x)\frac{t^\alpha}{\alpha} - (h^2 + nh + h)x\frac{t^\alpha}{\alpha} \\
&\qquad + (2nh^2 + 2h^3)(e^x - 1)\frac{t^{2\alpha}}{2\alpha^2} - h^2\frac{t^{2\alpha}}{2\alpha^2} + h^3 e^x \frac{t^{3\alpha}}{6\alpha^3}, \\
&\vdots
\end{aligned} \tag{20}$$

The series solution of Eq.(15) by q-HAM can be considered as

$$u(x,t;n;h) = e^x + \sum_{m=1}^{\infty} u_m(x,t;n;h)\left(\frac{1}{n}\right)^m. \tag{21}$$

For special values of $h=-1$, $n=1$ and $\alpha=1$, we get the exact solution of Eq.(15) as

$$u(x,t) = e^{x-t} + xt - \frac{t^2}{2}. \tag{22}$$

Example 2. Consider the following fractional nonlinear inviscid Burgers' equation [23, 24]

$$CD_t^\alpha u(x,t) + u(x,t)u_x(x,t) = 0, \quad x \in R, \quad t > 0, \quad 0 < \alpha \le 1, \tag{23}$$

subject to the initial condition

$$u(x,0) = x, \quad x \in R. \tag{24}$$

We can use the nonlinear operator for our problem as

$$\mathcal{N}\left[\omega(x,t;q)\right] = CD_t^\alpha \omega(x,t;q) + \omega(x,t;q)\omega_x(x,t;q). \tag{25}$$

Following similar steps with Example 1, we construct the q-HAM iterations as below

$$
\begin{aligned}
&u_0(x,t) = x,\\
&u_1(x,t) = hx\frac{t^\alpha}{\alpha},\\
&u_2(x,t) = nhx\frac{t^\alpha}{\alpha} + h^2x\frac{t^\alpha}{\alpha} + 2h^2x\frac{t^{2\alpha}}{2\alpha^2},\\
&u_3(x,t) = \left(n^2h + 2nh^2 + h^3\right)x\frac{t^\alpha}{\alpha} + \left(4nh^2 + 4h^3\right)x\frac{t^{2\alpha}}{2\alpha^2} + h^3x\frac{t^{3\alpha}}{\alpha^3},\\
&\vdots
\end{aligned}
\tag{26}
$$

An appropriate solution of Eq.(23) in terms of h and n is given

$$u(x,t;n;h) = x + \sum_{m=1}^{\infty} u_m(x,t;n;h)\left(\frac{1}{n}\right)^m. \tag{27}$$

For special values of $h=-1$, $n=1$ and $\alpha=1$, we have the exact solution of Eq.(23) as

$$u(x,t) = \frac{x}{1-t}. \tag{28}$$

In Figures below, the plots of solution functions of Eq.(23) for different values of α, h, n, t and x are presented.

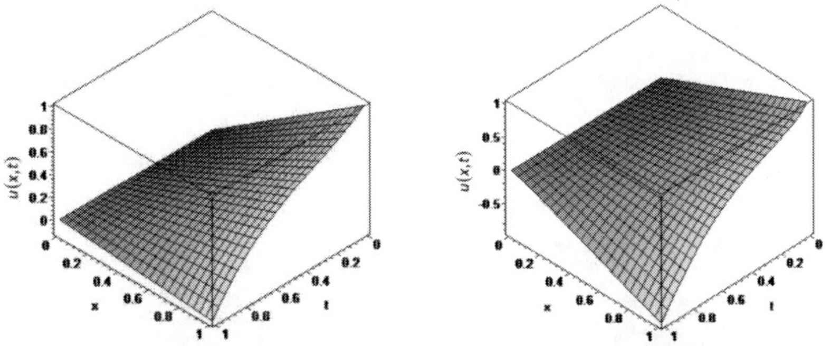

Fig.1. q-HAM Solutions of Eq.(23) with $\alpha = 0.75$ (left) $\alpha = 0.99$ (right)

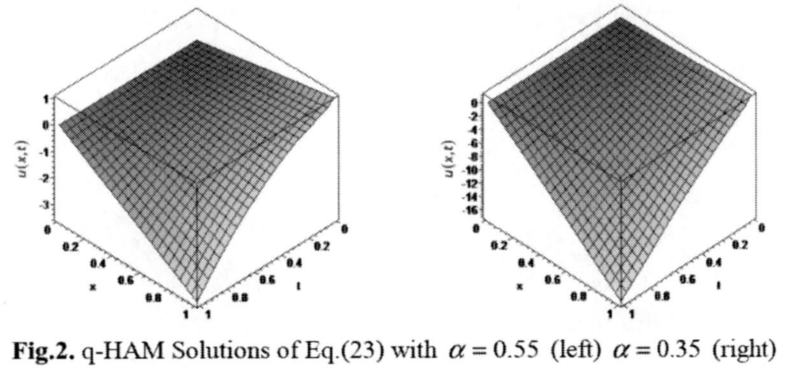

Fig.2. q-HAM Solutions of Eq.(23) with $\alpha = 0.55$ (left) $\alpha = 0.35$ (right)

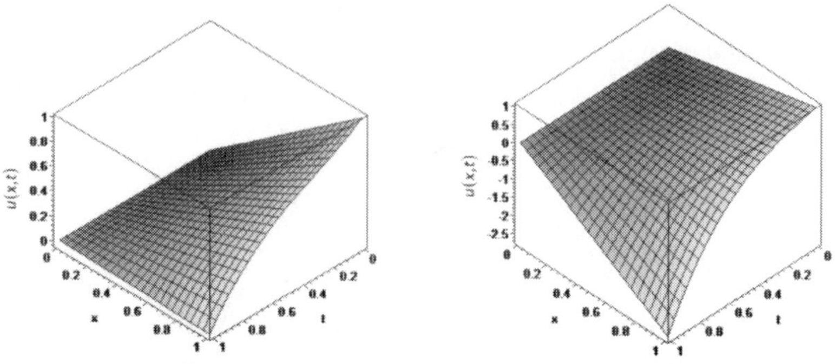

Fig.3. q-HAM Solutions of Eq.(23) for $n = 0.99$ (left) $n = 0.7$ (right)

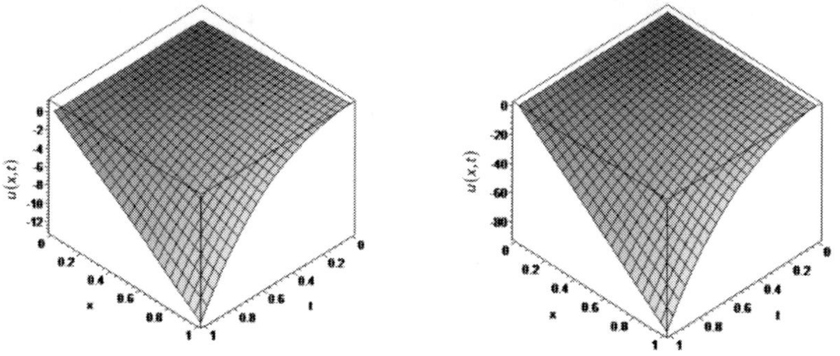

Fig.4. q-HAM Solutions of Eq.(23) for $n = 0.5$ (left) $n = 0.3$ (right)

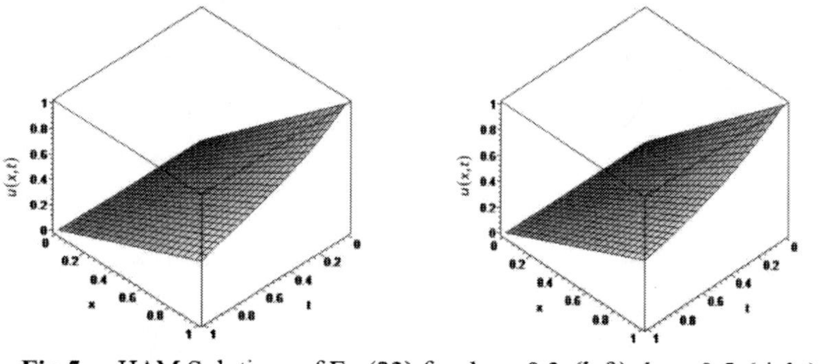

Fig.5. q-HAM Solutions of Eq.(23) for $h = -0.3$ (left) $h = -0.5$ (right)

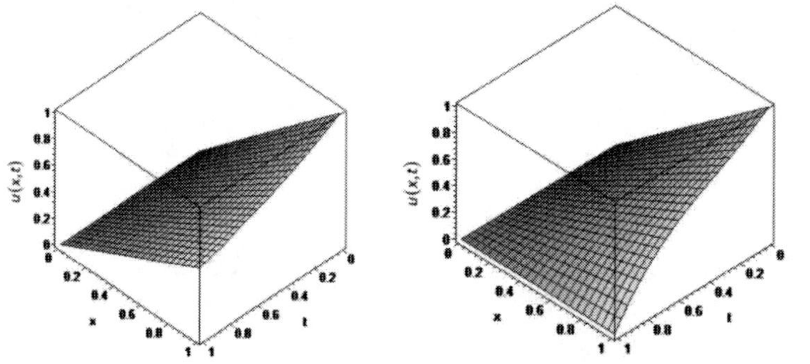

Fig.6. q-HAM Solutions of Eq.(23) for $h = -0.7$ (left) $h = -1$ (right)

4 Conclusions

In this study, we have considered the solutions of linear/nonlinear fractional PDEs by using the q-HAM that is a generalized form of HAM. We have redefined the suggested method with CDO to get the solutions of the fractional Cauchy problems and have applied it to the mentioned problems easily. Then we have verified the efficiencies and correctness of the recommended method by drawing the figures for different values of α, n, h, x and t. Moreover, the results show that the q-HAM solutions when $n = 1$, give HAM solutions and when the auxiliary parameter is equal to -1, give the classical HPM solutions. The successful results of the mentioned method show that the method is in complete coherent with the exact solutions even for considering a nonlinear fractional partial differential equation.

References

1. R. Hilfer, *Applications of fractional calculus in physics.* Applications of Fractional Calculus in Physics. Edited by HILFER R. Published by World Scientific Publishing Co. Pte. Ltd.,. ISBN# 9789812817747, (2000).
2. E. Bonyah, A. Atangana, M.A. Khan, *Modeling the spread of computer virus via Caputo fractional derivative and the beta-derivative.* Asia Pacific Journal on Computational Engineering, 4(1), p. 1. (2017).
3. H. Bulut, H.M. Baskonus, F.B.M. Belgacem, *The analytical solution of some fractional ordinary differential equations by the Sumudu transform method.* Abstract and Applied Analysis, (2013).
4. H.M. Baskonus, T. Mekkaoui, Z. Hammouch, H. Bulut, *Active control of a chaotic fractional order economic system.* Entropy, 17(8), pp. 5771-5783, (2015).
5. M. Yavuz, N. Özdemir, *New Numerical Techniques for Solving Fractional Partial Differential Equations in Conformable Sense. In: Ostalczyk P., Sankowski D., Nowakowski J. (eds) Non-Integer Order Calculus and its Applications. Lecture Notes in Electrical Engineering, vol 496. (2019), Springer, Cham*
6. M. Inc, M.T. Gencoglu, A. Akgül, *Application of Extended Adomian Decomposition Method and Extended Variational Iteration Method to Hirota-Satsuma Coupled KdV Equation.* Journal of Advanced Physics, 6(2), pp. 216-222, (2017).
7. M. Yavuz, N. Ozdemir, H.M. Baskonus, *Solutions of partial differential equations using the fractional operator involving Mittag-Leffler kernel.* The European Physical Journal Plus, 133(6), pp. 215, (2018).
8. M.G. Sakar, F. Erdogan, *The homotopy analysis method for solving the time-fractional Fornberg–Whitham equation and comparison with Adomian's decomposition method.* Applied Mathematical Modelling, 37(20-21), pp. 8876-8885, (2013).
9. M. Yavuz, N. Ozdemir, *Numerical inverse Laplace homotopy technique for fractional heat equations.* Thermal Science, 22(1), pp. 185-194, (2018).
10. N. Özdemir, M. Yavuz, *Numerical Solution of Fractional Black-Scholes Equation by Using the Multivariate Padé Approximation.* Acta Physica Polonica A, 132(3), 1050-1053, (2017).
11. J. Hristov, *Approximate solutions to fractional subdiffusion equations.* The European Physical Journal Special Topics, 193(1), pp. 229-243, (2011).
12. A. Yokuş, *Numerical Solutions of Time Fractional Korteweg--de Vries Equation and Its Stability Analysis.* Communications Faculty of Sciences University of Ankara Series A1 Mathematics and Statistics. 68(1), pp. 353-361, (2019).
13. A. Yokus, D. Kaya, *Numerical and exact solutions for time fractional Burgers' equation.* Journal of Nonlinear Sciences and Applications, 10(7), pp. 3419-3428, (2017).
14. R. Khalil, M. Al Horani, A. Yousef, M. Sababhehet, *A new definition of fractional derivative.* Journal of Computational and Applied Mathematics, 264, p. 65-70, (2014).

15. A. Atangana, D. Baleanu, A. Alsaedi, *New properties of conformable derivative.* Open Mathematics, 13(1), pp. 889-898, (2015).
16. T. Abdeljawad, *On conformable fractional calculus.* Journal of computational and Applied Mathematics, 279, pp. 57-66, (2015).
17. D. Avci, B.B. Iskender Eroglu, N. Ozdemir, *Conformable Heat Equation on a Radial Symmetric Plate.* Thermal Science, 21(2), pp. 819-826, (2017).
18. F. Evirgen, *Conformable fractional gradient based dynamic system for constrained optimization problem.* Acta Physica Polonica A, 132(3), pp. 1066-1069, (2017).
19. M. Yavuz, N. Özdemir, *A different approach to the European option pricing model with new fractional operator.* Mathematical Modelling of Natural Phenomena, 13(1), pp. 1-12, (2018).
20. M. Yavuz, *Novel solution methods for initial boundary value problems of fractional order with conformable differentiation.* An International Journal of Optimization and Control, 8(1), pp. 1-7, (2018).
21. O.S. Iyiola, G.O. Ojo, *On the analytical solution of Fornberg–Whitham equation with the new fractional derivative.* Pramana, 85(4), pp. 567-575, (2015).
22. S.-J. Liao, *An approximate solution technique not depending on small parameters:A special example.* International Journal of Non-Linear Mechanics, 30(3), pp. 371-380, (1995).
23. A. Tveito, R. Winther, *Introduction to partial differential equations: a computational approach.* Vol. 29, (Springer Science & Business Media 2004).
24. N.H. Asmar, *Partial differential equations with Fourier series and boundary value problems.* (Courier Dover Publications 2016).

Extensions of Hadamard Codes Defined on Rings

Mustafa Özkan[1,*], *Figen Öke*[2]

[1]Department of Mathematics, University of Trakya, Edirne, Turkey

[2]Department of Mathematics, University of Trakya, Edirne, Turkey

Abstract. In this study, some special matrices are constructed by choosing certain elements from finite rings and certain codes are written by using these matrices.

These codes are extended to a field. Moreover these codes are classified and more good codes are written.

1 Introduction

Structures of cyclic, constacyclic and quasi-cyclic codes over certain finite chain rings previously were worked . $(1+u)-$Constacyclic and Cyclic codes over the ring $\mathbb{F}_2+u\mathbb{F}_2$ were practiced by Qian J., Zhang L. and Zhu S. in [4]. These subjects were extended to the ring $\mathbb{F}_2+u\mathbb{F}_2+u^2\mathbb{F}_2$ by the same authors in [5]. Defined codes in $\mathbb{Z}_4$ ring were obtained by Krotov D. in [1] and [2]. Hadamard Code over $\mathbb{Z}_4$ in particular were worked by Krotov D. in [2].

We will deal with codes defined over finite rings and binary codes.

Main aim is to obtain new and good codes. An n . rank square matrix M that total ingredient 1 or -1 where $MM^t=nI$ is defined Hadamard matrix. Hadamard codes are called a code defined by Hadamard matrices. It says that using Hadamard matrices new codes could be obtained.

This study is a generalizations of results obtained in [7, 8, 9] for the finite chain rings over $\mathbb{F}_2$.

*Corresponding author: mustafaozkan@trakya.edu.tr, mustafaozkan22@icloud.com

2 Basics

Let R_m be a ring. C linear code of length n in $\mathbb{F}_2$ is a $\mathbb{F}_2$ - sub space of $\mathbb{F}_2^n$. C linear code of length n in R_m is a R_m - sub module of R_m^n.

Let length of C is n, it's number of elements is M and it's minimum distance is d. Therefore C be called $(n,M,d)\_$ code.

It is known that the ring $R_m = \mathbb{F}_2 + u\mathbb{F}_2 + \ldots + u^m\mathbb{F}_2$ with $m \geq 1,\ m \in \mathbb{Z}$ is isomorphic to the ring

$$\mathbb{F}_2[u]\Big/_{<u^{m+1}>} = \{x_0 + x_1.u + \ldots + x_m.u^m + <u^{m+1}> \mid\ x_i \in \mathbb{F}_2,\ 0 \leq i \leq m,\ m \geq 1,\ m \in \mathbb{Z}\}$$

in case $u^{m+1} = 0$. It is known that the $R_m = \mathbb{F}_2 + u\mathbb{F}_2 + \ldots + u^m\mathbb{F}_2$ where $u^{m+1} = 0$ is a ring

n. rank square matrix is called binary Hadamard matrix if it is gain from n. rank square Hadamard matrix M_n placing 0 instead of 1 and writing 1 instead of -1.

Let n be an even number. Write all rows of the matrix called foregoing as a vector where have length n. Let us create a set putting into these vectors and respectively their ingredient to back of these vectors. Then this vectors with length $2n$ are obtained. Write all this new vector as a codeword. If completions of codewords add to this set, the minimum distance of this code is n and Hadamard code included $4n$ elements.

The general Gray map is called as follows:

$$\Phi : R_m^n \longrightarrow \mathbb{F}_2^{2^m.n}$$

$$a_0 + a_1.u + \ldots + a_m.u^m \mapsto \Phi(a_0 + a_1.u + \ldots + a_m.u^m)$$

$$\begin{aligned}\Phi(a_0 + a_1.u + \ldots + a_m.u^m) = (&a_m, a_m + a_0, a_m + a_1, \ldots, a_m + a_{m-1}, a_m + a_0 + a_1, a_m + a_0 + a_2, \ldots,\\ &a_m + a_0 + a_{m-1}, a_m + a_1 + a_2, \ldots, a_m + a_1 + a_{m-1}, a_m + a_2 + a_3, \ldots, a_m + a_2 + a_{m-1}, \ldots,\\ &a_m + a_{m-2} + a_{m-1}, a_m + a_0 + a_1 + a_2, \ldots, a_m + a_0 + a_1 + a_{m-1}, \ldots,\\ &a_m + a_{m-3} + a_{m-2} + a_{m-1}, a_m + a_0 + a_1 + a_2 + a_3, \ldots,\\ &a_m + a_{m-4} + a_{m-3} + a_{m-2} + a_{m-1}, \ldots, \ldots, a_m + a_0 + a_1 + \ldots + a_{m-1})\end{aligned}$$

where $a_i \in \mathbb{F}_2^n$, for $0 \leq i \leq m$.

The homogeneous weight $w_{\text{hom}}(r)$ of $r \in R_m$ is given by

$$w_{\text{hom}}(r) = \begin{cases} 0 & ; r = 0 \\ 2^m & ; \ r = u^m \\ 2^{m-1} & ; \ otherwise \end{cases}$$

This expanded homogeneous weight function over R_m^n such that $w_{\text{hom}}(r) = \sum_{i=0}^{n-1} w_{\text{hom}}(r_i)$ for $r = (r_o, r_1, ..., r_{n-1}) \in R_m^n$. The homogeneous distance $d_{\text{hom}}(a,b)$ between any different vectors $a, b \in R_m^n$ is called as $w_{\text{hom}L}(a-b)$. The minimum homogeneous distance d_{hom} of C is called as for each $a,b \in C$, $a \neq b$ $d_{\text{hom}}(C) = \min\{d_{\text{hom}}(a,b)\}$.

Let C of length n is a code in $\mathbb{F}_2$ and $c = (c_o, c_1, ..., c_{n-1})$ be a codeword of C. The Hamming weight of C is defined as

$$w_H(c) = \sum_{i=0}^{n-1} w_H(c_i)$$

where $w_H(c_i) = 0$ if $c_i = 0$ and $w_H(c_i) = 1$ if $c_i = 1$. The minimum Hamming distance of C is defined as $d_H(C) = \min\{d_H(c,c')\}$ for any $c, c' \in C, c \neq c'$.

Therefore C of length n is a code in R_m and then $\Phi(C)$ is a binary code of length $2^m.n$.

3 Construction of Hadamard codes

In this chapter, generator matrices were created.

Formed every elements of one row of the matrix $G^{\alpha,\beta}$ from 1, select that the elements of the another rows from $\{0,1,u,...,u^m,1+u,...,1+u^m,...,1+u+...+u^m\}$ set if $\beta = 0$ and from $\{0,u^m\}$ set if $\alpha = 0$. $G^{\alpha,\beta}$ occured foregoing $\alpha + \beta + 1$ rows. Colums of $G^{\alpha,\beta}$ be lexicographically ordered.

Examples of $G^{\alpha,\beta}$ be given below :

$$G^{0,0}=[1]_{1x1}\ ,\ G^{0,1}=\begin{bmatrix}1 & 1\\ 0 & u^m\end{bmatrix}_{2x2},$$

$$G^{0,3}=\begin{bmatrix}1 & 1 & 1 & 1 & 1 & 1 & 1 & 1\\ 0 & 0 & 0 & 0 & u^m & u^m & u^m & u^m\\ 0 & 0 & u^m & u^m & 0 & 0 & u^m & u^m\\ 0 & u^m & 0 & u^m & 0 & u^m & 0 & u^m\end{bmatrix}_{4x8},$$

$$G^{1,0}=\begin{bmatrix}1 & 1 & 1 & 1 & . & . & . & 1 & . & . & . & 1\\ 0 & 1 & u & 1+u & . & . & . & 1+u^m & . & . & . & 1+u+...+u^m\end{bmatrix}_{2\times 2^{m+1}}.$$

Define the code $C^{\alpha,\beta}=\{\ (c_1,c_2).G^{\alpha,\beta}\ \big|\ c_1\in R_m^{\alpha+1}, c_2\in\mathbb{F}_2^{\beta}\}$ which has a generator matrix $G^{\alpha,\beta}$ where α,β are integers such that $\alpha,\beta\geq 0$. The length of $C^{\alpha,\beta}$ code is $n=2^{(m+1).\alpha+\beta}$ and parameters of code $C^{\alpha,\beta}$ be $(n,2^{m+1}.n,2^{m-1}.n)$.

Theorem 3.1 : If $C^{\alpha,\beta}$ is a code generated by the matrix $G^{\alpha,\beta}$ over R_m, it's image $\Phi(C^{\alpha,\beta})$ under the Gray map is the $(2^m.n,2^{m+1}.n,2^{m-1}.n)$_ Hadamard code over the field $\mathbb{F}_2$.

Proof : $C^{\alpha,\beta}$ code be obtained by the matrix $G^{\alpha,\beta}$ which has dimension $(\alpha+\beta+1)\times n$ is of $C^{\alpha,\beta}=\{\ (c_1,c_2).G^{\alpha,\beta}\ \big|\ c_1\in R_m^{\alpha+1},\ c_2\in\mathbb{F}_2^{\beta}\}$. The length of The length of $C^{\alpha,\beta}$ is $n=2^{(m+1).\alpha+\beta}$ and it has $2^{m+1}.n$ elements, i.e. $C^{\alpha,\beta}$ is a $(n,2^{m+1}.n,2^{m-1}.n)$ - code. Hence $\Phi(C^{\alpha,\beta})\subseteq\mathbb{F}_2^{2^m.n}$ and $\Phi(C^{\alpha,\beta})$ is a binary Hadamard code with $(2^m.n,2^{m+1}.n,2^{m-1}.n)$ parameters.

Theorem 3.2 : The $(C^{\alpha,\beta})^{\perp}$ is a $(n,\dfrac{(2^{m+1})^n}{2^{m+1}.n},4)$_ code and it's image

$\Phi((C^{\alpha,\beta})^{\perp})$ under the Gray map is a $(2^m.n,\dfrac{(2^{m+1})^n}{2^{m+1}.n},4)$_ code, in except the case

$\alpha=\beta=0$.

Proof : The generator matrix $G^{\alpha,\beta}$ of $C^{\alpha,\beta}$ is the parity-check matrix of the dual code $(C^{\alpha,\beta})^{\perp}$. The dual code of $(C^{\alpha,\beta})^{\perp}$ contains elements x of R_m^n satisfied $G^{\alpha,\beta}.x^T=0$.

It is seen that the number of vectors be $\frac{(2^{m+1})^n}{2^{m+1}.n}$, minimum weight of these vectors be 4.

Hence $(C^{\alpha,\beta})^\perp$ code be $(n, \frac{(2^{m+1})^n}{2^{m+1}.n}, 4)\_$ code. It is viewed that $\Phi((C^{\alpha,\beta})^\perp)$ has the parameters $(2^m.n, \frac{(2^{m+1})^n}{2^{m+1}.n}, 4)$.

4 Conclusion

In this paper; some special matrices are constructed to obtain new Hadamard codes using the elements of the finite chain ring $\mathbb{F}_2 + u\mathbb{F}_2 + ... + u^m\mathbb{F}_2$. Certain special codes are obtained using these special matrices as generator matrices.

References

1. Krotov, D. S.,Diskretn. Anal. Issled. Oper. Ser.1., **7, 4**. 78–90,(2000)
2. Krotov,D. S. ,Procs. of the International Workshop on Coding and Cryptography, Paris, 329-334,(2001)
3. Vermani, L. R., Chapman Hall, India., (1996)
4. J.F.Qian, L.N.Zhang, S.X. Zhu, Applied Mathematics Letters,**19**,820-823,(2006)
5. J.F.Qian, L.N.Zhang, S. X. Zhu,IEICE Trans.Fundamentals, **E89-a, no 6**,1863-1865,(2006)
6. W. C. Huffman and Vera Pless, Fundamentals of Error Correcting Codes, Cambridge, (2003)
7. M. Özkan and F.Öke , App Mathematics and Inf. Sci. 10 (2), 701-704.,(2016)
8. M. Ozkan, F. Oke, General Letters in Mathematics, 2(1), 110–118, (2017)
9. M. Ozkan, F. Oke, AIP Conf. Proc.,1926, 020035-1–020035-3,(2018).

On Gray Images of Constacyclic Codes

Mustafa Özkan[1,*], *Figen Öke*[2]

[1]Department of Mathematics, University of Trakya, Edirne, Turkey

[2]Department of Mathematics, University of Trakya, Edirne, Turkey

Abstract This paper deals with constacyclic codes over a new ring. Gray images of these codes over those rings are obtained. The relations between these codes and their Gray images are given.

1 Introduction

Cyclic and constacyclic codes over finite rings with two variables were investigated in many different studies. Transformations preserving distances from these rings to another finite ring or finite fields are studied. Construction of $(1+v)$-contacyclic codes over the ring was studied in [6] and Gray images of these codes are given. Cyclic codes, quasi-cyclic codes over $\mathbb{F}_2+u\mathbb{F}_2+v\mathbb{F}_2+uv\mathbb{F}_2$ and their dual were studied by different authors [3,6,7].

Using these studies, under the conditions $u^2=v^2=uv=0$, $(1+v)$-constacyclic codes over $\mathbb{F}_2+u\mathbb{F}_2+v\mathbb{F}_2$ were investigated in [8].

Under the conditions $u^2=v^3=uv=0$, the authors have shown in [1] the Gray images of the $(1+v)$-constacyclic codes over $\mathbb{F}_2+v\mathbb{F}_2+u\mathbb{F}_2+v^2\mathbb{F}_2$ are quasi-cyclic codes over the ring $R=\mathbb{F}_2+v\mathbb{F}_2+v^2\mathbb{F}_2$.

* Corresponding author: mustafaozkan@trakya.edu.tr

2 Preliminaries

Throughout this paper $R_{u,v} := \mathbb{F}_2 + v\mathbb{F}_2 + u\mathbb{F}_2 + u^2\mathbb{F}_2$ and $R_v := \mathbb{F}_2 + v\mathbb{F}_2$ under the conditions $u^3 = v^2 = uv = 0$ are two commutative local rings. Obviously R_v is a subring of $R_{u,v}$ whose R_v is a finite chain ring and $R_{u,v}$ is a finite non-chain ring. Moreover $R_{u,v}$ and R_v are vector spaces over $\mathbb{F}_2$ with basis $\{1; v; u; u^2\}$ and $\{1; v\}$, respectively.

The submodules of $(R_{u,v})^n$ (resp. $(R_v)^n$) are called linear codes of length n over $R_{u,v}$ (resp. R_v).

Definition: The Hamming weight over $\mathbb{F}_2$ is defined as;

$$w_H(m) = \begin{cases} 0, & m = 0 \\ 1, & m = 1 \end{cases}$$

Hamming weight of a code-word with length n, $m = (m_0, ..., m_{n-1}) \in \mathbb{F}_2^n$ is of the form $w_H(m) = \sum_{i=0}^{n-1} w_H(m_i)$.

The Lee weight $w_L : R_v \to \mathbb{N}$ is defined by letting $w_L(0) = 0$, $w_L(1) = w_L(1+v) = 1$, $w_L(v) = 2$.

Then $w_L(r) = \sum_{i=1}^{n-1} w_L(r_i)$ is satisfied for each element $r = (r_0, ..., r_{n-1}) \in R_v^n$.

Definition: The special weight over the ring $R_{u,v}$ as;

$$w_{Special}(k) := \begin{cases} 0 & ; \ k = 0 \\ 2 & ; \ k = 1+u, 1+v+u \\ 6 & ; \ k = u, 1, \ 1+u^2, u+u^2, 1+v+u+u^2 \\ 4 & ; \ otherwise \end{cases}$$

So the special weight of a codeword $k=(k_0,\dots,k_{n-1})\in R_{u,v}^n$ is of the form;

$$w_s(k)=\sum_{i=0}^{n-1} w_s(k_i).$$

Definition: The F_2- linear map $\Phi: R_{u,v}^n \to R_v^{4n}$ is given by : $\Phi(1)=e_1+ve_4$, $\Phi(v)=e_2+ve_4$, $\Phi(u)=e_3+ve_4$, and $\Phi(u^2)=e_4$, where $e_1:=(0,1,0,1)$, $e_2:=(0,1,1,0)$, $e_3:=(0,0,1,1)$, $e_4:=(1,1,1,1)$. The map $\Phi: R_{u,v}^n \to R_v^{4n}$ extends to $(R_{u,v})^n$ as follows :

$\Phi(c_1,\dots,c_n)=(\Phi(c_1),\dots,\Phi(c_n))$ for any $(c_1,\dots,c_n)\in(R_{u,v})^n$.

3 Constacyclic, Cyclic and Quasi-cyclic codes

A cyclic shift on $R_{u,v}^n$ is a permutation σ such that

$$\sigma(c_o,c_1,\dots,c_{n-1})=(c_{n-1},c_0,\dots,c_{n-2}).$$

A linear code C over $R_{u,v}$ of length n is said to be cyclic code if it is satisfied the equality $\sigma(C)=C$.

A $(1+u^2)$-constacyclic shift γ acts on $R_{u,v}^n$ as

$$\gamma(c_o,c_1,\dots,c_{n-1})=((1+u^2).c_{n-1},c_0,\dots,c_{n-2})$$

A linear code C over $R_{u,v}$ of length n is said to be $(1+u^2)$-constacyclic code if it is satisfied the equality $\gamma(C)=C$.

A cyclic shift on R_v^{4n} is a permutation τ such that

$$\tau(d_o,d_1,\dots,d_{4n-1})=(d_{4n-1},d_0,\dots,d_{4n-2}).$$

Let $D \subseteq R_v^{4n}$ is a linear code. If $\tau(D) = D$ then D is called a cyclic code over R_v.

A cyclic shift on R_v^{4n} is a permutation $\tau^{\otimes 2}$ such that $\tau^{\otimes 2}(c_o, c_1, \ldots, c_{4n-1}) = (c_{2n-1}, c_0, \ldots, c_{2n-2}, c_{4n-1}, c_{2n}, \ldots, c_{4n-2})$.

If $\tau^{\otimes 2}(D) = D$ then D is called a quasicyclic code of index 2 over R_v.

We have for $i \in \{1;2;3\}$, $\tau^{\otimes 2}(e_i) = e_i$ and $\sigma(e_4) = e_4$.

A $(1+v)$-constacyclic shift on R_v^{4n} is a permutation β such that

$$\beta(d_o, d_1, \ldots, d_{4n-1}) = ((1+v).d_{4n-1}, d_0, \ldots, d_{4n-2}).$$

Let $D \subseteq R_v^{4n}$ is a linear code. If $\beta(D) = D$ then D is called a $(1+v)$-constacyclic code over R_v.

4 Main Results

Lemma 4.1 : The equality $\Phi.\gamma = \sigma^{\otimes 2}.\Phi$ is satisfied.

Proof : $\gamma(k) = \gamma(k_0, \ldots, k_{n-1}) = ((1+u^2)k_{n-1}, k_0, \ldots, k_{n-2})$ is satisfied for each $k = (k_0, \ldots, k_{n-1}) \in R_{u,v}^n$, where $k_i = r_i + s_i u + t_i u^2$, $0 \le i \le n-1$.

Then it is obtained;

$$\Phi(\gamma(k)) = \Phi(r_{n-1} + s_{n-1}u + (r_{n-1} + t_{n-1})u^2, r_0 + s_0 u + t_0 u^2, \ldots, r_{n-2} + s_{n-2}u + t_{n-2}u^2)$$

$$= (r_{n-1} + t_{n-1}, t_0, t_1, \ldots, t_{n-2}, t_{n-1}, r_0 + t_0, \ldots, r_{n-2} + t_{n-2}, r_{n-1} + s_{n-1} + t_{n-1}, s_0 + t_0, \ldots, s_{n-1} + t_{n-1}, r_0 + s_0 + t_0, \ldots, r_{n-2} + s_{n-2} + t_{n-2})$$

On the other hand,

$$\Phi(k)=\Phi(k_0,k_1,\dots,k_{n-1})=\Phi(r_0+s_0u+t_0u^2,r_1+s_1u+t_1u^2,\dots,r_{n-1}+s_{n-1}u+t_{n-1}u^2)$$
$$=(t_0,t_1,\dots,t_{n-1},r_0+t_0,r_1+t_1,\dots,r_{n-1}+t_{n-1},s_0+t_0,s_1+t_1,\dots,s_{n-1}+t_{n-1},$$
$$r_0+s_0+t_0,r_1+s_1+t_1,\dots,r_{n-1}+s_{n-1}+t_{n-1})$$

Then it is seen that

$\sigma^{\otimes 2}\ (\Phi(c))=$

$\sigma^{\otimes 2}(t_0,t_1,\dots,t_{n-1},r_0+t_0,r_1+t_1,\dots,r_{n-1}+t_{n-1},s_0+t_0,s_1+t_1,\dots,s_{n-1}+t_{n-1},$
$r_0+s_0+t_0,r_1+s_1+t_1,\dots,r_{n-1}+s_{n-1}+t_{n-1})$

$=(r_{n-1}+t_{n-1},t_0,t_1,\dots,t_{n-2},t_{n-1},r_0+t_0,\dots,r_{n-2}+t_{n-2},r_{n-1}+s_{n-1}+t_{n-1},s_0+t_0,\dots,s_{n-1},$
$r_0+s_0+t_0,\dots,r_{n-2}+s_{n-2}+t_{n-2})$

$=\Phi(\gamma(k))$

Theorem 4.2.: C is a $(1+u^2)$-constacyclic code over $R_{u,v}$ with length n if and only if $\Phi(C)$ is a quasi-cyclic code of second order over R_v with length 4n.

Proof: If C is a $(1+u^2)$-constacyclic code with n length over the ring $R_{u,v}$, $\gamma(C)=C$. Then $\Phi(\gamma(C))=\Phi(C)$

Using the equality from the previous lemma the equality is obtained:

$\sigma^{\otimes 2}(\Phi(\gamma(C))=\Phi(\gamma(C))=\Phi(C)$.

Then we have the code $\Phi(C)$ is a quasi-cyclic code of length 4 with second order.

Conversely; if $\Phi(C)$ is a quasi-cyclic code over R_v with length 4n and of second order, $\sigma^{\otimes 2}(\Phi(C))=\Phi(C)$ is satisfied.

Using the previous lemma the equality

$\Phi(\gamma(C))=\sigma^{\otimes 2}(\Phi(C))=\Phi(C)$ is obtained.

Since Φ is injective, using the above equality; it is seen that $\gamma(C)=C$. It means that C is a $(1+u^2)$-constacyclic code with length n over the ring $R_{u,v}$.

5 Conclusion

In this paper, $(1+u^2)$- constacyclic codes over a ring isomorphic to $R_{u,v}$ are studied, and a new weight function on $R_{u,v}$ is defined.

References

1. P. Udomkavanich, S. Jitman, Int.J.Contemp. Math. Sciences,, **4(26)**, 1265–1272, (2009)
2. W. C. Huffman, V. Pless, Cambridge, (2003)
3. M. Ozkan, F. Oke, AIP Conf. Proc., **1926**, 020034-1–020034-3, (2018)
4. S. Roman, Graduate Texts in Mathematics, Springer Verlag, (1992)
5. M. Özkan, F.Öke, Mathematical Sciences and Applications E-Notes, **4(1)**, 40-44, (2016)
6. M. Özkan, F. Öke, Palestine Journal of Mathematics, **6(2)**, 241-245, (2017)
7. J.F.Qian, L.N.Zhang, S.X Zhu, Applied Mathematics Letters, **19**, 820-823, (2006)
8. Xu Xioafang, Computer Engineering and Applications, **49(12)**,77-79, (2013)

Comparison of Different Algorithms in the Radiotherapy Plans of Breast Cancer

Yonca Yahşi Çelen[1]* *and Atilla Evcin*[2]

[1]Department of Radiation Oncology, University of Kocatepe, 03200, Afyon, Turkey

[2]Department of Materials Science and Engineering, University of Kocatepe, 03200, Afyon, Turkey

Abstract. It is aimed to evaluate portal dosimetry results of planned breast cancer patients with intensity-modulated radiotherapy (YART) of Anisotropic Analytical Algorithm (AAA) and Pencil Beam Convolution (PBC) dose calculation algorithms. The plans of 10 treated patients will receive 6 MV photon energy and a total of 25 fractions of 50 Gray dose using the inverse YART technique, which is reverse planned in the Eclipse (ver.13.6) treatment planning system with Varian Trilogy Linear Accelerator prescribing. For each plan, dose was calculated after optimization using PBC and then AAA algorithms. The quality controls of the plans were made using the Electronic Portal Imaging Device (EPID) by creating individual verification plans for each algorithm. In addition, the maximum and average dose values in the target volume were compared in inverse YART plans calculated using PBC and AAA. When treatment plans generated by AAA and PBC dose calculation algorithms are analyzed using EPID, for the PBC algorithm, the mean values of γ_{Area} and γ_{Avg} are 98.15 ± 1.07, 0.40 ± 0.048 and 98.72 ± 1.13, 0.37 ± 0.051, respectively, for the AAA algorithm. The PTV D_{max} value for the PBC algorithm is 109.3 ± 1.09 and the D_{Avg} value is 101.7 ± 0.51. For the AAA algorithm, the PTV D_{max} value is 110.6 ± 1.12 and the D_{Avg} value is 102.9 ± 0.62. When the mean values of portal dosimetry γ_{Area} and γ_{Avg} evaluated using PBC and AAA algorithms were compared, the differences between the algorithms were not statistically significant ($p > 0.05$). Differences between the algorithms for PTV D_{max} and D_{Avg} values are not statistically significant ($p > 0.05$).

1 Introduction

Cancer disease is one of the most common health problems in recent years. Many patients with cancer have radiotherapy with or without chemotherapy. Radiotherapy is a method used in the treatment of cancer and some benign diseases using ionizing radiation. The success of radiation therapy is directly related to how well we can maintain robust tissue in addition to controlling tumor volume by irradiating the target.

*Corresponding author: yoncayashsi@gmail.com

External radiotherapy is calculated by dose calculation algorithms found in the treatment planning system (TPS), and the dose calculated by the system's loaded beam data is considered to be close to the actual dose distribution.
The calculated dose, the slight or even different occurrence from the dose administered to the patient, is related to the inhomogeneous dosimetry methods of the dose calculation algorithms.
In general, treatment plans are evaluated on the basis of dose calculations. Some comparisons between calculation algorithms may be found in the literature on breast irradiation, for PBC and AAA [1,2,3]. The calculations using the PBC algorithm are known to be inaccurate in regions of electronic non-equilibrium, such as in air cavities or in build-up regions, and inhomogeneity corrections due to the lung media or irregular body contour, which are major issues in breast RT [4,5,6]. Monte Carlo (MC) simulation methods and techniques are widely reported for their dose calculation accuracy. Applying MC techniques to dose calculations in RT has therefore the potential to decrease the uncertainties when compared to analytical/conventional treatment planning algorithms, regardless the beam geometry and target composition [7].
Accurate dose calculation algorithms play an important role in the decrease of global uncertainty and to achieve clinically effective dose distribution for tumor control [8]. Several commercial radiotherapy treatment planning systems (TPS) are implemented with photon beam models that rely on semi-empirical methods, using pencil beam algorithms (PBC) to calculate the absorbed dose and some approximations to correct for tissue inhomogeneities [9].
The aim of this study is to evaluate the results of portal dosimetry of dose calculation algorithms of AAA and PBC dosing in breast cancer patients planned with YART.

2 Fundamental Properties of the Method

As shown in figure 1, the plans of 10 treated patients were prepared by using a Varian Trilogy Linear Accelerator device in the Eclipse (ver.13.6) treatment planning system to prescribe 6 MV photon energy and a total of 25 fractions and 50 Gray doses using the inverse planned intensity radiotherapy (Inverse YART) technique. As shown in figure 2, for each plan, dose was calculated after optimization using PBC and then AAA algorithms. The quality controls of the plans were made using the Electronic Portal Imaging Device (EPID) by creating individual verification plans for each algorithm. Gamma analysis was performed based on 3% dose difference (DD), 3 mm distance adaptation (DTA) and 95% data transfer criteria. A 10% area of isodose was filtered in order to regulate the isodose differences resulting from fiber movements in the off-field low dose regions. In addition, the maximum and average dose values in the target volume were compared in inverse IMRT plans calculated using PBC and AAA.

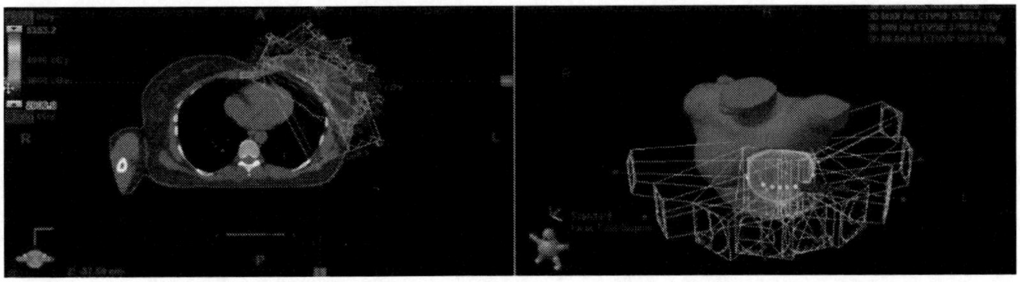

Fig.1. Treatment areas of planned breast cancer patient with YART technique.

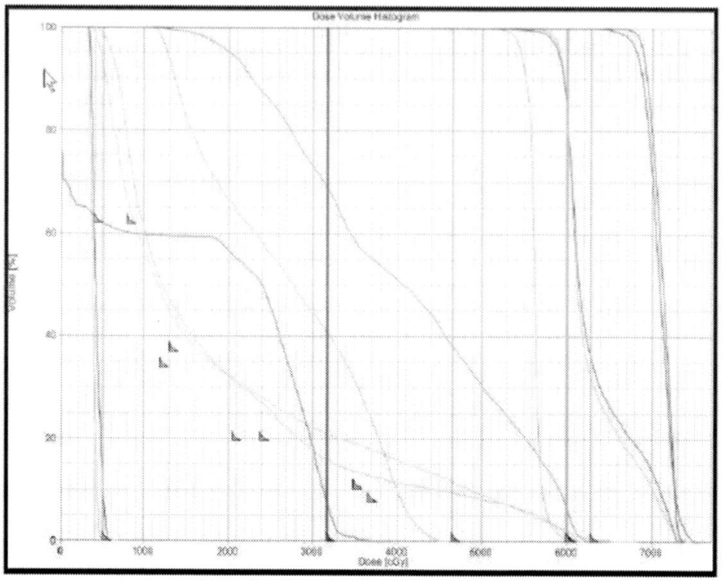

Fig.2. Dose-volume histogram during optimization

3 Application

When the treatment plans generated by AAA and PBC dose calculation algorithms are analyzed using EPID, For the PBC algorithm, as shown in table 1,the mean values of ɣArea and ɣAvg are 98.15 ± 1.07 0.40 ± 0.048 and 98.72 ± 1.13 0.37 ± 0.051, respectively, for the AAA algorithm. The PTV Dmax value for the PBC algorithm is 109.3 ± 1.09 and the D_{Avg}value is 101.7 ± 0.51. For the AAA algorithm, the PTV Dmax value is 110.6 ± 1.12 and the D_{Avg} value is 102.9 ± 0.62.

Table 1. Measurement results of PBC and AAA dose calculation algorithm data with EPID

	PBC	AAA
ɣArea	98,15±1,07	98,72±1,13
ɣAvg	0,40±0,048	0,37±0,051
PTV D_{max}	109,3±1,09	110,6±1,12
PTV D_{Avg}	101,7±0,51	102,9±0,62

4 Conclusion

Modern computerized planning systems and radiotherapy plans can be created with the dose calculation algorithms closest to the patient's satisfaction. These algorithms can be correction / measurement based, model based or Monte Carlo based. Percent dose, dose profiles and yield factor measurements should be made in water phantom for regular treatment areas under reference conditions to calculate doses in photon energies at Eclipse TPS. Some correction factors such as patient's illness, lack of tissue, tissue inhomogeneity are found by joining the calculator. Dose calculation algorithms use dose calculations and dose profiles taken at various depths by interpolating between deep doses measured in water phantom (8).
Research and development studies are continuing to obtain the current and modern state of the algorithm. AAA requires basic physical parameters to characterize the beam to obtain a dose calculation. These parameters are created by introducing to the TPS the beam data measured by the user for the configuration. The functional form of the basic physical parameters is activated by analytic convolution in AAA.
The AAA algorithm has significant advantages in terms of dose modeling in heterogeneous tissues. In dosing calculations, dose calculations also add to the dose distribution the effects of bundle and bunch-limiting accessories such as block, physical wedge filter, dynamic wedge filter, multi-leaf collimator, tray, and bolus and treatment table (9).

Treatment areas of breast cancer patients radiotherapy include different density tissues such as lung, fat, muscle. Inhomogeneity correction is performed to obtain the dose distribution closest to the truth in radiotherapy plans containing inhomogeneous tissues. Inhomogeneity correction with different dose calculation algorithms such as PBC, AAA and Acuros XB is done by different methods. The PBC algorithm performs inhomogeneity correction using correction methods such as Batho, Modified Batho and Etar. The AAA algorithm involves dose distribution by separately calculating contributions from primary photons, extra focal photons, and contaminant electron sources.
When the mean values of portal dosimetry, ɣArea and ɣAvg evaluated using PBC and AAA algorithms were compared, the differences between the algorithms were not statistically significant ($p> 0,05$). Differences between the algorithms for PTV Dmax and D_{Avg} values are not statistically significant ($p> 0.05$).

https://doi.org/10.1051/itmconf/20182201048

References

1. Ahnesjö, M. Saxner, A. Trepp, Med Phys. **19(2)**, 263e73 (1992)
2. S. Yoo, Q. Wu, J. O'Daniel, J. Horton, F. Yin, Radiother Oncol **103**, 172e7 (2012)
3. R. Chakarova, M. Gustafsson, A. Bäck, N. Grugge, A. Palm, A. Lindberg, Radiother Oncol, **102**, 102e7 (2012)
4. T. Knöos, A. Ahnesjö, P. Nilsson, P. Weber, Phys. Med. Biol., **40**, 1411e20 (1995)
5. V. Panettieri, P. Barsoum, M. Westermark, L. Brualla, I. Lax, Radiother Oncol, **93**, 94e101 (2009)
6. S. Almberg, T. Lindmo, J. Frengen, Radiother Oncol, **100**, 259e64 (2011)
7. N. Reynaert, S. Marck, D. Schaar, W. Van der Zee, C. Van Vliet-Vroegindeweij, M. Tomsej, Rad. Phys. Chem., **76**, 643e86 (2007)
8. N. Papanikolaou, J. Battista, A. Boyer, C. Kappas, E. Klein, T. R. Mackie, Medical Physics Publishing, 2004.
9. M. Siggel, P. Ziegenhein, S. Nill, U. Oelfke, Phys. Med., **28(4)**, 273e80(2012)

ITM Web of Conferences **22**, 01049 (2018) https://doi.org/10.1051/itmconf/20182201049
CMES-2018

Differentiation of Using Flattening Filter Free Energy in Vmat Plans for Prostate Cancer

Yonca Yahşi Çelen[1]* *and Atilla Evcin*[2]

[1]Department of Radiation Oncology, University of Kocatepe, 03200, Afyon, Turkey

[2]Department of Materials Science and Engineering, University of Kocatepe, 03200, Afyon, Turkey

Abstract. It is aimed to compare the values of conformity index (CI), homogeneity index (HI), monitor unit (MU) of volumetrically adjusted arthritis therapy (VMAT) plans using 10 prostate cancer patients with flattened filter (FF) and without flattening filter (FFF). In the study, treatment plans were prepared using 6 FF and 6 FFF in the Eclipse (ver.13.6) treatment planning system with Varian Trilogy Linear Accelerator. When planning was completed, CI averaged 0.87, HI averaged 0.44 and MU values were found to be 591 ± 26.8, 650 ± 33.06, respectively. When the PTV coverage, CI, HI and MU comparisons were made as a result of planning, there was no significant difference when comparing VMAT plans in FFF and FF energies. When we compare the MU values, the MU increase is seen when the straightening filter is removed. In both energy modes, good homogeneity in PTV was achieved with conventional francitation and close dose rates. No significant advantages and disadvantages of the unfiltered energy mode were observed in the assessment of plan quality in terms of CI, HI.

1 Introduction

Cancer disease is one of the most common health problems in recent years. Many patients with cancer have radiotherapy with or without chemotherapy. Radiotherapy is a method used in the treatment of cancer and some benign diseases using ionizing radiation.
The main goal in radiotherapy planning is to minimize the exposure of intact tissues to radiation while the target volume gives the necessary dose. Providing tumor control on this side also improves the quality of life as well as the prolongation of the life span (1,2).
Prostate cancer is the most common cancer in men. Standard therapy Radiotherapy is performed locally and locally advanced. Due to the widespread use of prostate specific antigen (PSA) screening and the development of an ultrasound-guided biopsy technique, diagnosis can be made at an earlier stage than in the past (3). By applying RT and surgery, the disease at the local-advanced stage can be controlled at a high rate

*Corresponding author: yoncayashsi@gmail.com

In recent years, advances in radiotherapy have resulted in conventional radiotherapy being replaced by Intensity Modulated Radiotherapy (IMRT) and Volumetric Arthritis Therapy (VMAT) techniques. These techniques provide better conformity to the target area and allow for the least possible intake of risky organs.
With IMRT and VMAT techniques, dose distribution can be obtained within the target volume with the radiotherapy fields or arc-based fixed control points arranged at different densities applied using non-uniform beam bundles. Multiple beam combinations at different intensities can be optimized to produce higher tumor control, lower normal tissue side effects. (4).
As shown in figure 3, another development in the technology is the use of standard linear accelerators without flat filter (FFF) heat. Expenditures are expected to decline in FFFs because of the reduced head scatter of accelerators. Thus, it is possible to reduce the dose of critical organ which is overdosed due to not being scattered, and the dose rate is higher treatment is performed in a shorter time.
It is aimed to compare the values of conformity index (CI), homogeneity index (HI), and monitor unit (MU) of volumetrically adjusted arc therapy (VMAT) plans using 10 prostate cancer patients with flattened filter (FF) and without flattening filter (FFF)

2 Fundamental Properties of the Method

CT images of 10 selected prostate cancer patients were scanned with 3 mm cross section and transferred to TPS. Using these CT images, target structures were plotted by gross tumor volume (GTV) and clinical target volume (CTV), planned target volume (PTV), and critical organs of radiation oncology. Patient plans were prepared using Varian Trilogy Linear Accelerator with 6 MV FF and 6 MV FFF in Eclipse (ver.13.6) treatment planning system. As shown in figure 1, two-tiered VMAT plans have been adopted as standard. Two backed VMAT plans with gantry turns in opposite directions were created. The plans were aimed at minimizing the leakage and "tongue and groove" effect from the MLCs with 30 ° and 330 ° collimator angles on opposite sides of the arcs. 6FF and 6FFF plans were made for each patient. As shown in figure 2, after optimization and calculations, the maximum doses of target and critical organs of each plan were examined with the help of DVH. The target and OAR maximum dose values, homogeneity and conformality of PTV and MU values were compared for FF and FFF plans.
Conformity index (CI) was developed as part of dosimetric analysis. When CI is equal to 1, ideal conformfality is achieved. The conformity index greater than 1 means that the irradiated volume is larger than the target volume and contains healthy tissue. If the conformity index is less than 1, only a part of the target volume is irradiated. The conformality index equation is given below (5).
$CI = VRI / TV$
VRI = irradiated volume (cc), TV = tumor volume (cc)
The conformity index values of the target volume of the PTV in the plans made using this equation are calculated and compared for FFF and FF beams.
The dose homogeneity characterizes the absorptive dose distribution in the target volume. Homogeneity index equal to or close to 0 indicates that the absorptive dose distribution is homogeneous. The HI equation is given below (6).$HI= (D\%2 - D\%98) / D\%50$
D 2% = target dose of 2% volume, D 98% = target dose of 98% volume,
D 50% = the dose of the target 50% of the volume
Using this equation, the homogeneity index values of the target volume of the PTVs are calculated for the FFF and FF beams.

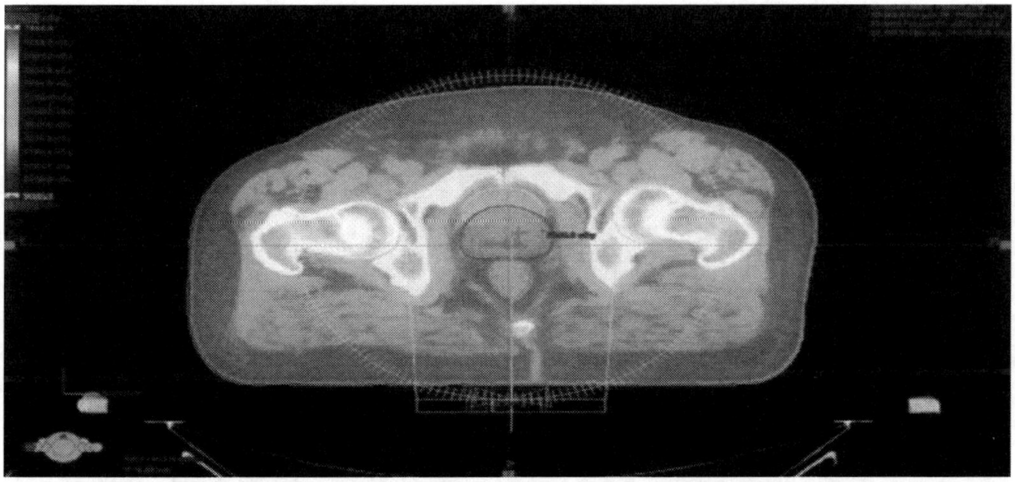

Fig.1. Treatment areas of planned prostat cancer patient with VMAT technique.

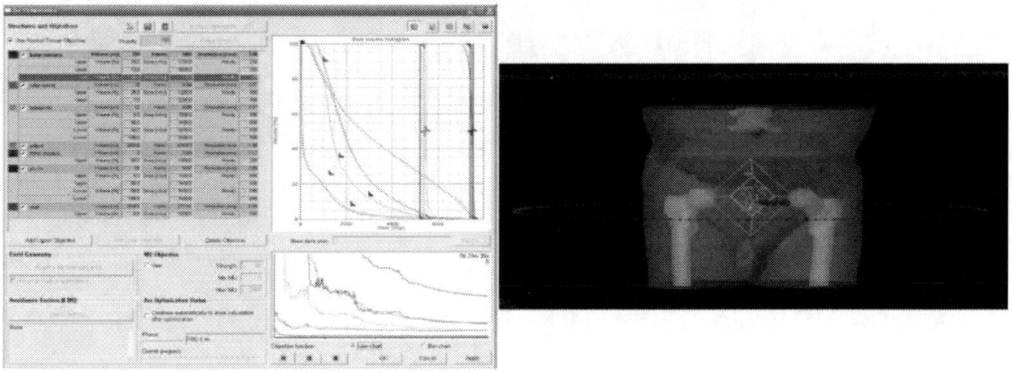

Fig.2. Optimization window and dose volume histogram

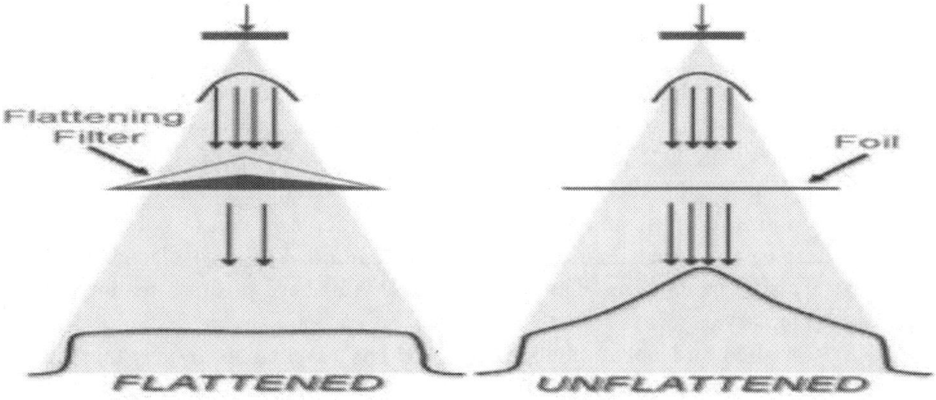

Fig.3. Physical structure in FFF energy

3 Application

As shown in figure 4, when the plans were completed, the mean CI for the FF and FFF plans was 0.87, the mean HI was 0.44, and the MU values were found to be 591 ± 26.8, 650 ± 33.06, respectively. As a result of planning, when PTV coverage, CI, HI and MU comparisons were made, there was no significant difference when comparing VMAT plans for FFF and FF energy. When we compare the MU values, the MU increase is seen when the straightening filter is removed.

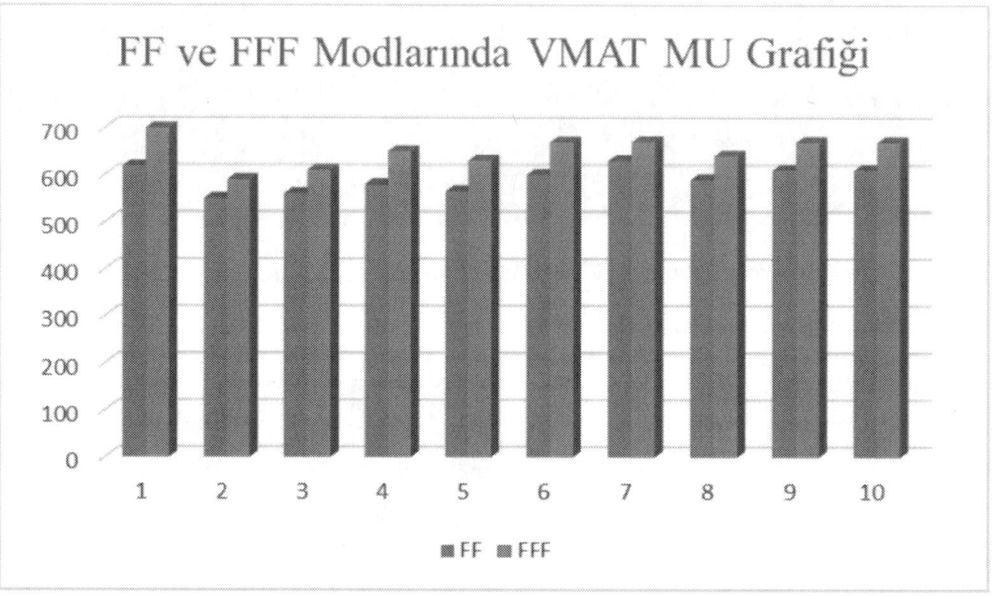

Fig.4. VMAT MU graph in FF and FFF modes

4 Conclusion

There has recently been an interest in the clinical use of liners (FFF) where there is no straightening filter. As a result of removing the smoothing filter; significant increase in dose rate, softening in photon flux (thus decreased in penetration), reduction in head scatter (thus reduction in peripheral dose) and uneven open beam profile. Other important features of the FFF linacs are that the surface dose and the maximum dose shift slightly to the surface. Increased dose rate for FFF linac; the energy of the photon beam, the linac maker, and the field size. The dose rate factor varies from 2 to 4 depending on these conditions. With increased dose rate, the duration of treatment is reduced, the likelihood of target motion is minimized, treatment volume may increase, and potentially lead to radiobiological benefits for some tumors (7). In addition, FFF linear accelerators usually have fewer neutron generations. Uncorrected rays with these positive features can reduce the risk of secondary cancer (8).
When FF is applied and not implemented, PTV has reached its target dose and critical organs at risk have been obtained within the initially defined dose limits.

Compared with the dose distribution effect of the use of the smoothing filter, the hot doses in the PTV target volumes were found to be significantly larger than those applied in the FF-unplanned schemes; but this difference is limited to 1%.
Zwahlen et al. In patients with prostate cancer, using 10 MV scans, it was reported that in the plans they performed with the VMAT technique, target homogeneity increased not 1% but not significantly when FF was not applied. Zwahlen et al. (9) reported an increase of 24.7% in the total MU value in the plans made with non-FF beams.
Ong et al. reported an 8.3% increase in the total MU value in the plans made with non-FF radiation. (10) The increase in the total MU value in this study is due to Zwahlen et al. and Ong et al. However, because Ong et al. Finds a lesser increase, the target volume may be smaller than this study. The reason for this increase in the total MU value is that the reduction of the flattening filter in unfused beams reduces the dose to the edge of the field relative to the FF rays; so that the FFF rays can be further modulated by the IMRT fields to increase the MU value so that the different dose levels occurring between the central region and the peripheral region can be homogenized. This increase in MU will increase the normal tissue integral dose (NTID). The increase in MU was found to be an expected result in order to make the dose levels homogeneous.
In both energy modes, good homogeneity in PTV was achieved with conventional francitation and close dose rates. No significant advantages and disadvantages of the unfiltered energy mode were observed in the assessment of plan quality in terms of CI, HI. When we compare the MU values, the MU increase is seen when the straightening filter is removed.

References

1. K. S. Chao, C. A. Perez , L. W. Brady, Nobel Tıp Kitabevleri, İstanbul, (2004)
2. F. M. Khan, The Physics of Radiation Therapy, **3**, (2003)
3. C. Öbek, Türkiye Kanserle Savaş Vakfı Onkolojide Güncel Yaklaşım Toplantıları, 23-29 (2011)
4. A. Çakır, F. Ağaoğlu, B. Şahin, M. Okutan, Y. Dizdar, E. Darendeliler, Türk Onkoloji Dergisi **24(4)**, 166-171 (2009)
5. L. Feuvret, G. Noel, J. J. Mazeron, P. Bey, Rad.. Oncol. Biol. Phys **64(2)**, 333-342 (2006)
6. International Comission on Radiation Units and Measurements (ICRU). Report 83 Prescribing, Recording and Reporting Photon Beam Intensity Modulated Radiation Therapy (IMRT) 2010.
7. P. Tsiamas, J. Seco, Z. Han, M. Bhagwat, J. Maddox, C. Kappas, K. Theodorou, M. Makrigiorgos, K. Marcus, P. Zygmanski, Med Phys **38(5)**, 2342-2352 (2011)
8. Y. Wang, M. K. Khan, J. Y. Ting, S. B. Easterling, Rad.. Oncol. Biol. Phys, **83(2)**, 281-285 (2012)
9. D. R. Zwahlen, S. Lang, J. Hrbacek, C. Glanzmann, S. Kloeck, Y. Najafi, T. Streller, G. Studer, K. Zaugg, Rad. Oncol. Biol. Phys **83(5)**, 1655-1660 (2012)
10. C. L. Ong, W. F. Verbakel,M. Dahele, J. P. Cuijpers, B. J. Slotman, Rad.. Oncol. Biol. Phys. **83(1)**, 137-143 (2012)

On The Solutions of Some Difference Equations Systems and Analytical Properties

Münevver Tuz[1]

[1]Department of Mathematics, University of Fırat, Elazığ, Turkey

maydin23@yahoo.com

Absract. In this study, we investigated the global asymptotic behaviors of their solutions by taking the second-order difference equation system. According to the given conditions, we obtained some asymptotic results for the positive balance of the system. We have also worked on q-fast changing functions. Such functions form the class of q-Caramate functions. We have applied q-Caramate functions to linear q-difference equations and We have also learned about the asymptotic behavior of solutions. In addition, we have studied the problem of initial and boundary value for the q-difference equation

1 Introduction

Difference equations (ordinary differential and q-differential) since ancient times are equations of interest to mathematicians and physicists. These equations solve the differential equations by shaping them into mathematical models of many other practical phenomena. Difference equations can be easily algorithmized and solved on a computer they are very suitable for.

Studies on q-difference equations appeared already at the beginning of the last century in intensive works especially by Adams[1], Atkinson[2], Bairamov[3,4], Jackson[5], Elaydi[6], Guseinov[7,8], Carmichael, Trjitzinsky and others. Unfortunately, from the thirties up to the beginning of the eighties, only nonsignificant interest in the area was observed. Since years eighties, an intensive and somewhat surprising interest in the subject reappeared in many areas of mathematics and applications including mainly new difference calculus and orthogonal polynomials, q-combinatorics, q-arithmetics, q-integrable systems and variational q calculus.

2 Preliminaries

We have investigated the global asymptotic behavior of the solutions of the system of the difference equation

$$k_n \frac{d^2 y_n(t)}{dt^2} = h_n \frac{dy_n(t)}{dt} + a_{n-1} y_{n-1}(t) + b_n y_n(t) + q a_n y_{n+1}(t) \tag{1}$$

$$n \in \{0,1\ldots,N-1\},\ t \geq 0,$$

where $N \geq 2$ is a fixed integer and $q > 1$ is a fixed real number.

$$y_{-1}(t) = 0,\ y_n(t) + \rho y_{n-1}(t) = 0,\ t \geq 0 \tag{2}$$

$$y(0) = w_n,\ \frac{dy_n(t)}{dt} = \kappa_n,\ n \in \{0,1,\ldots,N-1\}. \tag{3}$$

$\{y_n(t)\}_{n=1}^{N}$ find the solution and k_n, h_n, a_n, b_n real parametres, $w_n, \kappa_n\ n = (0,1,\ldots,N-1)$ complex, equation (2) is the real number of the limit condition ρ, meets the following conditions.

If a solution to the problem (1), (3) $\{y_n(t)\}_{n=1}^{N}$, (2) boundary conditions taking into account, they are obtained,

$$k_0\frac{d^2y_n(t)}{dt^2} = h_0\frac{dy_0(t)}{dt} + b_0y_0(t) + qa_0y_1(t), \tag{4}$$

$$k_n\frac{d^2y_n(t)}{dt^2} = h_n\frac{dy_n(t)}{dt} + a_{n-1}y_{n-1}(t) + b_ny_n(t) + qa_ny_{n+1}(t), \tag{5}$$

$$k_{N-1}\frac{d^2y_{N-1}(t)}{dt^2} = h_{N-1}\frac{dy_{N-1}(t)}{dt} + a_{N-2}y_{N-2}(t) + (b_{N-1} - hqa_{N-1})y_{N-1}(t) \tag{6}$$

This system, together with the initial conditions (3),

$$y(t) = \begin{bmatrix} y_0(t) \\ y_1(t) \\ \vdots \\ y_{N-1}(t) \end{bmatrix},\quad w = \begin{bmatrix} w_0 \\ w_1 \\ \vdots \\ w_{N-1} \end{bmatrix},\quad \kappa = \begin{bmatrix} \kappa_0 \\ \kappa_1 \\ \vdots \\ \kappa_{N-1} \end{bmatrix},$$

$$K = \begin{bmatrix} k_0 & \cdots & 0 \\ \vdots & \ddots & \vdots \\ 0 & \cdots & k_{N-1} \end{bmatrix},\quad H = \begin{bmatrix} h_0 & \cdots & 0 \\ \vdots & \ddots & \vdots \\ 0 & \cdots & h_{N-1} \end{bmatrix}$$

$$M = \begin{bmatrix} b_0 & qa_0 & 0 & 0 & \cdot & \cdot & \cdot & 0 & 0 \\ a_0 & b_1 & qa_1 & 0 & \cdot & \cdot & \cdot & 0 & 0 \\ 0 & a_1 & b_2 & qa_2 & \vdots & \vdots & \vdots & 0 & 0 \\ \vdots & \vdots & & \vdots & & & & & \\ 0 & 0 & 0 & 0 & 0 & 0 & 0 & a_{N-2} & b_{N-1} - pa_{N-1} \end{bmatrix}, \tag{7}$$

$$K\frac{d^2u(t)}{dt^2} = H\frac{dy_n(t)}{dt} + Mu(t), \quad 0 \le t < \infty \tag{8}$$

$$y(0) = w, \ \frac{dy(0)}{dt} = \kappa \tag{9}$$

we can write. Thus, the initial boundary value problem (1) - (3), the initial value problem (8), (9) the problem becomes equivalent [9]. Let

$$y(t) = e^{\alpha t}u \tag{10}$$

be a non-trivial solution of the equation, where α is the complex constant, $u = \{u_n\}_{n=0}^{N-1} \epsilon C^N$. Equation (10), (8) as well, they are obtained

$$(\lambda^2 K - \lambda H - M)\, u = 0 \tag{11}$$

Let us record (11) the equation

$$(\alpha^2 k_n - \alpha h_n - b_n)\, u_n - a_{n-1}u_{n-1} - qa_n u_{n+1} = 0,\ n = 0,1,\dots N-1 \tag{12}$$

$$u_{-1} = 0,\ u_N + hu_{N-1} = 0 \tag{13}$$

it is equivalent to the boundary value problem. If the solution of the problem (12), (13) is $\{u_n\}_{n=-1}^{N}$, the $u = \{u_n\}_{n=0}^{N-1}$ vector provides the equation (11) .

3 Asymptotics of regular solutions of q-difference equations

For $1 > q > 0$, we define the q- derivate of a real valued function u as

$$D_q y(t) = \frac{y(t) - y(qt)}{(1-q)t},\ D_q y(0) = \lim_{t\to 0} D_q y(t). \tag{14}$$

The higher order q – derivatives are given by

$$D_q^0 y(t) = y(t)\,,\ D_q^n y(t) = D_q D_q^{n-1} y(t),\ n\epsilon N. \tag{15}$$

we will describe asymptotics of solutions of q-difference equations. A q-difference equation for a sequence (y(1), y(2), y(3), . . .) of smooth functions of q has the form:

$$\sum_{j=0}^{d} b_j(q^k, q) y(k+j, q) = 0 \tag{16}$$

where $bj\ (v, u)$ are smooth functions and $y(k,q) = y(k)(q)$.
In the usual analytic theory of q-difference equations, q is a complex number inside or outside the unit disk.With this in mind, ϙ-difference equations are obtained from q-difference equations by the substitution $q = e^{2\pi i a/n}$ where ϙ is a small nonnegative real number, that plays the role of Planck's constant.[10]. The characteristic polynomial of the q-difference equation (16) is

$$P(v,\lambda) = \sum_{j=0}^{d} bj\ (v,1)\lambda^j$$

We will say that equation (16) is regular if

$$Dsc_\lambda P(v,\lambda)\ b0(v,1)\ bd(v,1) \neq 0$$

for all $v \in S^1$, where $Dsc_\lambda P(v,\lambda)$ is the discriminant of $P(v,\lambda)$, which is a polynomial in the coefficients of $P(v,\lambda)$. Let $\lambda_1(v),\dots,\lambda d(v)$ denote the roots of the characteristic polynomial, which we call the eigenvalues of eq. (16). It turns out that eq.(16) is regular iff the eigenvalues $\lambda_1(v), \ldots, \lambda_d(v)$ never collide and never vanish, for every $v \in S^1$. Moreover, it follows by the implicit function theorem that the roots are smooth functions of v $\in S^1$. Let $S^1 = \cup_{p\in P} Ip$ denote a partition of S^1 into a finite union of closed arcs (with nonoverlapping interiors), such that the magnitude of the eigenvalues does not change in each arc. In other words, for each $p \in P$, there is a permutation σ_p of the set {1, . ., d} such that

$$|\lambda_{\lambda\sigma p(1)}(v)| \geq |\lambda_{\lambda\sigma p(1)}(v)| \geq \cdots \geq |\lambda_{\lambda\sigma p(1)}(v)| \text{ for all } v \in Ip.$$

The following definition introduces a locally fundamental set of solutions of q-difference equations.
Fix a partition of I as above. A set $\{\psi_1, \ldots, \psi_d\}$ is a locally fundamental set of solutions of eq.(16) iff for every solution ψ for every $p \in P$ and for every $m = 1,\dots,d$ there exist smooth functions c p m such that

$$\psi(k,q) = c_1^p(q)\psi\sigma_p(1)(k,q) + \cdots + c_d^p(q)\psi\sigma_p(d)(k,q) \text{ for all } (k, q) : q^k \in I_p.$$

Theorem 1.([11]) Assume that eq.(16) is regular. Then, there exists a locally fundamental set of solutions $\{\psi_1, \ldots, \psi_d\}$ such that

- For every $m = 1,\dots,d$ and (k,q) such that $q^k \in I_p$ we have
  $\psi_m(k, e^{2\pi i a/n}) = \exp(\frac{n}{a}\Phi_m(\frac{ka}{n}, \frac{a}{n}))$.
- For some smooth functions Φ_m with uniform (with respect to $x \in I = [0,1]$) asymptotic expansion

$$\Phi m(x,\epsilon) \sim \sum_{s=0}^{\infty} \Phi m,s\,(x)\epsilon^s \tag{17}$$

where $\varphi_{m,s} \in C^\infty(I)$ for all s

- and leading term
  $\varphi_{m,0}(x) = \int_0^x \log(\lambda_m(e^{2\pi i t}))dt$

where we have chosen a branch for the logarithm of λ_m.

Definition 1. ([12]) We say that a formal series $\hat{\psi}$ is a solution to (8) iff

$$\frac{1}{\hat{\psi}(x,\epsilon)}\sum_{j=0}^{d} a_j(x,\epsilon).\hat{\psi}(x+j\epsilon,\epsilon) = 0\epsilon\ C^\infty(I)[[\epsilon]] \tag{18}$$

It is easy to see that if $\hat{\psi}$ is a formal solution to (8), then the leading term φ_0 satisfies the equation

$$exp(\varphi'0\,(x)) = \lambda(x)$$

where λ(x) is an eigenvalue of (8).
The hypergeometrical basic series:

$$_2\Phi_1(a,b,c,q,t) = \sum_{n\geq 0}\frac{(a;q^{-1})_n(b;q^{-1})_n}{(c;q^{-1})_n(q^{-1};q^{-1})_n}t^n \tag{19}$$

is a convergent series, and it is the solution of the q-difference equation rational linear order two,

$$\sigma_q^2 y - \lambda \sigma_q y + \mu y = 0 \text{ with } \begin{cases} \lambda = \frac{(a+b)t-(1+\frac{c}{q})}{abt-\frac{c}{q}} \\ \mu = \frac{t-1}{abt-\frac{c}{q}} \end{cases}.$$

It's a q-analogue (and a deformation) of Euler Gauss's classical hypergeometric symmetry, ${}_2F_1(\alpha,\beta,\gamma,t) = \sum_{n\geq 0} \frac{(\alpha)_n(\beta)_n}{(\gamma)_n(1)_n} t^n$

We find this last one by continuous limit (or "confluence") of coefficients. when $q \rightarrow 1$, with the constraints $a = q\alpha, b = q\beta, c = q\gamma$. We'll see further in which precise direction one can confuse the functional equation (1) it self to a differential equation

$$\sigma_q^2 y - \tilde{\lambda} \sigma_q y + \tilde{\mu} y = 0 \text{ with } \begin{cases} \tilde{\lambda} = \frac{(\alpha+\beta)t+(1-\gamma)}{1-t} \\ \tilde{\mu} = \frac{\alpha\beta t}{1-t} \end{cases}.$$

it admits for singularities 0, 1 and ∞ only and that these singularities are regular. The Frobenius Method provides fundamental systems of solutions in 0 and in ∞:

$$\begin{cases} {}_2F_1(\alpha,\beta,\gamma,t) \\ t^{1-\gamma}\, {}_2F_1(\alpha-\gamma+1,\beta-\gamma+1,2-\gamma,t) \end{cases}$$

$$\begin{pmatrix} (1/t)^{\alpha}\, {}_2F_1(\alpha,\alpha-\gamma+1,\alpha-\beta+1,\frac{1}{t}) \\ (1/t)^{\beta}\, {}_2F_1(\beta,\beta-\gamma+1,\beta-\alpha+1,\frac{1}{t}) \end{pmatrix}$$

The analytical extension of a fundamental system solutions in the neighborhood of a singularity still provides a fundamental system solutions and therefore expresses itself from another fundamental system by an array of constants.

To determine the behavior of the solutions we give the following lemmas.

Lemma 1. Define the function f : (0, ∞) → R by

$$f(t) = \Phi\left(\frac{t}{q} - \frac{1}{q}\right) - \Phi\left(1 - \frac{1}{t}\right).$$

Then $t \rightarrow f(t)$ is strictly increasing for $t > q^{1-1/\alpha}$ and strictly decreasing for $0 < t < q^{1-1/\alpha}$.

Proof. The statement follows from the equality

$$f'(t) = \frac{\alpha-1}{(t-1)^2}\left(q\left|\frac{t-1}{q}\right|^{\alpha} - \left|\frac{t-1}{t}\right|^{\alpha}\right).$$

Lemma 2. For $u \neq 0$ define the operator P by

$$P[u](t) = \Phi\left(\frac{u(q^2t)}{qu(qt)} - \frac{1}{q}\right) - \Phi\left(1 - \frac{u(t)}{u(qt)}\right).$$

Then

$$Dq(\Phi(Dqu(t))) = \frac{\Phi(u(qt))}{(q-1)\,\alpha^{\alpha_t}\alpha} P[u](t)$$

and equation (6) can be written as $P[u](t) = (q-1)\,\alpha^{\alpha_t \alpha} p(t)$ for $u \neq 0$.

Proof. From the q-derived formula

$$Dq(\Phi(D_q u(t))) = Dq\left(\Phi\left(\frac{u(qt)-u(t)}{(q-1)t}\right)\right)$$

$$=\frac{1}{(q-1)t}\left(\Phi\left(\left(\frac{u(q^2t)-u(qt)}{(q-1)qt}\right)-\Phi\left(\frac{u(qt)-u(t)}{(q-1)t}\right)\right)\right.$$

$$=\frac{1}{(q-1)\,a^{\alpha}t^{\alpha}}\left(\Phi\left(\left(\frac{u(qt)}{q}\left(\frac{u(q^2t)}{uqt}-1\right)\right)-\Phi\left(u(qt)\left(1-\frac{u(t)}{uqt}\right)\right)\right)\right)=\frac{\Phi(u(qt))}{(q-1)\,a^{\alpha}t^{\alpha}}P[u](t).$$

4 Solving the Initial Boundary Value Problem

In this section (4), (5) under the conditions of (1) - (3) the initial limit value the problem will be solved.

Theorem 2. (1) - (3) initial boundary value problem C^N arbitrary $\{f_n\}_{n=0}^{N-1}$ and g=$\{g_n\}_{n=0}^{N-1}$ for g g vectors

$$u_n(t)=\sum_{j=1}^{2N}a_je^{\lambda_jt}\varphi_n(\lambda_j)\qquad (n=-1,0,1,\ldots N)\qquad (20)$$

a unique $\{u_n(t)\}_{n=-1}^{N}$has the solution. Where $\lambda_1,\lambda_2,\ldots\lambda_{2N}$ the eigenvalues of the problem (11) and $\{\varphi_n(\lambda)\}_{n=-1}^{N}$ equation(12) the q- difference equation (13) is the solution that satisfies the initial conditions.

Proof. (20) formula gives a solution to the problem (1) - (3). Now that this solution is unique. We'll prove it. For this, the problems (1) - (9), which are equivalent to the problem (1) - (3) the problems

$$\frac{du(t)}{dt}=v(t)$$

$$\frac{dv(t)}{dt}=C^{-1}Rv(t)+C^{-1}Jv(t),$$

$$u(0)=f,\ v(0)=g$$

and the last problem is

$$A=\begin{bmatrix}0 & 1\\ C^{-1}J & C^{-1}R\end{bmatrix},\qquad w(t)=\begin{bmatrix}u(t)\\ v(t)\end{bmatrix},\ w^0=\begin{bmatrix}f\\ g\end{bmatrix},\ \text{to be}$$

$$\frac{dw(t)}{dt}=Aw(t),0\le t\le\infty,\qquad (21)$$

$$w(0)=w^0.\qquad (22)$$

Thus, problem (8), (9); the problem (21), (22) becomes equivalent. That is, if $u(t)$ If the solution of the vector function (8), (9) is the solution, then

$$w(t)=\begin{bmatrix}u(t)\\ u'(t)\end{bmatrix}$$

the vector function (21) is the solution to the problem (22) and, conversely,

$$w(t)=\begin{bmatrix}u(t)\\ v(t)\end{bmatrix}$$

if the vector function (21) is the solution of (22) then $v(t)=u'(t)$ and $u(t)$ is the solution of the vector function (8), (9). So (21), (22) the problem is that for the first order linear differential equation system .initial value problem and it is known that this problem is the only solution. So the theory is complete.

5 Conclusion

The examination of difference equations is indisputable in physics and mathematics. While solving the equations by discretizing the differential equations from one side, on the other hand, as mathematical models of many practical phenomena we also study the second - order self - adjoint linear homogeneous q - difference equations. We examined the solutions under the initial conditions and gave them their general properties.
We also examined the global asymptotic behavior of solutions by applying the q-Caramate functions to the q-difference equations. These equations are a wave equation type with the second - order q - difference derivative and therefore bear a physical meaning.

References

1. C. R. Adams, Annals of Mathematics, 2nd Ser. 30, 195 – 205 (1929)
2. F. V. Atkinson, Academic Press, New York (1964)
3. E. Bairamov ,Ö. Çakar, A.O. Çelebi, J. Math. Anal. Appl., 216, 303 – 320. (1997)
4. E. Bairamov ,Ö. Çakar, A.M. Krall, J. Diff. Equations, 151, 268 – 289. (1999)
5. F.H. Jackson, Amer. J. Math. 32, 305 – 314.(1910)

6. S. Elaydi, Springer - Verlag, New York (1996)
7. G.S. Guseinov , G. Oturanç, Turkish J. Math. 21, 409 – 422 (1997)
8. G.S. Guseinov, M.Terziler, Linear Algebra and Its Applications, 259, 329 – 346.(1997)
9. A. Hüseynov, Ankara University Inst. of Science and Tec.(2006)
10. R. Kashaev, Modern Phys. Lett. A 39 , 269–275. (1997).
11. O. Costin, R. Costin, SIAM J. Math. Anal. 27, 110–134. (1996)
12. S. Garoufalidis, J.S. Geronimo, arXiv:math/0405331 [math.QA], (2006)

Evaluation of Car Performances Using Data Envelopment Analysis

Yunus GÜRAL[1,*] *and Ayşe Turan BUĞATEKİN*[1]

[1]Department of Statistics, University of Fırat, Elazığ, Turkey

Abstract. Data Envelopment Analysis (DEA) is a nonparametric method used to examine the relative efficiencies of Decision Making Units (DMUs) on conditions where there are multiple inputs and multiple outputs. As in all sectors, it is very important for the automotive sector to operate effectively. Therefore, it is also important to measure the efficiency and find the source of the inefficiency. In this study, the performances of the DMU of the automobiles will examine using Data Envelopment Analysis. In this direction, it is aimed to assist consumers in purchasing by calculating the relative efficiencies of automobile models, determining effective and ineffective DMUs according to the wishes of the consumers. Sales price and fuel consumption are determined as input variables; maximum speed, cylinder volume, horsepower, maximum torque, luggage volume, acceleration time from 0 to 100 km are determined as output variables.

1 Introduction

In the automotive sector many brands and models are sold to consumers. The features of the automobile are different in these models. In addition to the features of the automobile price, fuel consumption, comfort and safety are also influential in making decisions for the consumers. These criteria have a great influence on consumers' choice.

DEA was first introduced in 1957 by Farrell in the literature and in the following years different researchers tried to be developed. In 1978 Charnes, Cooper and Rhodes studied it in detail [2].

The DMUs should be similar, since measurements are made relatively in the Data Envelopment Analysis. DEA is a method that uses similar inputs to obtain similar outputs.

In this study, the most selling car models in Turkey in 2017 are included. According to the wishes of the consumers it is aimed to assist consumers in purchasing by calculating the relative efficiencies of automobile models and determining effective and ineffective DMUs.

* Corresponding author: ygural@firat.edu.tr

2 Data Envelopment Analysis

DEA is a linear programming technique for measuring the relative efficiency of DMUs with multiple inputs and multiple number of outputs. DEA is used in many areas such as Health, Sports, Finance, Tourism, Marketing, Education [1,3]. The advantage of the DEA over other non-parametric methods is that multiple outputs can be measured in the production process. DEA identifies effective and ineffective DMUs. It also provides information on the amount and the source of ineffectiveness of DMUs.

2.1 Implementation Objectives of DEA

- Determine the amount and source of inefficiency of ineffective units,
- Creation of a reference set,
- Improving resources by reference set,
- Evaluation of the appropriateness of the administrations.

2.2 Charnes, Cooper, Rhodes (CCR)

CCR is the basic DEA model developed by Charnes, Cooper, Rhodes. Input-oriented CCR is concerned with how much input will be reduced to achieve the desired output. The output-oriented CCR is concerned with how much output level will increased by the amount of input available.

2.2.1 Output-oriented CCR (CCR-O)

CCR-O is the model that calculates how much increase the amount of output without changing the input level. Output-oriented CCR model [3] as follows

$$Max\beta_k + \varepsilon \sum_{r=1}^{l} T_r^- + \varepsilon \sum_{i=1}^{m} T_i^+$$

$$T_r^- + \sum_{j=1}^{t} \lambda_{jk} X_{rj} = X_{rk}$$

$$-T_i^+ + \sum_{j=1}^{t} \lambda_{jk} Y_{ij} - \beta_k Y_{ik} = 0$$

$$\lambda_{jk}, T_r^-, T_i^+ \geq 0$$

If $\beta^* = 1$, DMU is efficient. If $\beta^* > 1$ or slack variables are not zero, then DMU is inefficient.

For the full efficiency of the DMU, $\beta^* = 1$ and all slack variables must be zero. If $\beta^* = 1$ but one of the slack variables is different from zero, it is called weak efficient [4].

For an ineffective DMU to be effective, target values are calculated using the following formula.

$$X_{rk} = X_{rk} - T_r^{-*} \quad ; r = 1, \dots , l \tag{1}$$

$$Y_{ik} = \beta_k^* Y_{ik} + T_i^{+*} \; ; i = 1, \dots , m \tag{2}$$

Table 1. Efficiency Values of car model.

Segment C Sedan	Efficiency	Segment C HB	Efficiency	Segment C Suv	Efficiency
Fiat Egea	1	VW Golf	1,01723	Nissan Qashqai	1
Renault Megane	1	Seat Leon	1	Dacia Duster	1
Toyota Corolla	1	Opel Astra 1.6 Cdti Enjoy	1	VW Tiguan	1
Ford Focus	1	Renault Megane	1	Hyundai Tucson	1
Opel Astra	1	Fiat Egea	1,00306	Toyota C-Hr	1
Honda Civic	1,01163	Toyota Auris	1	Peugeot 3008	1
VW Jetta	1	Audi A3 Sportback	1	Renault Kadjar	1
Skoda Octavia	1	Bmw 1 serial 116d	1	Ford Kuga	1
Audi A3	1	Mercedes A Serial A180	1,0192	Honda CR-V	1
Mercedes CLA 180	1,02588	Hyundai İ30	1	Kia Sportage	1
Mazda 3	1,02173	Nissan Pulsar	1	Mercedes GLA 200	1
Mitsubishi Lancer	1,05338	Peugeot 308	1	Seat Ateca	1,022691
		Skoda Rapid Spaceback	1	Bmw X1	1
		Honda Civic	1	Subaru Forester 2	1,048195
		Ford Focus	1	Audi Q3	1
		Volvo V40	1	Subaru XV	1,058605
		Volvo V40 Cross Country	1,0198	Mini Countryman	1,030307
		Citroen C4	1,0521	Range Rover Evoque	1
		Alfa Romeo Giulietta	1	Skoda Yeti	1
		Kia Cee'd	1	Mitsubishi ASX	1,026402
				Toyota RAV4	1
				Mazda CX-5	1

Table 2. Efficiency Values of car model.

Segment B	**Efficiency**	**Segment D**	**Efficiency**
Toyota Yaris	1	VW Passat	1
VW Polo	1	Mercedes C180	1
Honda Jazz	1	Skoda Superb	1
Kia Rio	1,0317	Opel Insignia	1
Ford Fiesta	1	Audi A4	1,00645
Hyundai i20	1,0254	Bmw 318d	1
Renault Clio	1	Renault Talisman	1
Citroen C3	1	Ford Mondeo	1,02004
Opel Corsa	1,04848	Volvo S60	1
Peugeot 208	1	Peugeot 508	1
Dacia Sandero	1	Toyota Avensis	1
Nissan Micra	1		
Skoda Fabia	1		
Mini Cooper	1		
Audi A1 Sportback	1		
Suzuki Baleno	1,0255		

In Turkey best-selling car models in 2017 were included in the application. According to Table 1 and 2, twenty of these car models are not effective. Target values for ineffective automobile models can be found using (1) and (2).

3 Conclusion

Relative efficiency measurement by Data Envelopment Analysis has many application fields in many sectors such as education, bank and health. However, unlike these sectors, it is noteworthy that in the automotive sector there are very few studies in which the relative efficiencies of automobile models are measured using DEA, especially from a consumer point of view. In addition, DEA based decision support system has been created to help automobile consumers make purchasing decisions. It is believed that this aspect of the work contributes to the literature and thus may be a guide to similar work in the future.

References

1. Aktaş , H., An Application of Data Envelopment Analysis with financial ratios assess the performance of banks , Yönetim ve Ekonomi, Celal Bayar Üniversitesi, İ.İ.B.F. Dergisi, 5, 27-40, 1999.
2. Charnes, A., Cooper, W.W., Rhodes, E., Measuring the efficiency of decision making units. European Journal of Operational Research 2, 429–444, 1978.
3. Erpolat, S., Veri Zarflama Analizi (Ağırlık Kısıtlamasız, Ağırlık Kısıtlamalı, Şans Kısıtlı, Bulanık) Türkiye'deki Özel Bütçeli Diğer İdarelerin Etkinlik Analizi, Evrim Yayınevi, İstanbul, 2011.
4. Kutlar, A. ve Babacan, A., A Comparison Of The Effectivity Of Cumhuriyet University With State Universities: An Applicatoin Of Dea Technique, Kocaeli Üniversitesi Sosyal Bilimler Enstitüsü Dergisi, 15(1), 148-172, 2008.

https://doi.org/10.1051/itmconf/20182201052

Lacunary Statistical Convergence of Order alpha and Lacunary Strongly Summable Sequences of Order alpha of Generalized Difference Sequences

Mikail Et[1,*], *Hacer Şengül*[2], and *Muhammed Çınar*[3]

[1]Department of Mathematics, Fırat University, Elazığ, Turkey

[2]Faculty of Education, Harran University, Osmanbey Campus, Şanlıurfa, Turkey

[3]Department of Mathematic Education, Muş Alparslan University, Muş, Turkey

Abstract. The notion of the α th order Δ_i^m − lacunary statistical convergence and α th order lacunary strongly $\left(\Delta_i^m, p\right)$ − summable sequences was introduced by Altınok et al. [1]. They also gave significant inclusion correlation associated with the aforementioned sequence spces. In this paper, our aim is to exploit some other important relations between the given notions.

1 Introduction

The idea of statistical convergence was firstly determined by Fast [12] and redefined independently by Schoenberg [28]. The concept of the density of the set of positive integers $\mathbb{N}$ was used as the main motivation to create this idea. For any subset F of the $\mathbb{N}$, the density of F is defined as

$$\beta(F) = \lim_{n\to\infty} \tfrac{1}{n} \textstyle\sum_{l=1}^{n} \omega_F(l)$$

where ω_F is the characteristic function of F. It is needless to say that this definition is valid only for the existence of the limit. Let $y = (y_l)$ be a given any sequence then it is statistically converged to R, if for every $\varepsilon > 0$, $\beta\{l \in N : |y_l - R| \geq \varepsilon\} = 0$. Furthermore, the relation and some application of this definition with the summabilitiy theory was presented by using the perspective of the sequence space by Çınar et al. ([3], [6], [30]), Connor [5], Fridy [14], Et et al. ([2], [11], [30]), Işık et al. ([17], [18], [19], [20], [21])). Gadjiev and Orhan [16] stated firstly the order of the sequence's statistical

* Corresponding author: mikailet68@gmail.com

convergence. Based on this research, Çolak [4] studied the α th order of the statistical convergence and α th order of the strong $p-$Cesàro summability.

The lacunary sequence is defined by $h_k = (l_k - l_{k-1}) \to \infty$ as $k \to \infty$, where $\gamma = (l_k)$ is an increasing sequence of integer. The interval denoted by γ is determined by $I_k = (l_{k-1}, l_k]$ and q_k is the abbreviation of the ratio of $\frac{l_k}{l_{k-1}}$ ([13], [15], [20], [29]). The introduction of the difference sequence space and its modification and extension were recently conducted by Kızmaz [23] and Et and Çolak [7]. Tripathy et al. [31] obtained the generalized shape of the difference sequences by considering the former difference sequences in the following form

$$Y(\Delta_i^m) = \{y = (y_l) \; : \; (\Delta_i^m y_l) \in Y\}$$

where Y is regarded as a sequence, i and m are any two arbitrary positive integers, and $\Delta_i^m y_l = \sum_{w=0}^{m} (-1)^w \binom{m}{w} y_{l+iw}$.

We rather choose to take $Y(\Delta^m)$ in lieu of $Y(\Delta_i^m)$ and $\Delta^m y$ in lieu of $\Delta_i^m y$ if we assume that $i = 1$. Other well known studies can be found in the following papers ([8], [9], [10], [17], [22], [25]).

2 Main Results

This section is the core of the study since it contains main results ans theorems.

Definition 2.1. [1] Let $\gamma = (l_k)$ be a lacunary sequence such that m and i are choosen as arbitrarily non-negative integers and $0 < \alpha \leq 1$ be given. If a real number R is defined as follows

$$\lim_{k\to\infty} \frac{1}{h_k^\alpha} \left| \left\{ l \in I_k \; : \; \left| \Delta_i^m y_l - R \right| \geq \varepsilon \right\} \right| = 0,$$

then the sequence $y = (y_l)$ is said to be $\Delta_i^m -$ lacunary statistically convergent of order α (or $S_\gamma^\alpha(\Delta_i^m) -$ convergent).

Thus it is obvious that $S_\gamma^\alpha(\Delta_i^m) - \lim y_l = R$. The set of all $\Delta_i^m -$ lacunary statistically convergent sequences will be denoted by $S_\gamma^\alpha(\Delta_i^m)$.

$\Delta_i^m -$lacunary statistical convergence is well defined for $0 < \alpha \leq 1$, but it is not well defined for $\alpha > 1$ in general. It is evidently true that every $\Delta^m -$convergent sequence is lacunary $\Delta_i^m -$ statistically convergent of order α $(0 < \alpha \leq 1)$, but the converse does not hold.

Definition 2.2. [1] Let $\gamma = (l_k)$ be a lacunary sequence such that m and i are choosen as arbitrarily non-negative integers and $\alpha, p > 0$ be given. If a real number R is defined as follows

$$\lim_{k\to\infty}\frac{1}{h_k^{\alpha}}\sum_{l\in I_k}\left|\Delta_i^m y_l - R\right|^p = 0,$$

then a sequence $y=(y_l)$ is said to be strongly $w_\gamma^\alpha(\Delta_i^m, p)-$summable (or lacunary strongly $(\Delta_i^m, p)-$summable of order α).

Thus it is obvious that $w_\gamma^\alpha(\Delta_i^m, p)-\lim y_l = R$. The set of all strongly $w_\gamma^\alpha(\Delta_i^m, p)-$summable sequences of order α will be denoted by $w_\gamma^\alpha(\Delta_i^m, p)$. If we take $\gamma=(2^k)$ then we write $w^\alpha(\Delta_i^m, p)$ instead of $w_\gamma^\alpha(\Delta_i^m, p)$.

For the following theorems we omit to give detailed proof since it is easily proved by using standard methods.

Theorem 2.3. Let $0<\alpha\le 1$ and $\gamma=(l_k)$ be a lacunary sequence. If $\limsup_k \frac{l_k}{l_{k-1}^{\alpha}}<\infty$, then $w_\gamma^\alpha(\Delta_i^m, p)\subset w^\alpha(\Delta_i^m, p)$.

Theorem 2.4. Let $\gamma=(l_k)$ and $\gamma'=(s_k)$ be two lacunary sequences such that $I_k\subset J_k$ for all $k\in\mathbb{N}$ and let α and c be fixed real numbers such that $0<\alpha\le c\le 1$,

(i) If

$$\liminf_{k\to\infty}\frac{g_k^{\alpha}}{\ell_k^{c}}>0 \qquad (1)$$

then $S_{\gamma'}^{\beta}(\Delta_i^m)\subseteq S_\gamma^\alpha(\Delta_i^m)$,

(ii) If

$$\lim_{k\to\infty}\frac{\ell_k}{g_k^{c}}=1 \qquad (2)$$

then $S_\gamma^\alpha(\Delta_i^m)\subseteq S_{\gamma'}^{c}(\Delta_i^m)$, where $I_k=(l_{k-1},l_k]$, $J_k=(s_{k-1},s_k]$, $g_k=l_k-l_{k-1}$ and $\ell_k=s_k-s_{k-1}$.

Theorem 2.5. Let $\gamma=(l_k)$ and $\gamma'=(s_k)$ be two lacunary sequences such that $I_k\subseteq B_k$ for all $k\in\mathbb{N}$, α and c be fixed real numbers such that $0<\alpha\le c\le 1$ and $0<p<\infty$. Then we get

(i) If (1) holds then $w_{\gamma'}^{\beta}(\Delta_i^m, p)\subset w_\gamma^\alpha(\Delta_i^m, p)$,

(ii) If (2) holds and $y\in\ell_\infty(\Delta_i^m)$ then $w_\gamma^\alpha(\Delta_i^m, p)\subset w_{\gamma'}^{\beta}(\Delta_i^m, p)$.

Theorem 2.6. Let $\gamma=(l_k)$ and $\gamma'=(s_k)$ be two lacunary sequences such that $I_k\subseteq B_k$ for all $k\in\mathbb{N}$, α and c be fixed real numbers such that $0<\alpha\le c\le 1$ and $0<p<\infty$. Then

(i) Let (1) holds, if a sequence is strongly $w_{\gamma}^{\beta}\left(\Delta_i^m, p\right)-$ summable to R, then it is $S_{\gamma}^{\alpha}\left(\Delta_i^m\right)-$ convergent to R,

(ii) Let (2) holds, if a Δ_i^m- bounded sequence is $S_{\gamma}^{\alpha}\left(\Delta_i^m\right)-$ convergent to R then it is strongly $w_{\gamma}^{\beta}\left(\Delta_i^m, p\right)-$ summable to R.

Theorem 2.7. If $y \in w^{\alpha}\left(\Delta_i^m, p\right) \cap w_{\gamma}^{\alpha}\left(\Delta_i^m, p\right)$ and $\limsup_k \frac{l_k}{l_{k-1}^{\alpha}} < \infty$, then $w_{\gamma}^{\alpha}\left(\Delta_i^m, p\right) - \lim y_l = w^{\alpha}\left(\Delta_i^m, p\right) - \lim y_l = R$.

Proof. Let $w_{\gamma}^{\alpha}\left(\Delta_i^m, p\right) - \lim y_l = R$ and $w^{\alpha}\left(\Delta_i^m, p\right) - \lim y_l = R^{'}$, and suppose that $R \neq R^{'}$. Since $\limsup_k \frac{l_k}{l_{k-1}^{\alpha}} < \infty$, from Theorem 2.3 we already know that $w_{\gamma}^{\alpha}\left(\Delta_i^m, p\right) \subset w^{\alpha}\left(\Delta_i^m, p\right)$. Since $\left(\Delta_i^m y - R^{'}\right) \in w_{\gamma,0}^{\alpha}\left(\Delta_i^m, p\right)$, it follows that $\left(\Delta_i^m y - R^{'}\right) \in w_0^{\alpha}\left(\Delta_i^m, p\right)$ and therefore $\frac{1}{t^{\alpha}}\sum_{b=1}^{t}\left|\Delta_i^m y_b - R^{'}\right| \to 0$. Then we have

$$\frac{1}{t^{\alpha}}\sum_{b=1}^{t}\left|\Delta_i^m y_b - R^{'}\right| + \frac{1}{t^{\alpha}}\sum_{b=1}^{t}\left|\Delta_i^m y_b - R\right| \geq \frac{1}{t^{\alpha}}\left|R - R^{'}\right| > 0,$$

and hence $R = R^{'}$.

3 Results Related to Modulus Function

Nakano developed the mathematical background of the notion of the modulus [26]. Then this notion is used for the construction of some sequence spaces by Maddox [24] and Ruckle [27]. Now, we present the inclusion relations among the sets of lacunary strongly $\left(\Delta_i^m, p\right)-$summable sequences of order α and Δ_i^m-lacunary statistically convergent sequences of order α with respect to the modulus function f.

Definition 3.1. Let f be a modulus function, $p = \left(p_l\right)$ be a sequence of strictly positive real numbers and $\alpha > 0$. Now we define

$$w_{\gamma}^{\alpha}\left(\Delta_i^m, (p), f\right) = \left\{ y = \left(y_l\right) : \lim_{k\to\infty} \frac{1}{g_k^{\alpha}} \sum_{l \in I_k} \left[f\left(\left|\Delta_i^m y_l - R\right|\right)\right]^{p_l} = 0, \text{ for some } R \right\}.$$

If $y \in w_{\gamma}^{\alpha}\left(\Delta_i^m, (p), f\right)$, then we say that y is strongly $w_{\gamma}^{\alpha}\left(\Delta_i^m, (p), f\right)-$summable with respect to the modulus function f. If $p_l = p$ for all $l \in \mathbb{N}$ and $f(y) = y$ we shall write $w_{\gamma}^{\alpha}\left(\Delta_i^m, p\right)$ instead of $w_{\gamma}^{\alpha}\left(\Delta_i^m, (p), f\right)$ and in the special case $f(y) = y$ we shall write $w_{\gamma}^{\alpha}\left(\Delta_i^m, (p)\right)$ instead of $w_{\gamma}^{\alpha}\left(\Delta_i^m, (p), f\right)$.

In the following theorems we shall assume that the sequence $(p_l) = p$ is bounded and $0 < g = \inf_l p_l \le p_l \le \sup_l p_l = G < \infty$.

Theorem 3.2. Let m and i be two non-negative integers, $\gamma = (l_k)$ be a lacunary sequence, α and c be fixed real numbers such that $0 < \alpha \le c \le 1$ and f be a modulus function, then $w_\gamma^\alpha(\Delta_i^m,(p),f) \subset S_\gamma^c(\Delta_i^m)$.

Proof. Let $y \in w_\gamma^\alpha(\Delta_i^m,(p),f)$ and let $\varepsilon > 0$ be given and $\sum_1$ and $\sum_2$ denote the sums over $l \in I_k$, $|\Delta_i^m y_l - R| \ge \varepsilon$ and $l \in I_k$, $|\Delta_i^m y_l - R| < \varepsilon$, respectively. Since $g_k^\alpha \le g_k^c$ for each k we may write

$$\begin{aligned}\frac{1}{g_k^\alpha}\sum_{l\in I_k}\left[f\left(\left|\Delta_i^m y_l - R\right|\right)\right]^{p_l} &= \frac{1}{g_k^\alpha}\left[\Sigma_1\left[f\left(\left|\Delta_i^m y_l - R\right|\right)\right]^{p_l} + \Sigma_2\left[f\left(\left|\Delta_i^m y_l - R\right|\right)\right]^{p_l}\right]\\ &\ge \frac{1}{g_k^c}\left[\Sigma_1\left[f\left(\left|\Delta_i^m y_l - R\right|\right)\right]^{p_l} + \Sigma_2\left[f\left(\left|\Delta_i^m y_l - R\right|\right)\right]^{p_l}\right]\\ &\ge \frac{1}{g_k^c}\left|\left\{l \in I_k \ : \ \left|\Delta_i^m y_l - R\right| \ge \varepsilon\right\}\right| \ \min([f(\varepsilon)]^g,[f(\varepsilon)]^G).\end{aligned}$$

Hence $y \in S_\gamma^c(\Delta_i^m)$.

Theorem 3.3. Let $\gamma = (l_k)$ be a lacunary sequence, m and i be two non-negative integers and $0 < \alpha \le 1$. If $\lim_{k\to\infty} \frac{g_k}{g_k^\alpha} = 1$ and the modulus f is bounded, then $S_\gamma^\alpha(\Delta_i^m) \subset w_\gamma^\alpha(\Delta_i^m,(p),f)$.

Proof. Let $y \in S_\gamma^\alpha(\Delta_i^m)$ and suppose that f is bounded and $\varepsilon > 0$ be given. Since f is bounded there exists an integer K such that $f(y) \le K$, for all $y \ge 0$. Then we may write

$$\begin{aligned}\frac{1}{g_k^\alpha}\sum_{l\in I_k}\left[f\left(\left|\Delta_i^m y_l - R\right|\right)\right]^{p_l} &= \frac{1}{g_k^\alpha}\left(\Sigma_1\left[f\left(\left|\Delta_i^m y_l - R\right|\right)\right]^{p_l} + \Sigma_2\left[f\left(\left|\Delta_i^m y_l - R\right|\right)\right]^{p_l}\right)\\ &\le \max\left(K^g,K^G\right)\frac{1}{g_k^\alpha}\left|\left\{l \in I_n \ : \ \left|\Delta_i^m y_l - R\right| \ge \varepsilon\right\}\right|\\ &\quad + \frac{g_k}{g_k^\alpha}\max\left(f(\varepsilon)^g, f(\varepsilon)^G\right).\end{aligned}$$

and so $w_\gamma^\alpha(\Delta_i^m,(p),f)$.

Theorem 3.4. Let $\gamma=(l_k)$ be a lacunary sequence, m and i be two non-negative integers and $0<\alpha\leq 1$. If $y=(y_l)$ is strongly $w_\gamma^\alpha(\Delta_i^m,(p),f)-$ summable to R with respect to the modulus function f and $\lim p_l>0$, then $w_\gamma^\alpha(\Delta_i^m,(p),f)-\lim y_l=R$ unique.

Proof. Omitted.

Theorem 3.5. Let $\gamma=(l_k)$ be a lacunary sequence, m and i be two non-negative integers and $0<\alpha\leq 1$. The sequence spaces $w_\gamma^\alpha(\Delta_i^m,p), S_\gamma^\alpha(\Delta_i^m)$ and $w_\gamma^\alpha(\Delta_i^m,(p),f)$ are neither solid nor symmetric, nor sequence algebras for $m\geq 1$.

Proof. Let $\gamma=(2^k)$ and $i=1$, then $(y_l)=(l^{m-1})\in w_\gamma^\alpha(\Delta_i^m,p)$ but $(\alpha_l y_l)\notin w_\gamma^\alpha(\Delta_i^m,p)$ when $\alpha_l=(-1)^l$ for all $l\in\mathbb{N}$. Hence $w_\gamma^\alpha(\Delta_i^m,p)$ is not solid. The other cases can be proved on considering similar examples.

References

1. H. Altınok, M. Et, M. Işık, AIP Conference Proceedings **1926**, 020004 (2018); https://doi.org/10.1063/1.5020453
2. H. Altınok, M. Et, R. Çolak, Iran. J. Fuzzy Syst. **11(5)**, 39–46, (2014)
3. M. Çınar, M. Karakaş, M. Et, Fixed Point Theory Appl. **2013:33**, 11 pp (2013)
4. R. Çolak, Modern Methods in Analysis and Its Applications, New Delhi, India : Anamaya Pub, **2010**, 121–129 (2010)
5. J.S. Connor, Analysis **8,** 47-63 (1988)
6. M. Et, M. Çınar, M. Karakaş, J. Inequal. Appl., **2013:204**, 8 pp (2013)
7. M. Et, R. Colak, Soochow J. Math. **21(4)**, 377–386 (1995)
8. M. Et, H. Altınok, Y. Altın, Appl. Math. Comput. **154(1)**, 167–173 (2004)
9. M. Et, A. Alotaibi, S.A. Mohiuddine, The Scientific World Journal, 535419 (2014)
10. M. Et, M. Mursaleen, M. Işık, Filomat **27(5)**, 789–796 (2013)
11. M. Et, B.C. Tripathy, A.J. Dutta, Kuwait J. Sci. **41(3)**, 17–30 (2014)
12. H. Fast, Colloq. Math. **2**, 241–244 (1951)
13. A.R. Freedman, J.J. Sember M. Raphael, Proc. Lond. Math. Soc. **37(3)**, 508–520 (1978)
14. J.A. Fridy, Analysis, **5,** 301–313 (1985)
15. J.A. Fridy, C. Orhan, Pacific J. Math. **160**, 43–51 (1993)
16. A.D. Gadjiev, C. Orhan, Rocky Mountain J. Math. **32(1)**, 129–138 (2002)
17. M. Işık, J. Inequal. Appl. **2010**, Art. ID 457892, 7 pp (2010)
18. M. Işık, K.E. Akbaş, J. Inequal. Spec. Funct. **8(4)**, 57–64 (2017)
19. M. Işık, Math. Slovaca **61(5)**, 779–788 (2011)
20. M. Işık, K.E. Et, AIP Conference Proceedings **1676**, 020045 (2015); doi: http://dx.doi.org/10.1063/1.4930471.
21. M. Işık, K.E. Akbaş, ITM Web of Conferences **13**, 01024 (2017); doi: 10.1051/itmconf/20171301024
22. M. Karakaş, M. Et, V. Karakaya, Acta Math. Sci. Ser. B (Engl. Ed.) **33(6)**, 1711–1720 (2013)

23. H. Kızmaz, Canad. Math. Bull. **24(2)**, 169–176 (1981)
24. I.J. Maddox, Math. Proc. Camb. Philos. Soc. 1**00**, 161–166 (1986)
25. M. Mursaleen, R. Çolak, M. Et, J. Inequal. Appl. **2007**, Art. ID 86757, 6 pp (2007)
26. H. Nakano, Proc. Japan Acad. **27**, 508–512 (1951)
27. W.H. Ruckle, Canad. J. Math. **25**, 973–978 (1973)
28. I.J. Schoenberg, Amer. Math. Monthly **66**, 361–375 (1959)
29. H. Şengül, M. Et, Acta Math. Sci. Ser. B Engl. Ed. **34(2)**, 473–482 (2014)
30. H.M. Srivastava, M. Et, Filomat **31(6)**, 1573–1582 (2017)
31. B.C. Tripathy, A. Esi, B.K. Tripathy, Soochow J. Math. **31(3)**, 333–340 (2005)

Optical Solitons and Other Solutions to the (2+1)-Dimensional Cubic Nonlinear Schrödinger Equation with Fractional Temporal Evolution

Sibel Sehriban Atas[1 *], *Tukur Abdulkadir Sulaiman*[1,2], and *Hasan Bulut*[1,3]

[1]Department of Mathematics, Firat University, Elazig, Turkey

[2]Department of Mathematics, Federal University Dutse, Jigawa, Nigeria

[3]Department of Mathematics Education, Final University, Girne, Cyprus

Abstract. In this study, the (2+1)-dimensional cubic nonlinear Schrödinger equation with fractional temporal evolution is investigated by using the extended sinh-Gordon equation expansion method. The idea of conformable fractional derivative is used in transforming the complex nonlinear partial differential equation to nonlinear ordinary differential equation. Dark, bright, mixed dark-bright, singular, mixed singular solitons and singular periodic wave solutions are successfully reached. The parametric conditions for the existence of valid solitons are given. The 2D and 3D graphics to some of the reported solutions are plotted.

1 Introduction

Nonlinear Shrodinger's type equations (NLSEs) are special kind of nonlinear evolution equations expressing several complex nonlinear aspect in the field of nonlinear sciences such as the optical fiber, fluid mechanics, plasma physics, biology, hydrodynamics and so on. Several analytical techniques have been used to find the solutions of various NLSEs [1-12].

In this study, the (2+1)-dimensional cubic nonlinear Schrödinger equation with fractional temporal evolution [13] is going to be soughted by using the extended sinh-Gordon equation expansion method (ShGEEM) [14-17].

The (2+1)-dimensional cubic nonlinear Schrödinger equation with fractional temporal evolution is given by [13]

$$iD_t^\alpha \psi + c\left(\psi_{xx} + \psi_{yy}\right) + a|\psi|^2 \psi + b\psi = 0,\ \alpha \in (0,1), \tag{1.1}$$

where a,b,c are nonzero constant.

When $\alpha = 1$, we have the original (2+1)-dimensional cubic nonlinear Schrödinger equation [13].

* Corresponding author: sibel.s.atas@gmail.com

2 The Conformable Fractional derivative

In this section, we give some basic definition, properties and theorem about the conformable fractional derivative.

Definition 1. Let $g:(0,\infty)\to\mathbb{R}$, then the conformable fraction derivative of g of order α is defined as here, some basic properties of conformable fractional derivative are presented [18].

1. $T_\alpha(bg+ch)=bT_\alpha(g)+cT_\alpha(h), b,c\in\mathbb{R}$,
2. $T_\alpha(t^\lambda)=\lambda t^{\lambda-\alpha}, \lambda\in\mathbb{R}$,
3. $T_\alpha(gh)=gT_\alpha(h)+hT_\alpha(g)$,
4. $T_\alpha\left(\frac{g}{h}\right)=\frac{hT_\alpha(g)+gT_\alpha(h)}{h^2}$,
5. If g is differentiable, then $T_\alpha(g)(t)=t^{1-\alpha}\frac{dg}{dt}$.

Theorem 1. Let $g,h:(0,\infty)\to\mathbb{R}$ be differentiable functions and also α differentiable, then the following rule holds:

$$T_\alpha(g\circ h)(t)=t^{1-\alpha}h'(t)g'(h(t)). \tag{2.1}$$

3 Application

In this section, the application of the extended ShGEEM to the (2+1)-dimensional cubic nonlinear Schrödinger equation with fractional temporal evolution is presented.

Consider the complex time-fractional travelling wave transformation

$$\psi(x,y,t)=\psi(\xi)e^{i\theta}, \theta=px+qy+r\frac{t^\alpha}{\alpha}, \xi=\mu\left(x+y-k\frac{t^\alpha}{\alpha}\right).$$

(3.1)

Inserting Eq. (3.1) into Eq. (1.1), we get the following NODE:

$$\left(b-c\left(p^2+q^2\right)-r\right)\psi+a\psi^3+2c\mu^2\psi''=0 \tag{3.2}$$

from the real part and the relation $k=2c(p+q)$ from the imaginary part.

By the extended ShGEEM, the solutions of any given nonlinear partial differential equation are assumed to be of the forms [14]

$$\psi(\theta)=\sum_{j=1}^{m}\left[b_j\sinh(\theta)+a_j\cosh(\theta)\right]^j+a_0 \tag{3.3}$$

$$\psi(\xi)=\sum_{j=1}^{m}\left[\pm ib_j \operatorname{sec}h(\xi)+a_j \tan h(\xi)\right]^j+a_0 \tag{3.4}$$

and

$$\psi(\xi)=\sum_{j=1}^{m}\left[\pm b_j \operatorname{csc}h(\xi)+a_j \cot h(\xi)\right]^j+a_0, \tag{3.5}$$

respectively.

Balancing the terms ψ^3 and ψ'' in Eq. (3.2), yields $m=1$.

With $m=1$, Eqs. (3.3)-(3.5) take the forms

$$\psi(\theta)=b_1\sinh(\theta)+a_1\cosh(\theta)+a_0 \tag{3.6}$$

$$\psi(\xi)=\pm ib_1 \operatorname{sec}h(\xi)+a_1\tan h(\xi)+a_0 \tag{3.7}$$

and

$$\psi(\xi)=\pm b_1 \operatorname{csc}h(\xi)+a_1\cot h(\xi)+a_0, \tag{3.8}$$

where $i=\sqrt{-1}$ and $\theta'=\sinh(\theta)$ [14].

Inserting Eq. (3.6) and its second derivative into Eq. (3.2) gives a polynomial in powers of hyperbolic functions. Summing each coefficient of the hyperbolic functions of the same power and equating each summation to zero, gives a group of algebraic equations. The system of algebraic equations is simplified to secure the values of the parameters involved. For each case, inserting the obtained values of the parameters into each of Eqs. (3.6) and (3.8), yields the solutions to Eq. (1.1). Simplifying the group of algebraic equation, we get the values of the parameters involved.

Case 1: When

$$a_0=0, a_1=-\frac{2\sqrt{-c}\mu}{\sqrt{a}}, b_1=0, q=\frac{\sqrt{b-r-c\left(q^2+4\mu^2\right)}}{\sqrt{c}}, \tag{3.9}$$

we have the following dark and singular

soliton: $$u_1(x,t)=-\frac{2\sqrt{-c}\mu\left(\tanh\left[\left(x+y-\frac{kt^\alpha}{\alpha}\right)\mu\right]\right)}{\sqrt{a}}e^{i(px+qy+r\frac{t^\alpha}{\alpha})}, \tag{3.10}$$

$$u_2(x,t) = -\frac{2\sqrt{-c}\mu\left(\coth\left[\left(x+y-\frac{kt^\alpha}{\alpha}\right)\mu\right]\right)}{\sqrt{a}} e^{i(px+qy+r\frac{t^\alpha}{\alpha})}, \tag{3.11}$$

where $ac<0$ and $c\left(b-r-c\left(q^2+\mu^2\right)\right)>0$ for valid soliton.

Case 2: When

$$a_0 = 0, a_1 = 0,\ b_1 = \frac{2\sqrt{-c}\mu}{\sqrt{a}}, q = -\frac{\sqrt{b-r-c\left(q^2-2\mu^2\right)}}{\sqrt{c}}, \tag{3.12}$$

we have the following bright and singular

soliton: $$u_3(x,t) = -\frac{2\sqrt{-c}\mu\left(\operatorname{sech}\left[\left(x+y-\frac{kt^\alpha}{\alpha}\right)\mu\right]\right)}{\sqrt{a}} e^{i(px+qy+r\frac{t^\alpha}{\alpha})}, \tag{3.13}$$

where $c\left(b-r-c\left(q^2+\mu^2\right)\right)>0$ for valid soliton.

$$u_4(x,t) = \frac{2i\sqrt{c}\mu\left(\operatorname{csc}h\left[\left(x+y-\frac{kt^\alpha}{\alpha}\right)\mu\right]\right)}{\sqrt{a}} e^{i(px+qy+r\frac{t^\alpha}{\alpha})}, \tag{3.14}$$

where $c\left(b-r-c\left(q^2+\mu^2\right)\right)>0$ for valid soliton.

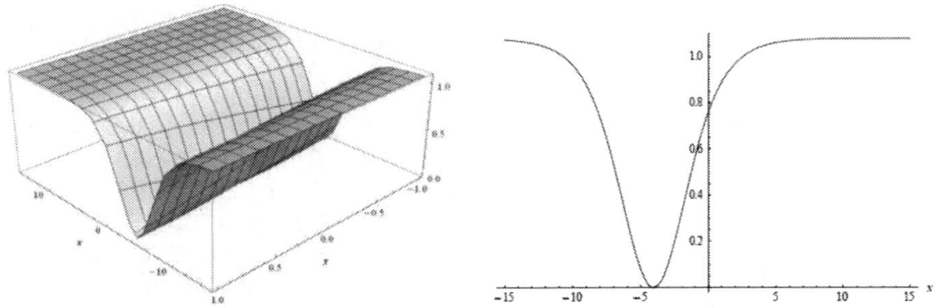

Figure 1. The 3D and 2D surfaces of Eq. (3.10) under the values of $c=-3, a=1, k=-0.75, \mu=0.3, \alpha=0.2, -15<x<15, -1<y<1$ and $t=0.5$ for the 2D surface.

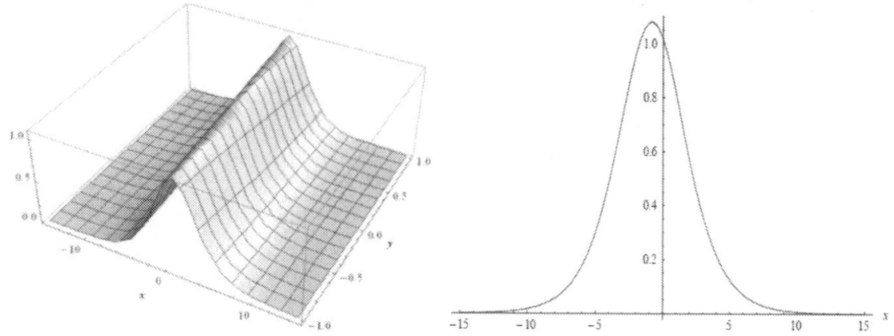

Figure 2. The 3D and 2D surfaces of Eq. (3.13) under the values of $c=-3, a=1, k=-0.003, \mu=0.3, \alpha=0.2, y=0.8, -15<x<15, -1<y<1$ and $t=0.5$ for the 2D surface.

4 Conclusion

In this work, the extended sinh-Gordon equation expansion method is used in extracting various soliton and other solutions such as the dark, bright and singular solutions to the (2+1) dimensional cubic nonlinear Schrödinger equation with fractional temporal evolution. Using suitable values of parameters, the 2- and 3-dimensional surfaces to some of the reported solutions are plotted. The reported solutions in this study may be useful in explaining the physical meaning of several nonlinear models arising in the different fields of nonlinear sciences. The extended ShGEEM provides a powerful mathematical tool for securing family of wave solutions to different kind of nonlinear models.

References

1. H. Bulut, T.A. Sulaiman, B. Demirdag, Nonlinear Dyn **91(3)**, 1985-1991 (2018)
2. S. Duran, M. Askin, T.A. Sulaiman, IJOCTA **7(3)**, 240-247 (2017)
3. H. Bulut, H.A. Isik, T.A. Sulaiman, ITM Web of Conferences **13**, 01019 (2017)
4. H. Bhrawy, M. A. Abdelkawy, A. Biswas. Indian J. Phys **87**,1125–1131 (2013)
5. H.M. Baskonus, H. Bulut, Waves in Rnd. and Comp. Med **26**, 189–196 (2016)
6. F. Özpinar, H. M. Baskonus and H. Bulut, Entropy **17(12)**, 8267–8277 (2015)
7. H.M. Baskonus, H. Bulut, Waves in Rnd. and Comp. Med **25(4)**, 720-728 (2015)
8. F. Belgacem, A.A. Karaballi, and S.L. Kalla, Math. Prob. Eng. **3-4**,103 (2003)
9. A. Yokus, H.M. Baskonus, T.A. Sulaiman and H. Bulut, Numer Methods Partial Differential Eq. **34**, 211-227 (2018)
10. A. Yokus, T.A. Sulaiman and H. Bulut, Opt Quant Electron **50**, 31 (2017)
11. H. Bulut, T.A. Sulaiman, H.M. Baskonus and A.A. Sandulyak, Optik **135**, 327-336 (2017)
12. H. Bulut, T.A. Sulaiman, H.M. Baskonus and T. Akturk, Opt Quant Electron **50**, 134 (2018)

13. N. Taghizadeh, M. Mirzazadeh, , International Journal of Applied Mathematics and Computation, **5(3)** 12-16 (2013)
14. H.M. Baskonus, T.A. Sulaiman, H. Bulut, T. Akturk, Superlattices Microstruct, **115** 19–29 (2018)
15. C. Cattani, T.A. Sulaiman, H.M. Baskonus and H. Bulut, Opt Quant Electron **50** 138, (2018)
16. C. Cattani, T.A. Sulaiman, H.M. Baskonus and H. Bulut, Eur. Phys. J. Plus **133**, 228 (2018)
17. H.M. Baskonus, T.A. Sulaiman and H. Bulut, Opt Quant Electron **50**, 253 (2018)
18. R. Khalil, M. Al Horani, A. Yousef, M. Sababheh, Journal of Computational and Applied Mathematics, **264** 65-70 (2014)

Analysis of Repairable k-out-of-n System Model Using Inter-Arrival Failure Times

Mehmet GÜRCAN [1] , *Yunus* GÜRAL [1], and *Gökhan* GÖKDERE [1,*]

[1]Department of Statistics, University of Fırat, Elazığ, Turkey

Abstract. Examination of repairable k-out-of-n system which has an important place in systems is an important application area of reliability theory. In general, the reliability analysis of a repairable system is based on the distribution of the lifetimes of the components in the system. In this paper, the reliability analysis of the repairable k-out-of-n system is made through the inter-arrival failure times of the components in the system.

1 Introduction

A working system is defined as a dynamic structure. When the system is observed the time of observation forms the section of the working system. In systems with dynamic structure, the most important problem is the moment when the system should be examined. This problem can be solved with the help of stochastic processes with continuous parameters. In general, a dynamic system is observed at a time when it changes its states. Since the system which is examined is a repairable system, it needs to be considered at failure times of the components.

As the system is examined at failure times of the components, the random variable that is important to us is the number of failed components in the repair. Let $X(t)$ denotes the number of failed components in the repair at time t, $t \geq 0$. Note that, the distribution of the time between successive failures of the components may be different from the exponential distribution. Therefore, $X(t)$ is not a Markov or semi-Markov process. However, a semi-Markov process can be obtained with the help of the $X(t)$, which represents the times of failure of the system. Let t_k represents the moment that the k-th failure takes place. At the end of the t_k, let $X(t_k - 0) = X_k$ denotes the number of failed components in the repair, excluding the last failed component. Then, the process $\{\xi(t), t \geq 0\}$ with $\xi(t) = X_k$, $t_k \leq t < t_{k+1}$ becomes a semi-Markov process. The state space of this process becomes $\{0,1,2, \dots, k\}$ in the repairable k-out-of-n system. Also, $t_1^k, t_2^k, \dots$ are the times of arrival to the case of k.

* Corresponding author: g.g.gokdere@gmail.com

Thus, in the process $\xi(t)$, the first failure time of the repairable k-out-of-n system and the elapsed time between successive failure times are reduced to the finding the distributions of the first arrival time and return time to the case of k. Note that, after the first failure of the system, it is being repaired and starting to work.
Assume that the repair time in the system is random variable with exponential distribution with parameter μ. Let, F be the cumulative distribution function of the times between successive failure times of the components. The function $a_r(s)$, $r = 0,1, \dots$ $s \geq 0$ is defined by

$$a_r(s) = \int_0^{\infty} \frac{(\mu t)^r}{r!} e^{-(s+\mu)t} dF(t).$$

Let $f(s)$ be the Laplace transformation of $F(x)$. Determination d_k is defined as

$$d_k = \det \begin{bmatrix} 1 & -a_0 & & & & \\ 1 & 1-a_1 & -a_0 & & & \\ \vdots & \vdots & \vdots & & & \\ 1 & -a_{k-2} & -a_{k-3} & \cdots & 1-a_1 & -a_0 \\ 1 & -a_{k-1} & -a_{k-2} & \cdots & -a_2 & 1-a_1 \end{bmatrix}$$

where

$$a_r = a_r(0) = \int_0^{\infty} \frac{(\mu t)^r}{r!} e^{-\mu t} dF(t).$$

Let T_{0k} be the first failure time of the system and T_{kk} be the time between successive failures. Then we have

$$ET_{0k} = \frac{d_k}{\alpha^k} m$$

$$ET_{kk} = \left(\frac{d_k}{\alpha^k} - \frac{d_{k-1}}{\alpha^{k-1}}\right) m,$$

where

$$m = -f'(0)$$

and

$$\alpha = f(\mu) = \int_0^{\infty} e^{-\mu t} dF(t).$$

Here, m is the average time between failures of the components and α is the probability that the repair time is greatest among the failures.

 https://doi.org/10.1051/itmconf/20182201054

2 Illustrative Example

As an illustration, consider a repairable 10-out-of-n system. Assume that the distribution of components is exponential distribution with the parameter $\mu = 0.025$. The distribution of the times between successive failures of the system is given in Table 1.

Table 1. The distribution of the times between successive failures of the system

No	$t_{k+1} - t_k$	$F(t)$
1	12	0.01
2	20	0.085
3	35	0.21
4	45	0.27
5	50	0.42
6	55	0.57
7	60	0.78
8	65	0.905
9	72	1

From the above table we can get the following results:

1. The mean of the times between successive failures is $m = 50.21$.

2. $d_{10} = 0.0188$, $d_9 = 0.0358$ and $\alpha = 0.2876$.

3. The mean time to first failure of the system is $ET_{0k} = 2.4 \times 10^5$.

4. The mean time between successive failures is $ET_{kk} = 1.1 \times 10^5$.

Conclusions

Studies related to the reliability of the systems are carried out using the distribution of the life time of the components. In this paper, we have studied repairable k-out-of-n system under the assumption that elapsed time between components is known. We have obtained two important results. In the first of the results, the mean time until the moment of failure can be calculated. In the second result, it is possible to calculate the mean running time between successive failures. These results are very important for preventive maintenance of the repairable k-out-of-n system. Also, the number of the components in the system has no importance since the distribution of the lifetime of the components are not consideration in the examination.

References

1. F.S.Q. Alves, H.C. Yehia, L.A.C. Pedrosa, F.R.B. Cruz and L. Kerbache, Upper bounds on performance measures of heterogeneous M/M/c queues, Mathematical Problems in Engineering, Article ID 702834 (2011).
2. J.B. Atkinson, Two new heuristics for the GI/G/n/0 queueing loss system with examples based on the two-phase Coxian distribution, Journal of the Operational Research Society, **60**, 818-830 (2009).
3. S.L. Brumelle, A generalization of Erlang's loss system to state dependent arrival and service rates, Mathematics of Operations Research, **3**, 10-16 (1978).
4. E. Çinlar, Markov renewal theory, Advances in Applied Probability, **12**, 123-187 (1969).
5. S. Halfin, Distribution of the interoverflow time for the GI/G/1 loss system, Mathematics of Operations Research, **6**, 563-570 (1981).
6. H.O. Isguder and C.C. Celikoglu, Minimizing the loss probability in GI/M/3/0 queueing system with ordered entry, Scientific Research and Essays, **7**, 963-968 (2012).
7. R. Pyke, Markov renewal processes: Definitions and preliminary properties, Annals of Mathematical Statistics, **32**, 1231-1242 (1961a).
8. R. Pyke, Markov renewal processes with finitely many states, Annals of Mathematical Statistics, **32**, 1243-1259 (1961b).
9. W.L. Smith, Regenerative stochastic processes, Proceedings of the Royal Society of London, Series A, Mathematical and Physical Sciences, **232**, 6-31 (1955).

A Survey on Photo Forgery Detection Methods

Bilgehan Gurunlu[1,*], *Serkan Ozturk*[2]

[1]Department of Informatics, Kahramanmaras Sutcu Imam University, K.Maras, Turkey

[2] Department of Computer Engineering, Erciyes University, Kayseri, Turkey

Abstract. In recent years, digital image forgery detection has become one of the hardest studying area for researchers investigations in the field of information security and image processing. Image forgery detection methods can be divided into two extensive groups such as Active methods and Passive (Blind) methods. Active methods have been used data hiding techniques like watermarking and digital signatures. Passive forensic methods (or Blind) use image statistics or they investigate the attributes of the image to determine the forgeries. Passive detection techniques are also split into three branches; image splicing, image retouching, copy-move. Such image forgery detection methods are focus of this paper.

1 Introduction

Widespread of the digital cameras and the image editing tools like Adobe Photoshop, Microsoft Paint that gives for people doctored images for the bad aims. Increasing in image forensics has given boosting to special techniques for the detection of changing image. Image Forgery Detection is a new studying area which goals confirm the authenticity of image by collected information their feature. Diverse methods have been evolved to handle with changing image and forgery because of provide the origin of the image. We will evaluate different groups of algorithm. Unfortunately, we have not been able to review some important study here.

Digital image forgery methods can be classified two main categories. One of them active methods and the other passive(blind) methods[2]. Active methods must be have pre-embedded information, such as digital watermarking and steganography. Digital watermarking is a method embedding secret information in the data, can be divided two categories, visible and invisible [1]. Steganography's example of digital signatures.

Passive methods can be divided five group.

- Pixel Based (copy-move, resampling, splicing, statistical)
- Format Based (JPEG quantization, double JPEG, JPEG blocking) [2]
- Camera Based (chromatic aberration,color filter array,camera response, sensor noise)
- Physics Based (light direction(2-D), light direction(3-D), light environment) [3]

* Corresponding author: gurunlu@gmail.com

- Geometric Based (principal point, metric measurements) [4]

We evaluated the study on copy-move (or cloning) in this survey. Copy-move forgery detection methods are following three groups.

- Brute Forces [5]
- Block-Base Techniques
- Keypoint Based Techniques

Brute force techniques is based exhaustive search and auto correlation methods [6]. Block based techniques use like this algorithms; DCT(Discrete Cosine Transform)[7], PCA(Principle Component Analysis) [8], SVD(Singular Value Decomposition)[8], DWT (Discrete Wavelet Transform)[10]. Keypoint based techniques use like this algorithms; SIFT(Scale Invariant feature transform), SURF(Speeded-up Robust features). An Digital forgery example of a newspaper report on Saddam and Bill Clinton photographs is shown in Figure 1.

INTERNET

Spies turn to high-tech info ops

PCs, Internet used for manipulating images, public opinion

Altered Egos
The top photo, of a fictitious meeting between President Clinton and Saddam Hussein, was generated by digitally altering and combining elements from three photos, shown below the altered image. Intelligence agencies plan to use similar techniques to create images and then disseminate them via the Internet in their efforts to influence events in foreign countries, such as Iraq.

BY DANIEL VERTON

Federal intelligence agencies are studying ways to use computers and the Internet, rather than just leaflets and radio broadcasts, to shape and disseminate information designed to sway public opinion in the world's hot spots.

As part of its so-called "perception management" program, the intelligence community has for decades created misinformation to trigger political change without direct political or military involvement in countries where the United States has vested interests, such as Iraq and North Korea.

Acting on congressional recommendations to bolster research and development in information technology, intelligence agencies are turning to PCs to develop more sophisticated means of manipulating and delivering digital photos, video clips and recorded sound to portray fictitious events in hopes of provoking desirable outcomes.

Fig.1. Digital Forgery Example [5]

2 Basic Steps of the Methods

Although there are a large number of proposed copy-move fraud detection methods, all the methods applies the general flow chart given in Figure 2. As seen in the flow chart, there are two alternatives techniques after pre-processing: block based methods and keypoint based methods. In both methods pre-processing steps are applied. For example; many methods studying on gray level images and so color channels must be combined.

Step 1. Image Pre-Processing; is the first step. Methods change some specific details on the image like image filtering, DCT coefficients, RGB to gray level before the feature extraction step [9].

Step 2. Feature Extraction; in the block based method the image is divided into sub-blocks in the form of a rectangle. For each sub-block, the feature vectors are computed. Then similar attribute vectors are matched. On the other hand, in feature based methods, only image regions with high entropy value are determined without any image subdivision. These regions are called "key points" and the feature vectors of these regions subtracted [10].

Step 3. Feature Matching; similarity between two feature is marked for a duplicated image. Some methods uses lexicographic sorting and others best-bin-first search(kd-tree algorithm) in determing the similar features vectors.

Step 4. Filtering; have been committed for decrease the possibility of the false positive matches. There are many distance algorithm with the near intensities. Some research proposed Euclidean distance[11], and someone correlation coefficient[12].

Step 5. Pro-Processing; this is last step of all the methods and its optional. The goal of this last step is to protect only those blocks(or keypoint) that share common feature. When considering a set of mappings for the region, it is expected that the source and target blocks(or keypoint) of these mappings are close to each other.

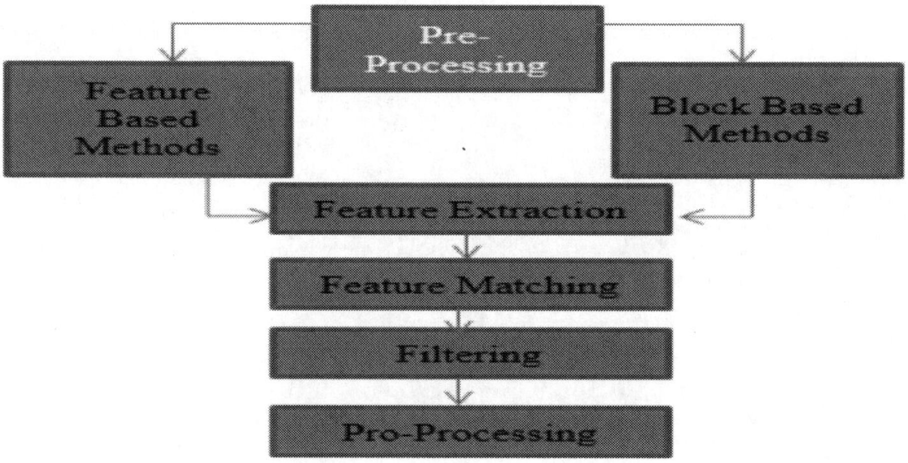

Fig.2. Image Forgery Detection Flow-Chart

3 Comparison of the Related Work

In this subdivision of manuscript, we supply some experimental solutions to describe the demonstration of the analytic algorithm suggested.

Author(s)	Techniques	Benefits	Drawbacks
[13]	Coefficient map and threshold. DWT and segmentation	A good result for diverse condition, copy-move and transformation	Execution time
[14]	Transform matrix. Segmentation	Give preferable result using SIFT	

[15]	Histogram of Gradient	High accurate ratio against to rotation, color reduction and contrast change attacks. Attribute size and calculation cost more effective	It needs to be developed and improved against large scale rotation, JPEG compression and scaling attacks
[16]	Binary DCT	Strong against contrast change, noises, JEG compression applied it also works on images at the same time	Like scaling and rotating weak against attacks
[17]	SVD and DCT	Low dimension. Multiple fraud detection. Robust to Gaussian blurring, JPEG compression and other operation.	Solely tested post-processing step
[18]	Fast Walsh-Hadamard Transform(FWHT)	Great correctness	Bad accuracy if the image transform
[19]	SURF and Nearest Neighbour Ratio(NNR)	Good to rotation, scaling and JPEG compression	Copy-move pixel are not localized
[20]	Convolution Neural Network (CNN)	Learned automatically	Copy-move pixel are not localized

4 Conclusion

In this paper, a brief survey of the image forgery detection(IFD) methods will be help the researcher in this studying area. We have mostly surveyed publication on IFD between 2013 and 2018. We provided a wide review of up-to-date copy move forgery methods. The methods are classifying two groups; like block base and keypoint based. Although both methods are superior to themselves, there have been more recent publications on block based methods last five years.

References

1. V. Aslantaş, Ş.Özer, S.Öztürk, "Improving the performance of DCT-based fragile watermarking using intelligent optimization algorithms", Optics Communications, vol.282, pp.2806-2817, (2009).
2. M. D. Ansari, S. P. Ghrera, V. Tyagi, "Pixel-Based Image Forgery Detection A Review", IETE Journal of Education, 55(1), 40-46, (2014).
3. H. Farid, "A survey of image forgery detection", IEEE Signal Processing Magazine, 26(2), 16-25, (2009).

4. M.K. Tohnson, H. Farid, "Metric measurements on a plane from single image" Dept. Comput.Sci., Darmouth College, Tech. Rep. TR2006-576, (2006).
5. J. Fridrich, D. Soukal, J. Luka, "Detection of copy-move forgery in digital iamges", Digital Forensic Research Workshop, pp.6-8, (2003).
6. N.K. Gill, R. Garg, A. Doger, "A review paper on digital image forgery detection techniques", 8th ICCCNT 2017, pp.1-7, (2017).
7. E. Ardizzone, A. Bruno, G. Mazzola," Copy-move forgery detection via texture description" ACM Workshop on multimedia in forensics, security and intelligence, pp.59-64, (2010).
8. B. Soloria, A. K. Nandi," Automated detection and localization of duplicated regions affected by reflection, rotation and scaling in image forensics", International Journal of Signal Processing, pp.1759-1770, (2011).
9. A. Makandar, B. Halalli," A review on preprocessing techniques for digital mammography images", International Journal of Computer Applications, pp.0875-887, (2015).
10. I. Amerini, L.Ballan, Caldelli, Caldelli, A. D. Bimbo, "A SIFT-based forensic method for copy-move attack detection and transformation recovery". IEEE Transactions on Information Forensics and Security, 6(3), 1099–1110, (2011).
11. S. Ryu, M. Lee, H. Lee, "Detection of Copy-Rotate-Move Forgery using Zernike Moments," in Information Hiding Conference, Jun. 2010, pp. 51–65. (2010).
12. S. Bravo-Solorio, A. K. Nandi, "Exposing Duplicated Regions Affected by Reflection, Rotation and Scaling," in International Conference on Acoustics, Speech and Signal Processing, May 2011, pp. 1880–1883. (2011).
13. C. M. Pun, X. C. Yuan, X.L. Bi. "Image Forgery Detection Using Adaptive Over-Segmentation and Feature Points Matching." IEEE, (2015).
14. L. Jian. "Segmentation-based Image Copy-move Forgery Detection Scheme." IET, (2015).
15. J.C. Lee, C.P.Chang, W.K. Chen. "Detection of copy–move image forgery using histogram of orientated gradients. Information Sciences, Informatics and Computer Science Intelligent Systems Applications, 3(21), 250-262, (2015).
16. S.Kumar, J. V. Desai, S. Mukherjee, "Copy Move Forgery Detection in Contrast Variant Environment using Binary DCT Vectors". International Journal of Image, Graphics and Signal Processing (IJIGSP), 7(6), 38-44, (2015).
17. J.Zhao, J.Guo, "Passive Forensics for Copy-Move Image Forgery Using a Method Based on DCT and SVD". ForensicSci.Int.233,158–166.(2015).
18. B.Yang, X.Sun, X.Chen, J.Zhang, X.Li, "An Efficient Forensic Method for Copy – Move Forgery Detection Based on DWT-FWHT", RadioEng.22,1098–1105. (2013).
19. E.Silva, T. Carvalho, A. Ferreira, A. Rocha, "Going Deeper into Copy-Move Forgery Detection: Exploring Image Tell tales via multi-scale analysis and voting processes. J Visual Commun Image Represent.;29:16–32, (2015).
20. J. Chen, X. Kang, Y. Liu, ZJ. Wang,"Median Filtering Forensics Based on Convolutional Neural Networks. IEEE Signal Process Lett. 22:1849–1853, (2015).

Investigation of Dark and Bright Soliton Solutions of Some Nonlinear Evolution Equations

Seyma Tuluce Demiray[1,*] and *Hasan Bulut*[1]

[1] Department of Mathematics, Fırat University, Elazığ, Turkey

Abstract. In this paper, generalized Kudryashov method (GKM) is used to find the exact solutions of (1+1) dimensional nonlinear Ostrovsky equation and (4+1) dimensional Fokas equation. Firstly, we get dark and bright soliton solutions of these equations using GKM. Then, we remark the results we found using this method.

1 Introduction

During the past years, soliton solutions are considerably important issue in biophysics, geophysical sciences, chemical kinematics, optical fibers, technology of space, electricity, elastic media and several topics in nonlinear sciences.

In recent years, most authors have presented several methods to find the soliton solutions of NLEEs such as (G'/G)-expansion method [1], exp-function method [2], the tanh method [3] and many more. In this work, GKM [4] will be used to find exact solutions of (1+1) dimensional nonlinear Ostrovsky equation.

Firstly, we consider the following (1+1) dimensional nonlinear Ostrovsky equation,

$$uu_{xxt} - u_x u_{xt} + u^2 u_t = 0. \tag{1}$$

Eq. (1) is a model for weakly nonlinear surface and internal waves in a rotating ocean. This equation has been introduced by Vakhnenko and Parkers [5]. They have found completely integrable of Eq. (1) by inverse scattering method [6]. Then, some authors have used various methods to find travelling wave solutions of this equation [7-11].

Secondly, we handle (4+1) dimensional Fokas equation [12-18],

$$4u_{tx} - u_{xxxy} + u_{xyyy} + 12u_x u_y + 12uu_{xy} - 6u_{zw} = 0. \tag{2}$$

This equation has been obtained by Fokas by expanding the Lax pairs of the integrable Kadomtsev–Petviashvili (KP) and Davey–Stewartson (DS) equations to some higher-dimensional nonlinear wave equations [12]. The importance of Eq. (1) suggests that the idea of complexifying time can be considered in the context of modern field theories via the existence of integrable nonlinear equations in four spatial dimensions involving complex time [12].

The layout of this paper is organized as follows. In Sec. 2, we give basic facts of GKM. In Sec. 3, we implement GKM to (1+1) dimensional nonlinear Ostrovsky equation

* Corresponding author: seymatuluce@gmail.com

and (4+1) dimensional Fokas equation. In Sec. 4, we report the results we obtained using this method.

2 Basic Facts of GKM

Step 1. We consider PDE as follows:

$$P\left(u, u_t, u_x, u_{xx}, u_{xxx}, \cdots\right) = 0. \tag{3}$$

Then, we get traveling transformation

$$u\left(x,t\right) = u\left(\xi\right), \xi = x - ct, \tag{4}$$

where c is arbitrary constant. Eq.(3) was reduced to a nonlinear ordinary differential equation:

$$N(u, u', u'', u''', \cdots) = 0, \tag{5}$$

where the prime denotes differentiation with respect to ξ.

Step 2. Suggest that the exact solutions of Eq.(5) can be found as follows;

$$u\left(\xi\right) = \frac{\sum_{i=0}^{N} a_i Q^i\left(\xi\right)}{\sum_{j=0}^{M} b_j Q^j\left(\xi\right)} = \frac{A\left[Q(\xi)\right]}{B\left[Q(\xi)\right]}, \tag{6}$$

where Q is $\frac{1}{1 \pm e^{\xi}}$. We highlight that the function Q is solution of Eq. (7)

$$Q_\xi = Q^2 - Q. \tag{7}$$

Taking Eq.(6) into account, we gain

$$u'\left(\xi\right) = \frac{A'Q'B - AB'Q'}{B^2} = Q'\left[\frac{A'B - AB'}{B^2}\right] = \left(Q^2 - Q\right)\left[\frac{A'B - AB'}{B^2}\right], \tag{8}$$

$$u''\left(\xi\right) = \frac{Q^2 - Q}{B^2}\left[\left(2Q-1\right)\left(A'B - AB'\right) + \frac{Q^2 - Q}{B}\left[B\left(A''B - AB''\right) - 2B'A'B + 2A\left(B'\right)^2\right]\right], \tag{9}$$

$$\begin{aligned} u'''\left(\xi\right) = {} & \left(Q^2 - Q\right)^3\left[\frac{\left(A'''B - AB''' - 3A''B' - 3B''A'\right)B + 6B\left(AB'' + B'A'\right)}{B^3} - \frac{6A\left(B'\right)^3}{B^4}\right] \\ & + 3\left(Q^2 - Q\right)^2\left(2Q - 1\right)\left[\frac{B\left(A''B - AB''\right) - 2B'A'B + 2A\left(B'\right)^2}{B^3}\right] \\ & + \left(Q^2 - Q\right)\left(6Q^2 - 6Q + 1\right)\left[\frac{A'B - AB'}{B^2}\right]. \end{aligned} \tag{10}$$

Step 3. The solution of Eq.(5) can be expressed as follows:

$$u\left(\xi\right) = \frac{a_0 + a_1 Q + a_2 Q^2 + \cdots + a_N Q^N + \cdots}{b_0 + b_1 Q + b_2 Q^2 + \cdots + b_M Q^M + \cdots}. \tag{11}$$

To compute the values M and N in Eq.(11) that is the pole order for the general solution of Eq.(5), we develop comparably as in the classical Kudryashov method on balancing the highest order nonlinear terms in Eq.(5) and we can establish a relation between M and N. We can find values of M and N.

Step 4. Substituting Eq.(6) into Eq.(5) ensures a polynomial $R(Q)$ of Q. Equating the coefficients of $R(Q)$ to zero, we get a system of algebraic equations. Solving this system, we can identify c and the coefficients $a_0, a_1, a_2, \cdots, a_N, b_0, b_1, b_2, \cdots, b_M$. Thus, we gain the exact solutions to Eq.(5).

3 Application of GKM to (1+1) Dimensional Nonlinear Ostrovsky Equation and (4+1) Dimensional Fokas Equation

In order to solve Eq. (1), we use the travelling wave transform

$$u(x,t) = u(\xi), \xi = x - ct, \tag{12}$$

where c is arbitrary constant.
Substituting Eq. (13) into Eq. (1),

$$u_t = -cu', \; u_x = u', u_{xt} = -cu'', u_{xxt} = -cu''', \tag{13}$$

we get the following equation

$$3uu'' - 3(u')^2 + u^3 = 0. \tag{14}$$

After Eq. (1) has been turned into Eq. (14), Substituting Eqs. (6) and (9) into Eq. (14) and balancing the highest order nonlinear terms of uu'' and u^3 in Eq. (14), then the following relation is acquired

$$2N - 2M + 2 = 3N - 3M \Rightarrow N = M + 2. \tag{15}$$

The exact solution of Eq.(1) is obtained as follows;

$$a_0 = 0, \; a_1 = -a_2 + 6b_1, \; a_3 = -6b_1, \; b_0 = -\frac{a_2}{6} + b_1, \tag{16}$$

Substituting Eq. (16) into Eq. (11), we get dark soliton solutions of Eq.(1)

$$u_1(x,t) = \frac{3}{2}\left(1 - \left[\tanh\left(\frac{x-ct}{2}\right)\right]^2\right), \tag{17}$$

$$u_2(x,t) = \frac{3}{2}\left(1 - \left[\coth\left(\frac{x-ct}{2}\right)\right]^2\right), \tag{18}$$

In an effort to find travelling wave solutions of the Eq. (2), we get the transformation via the wave variables

$$u(x,y,z,w,t) = u(\xi), \xi = \alpha x + \beta y + \gamma z + \delta w + \varepsilon t, \tag{19}$$

where $\alpha, \beta, \gamma, \delta$ and ε are arbitrary constants.

Substituting Eq. (20) into Eq. (2),

$$\begin{aligned} u_x &= \alpha u', u_y = \beta u', u_{tx} = \alpha\varepsilon u'', \; u_{xy} = \alpha\beta u'', \\ u_{zw} &= \gamma\delta u'', u_{xxxy} = \alpha^3 \beta u^{(4)}, u_{xyyy} = \alpha\beta^3 u^{(4)}, \end{aligned} \tag{20}$$

we get the following equation

$$(\alpha\beta^3 - \alpha^3\beta)u'' + (4\alpha\varepsilon - 6\gamma\delta)u + 6\alpha\beta u^2 = 0. \tag{21}$$

After Eq. (2) has been turned into Eq. (21), Substituting Eqs. (6) and (9) into Eq. (21) and balancing the highest order nonlinear terms between u'' and u^2 in Eq. (21), then the following relation is acquired

$$N - M + 2 = 2N - 2M \Rightarrow N = M + 2. \tag{22}$$

The exact solution of Eq.(2) is obtained as follows;

Case 1.

$$\begin{gathered} a_0 = 0,\ a_1 = \left(-\alpha^2 + \beta^2\right)b_0, a_2 = \left(\alpha^2 - \beta^2\right)\left(b_0 - b_1\right), \\ a_3 = \left(\alpha^2 - \beta^2\right)b_1,\ \varepsilon = \frac{\alpha\beta^3 - \alpha^3\beta + 6\gamma\delta}{4\alpha}, \end{gathered} \tag{23}$$

Inserting Eq. (23) into Eq. (11), we get bright soliton solutions of Eq.(2)

$$u_1\left(x,y,z,w,t\right) = \frac{\left(-\alpha^2 + \beta^2\right)}{4}\left[\sec h^2\left(\frac{\alpha x + \beta y + \gamma z + \delta w + \varepsilon t}{2}\right)\right], \tag{24}$$

$$u_2\left(x,y,z,w,t\right) = \frac{\left(\alpha^2 - \beta^2\right)}{4}\left[\csc h^2\left(\frac{\alpha x + \beta y + \gamma z + \delta w + \varepsilon t}{2}\right)\right]. \tag{25}$$

Case 2.

$$\begin{gathered} a_0 = \frac{1}{6}\left(\alpha^2 - \beta^2\right)b_0,\ a_1 = \frac{1}{6}\left(\alpha^2 - \beta^2\right)\left(-6b_0 + b_1\right), a_2 = \left(\alpha^2 - \beta^2\right)\left(b_0 - b_1\right), \\ a_3 = \left(\alpha^2 - \beta^2\right)b_1,\ \varepsilon = \frac{-\alpha^3\beta + \alpha\beta^3 + 6\gamma\delta}{4\alpha}, \end{gathered} \tag{26}$$

Embedding Eq. (26) into Eq. (11) , we get dark soliton solution of Eq.(2)

$$u_3\left(x,y,z,w,t\right) = \frac{\left(\alpha^2 - \beta^2\right)}{6}\left(-\frac{1}{2} + \frac{3}{2}\left[\tanh\left(\frac{\alpha x + \beta y + \gamma z + \delta w + \varepsilon t}{2}\right)\right]^2\right), \tag{27}$$

$$u_4\left(x,y,z,w,t\right) = \frac{\left(\alpha^2 - \beta^2\right)}{6}\left(-\frac{1}{2} + \frac{3}{2}\left[\coth\left(\frac{\alpha x + \beta y + \gamma z + \delta w + \varepsilon t}{2}\right)\right]^2\right). \tag{28}$$

Remark. If we compare the exact solutions of Eq. (1) and Eq. (2) reported by the other authors, our solution (17) is a similar solution with the solution (4.15) in [7]. Also, our solution (24) is a similar solution with the solution (10) in [15]. To our knowledge, other solutions of Eq. (1) we reported here are new and are not trackable in the former literature.

4 Conclusion

In this paper, we gain dark soliton solutions of (1+1) dimensional nonlinear Ostrovsky equation and (4+1) dimensional Fokas equation using GKM. Applications show that this method can successfully solve this equation. Thus, we can conclude that not only this method has a significiant role in observing NLEEs but also it is highly strong with regard to yielding analytical solutions of NLEEs.

References

1. M. Wang, X. Li, J. Zhang, Phys. Lett. A, **372,** 417-423 (2008).
2. J. H. He, X. H. Wu, Chaos Soliton. Fract., **30,** 700-708 (2006).
3. E. Fan, Phys. Lett. A, **277,** 212-218 (2000).
4. S. Tuluce Demiray, H. Bulut, Opt. Quant. Electron., **49(117)**, 1-8 (2017).
5. V. O. Vakhnenko, E. J. Parkers, Nonlinearity, **11,** 1457-1464 (1998).
6. V. O. Vakhnenko, E. J. Parkers, Chaos Soliton. Fract., **13,** 1819-1826 (2003).
7. E. Yusufoglu, A. Bekir, Appl. Math. Comput., **186,** 256-260 (2007).
8. C. Köroğlu, T. Öziş, Computers and Mathematics with Applications, **58,** 2142-2146 (2009).
9. E. M. E. Zayed, S. Al-Joudi, International Journal of Nonlinear Science, **8(4),** 435-447 (2009).
10. J. Lee, R. Sakthivel, Commun. Math. Sci., **7(4),** 1053-1062 (2009).
11. S. Marzan, F. Farhana, Md. T. Ahmed, K. Khan, M. A. Akbar, Global Journal of Scence Frontier Research (F), **13(9),** 45-53 (2013).
12. A. S. Fokas, Phys. Rev. Lett., **96(19),** Article ID:190201 (2006).
13. Z. Z. Yang, Z. Y. Yan, Commun. Theor. Phys. (Beijing,) **51(5),** 876-880 (2009).
14. J. Lee, R. Sakthivel, L. Wazzan, Modern Phys. Lett. B, **24(10),** 1011-1021 (2010).
15. H. Kim, R. Sakthivel, Rep. Math. Phys., **70(1),** 39-50 (2012).
16. Y. He, Math. Probl. Eng., **2014,** Article ID: 972519, 11 pages (2014).
17. S. Zhang, C. Tian, W. Y. Qian, Pramana J. Phys., **86(6),** 1259-1267 (2016).
18. M. O. Al-Amr, S. El-Ganaini, Comput. Math. Appl., **74**, 1274-1287 (2017).

Dynamic Reliability Evaluation of Linear Consecutive k-out-of-n: F System With Multi-State Components

Gökhan GÖKDERE [1,*] and *Mehmet* GÜRCAN [1]

[1]Department of Statistics, University of Fırat, Elazığ, Turkey

Abstract. In engineering applications, analyzing a technical system vary according to the operating principles of the system. In some situations, the status of the system is a function of stresses which act on the system and cause degradation. In order to efficiently analysis the reliability of a system which operates under stress, assigning the various states to the components depending on their operating performance is very important. In this paper, we have investigated the linear consecutive k-out-of-n: F system and assigned multiple states to its components. Due to the reason, the operating performance of the components can easily be controlled. Apart from that the reliability of the system depending on the states of its components can be calculated at any time interval. In the numerical example, the states of the components and the reliability calculation of the system at specific time intervals are shown clearly.

1 Introduction

A linear consecutive k-out-of-n:F (Lin/Con/k/n:F) system consists of n linearly arranged components such that the system fails if and only if at least k consecutive components fail. The first report for the Lin/Con/k/n:F system was presented by [1].

For reliability analysis and engineering applications, depending on the operating environment of the system, it may be exposed to some stresses and the system may respond to these strength. In this process, which is called the stress-strength model in the literature, the reliability of the system is the probability that the system is strong enough to counteract the stresses applied on it. The stress-strength model remains important since it is a viable model in engineering application, and many studies have been done in the literature on the analysis of systems and components using this model. [2, 3, 4, 5].

* Corresponding author: g.g.gokdere@gmail.com

Theoretically, in the stress-strength model, the stress and strength concepts are considered as random variables that do not change with time. However, in engineering applications, the status of the stress and strength variables may improve or deteriorate over time. As a result, the status of these variables may change over time. Therefore, it will be more meaningful to consider the stress-strength model depending on time. Regarding this issue, [6] defined the random lifetime, T, of the system as

$$T = \inf\{t: t \geq 0, Y(t) \leq X(t)\}, \tag{1}$$

where $X(t)$ and $Y(t)$ are the stress and strength random variables at time t, respectively. From (1) and for $(0, t_0]$, the reliability of the system is $R(t_0) = P(T > t_0)$.
Depending on the location of the system and external factors, the system may be exposed to more than one stress. Although the system is not completely degraded as a result of the stresses, there may be various degradation levels in system performance. Such a situation corresponds to a multi-state system [5]. In general, two different system structures are studied, namely binary system and multi-state system. In a binary system, the definition domains of the states of the system and its components are $\{0,1\}$. In a multi-state system, the definition domains of the states of the system and its components are $\{0,1,2,\dots,M\}$. In engineering applications, the multi-state system has a more flexible structure than the binary system when the binary system and the multi-state system are compared. In literature, much attention has been paid to multi-state system modeling [7, 8, 9, 10, 11].
The failure rate $\lambda(t)$ which is corresponds to the absolutely continuous cumulative distribution function $F(t)$, is represented by

$$\lambda(t) = \frac{f(t)}{\bar{F}(t)}, \tag{2}$$

and defines the probability that an operating component will fail in the next sufficiently small unit interval of time. The failure rate plays an exceptional role in reliability engineering that mostly deals with positive random variables [12].
Clearly, as $F(t)$ is absolutely continuous, the probability distribution function $f(t) = \frac{d}{dt}F(t)$ exists almost everywhere. Then, from (2), we have main exponential formula

$$F(t) = 1 - \exp\left(-\int_0^t \lambda(u)du\right), \tag{3}$$

where $\lambda(t)$ is the corresponding failure rate. Equation (3) shows that the failure rate $\lambda(t)$ uniquely defines the absolutely continuous cdf $F(t)$.
In this paper, we aim to propose a new method for computing the dynamic reliability of the Lin/Con/k/n: F system depending on the states of its components. A method is presented for the case in which the system consists of n independent components whose deteriorating strengths are independent and identically distributed and these components are subjected to two random stresses.

2 Proposed Method

Suppose that the Lin/Con/k/n:F system consists of n independent components whose deteriorating strengths $Y_i(t) \geq 0$, $i = 1, 2, \dots, n$ are stochastically decreasing in time and independent and identically with continuous cumulative distribution function $G_{i(t)}(x) = P\{Y(t) \leq x\}$. Also assume that, these components are subjected to two types of stresses X_l, $l = 1, 2$ and stresses remain fixed over time with cdf $F_l(x) = P\{X_l \leq x\}$. Let $EY_i(t)$ and EX_l are the finite mean of $Y_i(t)$ and X_l, respectively.

Using the finite mean of $Y_i(t)$ and X_l the operation states of the components, depending on the number of stresses, can be defined as follows

$$S_i(t):\begin{cases}3, & EX_1+EX_2<EY_i(t)\\ 2, & EX_2<EY_i(t)\\ 1, & EX_1<EY_i(t)\\ 0, & EY_i(t)<EX_1\end{cases},$$

where $S_i(t)$ is the state of the component i at time t and $EX_1 < EX_2$. The states 0 and 3 represent completely failed and perfectly working states, respectively and the others degraded states.

There is no doubt that the operation time in the possible states of the system will be reduced as time progresses, due to reduction in the operation performance of the components which work under stresses. This case causes a problem such that analyzing how to reduce of the operation performance of the system in the course of time. Hence, it is necessary considering as a dynamic structure of the system and also making examination at a certain time interval for the reliability of the system. Therefore, calculation of efficient operation performance of the system at a certain time periods is important.

Let, $Z_{i,j}(t)\geq 0$ denote the performance rate of the component i in state j at time t and $p_{i,j}(t)$ be the operating performance probability of the component i in state j at time t, $1\leq i\leq n$, $1\leq j\leq 3$. Then,

$$p_{i,j}(t)=P\{Z_{i,j}(t)>t+h|Z_{i,j}(t)>t\}=\frac{P\{Z_{i,j}(t)>t+h\}}{P\{Z_{i,j}(t)>t\}}=\frac{1-H_i(t+h)}{1-H_i(t)} \tag{4}$$

where $H_i(t)=P\{Z_{i,j}(t)\leq t\}$, $i=1,2,\dots,n$. Now using (3) in (4), we can obtain the dynamic operation performance probabilities as follows

$$p_{i,j}(t)=\exp\left(-\int_t^{t+h}\lambda_{i,j}(u)du\right). \tag{5}$$

where $\lambda_{i,j}(t)$ is the corresponding failure rate for the component i in state j at time t. The failure rate for the component i in state "3", "2" and "1" at time t is as follows, respectively.

$$\lambda_{i,3}(t)=\frac{1}{(EY_i(t)-EX_1-EX_2)+(EY_i(t)-EX_2)+(EY_i(t)-EX_1)}\psi_i^{(3)},$$

where $\psi_i^{(3)}$ is an indicator function such that

$$\psi_i^{(3)}=\begin{cases}1, & \text{if event } EY_i(t)-EX_1-EX_2>0 \text{ occurs at time } t\\ 0, & \text{if event } EY_i(t)-EX_1-EX_2>0 \text{ does not occurs at time } t.\end{cases}$$

$$\lambda_{i,2}(t)=\frac{1}{(EY_i(t)-EX_2)+(EY_i(t)-EX_1)}\psi_i^{(2)},$$

where $\psi_i^{(2)}$ is an indicator function such that

$$\psi_i^{(2)}=\begin{cases}1, & \text{if event } EY_i(t)-EX_2>0 \text{ occurs at time } t\\ 0, & \text{if event } EY_i(t)-EX_2>0 \text{ does not occurs at time } t.\end{cases}$$

and

$$\lambda_{i,1}(t)=\frac{1}{(EY_i(t)-EX_1)}\psi_i^{(1)},$$

where $\psi_i^{(1)}$ is an indicator function such that

$$\psi_i^{(1)}=\begin{cases}1, & \text{if event } EY_i(t)-EX_1>0 \text{ occurs at time } t\\ 0, & \text{if event } EY_i(t)-EX_1>0 \text{ does not occurs at time } t.\end{cases}$$

3 Conclusions

The evaluation of the operating performance of the system which works under varies stresses is very important for investigation of the technical system. For this reason, the correct evaluation depending on number of stresses plays an important role for determining the reliability of the system. In case of more than one stress, the expected values of the stresses are arranged in descending order. At the beginning when the system starts working, the expected value of strength of each component is greater than sum of all stresses. As the process continuous, the expected value of the strength for each component reduces within the operating time until reaches the minimum expected stress. For each process, we have assigned a state. Finally using the states of the components, the reliability of the system is determined successively without assigning any state to the system.

References

1. Kontoleon, J. M. (1980). Reliability determination of a r-successive-out-of-n:F system. *IEEE Trans. Reliab.*, 29, 437.
2. Chandra, S., & Owen, D. B. (1975). On estimating the reliability of a component subject to several different stresses (strengths). *Naval Res. Log. Quart.*, 22, 31-40.
3. Kotz, S., Lumelskii, Y., & Pensky, M. (2003). *The Stress-Strength Model and its Generalizations. Theory and Applications*. Singapore: World Scientific.
4. Nadarajah, S., & Kotz, S. (2006). Reliability for some bivariate exponential distributions. *Mathematical Problems in Engineering*, 2006, 1-14.
5. Eryılmaz, S., & İşçioğlu, F. (2011). Reliability evaluation for a multi-state system under stress-strength setup. *Commun. Statist. Theor. Meth.*, 40, 547-558.
6. Basu, A. P., & Ebrahimi, N. (1983). On the reliability of stochastic systems. *Statist. Probab. Lett.*, 1, 265-267.
7. El-Neweihi, E., Proschan, F. & Sethuraman, J. (1978). Multi-state coherent system. *Journal of Applied Probability*, 15, 675-688.
8. Hudson, J. C., & Kapur, K. C. (1983). Reliability analysis for multistate systems with multistate components. *IIE Transactions*, 15, 127-135.
9. Ebrahimi N. (1984). Multistate reliability models. *Naval Research Logistics Quarterly*, 31, 671-680.
10. Brunelle R. D., & Kapur, K. C. (1999). Review and classification of reliability measures for multistate and continuum models. *IIE Transactions*, 31, 1171-1180.
11. Kuo, W., & Zuo, M. J. (2003). *Optimal Reliability Modeling: Principles and Applications*. New York: John Wiley & Sons.
12. Finkelstein, M., & Cha, J. H. (2013). *Stochastic modelling for reliability. shocks, burn-in and heterogeneous populations*. London: Springer

(λ,μ) Statistical Convergence of Order γ for Double Sequences of Functions

Muhammed Çınar[1,*], *Dilan* Şeker[2], and *Mikail* Et[3]

[1,2]Department of Mathematic Education, University of Muş Alparslan, Muş, Turkey

[3] Department of Mathematics, University of Fırat, Elazığ, Turkey

Abstract. In this present work, we investigate the concepts of Δ^r – statistical convergence of order $\widetilde{\gamma}$ for double sequences of real valued functions. Also some definitions we given for Δ^r – statistical Cauchy sequence for double sequences of real valued functions and some relations between $S^{\widetilde{\gamma}}_{(\lambda,\mu)}(\Delta^r, f)$ statistical convergence and strong $\left[V^{\widetilde{\gamma}}, \lambda, \mu\right]_q (\Delta^r, f)-$ summability are given.

1 Introduction

Zygmund [20] defined statistical convergence in 1935. After that defined statistical convergence for sequences of real numbers was given by Steinhaus [19] and Fast [6] later reintroduced by Schoenberg [14] independently. Then, in terms of the sequence spaces is examined and correlated with the summability theory by Connor [9], Çinar et. al. [1], Çolak [2], Et et. al. ([3],[4],[15],[16]), Fridy [7], Mursaleen [12], and many others [8],[10],[13],[17],[18]). Statistical convergence is related to density of subsets of the set ℕ of natural numbers. The density of a subset D of ℕ is defined by

$$d(D) = \lim_{n\to\infty} \frac{1}{n}\sum_{k=1}^{n} \chi_D(K)$$

provided that the limit texists. x = (x_n) sequences of real numbers is statistically convergent to ξ if for every ε > 0, $d(\{n \in \mathrm{N} : |x_n - \xi| \geq \varepsilon\}) = 0$

Pringsheim[21] defined the convergence of double sequences in 1900 as follows :

$x = (x_{nm})$ be a double sequences , for every $\varepsilon > 0$ there exists $N \in \mathbb{N}$ such that $|x_{nm} - \xi| < \varepsilon$ whenever $n, m > N$. In this situation we write $P - \lim x = \xi$.

---

* Corresponding author: muhammedcinar23@gmail.com

A double sequence $x=(x_{nm})_{n,m=0}^{\infty}$ can be defined as bounded if there exists a positive real number T such that $|x_{nm}|<T$ for all n and m , i.e. $\|x\|=\sup_{n,m\geq 0}|x_{nm}|<\infty$.

Let $K\subseteq \mathbb{N}x\,\mathbb{N}$ and $K(r,s)=\{(n,m)\;:\;n\leq r, m\leq s\}$. The double natural density of K is defined by

$$d^2(K)=P-\lim_{r,s}\frac{1}{rs}|K(r,s)|,\ \text{if the limit exists.}\ [12]$$

The idea of difference sequences was given Kızmaz [11] and were generalized by Et and Çolak [5].

Throught this work unless said otherwise by for all $n,m\in \mathrm{N}_{n_0}$ we mean for all $n,m\in \mathrm{N}_{n_0}$ except finite numbers of positive integers where $\mathrm{N}_{n_0}=\{n_0,n_0+1,n_0+2,...\}$ for some $n_0\in\mathbb{N}=\{1,2,3,...\}$

2 Main Result

Definition 1 Let $\widetilde{\gamma}\in(0,1]$ be given. A double sequence of functions $\{f_{kl}\}$ is said to be $\Delta^r-(\lambda,\mu)$ statistically convergent of order $\widetilde{\gamma}$ to the function f on a set D if for every $\varepsilon>0$ and for every $x\in D$

$$\lim_{uv}\frac{1}{\lambda_u^s\mu_v^t}\left|\left\{(k,l)\in I_u x J_v\;:\;\left|\Delta^r f_{kl}(x)-f(x)\right|\geq\varepsilon\right\}\right|=0.$$

where $I_u=[u-\lambda_u+1,u], J_v=[v-\mu_v+1,v]$ Then we can write $S_{(\lambda,\mu)}^{\widetilde{\gamma}}-\lim\Delta^r f_{kl}(x)=f(x)$ on D. The function f is $\Delta^r-(\lambda,\mu)$ statistical limit of order $\widetilde{\gamma}$ of the sequence $\{f_{kl}\}$. The set of all $\Delta^r-(\lambda,\mu)$ statistically convergent sequences of functions order $\widetilde{\gamma}$ will be denoted by $S_{(\lambda,\mu)}^{\widetilde{\gamma}}(\Delta^r,f)$ Then we write shortly $\Delta^r f_{kl}(x)\xrightarrow{S_{(\lambda,\mu)}^{\widetilde{\gamma}}}f(x)$.

$\Delta^r-\ (\lambda,\mu)$ statistical convergence of order $\widetilde{\gamma}$ is well defined for $\widetilde{\gamma}\in(0,1]$, but is not well defined for $\widetilde{\gamma}\succ 1$. $\{f_{kl}\}$ sequence of functions defined as follows

$$f_{kl}(x)=\begin{cases}1 & k+l=2n\\ x^{k+l} & k+l\neq 2n\end{cases}\qquad n=1,2,3...,x\in\left[0,\tfrac{1}{2}\right]$$

For $\widetilde{\gamma}\succ 1$, $S_{(\lambda,\mu)}^{\widetilde{\gamma}}-\lim\Delta f_{kl}(x)=2$ and $S_{(\lambda,\mu)}^{\widetilde{\gamma}}-\lim\Delta f_{kl}(x)=-2$ which is impossible.

Theorem 2 Let $\{f_{kl}\}$ and $\{g_{kl}\}$ are two double sequences of real valued functions defined on a set D and $\widetilde{\gamma}\in(0,1]$.

(i) If $\Delta^r f_{kl}(x) \xrightarrow{S^{\tilde{\gamma}}_{(\lambda,\mu)}} f(x)$ and $c \in \mathrm{R}$, then $c\Delta^r f_{kl}(x) \xrightarrow{S^{\tilde{\gamma}}_{(\lambda,\mu)}} cf(x)$,

(ii) If $\Delta^r f_{kl}(x) \xrightarrow{S^{\tilde{\gamma}}_{(\lambda,\mu)}} f(x)$ and $\Delta^r g_{kl}(x) \xrightarrow{S^{\tilde{\gamma}}_{(\lambda,\mu)}} g(x)$, then $\Delta^r f_{kl}(x) + \Delta^r g_{kl}(x) \xrightarrow{S^{\tilde{\gamma}}_{(\lambda,\mu)}} f(x) + g(x)$.

Theorem 3 $\tilde{\gamma}, \tilde{\beta} \in (0,1]$ are given such that $\tilde{\gamma} \le \tilde{\beta}$ then $S^{\tilde{\gamma}}_{(\lambda,\mu)}(\Delta^r, f) \subseteq S^{\tilde{\beta}}_{(\lambda,\mu)}(\Delta^r, f)$ which is proper.

Proof. It's easy to show that inclusion is provided. To show that the inclusion is strict define a double sequence $\{f_{kl}\}$ by

$$f_{kl}(x) = \begin{cases} 1 & k,l = n^2 \\ \frac{k^2 l^2 x}{1+k^3 l^3 x^2} & k,l \ne n^2 \end{cases}.$$

Then $S^{\tilde{\beta}}_{(\lambda,\mu)} - \lim \Delta f_{jk}(x) = 0$ for $\tilde{\beta} \in (\frac{1}{2}, 1]$, but $f(x) \notin S^{\tilde{\gamma}}_{(\lambda,\mu)}(\Delta, f)$ for $\tilde{\gamma} \in (0, \frac{1}{2}]$.

Corollary 4 If $\{f_{kl}\} \in S^{\tilde{\gamma}}_{(\lambda,\mu)}(\Delta^r, f)$, then $\{f_{kl}\} \in S_{(\lambda,\mu)}(\Delta^r, f)$

Definition 5 Let $\tilde{\gamma} \in (0,1]$. The sequence $\{\Delta^r f_{kl}\}$ is a (λ,μ) statistically Cauchy sequence of order $\tilde{\gamma}$ provided that for every $\varepsilon > 0$ there are two numbers $N(=N(\varepsilon)), M(=M(\varepsilon))$ such that

$$\left|\Delta^r f_{kl}(x) - \Delta^r f_{NM}(x)\right| < \varepsilon \quad a.a.(k,l)\ (\tilde{\gamma}) \text{ and for each } x \in D$$

i.e.

$$\lim_{uv} \frac{1}{\lambda_u^s \mu_v^t} \left|\left\{(k,l) \in I_u x J_v \ : \ \left|\Delta^r f_{kl}(x) - \Delta^r f_{NM}(x)\right| \ge \varepsilon \right\}\right| = 0.$$

for each $x \in D$.

Definition 6 Suppose that $\tilde{\gamma} \in (0,1]$ and u is a positive real number. A double sequence of functions $\{\Delta^r f_{kl}\}$ is strongly $(V^{\tilde{\gamma}}, \lambda, \mu)_q$ summable of order $\tilde{\gamma}$ (or $[V^{\tilde{\alpha}}, \lambda, \mu]_q (\Delta^r, f)-$ summable) if there is a function f such that

$$\lim_{uv\to\infty} \frac{1}{\lambda_u^s \mu_v^t} \sum_{k \in I_u} \sum_{l \in J_v} \left|\Delta^r f_{kl}(x) - f(x)\right|^q = 0.$$

Thus it can be written $[V^{\tilde{\gamma}}, \lambda, \mu]_q - \lim \Delta^r f_{kl}(x) = f(x)$ on D. The set of all strongly $(V^{\tilde{\gamma}}, \lambda, \mu)_q (\Delta^r, f)-$ summable double sequences of functions of order $\tilde{\gamma}$ is shown by $[V^{\tilde{\gamma}}, \lambda, \mu]_q (\Delta^r, f)$. We can write shortly $\Delta^r f_{kl}(x) \xrightarrow{[V^{\tilde{\gamma}}, \lambda, \mu]_q} f(x)$.

Teorem 7 Assume that q is a positive real number and $\widetilde{\gamma},\widetilde{\beta}\in(0,1]$ such that $\widetilde{\gamma}\le\widetilde{\beta}$. Then $[V^{\widetilde{\gamma}},\lambda,\mu]_q(\Delta^r,f)\subseteq[V^{\widetilde{\beta}},\lambda,\mu]_q(\Delta^r,f)$ is proper for some $\widetilde{\gamma}=(s,t)$ and $\widetilde{\gamma}=(u,v)$ such that $\widetilde{\gamma}\prec\widetilde{\beta},\ s,t,u,v\in R$.

Proof. It's easy to show that inclusion is provided.. To show that the inclusion is strict define a double sequence $\{f_{kl}\}$ by

$$f_{kl}(x)=\begin{cases}\frac{klx}{1+klx} & k,l=n^2\\ 0 & k,l\neq n^2\end{cases}\quad x\in[1,2]$$

the sequence $\{\Delta f_{kl}\}$ is strongly $(V^{\widetilde{\beta}},\lambda,\mu)_q$ summable of order $\widetilde{\beta}$, for $\widetilde{\beta}\in(\frac{1}{2},1]$, the sequence $\{\Delta f_{kl}\}$ is not strongly $(V^{\widetilde{\gamma}},\lambda,\mu)_q$ summable of order $\widetilde{\gamma}$, for $\widetilde{\gamma}\in(0,\frac{1}{2}]$ where $\lambda_u=u,\quad \mu_v=v$.

Corollary 8 Let $\widetilde{\gamma},\widetilde{\beta}\in(0,1]$ and q be a positive real number. Then

(i) if $\widetilde{\gamma}\cong\widetilde{\beta}$, then $[V^{\widetilde{\gamma}},\lambda,\mu]_q(\Delta^r,f)\subseteq[V^{\widetilde{\beta}},\lambda,\mu]_q(\Delta^r,f)$

(ii) $[V^{\widetilde{\gamma}},\lambda,\mu]_q(\Delta^r,f)\subseteq[V,\lambda,\mu]_q(\Delta^r,f)$ for each $\widetilde{\gamma}\in(0,1]$ and $0<u<\infty$.

Theorem 9 Let $\widetilde{\gamma},\widetilde{\beta}\in(0,1]$ such that $\widetilde{\gamma}\le\widetilde{\beta}$ and u be a positive real number. If a double sequence of functions $\{f_{kl}\}$ is in the $[V^{\widetilde{\gamma}},\lambda,\mu]_q(\Delta^r,f)$ then $\{f_{kl}\}$ is also in the $S^{\widetilde{\beta}}_{(\lambda,\mu)}(\Delta^r,f)$.

Corollary 10 Let $\widetilde{\gamma}\in(0,1]$ be given and and u be a positive real number. If $\{f_{kl}\}\in[V^{\widetilde{\gamma}},\lambda,\mu]_q(\Delta^r,f)$, then $\{f_{kl}\}\in S^{\widetilde{\gamma}}_{(\lambda,\mu)}(\Delta^r,f)$.

Theorem 11 Let $\lambda=(\lambda_u)$, $\mu=(\mu_v)$ $\sigma=(\sigma_u),\rho=(\rho_v)$ be four sequences in Λ such that $\lambda_u\le\sigma_u$, $\mu_v\le\rho_v$ for all $u,v\in\mathbb{N}_{n_o}$ $0<\widetilde{\gamma}\le\widetilde{\beta}\le1$ and $\{f_{kl}\}$ be a double sequence of real valued functions defined on a set D.

(i) If

$$\liminf_{uv\to\infty}\frac{\lambda_u^{\widetilde{\alpha}}\mu_v^{\widetilde{\alpha}}}{\sigma_u^{\widetilde{\beta}}\rho_v^{\widetilde{\beta}}}>0$$

then $S^{\widetilde{\beta}}_{(\lambda\mu)}(\Delta^r,f)\subseteq S^{\widetilde{\gamma}}_{(\sigma,\rho)}(\Delta^r,f)$,

(ii) If

$$\lim_{u\to\infty}\frac{\lambda_u^{\tilde{\alpha}}}{\sigma_u^{\tilde{\beta}}}=1,\lim_{v\to\infty}\frac{\mu_v^{\tilde{\alpha}}}{\rho_v^{\tilde{\beta}}}=1 \quad and \quad \lim_{u\to\infty}\frac{\sigma_u}{\sigma_u^{\tilde{\beta}}}=1,\lim_{v\to\infty}\frac{\rho_v}{\rho_v^{\tilde{\beta}}}=1$$

then $S_{(\lambda\mu)}^{\tilde{\gamma}}(\Delta^r, f)\subseteq S_{(\sigma,\rho)}^{\tilde{\beta}}(\Delta^r, f)$

References

1. M. Çınar, M. Karakaş, M. Et, Fixed Point Theory Appl. 2013:33, 11 pp. (2013)
2. R. Çolak, New Delhi, India, Anamaya Pub. 121—129 (2010)
3. M. Et, M. Çınar, M. Karakaş, J. Inequal. Appl. 2013:204, 8 pp. (2013)
4. M. Et, H. Şengül, , Filomat **28(8)**, 1593-1602 (2014),
5. M. Et and R. Colak, Soochow J. Math., **21(4)**, 377-386 (1995),
6. H. Fast, Colloq. Math. **2** 241-244 (1951)
7. J. Fridy, Analysis, **5** ,301-313 (1985)
8. A. Gökhan, M. Güngör, , Indian J. Pure Appl. Math., **33(9)**, 1379-1384 (2002)
9. J. S. Connor, Analysis, **8**, 47-63 (1988)
10. O. Duman and C. Orhan, Czechoslovak Math. J., 54(2), 413-422 (2004)
11. H. Kızmaz, Canad. Math. Bull., **24(2)**, 169-176 (1981)
12. M. Mursaleen, Osama H. H. Edely, J. Math. Anal. Appl., **288(1)**, 223--231(2003)
13. A. Pringsheim, Zur,Math. Ann., **53(3)**, 289—321 (1900)
14. I. J. Schoenberg, Amer. Math. Monthly, **66** 361-375 (1959)
15. M. Et, M. Mursaleen, M. Işık, Filomat, **27(5)**, 789-796 (2013)
16. M. Et, A. Alotaibi, S.A. Mohiuddine, Scientific World Journal, (2014)
17. M. Karakaş, M.Et, V. Karakaya, Acta Mathematica Scientia, **33(6)**, 1711-1720(2013)
18. H.M. Srivastava, M. Et, Fixed Point Theory and Applications, **31(6)**, 1573-1582 (2017)
19. H. Steinhaus, Sur la convergence ordinaire et la convergence asymptotique, Colloquium Mathematicum 2 (1951), 73-74.
20. A. Zygmund, Trigonometric Series, Cambridge University Press, Cambridge, UK, 1979
21. A. Pringsheim,. (German) *Math. Ann.* **53 (3)**, 289—321, 1900

On rough convergence variables of triple sequences

M. Kemal Ozdemir[1,*], *Ayhan* Esi[2,**], and *Ayten* Esi[2,***]

[1]Department of Mathematics, İnönü University, 44280, Malatya, Turkey
[2]Department of Mathematics, Adıyaman University, 02040, Adıyaman, Turkey

Abstract. Triple sequence convergence has an extremly important position in the basic theory of mathematics. The present manuscript contains four types of convergence concept of convergence almost surely, convergence incredibility, trust convergence in mean and convergence in distribution and discuss the relation ship among those and some mathematical properties of those new convergence.

1 Introduction

Initially Phu [8–10] has presented the concept for the rough convergence in a finite dimensional normed spaces. He also illustrated the fact that this set LIM_x^r is closed, bounded and convex; at the same time presented the concept for the rough Cauchy sequence. He again studied the connections among rough convergence and the other types of convergence and also the dependence for LIM_x^r on the roughness having degree of r.

Aytar [1] investigated rough statistical convergence and gave the definition for the set of rough statistical limit points of a sequence and also got two important statistical convergence criteria related to this set and show that it is convex and closed set. Furthermore, Aytar [2] examined that this r-limit set of the sequence equals to intersection of those sets and also that r-core of the sequence equals to the union of those sets. Dundar and Cakan [5] studied about rough ideal convergence and defined the set of rough ideal limit points for a given sequence.

In the present manuscript, we present the concept of rough convergence and the set of rough limit points of a triple sequence and get two rough convergence criteria related to this set, also show that the set is closed and convex and investigated the relations among the set of accumulation points and the set of rough limit points of a triple sequence.

A triple sequence (either complex or real) can also be described via a function defined as $x : \mathbb{N}^3 \to \mathbb{R}(\mathbb{C})$, in which the natural, real and complex numbers are denoted by $\mathbb{N}$, $\mathbb{R}$ and $\mathbb{C}$, respectively. The various types of concepts for triple sequence has been presented and examined initially by *Datta et al. [3]*, *Debnath et al. [4]*, *Esi et al. [6, 7]*, *Sahiner et al. [11]* and several others.

Consider (x_{mnk}) is a triple sequence of rough variables. In this paper we discuss some convergence notions of rough triple sequence: convergence almost surely (a.s.), convergence in credibility, trust convergence in mean and also in distribution and some mathematical properties among them.

*e-mail: kozdemir73@gmail.com
**e-mail: aesi23@hotmail.com
***e-mail: aytenesi@yahoo.com

2 Definitions and Preliminaries

Definition 2.1 *It is said that a triple sequence $x = (x_{mnk})$ is rough convergent (r-convergent) to l, denoted as $x_{mnk} \to^r l$, under the assumption that*

$$\forall \epsilon > 0, \exists\, i_\epsilon \in \mathbb{N} : m, n, k \geq i_\epsilon \Longrightarrow |x_{mnk} - l| < r + \epsilon, \tag{2.1}$$

or equivalently, if

$$\limsup |x_{mnk} - l| \leq r. \tag{2.2}$$

where r is known as the roughness of degree. When $r = 0$ is taken, the ordinary convergence of a triple sequence is obtained.

Definition 2.2 *Let a triple sequence of rough variables (x_{mnk}) is said to be convergent almost surely to the rough variable l iff there is a set $A \in \mathbb{A}$ with $Tr(A) = 1$ such that $\lim_{uvw \to \infty} |x_{mnk}(\lambda) - l(\lambda)| = 0$ for every $\lambda \in A$. Under these conditions, we can write $x_{mnk} \to^{a.s.} l$.*

Definition 2.3 *Let us assume that r is a real number which is non-negative and (x_{mnk}) be a triple sequence of rough variables. We say that the triple sequence (x_{mnk}) converges in trust to the rough variable l if $\lim_{mnk \to \infty} Tr\{|x_{mnk} - l| \geq r + \epsilon\} = 0$ for every $\epsilon > 0$. Under these conditions, we can also write $(x_{mnk}) \to_r^{Trv} l$.*

Definition 2.4 *Let r be any real number which is non-negative and (x_{mnk}) be a triple sequence of rough variables having finite expected values, the triple sequence (x_{mnk}) converges in mean to the rough variable l if $\lim_{mnk \to \infty} E[|x_{mnk} - l|] = 0$. Under these conditions, we can write $(x_{mnk}) \to^E l$.*

Definition 2.5 *Let us assume that $\Phi, \Phi_1, \Phi_2, \ldots,$ are the trust distributions of rough variables $l, l_1, l_2, \ldots,$ respectively. If the triple sequence (Φ_{mnk}) converges weakly to Φ, then we say that $(x_{mnk}) \to^{Dis} l$.*

3 Main Results

Theorem 3.1 *Let us r be a real number which is non-negative and (x_{mnk}) be a triple sequence of rough variables, then $(x_{mnk}) \to^{a.s.} l \Longleftrightarrow$ for every $\epsilon > 0$, we have*

$$\lim_{uvw \to \infty} Tr \left\{ \bigcup_{(m,n,k)=(u,v,w)} \{|x_{mnk} - l| \geq r + \epsilon\} \right\} = 0$$

Proof. For every $(m, n, k) \geq 1$ and $\epsilon > 0$, we describe

$$X = \left\{ \lambda \in \Lambda : \lim_{mnk \to \infty} x_{mnk}(\lambda) \neq l(\lambda) \right\},\ X = \{\lambda \in \Lambda : |x_{mnk}(\lambda) - l(\lambda)| \geq r + \epsilon\}.$$

It is clear that

$$X = \bigcup_{\epsilon > 0} \left(\bigcap_{(u,v,w)=1}^{\infty} \bigcup_{(m,n,k)=(u,v,w)}^{\infty} X_{mnk}(r + \epsilon) \right).$$

Note that $x_{mnk} \to_r^{a.s.} l$, if and only if $Tr(X) = 0$, i.e.,

$$x_{mnk} \to_r^{a.s.} l \Longleftrightarrow Tr \left\{ \bigcap_{(u,v,w)=1}^{\infty} \bigcup_{(m,n,k)=(u,v,w)}^{\infty} X_{mnk}(r + \epsilon) \right\} = 0$$

for every $\epsilon > 0$. Since,

$$\bigcup_{(m,n,k)=(u,v,w)}^{\infty} X_{mnk}(r+\epsilon) \downarrow \bigcap_{(u,v,w)=1}^{\infty} \bigcup_{(m,n,k)=(u,v,w)}^{\infty} X_{mnk}(r+\epsilon)$$

it is followed from the continuity of trust measure that

$$\lim_{uvw\to\infty} Tr\left\{\bigcup_{(m,n,k)=(u,v,w)}^{\infty} X_{mnk}(r+\epsilon)\right\} = Tr\left\{\bigcap_{(u,v,w)=1}^{\infty} \bigcup_{(m,n,k)=(u,v,w)}^{\infty} X_{mnk}(r+\epsilon)\right\}.$$

■

Theorem 3.2 *Let us assume that r is a real number which is non-negative and (x_{mnk}) be a triple sequence of rough variables. If $(x_{mnk}) \to_r^{a.s.} l$, then $(x_{mnk}) \to_r^{Tr} l$.*

Proof. It is followed from both the convergence a.s. and also Theorem 3.1 that

$$\lim_{uvw\to\infty} Tr\left\{\bigcup_{(m,n,k)=(u,v,w)}^{\infty} \{|x_{mnk} - l| \geq r+\epsilon\}\right\} = 0$$

for each $\epsilon > 0$. For every $(u, v, w) \geq 1$, since we have

$$\{|x_{uvw} - l| \geq r+\epsilon\} \subset \bigcup_{(m,n,k)=(u,v,w)}^{\infty} \{\{|x_{mnk} - l| \geq r+\epsilon\}\}$$

the theorem holds. ■

Theorem 3.3 *Let us assume that r be a real number which is non-negative and (x_{mnk}) be a triple sequence of rough variables. If the triple sequence $(x_{mnk}) \to_r^{mean} l$, then $(x_{mnk}) \to_r^{Tr} l$.*

Proof. For any given number $\epsilon > 0$

$$Tr\{|x_{mnk} - l| \geq r+\epsilon\} \leq \frac{E[|x_{mnk} - l|]}{r+\epsilon} \to 0 \text{ as } m,n,k \to \infty.$$

Thus $(x_{mnk}) \to_r^{Tr} l$.

Thus trust distribution $\Phi : \mathbb{R}^3 \to [0,1]$ of a rough variable l is described as

$$\Phi(x) = Tr\{\lambda \in \Lambda : l(\lambda) \leq x\} \text{ for all } x \in \mathbb{R}^3$$

i.e., $\Phi(x)$ is the trust having the rough variable $l \leq x$. ■

Theorem 3.4 *Suppose that rough variables are $l, l_1, l_2, \ldots,$. When triple sequence $(x_{mnk}) \to_r^{Tr} l$, then $(x_{mnk}) \to_r^{Dist} l$.*

Proof. Let us assume x is a prescribed continuity point in the distribution Φ. Again, On the other side, when $y > x$, we have

$$\{x_{mnk} \leq x\} = \{x_{mnk} \leq x, l < y\} \bigcup \{x_{mnk} \leq x, l > y\} \subset \{l < y\} \bigcup \{|x_{mnk} - l| \geq y - x\}$$

$$\Longrightarrow \Phi_{mnk}(x) \leq \Phi(y) + Tr\{|x_{mnk} - l| \geq y - x\}.$$

Since $(x_{mnk}) \to_r^{Tr} l$, we have

$$Tr\{\|x_{mnk} - l\| \geq y - x\} \to 0.$$

Therefore we have

$$\lim_{m,n,k\to\infty} \sup \Phi_{mnk}(x) \leq \Phi(y) \text{ for every } y > x.$$

If we let $y \to x$, then we have

$$\lim_{m,n,k\to\infty} \sup \Phi_{mnk}(x) \leq \Phi(x). \tag{3.1}$$

Similarly, on the way around, when $z < x$ is taken, we obtain

$$\begin{aligned} \{l \leq z\} &= \{l \leq z, x_{mnk} \leq x\} \bigcup \{l \leq z, x_{mnk} > x\} \\ &\subset \{x_{mnk} \leq x\} \bigcup \{\|x_{mnk} - l\| \geq x - z\} \\ &\Longrightarrow \Phi(z) \leq \Phi_{mnk}(x) + Tr\{\|x_{mnk} - l\| \geq x - z\}. \end{aligned}$$

Since $Tr\{\|x_{mnk} - l\| \geq x - z\} \to 0$, we get $\Phi(z) \leq \lim_{mnk\to\infty} \inf \Phi_{mnk}(x)$ for the condition $z < x$. If we let $z \to x$, then we have

$$\Phi(x) \leq \lim_{mnk\to\infty} \inf \Phi_{mnk}(x). \tag{3.2}$$

This is followed directly from Eqs. (3.1)-(3.2) we get $\Phi_{mnk}(x) \to \Phi(x)$. ■

Conclusion: This paper contributed to the research area of rough variable triple sequence were defined and then some mathematical properties of those new convergence are discussed.

References

[1] Aytar S., Rough statistical convergence, Numer. Funct. Anal. Optim. **29(3-4)**, 291-303 (2008).

[2] Aytar S., The rough limit set and the core of a real sequence, Numer. Funct. Anal. Optim. **29(3-4)**, 283-290 (2008).

[3] Datta, A. J.; Esi, A. and Tripathy, B. C., Statistically convergent triple sequence spaces defined by Orlicz function, J. Math. Anal. **4(2)**, 16-22 (2013).

[4] Debnath S., Sarma, B. and Das, B. C., Some generalized triple sequence spaces of real numbers, J. Nonlinear Anal. Optim. **6(1)**, 71–79 (2015).

[5] Dundar, E.; Cakan, C., Rough *I*-convergence, Demonstr. Math.**47(3)**, 638-651 (2014).

[6] Esi, A., On some triple almost lacunary sequence spaces defined by Orlicz functions, Research and Reviews: Discrete Mathematical Structures **1(2)**, 16-25 (2014).

[7] Esi, A. and Catalbas M. N., Almost convergence of triple sequences, Global Journal of Mathematical Analysis **2(1)**, 6–10 (2014).

[8] Phu, H. X., Rough convergence in normed linear spaces, Numer. Funct. Anal. Optim. **22(1-2)**, 199-222 (2001).

[9] Phu H. X., Rough continuity of linear operators, Numer. Funct. Anal. Optim. **23(1-2)**, 139–146 (2002).

[10] Phu H. X., Rough convergence in infinite dimensional normed spaces, Numer. Funct. Anal. Optimiz. **24(3-4)**, 285–301 (2003).

[11] Sahiner, A.; Gurdal, M. and Duden F. K., Triple sequences and their statistical convergence, Selcuk J. Appl. Math. **8(2)**, 49-55 (2007).

ITM Web of Conferences **22**, 01060 (2018) https://doi.org/10.1051/itmconf/20182201060
CMES-2018

Bernstein operator of rough $\mathfrak{I}$-core of triple sequences

M. Kemal Ozdemir[1,*], *Ayhan* Esi[2,**], and *Ayten* Esi[2,***]

[1]Department of Mathematics, İnönü University, 44280, Malatya, Turkey
[2]Department of Mathematics, Adıyaman University, 02040, Adıyaman, Turkey

Abstract. We introduce and study some basic properties of Bernstein-Stancu polynomials of rough $\mathfrak{I}$-convergent of triple sequence spaces and also study the set of all Bernstein-Stancu polynomials of rough $\mathfrak{I}$-limits of a triple sequence spaces and relation between analytic ness and Bernstein-Stancu polynomials of rough $\mathfrak{I}$-core of a triple sequence spaces.

1 Introduction

The notion of rough convergence has firstly been presented by Phu [10–12] in finite dimensional normed spaces. The author also illustrated that this set LIM_x^r is closed, convex and bounded; and he at the same time proposed the concept about the rough Cauchy sequence. The author again examined the connections among rough convergence and the other types of convergence and the dependence for LIM_x^r on the roughness having degree of r. Aytar [1] investigated rough statistical convergence and described rough statistical limit points set for a given sequence and had two important criteria about statistical convergence related to the set and proved the fact that the set is convex and closed. Meanwhile, Aytar [2] investigated that the r-limit set of the sequence equals to the intersection of those sets and that r-core of the sequence equals to the union of those sets. Dundar and Cakan [9] investigated rough ideal convergence and defined the set of rough ideal limit points of a sequence. The concept of $\mathfrak{I}$-convergence of a triple sequence spaces depended on the very structure for the ideal $\mathfrak{I}$ of the subsets of $\mathbb{N}^3$, where $\mathbb{N}$ denotes the natural numbers, is an expected result of the concept of convergence and statistical convergence.

In the present manuscript, we investigate several fundamental characteristics for rough $\mathfrak{I}$-convergence of a triple sequence spaces in three dimensional matrix spaces. We analyze the set of all rough $\mathfrak{I}$-limits consisting of a triple sequence spaces and at the same time the connection among analytic ness and rough $\mathfrak{I}$-core of a triple sequence spaces.

Let us assume that K is a subset consisting of the set consisting of positive integer numbers $\mathbb{N}^3$ and let us assume this set $K_{ik\ell} = \{(m,n,k) \in K : m \leq i, n \leq j, k \leq \ell\}$. Under these assumptions, the natural density for K can be denoted by $\delta(K) = \lim_{i,j,\ell\to\infty} \frac{|K_{ij\ell}|}{ij\ell}$, where $|K_{ij\ell}|$ stands for the number of elements in $K_{ij\ell}$.

First applied the notion of (p,q)-calculus in approximation theory and put forward the (p,q)-analogue of Bernstein operators. Later, depending on (p,q)-integers, several approximation results for Bernstein-Stancu, Bernstein-Kantorovich, (p,q)-Lorentz, Bernstein-Schurer, Bleimann, Butzer and Hahn operators etc.

*e-mail: kozdemir73@gmail.com
**e-mail: aesi23@hotmail.com
***e-mail: aytenesi@yahoo.com

In recent years, Khalid et al. have presented a very nice application using computer-aided geometric design and also applied those Bernstein basis to construct (p, q)-Bezier curves and surfaces depending upon (p, q)-integers, being a step other generalization of q-Bezier surfaces and curves.

Motivated by the above mentioned work on (p, q)-approximation and its application, in this paper we study statistical approximation properties of Bernstein-Stancu Operators depending on (p, q)-integers. Now we are going to remember several fundamental descriptions related to (p, q)-integers. For any $u, v, w \in \mathbb{N}$, the (p, q)-integer $[uvw]_{p,q}$ is described as

$$[0]_{p,q} := 0 \text{ and } [uvw]_{p,q} = \frac{p^{uvw} - q^{uvw}}{p - q} \text{ if } u, v, w \geq 1,$$

in which $0 < q < p \leq 1$. The (p, q)-factorial is described as

$$[0]_{p,q}! := 1 \text{ and } [uvw]!_{p,q} = [1]_{p,q}[2]_{p,q} \cdots [uvw]_{p,q} \text{ if } u, v, w \geq 1.$$

We can describe the (p, q)-binomial coefficient as

$$\binom{u}{m}\binom{v}{n}\binom{w}{k}_{p,q} = \frac{[u]!_{p,q}}{[m]!_{p,q}\,[u-m]!_{p,q}} \frac{[v]!_{p,q}}{[n]!_{p,q}\,[v-n]!_{p,q}} \frac{[w]!_{p,q}}{[k]!_{p,q}\,[w-k]!_{p,q}}$$

for all $u, v, w, m, n, k \in \mathbb{N}$ with $(u, v, w) \geq (m, n, k)$.

The equality about (p, q)-binomial expansion can be given as

$$(ax + by)^{uvw}_{p,q} = \sum_{m=0}^{u}\sum_{n=0}^{v}\sum_{k=0}^{w} p^{\frac{(u-m)(u-m-1)+(v-n)(v-n-1)+(w-k)(w-k-1)}{2}} q^{\frac{m(m-1)+n(n-1)+k(k-1)}{2}} \cdot$$
$$\binom{u}{m}\binom{v}{n}\binom{w}{k}_{p,q} a^{(u-m)+(v-n)+(w-k)} b^{m+n+k} x^{(u-m)+(v-n)+(w-k)} y^{m+n+k},$$

$$(x + y)^{uvw}_{p,q} = (x + y)(px + qy)\left(p^2x + q^2y\right)\cdots\left(p^{(u-1)+(v-1)+(w-1)}x + q^{(u-1)+(v-1)+(w-1)}y\right),$$
$$(1 - x)^{uvw}_{p,q} = (1 - x)(p - qx)\left(p^2 - q^2x\right)\cdots\left(p^{(u-1)+(v-1)+(w-1)} - q^{(u-1)+(v-1)+(w-1)}x\right), \text{ and}$$
$$(x)^{mnk}_{p,q} = x(px)\left(p^2x\right)\cdots\left(p^{(u-1)+(v-1)+(w-1)}x\right) = p^{\frac{m(m-1)+n(n-1)+k(k-1)}{2}}.$$

The Bernstein operator of order (r, s, t) is presented as follows

$$B_{rst}(f, x) = \sum_{m=0}^{r}\sum_{n=0}^{s}\sum_{k=0}^{t} f\left(\tfrac{mnk}{rst}\right)\binom{r}{m}\binom{s}{n}\binom{t}{k} x^{m+n+k}(1 - x)^{(m-r)+(n-s)+(k-t)}$$

in which f is a continuous (either complex or real valued) function on $[0, 1]$.

We can define (p, q)-Bernstein operators as

$$B_{rst,p,q}(f, x) = \frac{1}{p^{\frac{r(r-1)+s(s-1)+t(t-1)}{2}}} \sum_{m=0}^{r}\sum_{n=0}^{s}\sum_{k=0}^{t}\binom{r}{m}\binom{s}{n}\binom{t}{k} p^{\frac{m(m-1)+n(n-1)+k(k-1)}{2}} x^{m+n+k}.$$
$$\prod_{u_1=0}^{(r-m-1)}(p^{u_1} - q^{u_1}x)\prod_{u_2=0}^{(s-n-1)}(p^{u_2} - q^{u_2}x)\prod_{u_3=0}^{(t-k-1)}(p^{u_3} - q^{u_3}x)\cdot \tag{1.1}$$
$$f\left(\frac{p^{(r-m)}[m]_{p,q} + p^{(s-n)}[n]_{p,q}\,p^{(t-k)}[k]_{p,q}}{[r]_{p,q} + [s]_{p,q} + [t]_{p,q}}\right), x \in [0, 1]$$

Also, we have

$$(1-x)_{p,q}^{rst} = \sum_{m=0}^{r}\sum_{n=0}^{s}\sum_{k=0}^{t}(-1)^{m+n+k}\, p^{\frac{(r-m)(r-m-1)+(s-n)(s-n-1)+(t-k)(t-k-1)}{2}}\, q^{\frac{m(m-1)+n(n-1)+k(k-1)}{2}}\binom{r}{m}\binom{s}{n}\binom{t}{k}x^{m+n+k}.$$

$$S_{rst,p,q}(f,x) = \frac{1}{p^{\frac{r(r-1)+s(s-1)+t(t-1)}{2}}}\sum_{m=0}^{r}\sum_{n=0}^{s}\sum_{k=0}^{t}\binom{r}{m}\binom{s}{n}\binom{t}{k}p^{\frac{m(m-1)+n(n-1)+k(k-1)}{2}}x^{m+n+k}.$$

$$\prod_{u_1=0}^{(r-m-1)}(p^{u_1}-q^{u_1}x)\prod_{u_2=0}^{(s-n-1)}(p^{u_2}-q^{u_2}x)\prod_{u_3=0}^{(t-k-1)}(p^{u_3}-q^{u_3}x)\cdot \tag{1.2}$$

$$f\left(\frac{p^{(r-m)}[m]_{p,q}+p^{(s-n)}[n]_{p,q}\,p^{(t-k)}[k]_{p,q}+\eta}{[r]_{p,q}+[s]_{p,q}+[t]_{p,q}+\mu}\right), x\in[0,1]$$

Note that for $\eta = \mu = 0$, (p,q)-Bernstein-Stancu operators in Eq. (1.2) reduces into (p,q)-Bernstein operators. Also for $p = 1$, (p,q)-Bernstein-Stancu operators in Eq. (1.1) result in q-Bernstein-Stancu operators.

A triple sequence (either complex or real) may be described as a function $x : \mathbb{N}^3 \rightarrow \mathbb{R}(\mathbb{C})$, in which the complex, real and natural numbers are denoted by $\mathbb{C}$, $\mathbb{R}$ and $\mathbb{N}$, respectively. The various types of concepts for triple sequence has been first proposed and examined initially by *Datta et al. [7]*, *Debnath et al. [8]*, *Esi et al. [3–6]*, *Sahiner et al. [13, 14]*, *Subramanian et al. [15]* and several others.

It is said that a triple sequence $x = (x_{mnk})$ is triple analytic when

$$\sup_{m,n,k}|x_{mnk}|^{\frac{1}{m+n+k}} < \infty.$$

Λ^3 usually denotes all triple analytic sequences space.

2 Definitions and Preliminaries

For the rest of the manuscript $\mathbb{R}^3$ symbolizes the real three dimensional case having the metric. Take into consideration a triple sequence $x = (x_{mnk})$ such that $x_{mnk} \in \mathbb{R}^3$; $m, n, k \in \mathbb{N}^3$. The following definition are obtained:

Definition 2.1 *Assume that f is a continuous function on* $[0,1]$. *A triple sequence of Bernstein-Stancu polynomials* $\left(S_{rst,p,q}(f,x)\right)$ *is statistically convergent to (f,x) and denoted by* $S_{rst,p,q}(f,x) \rightarrow^{st-\lim x} (f,x)$,*when for every* $\epsilon > 0$, $d(A(\epsilon)) = 0$ *is obtained, in which* $A(\epsilon) = \left\{(m,n,k) \in \mathbb{N}^3 : \left|S_{rst,p,q}(f,x) - (f,x)\right| \geq \epsilon\right\}$.

Definition 2.2 *Assume that f is any given continuous function on* $[0,1]$. *It is said that a triple sequence of Bernstein-Stancu polynomials* $\left(S_{rst,p,q}(f,x)\right)$ *is statistically convergent to (f,x) denoted by* $S_{rst,p,q}(f,x) \rightarrow^{st-\lim x} (f,x)$, *providing that the set* $\left\{(m,n,k) \in \mathbb{N}^3 : \left|S_{rst,p,q}(f,x) - (f,x)\right| \geq \epsilon\right\}$, *with natural density zero for each* $\epsilon > 0$. *Under this condition, (f,x) is known as the statistical limit for the sequence of Bernstein-Stancu polynomials.*

Definition 2.3 *Assume that f is a given continuous function given on* $[0,1]$. *It is said that a triple sequence of Bernstein-Stancu polynomials* $\left(S_{rst,p,q}(f,x)\right)$ *in a metric space* $(X, |.,.|)$ *and* $r \geq 0$ *be a real number is r-convergent to (f,x), denoted by* $S_{rst,p,q}(f,x) \rightarrow^{r} (f,x)$, *if for any* $\epsilon > 0$ *there exists* $N_\epsilon \in \mathbb{N}^3$ *such that for all* $m,n,k \geq N_\epsilon$ *we have* $\left|S_{rst,p,q}(f,x) - (f,x)\right| < r+\epsilon$. *In this case* $S_{rst,p,q}(f,x)$ *is called an r-limit of (f,x).*

Remark 2.1 *We consider r-limit set* $S_{rst,p,q}(f,x)$ *which is denoted by* $LIM^{r}S_{rst,p,q}(f,x)$ *and is denoted as* $LIM^{r}S_{rst,p,q}(f,x) = \left\{f : S_{rst,p,q}(f,x) \to^{r} (f,x)\right\}$.

Definition 2.4 *It is assumed that* f *is a prescribed continuous function defined on* $[0,1]$. *It is said that a triple sequence of Bernstein-Stancu polynomials* $\left(S_{rst,p,q}(f,x)\right)$ *is r-convergent if* $LIM^{r}S_{rst,p,q}(f,x) \neq \phi$ *and r is known as a rough convergence degree of* $S_{rst,p,q}(f,x)$. *If* $r = 0$ *then it is ordinary convergence of triple sequence of Bernstein-Stancu polynomials.*

Definition 2.5 *Assume that* f *is a continuous function prescribed on* $[0,1]$. *It is said that A triple sequence of Bernstein polynomials* $(B_{mnk}(f,x))$ *in a metric space* $(X, |.,.|)$ *and* $r \geq 0$ *is a real number is r-statistically convergent to* (f,x), *symbolized by* $B_{mnk}(f,x) \to^{r-st_3} (f,x)$, *when for any* $\epsilon > 0$ *we obtain* $d(A(\epsilon)) = 0$, *in which*

$$A(\epsilon) = \left\{(m,n,k) \in \mathbb{N}^3 : |B_{mnk}(f,x) - (f,x)| \geq r + \epsilon\right\}.$$

When this happens, (f,x) *is known as r-statistical limit of* $B_{mnk}(f,x)$. *If* $r = 0$ *then it is ordinary statistical convergent of triple sequence of Bernstein polynomials.*

Definition 2.6 *It is said that a class* $\mathfrak{I}$ *of subsets consisting of a nonempty set X is an ideal in X providing that (i)* $\phi \in \mathfrak{I}$, *(ii)* $A, B \in \mathfrak{I} \Rightarrow A \bigcup B \in \mathfrak{I}$, *(iii)* $A \in \mathfrak{I}$, $B \subset A \Rightarrow B \in \mathfrak{I}$. $\mathfrak{I}$ *is known as a non-trivial ideal if* $X \notin \mathfrak{I}$.

Definition 2.7 *Assume that* f *is a continuous function on* $[0,1]$. *It is said that a triple sequence of Bernstein polynomials* $(B_{mnk}(f,x))$ *in a metric space* $(X, |.,.|)$ *and* $r \geq 0$ *be a real number is rough ideal convergent or* $r\mathfrak{I}$*-convergent to* (f,x), *denoted by* $B_{mnk}(f,x) \to^{r\mathfrak{I}} (f,x)$, *if for any* $\epsilon > 0$ *we have* $\left\{(m,n,k) \in \mathbb{N}^3 : |B_{mnk}(f,x) - (f,x)| \geq r + \epsilon\right\} \in \mathfrak{I}$. *In this case* $(B_{mnk}(f,x))$ *is called* $r\mathfrak{I}$*-convergent to* (f,x) *and a triple sequence of Bernstein polynomials* $(B_{mnk}(f,x))$ *is called rough* $\mathfrak{I}$*-convergent to* (f,x) *with r as roughness of degree. If* $r = 0$ *then it is ordinary* $\mathfrak{I}$*-convergent.*

Note 2.1 *Generally, it is assumed that* f *is a continuous function on* $[0,1]$. *A triple sequence of Bernstein polynomials* $(B_{mnk}(g,x))$ *is not* $\mathfrak{I}$*-convergent in usual sense and* $|B_{mnk}(f,x) - B_{mnk}(g,x)| \leq r$ *for all* $(m,n,k) \in \mathbb{N}^3$ *or* $\left\{(m,n,k) \in \mathbb{N}^3 : |B_{mnk}(f,x) - B_{mnk}(g,x)| \geq r\right\} \in \mathfrak{I}$ *for some* $r > 0$. *Then the triple sequence of Bernstein polynomials* $(B_{mnk}(f,x))$ *is* $r\mathfrak{I}$*-convergent.*

Note 2.2 *It is clear that* $r\mathfrak{I}$*-limit of a sequence* $B_{mnk}(f,x)$ *of Bernstein polynomial is not necessarily unique.*

Definition 2.8 *Let us consider* $r\mathfrak{I}$*-limit set of* $f(x)$, *symbolized by* $\mathfrak{I} - LIM^{r}B_{mnk}(f,x) = \left\{f : B_{mnk}(f,x) \to^{r\mathfrak{I}} (f,x)\right\}$, *then the triple sequence of Bernstein polynomials* $(B_{mnk}(f,x))$ *is said to be* $r\mathfrak{I}$*-convergent if* $\mathfrak{I} - LIM^{r}B_{mnk}(f,x) \neq \phi$ *and r is known as a rough* $\mathfrak{I}$*-convergence having degree of* $B_{mnk}(f,x)$.

Definition 2.9 *Assume that* f *is a continuous function on* $[0,1]$. *It is said that a triple sequence of Bernstein polynomials* $(B_{mnk}(f,x))$ *is* $\mathfrak{I}$*-analytic if there exists a real number* $M > 0$ *satisfying*

$$\left\{(m,n,k) \in \mathbb{N}^3 : |B_{mnk}(f,x)|^{1/m+n+k} \geq M\right\} \in \mathfrak{I}.$$

Definition 2.10 *It is said that a point* $L \in X$ *is an* $\mathfrak{I}$*-accumulation point and suppose that* f *is a continuous function on* $[0,1]$. *A Bernstein polynomials* $(B_{mnk}(f,x))$ *is a metric space* (X,d) *iff for each* $\epsilon > 0$ *the set*

$$\left\{(m,n,k) \in \mathbb{N}^3 : d\left(B_{mnk}(f,x), f(x)\right) = |B_{mnk}(f,x) - f(x)| < \epsilon\right\} \notin \mathfrak{I}.$$

The set consisting of all $\mathfrak{I}$-accumulation points of $(B_{mnk}(f,x))$ is symbolized by $\mathfrak{I}(\Gamma(B_{mnk}(f,x)))$.

Definition 2.11 *Suppose that f is a continuous function on $[0,1]$. It is said that a triple sequence of Bernstein polynomials $(B_{mnk}(f,x))$ is rough $\mathfrak{I}$-convergent if $\mathfrak{I}-LIM^r B_{mnk}(f,x) \neq \phi$. It is obvious that if $\mathfrak{I} - LIM^r B_{mnk}(f,x) \neq \phi$ for a triple sequence of Bernstein polynomials $(B_{mnk}(f,x))$ of real numbers, under these conditions we conclude $\mathfrak{I} - LIM^r B_{mnk}(f,x) = [\mathfrak{I}-\limsup B_{mnk}(f,x) - r, \mathfrak{I}-\liminf B_{mnk}(f,x)+r]$.*

Definition 2.12 *Assume that f is a continuous function on $[0,1]$. It is said that a triple sequence of Bernstein polynomials $(B_{mnk}(f,x))$ is rough $\mathfrak{I}$ − core $B_{mnk}(f,x)$ is defined to the closed interval $[+\infty, -\infty]$.*

3 Main Results

Theorem 3.1 *Assume that f is a continuous function on $[0,1]$. When a triple sequence of Bernstein polynomials $(B_{mnk}(f,x))$ of real numbers and $\mathfrak{I} \subset 2^{\mathbb{N}}$ be an admissible ideal, then we conclude that* $\operatorname{diam}(\mathfrak{I} - LIM^r B_{mnk}(f,x)) \leq 2r$. *Generally,* $\operatorname{diam}(\mathfrak{I} - LIM^r B_{mnk}(f,x))$ *has an upper bound.*

Proof. Assume that $\operatorname{diam}(LIM^r B_{mnk}(f,x))$. Then,

$$\exists B_{mnk}(p,x), B_{mnk}(q,x) \in LIM^r B_{mnk}(f,x) \ni: |B_{mnk}(p,x) - B_{mnk}(q,x)| > 2r.$$

Take $\epsilon \in \left(0, \frac{|B_{mnk}(p,x)-B_{mnk}(q,x)|}{2} - r\right)$. Because $B_{mnk}(p,x)$, $B_{mnk}(q,x) \in \mathfrak{I} - LIM^r B_{mnk}(f,x)$, we have $A_1(\epsilon) \in \mathfrak{I}$ and $A_2(\epsilon) \in \mathfrak{I}$ for every $\epsilon > 0$, where

$$A_1(\epsilon) = \left\{(i,j,k) \in \mathbb{N}^3 : |B_{mnk}(f,x) - B_{mnk}(p,x)| \geq r + \epsilon\right\} \text{ and}$$

$$A_2(\epsilon) = \left\{(i,j,k) \in \mathbb{N}^3 : |B_{mnk}(f,x) - B_{mnk}(q,x)| \geq r + \epsilon\right\}.$$

Using the properties $F(\mathfrak{I})$, we get $(A_1(\epsilon)^c \bigcap A_2(\epsilon)^c) \in F(\mathfrak{I})$. Thus we write,

$$\begin{aligned} |B_{mnk}(p,x) - B_{mnk}(q,x)| &\leq |B_{mnk}(f,x) - B_{mnk}(p,x)| + |B_{mnk}(f,x) - B_{mnk}(q,x)| \\ &< (r+\epsilon) + (r+\epsilon) < 2(r+\epsilon), \text{ for all } (m,n,k) \in A_1(\epsilon)^c \bigcap A_2(\epsilon)^c \end{aligned}$$

and this is an clear contradiction. Hence $\operatorname{diam}(LIM^r B_{mnk}(f,x)) \leq 2r$.

Now, consider a triple sequence of Bernstein polynomials of $(B_{mnk}(f,x))$ of real numbers so that $\mathfrak{I} - \lim_{mnk\to\infty} B_{mnk}(f,x) = f(x)$.

Let $\epsilon > 0$. So, it can be written that $\left\{(m,n,k) \in \mathbb{N}^3 : |B_{mnk}(f,x) - (f,x)| \geq \epsilon\right\} \in \mathfrak{I}$. Consequently, we got

$$\begin{aligned} |B_{mnk}(f,x) - B_{mnk}(p,x)| &\leq |B_{mnk}(f,x) - (f,x)| + |(f,x) - B_{mnk}(p,x)| \\ &\leq |B_{mnk}(f,x) - (f,x)| + r \leq r + \epsilon, \end{aligned}$$

for each $B_{mnk}(p,x) \in \bar{B}_r((f,x)) := \left\{B_{mnk}(p,x) \in \mathbb{R}^3 : |B_{mnk}(p,x) - (f,x)| \leq r\right\}$. Then, we get $|B_{mnk}(f,x) - B_{mnk}(p,x)| < r+\epsilon$ for each $(m,n,k) \in \{(m,n,k) \in \mathbb{N}^3 : |B_{mnk}(f,x) - (f,x)| < \epsilon\}$. Because the triple sequence of Bernstein polynomials of $B_{mnk}(f,x)$ is $\mathfrak{I}$-convergent to $f(x)$, we have

$$\left\{(m,n,k) \in \mathbb{N}^3 : |B_{mnk}(f,x) - (f,x)| < \epsilon\right\} \in F(\mathfrak{I}).$$

Therefore, we get $p \in \mathfrak{I} - LIM^r B_{mnk}(f, x)$. Consequently, we can write

$$\mathfrak{I} - LIM^r B_{mnk}(f, x) = \bar{B}_r((f, x)). \tag{3.1}$$

Since $\text{diam}\left(\bar{B}_r((f, x))\right) = 2r$, this can be usually shown that the upper bound $2r$ of the diameter of the set $\mathfrak{I} - LIM^r B_{mnk}(f, x)$ is not lower bound. ■

Theorem 3.2 *Assume that f is a continuous function prescribed on* $[0, 1]$. *Assume that a triple sequence of Bernstein polynomials $(B_{mnk}(f, x))$ of real numbers, $\mathfrak{I} \subset 3^{\mathbb{N}}$ is an admissible ideal. For a given arbitrary $(f, c) \in \mathfrak{I}(\Gamma_x)$, we have $|B_{mnk}(f, x) - (f, c)| \leq r$ for every $B_{mnk}(f, x) \in \mathfrak{I} - LIM^r B_{mnk}(f, x)$.*

Proof. If we suppose, on the contrariwise, that there is a point $(f, c) \in \mathfrak{I}(\Gamma_x)$ and $B_{mnk}(f, x) \in \mathfrak{I} - LIM^r B_{mnk}(f, x)$ so that $|B_{mnk}(f, x) - (f, c)| > r$. Define $\epsilon := \frac{|B_{mnk}(f,x)-(f,c)|-r}{3}$. Then

$$\left\{(m, n, k) \in \mathbb{N}^3 : |B_{mnk}(f, x) - (f, c)| < \epsilon\right\} \subseteq \left\{(m, n, k) \in \mathbb{N}^3 : |B_{mnk}(f, x) - (f, x)| \geq r + \epsilon\right\}. \tag{3.2}$$

Since $(f, c)\, c \in \mathfrak{I}(\Gamma_x)$, we have $\left\{(m, n, k) \in \mathbb{N}^3 : |B_{mnk}(f, x) - (f, c)| < \epsilon\right\} \notin \mathfrak{I}$. But from definition of $\mathfrak{I}$-convergence, since $\{(m, n, k) \in \mathbb{N}^3 : |B_{mnk}(f, x) - f(x)| \geq r + \epsilon\} \in \mathfrak{I}$, so by (3.2) we have $\left\{(m, n, k) \in \mathbb{N}^3 : |B_{mnk}(f, x) - (f, c)| < \epsilon\right\} \in \mathfrak{I}$, which contradicts the fact $(f, c) \in \mathfrak{I}(\Gamma_x)$. On the way around, if $(f, c) \in \mathfrak{I}(\Gamma_x)$ i.e., $\{(m, n, k) \in \mathbb{N}^3 : |B_{mnk}(f, x) - (f, c)| < \epsilon\} \notin \mathfrak{I}$, then $\{(m, n, k) \in \mathbb{N}^3 : |B_{mnk}(f, x) - (f, x)| \geq r + \epsilon\} \notin \mathfrak{I}$, which is an obvious contradiction to the fact $(f, x) \in \mathfrak{I} - LIM^r B_{mnk}(f, x)$. ■

Theorem 3.3 *Assume that f is a continuous function on* $[0, 1]$. *A triple sequence of Bernstein polynomials*

$$(B_{mnk}(f, x)) \rightarrow^{\mathfrak{I}} (f, x) \Leftrightarrow \mathfrak{I} - LIM^r B_{mnk}(f, x) = \bar{B}_r((f, x)).$$

Proof. *Necessity:* By Theorem 3.1.

Sufficiency: Let $\mathfrak{I} - LIM^r B_{mnk}(f, x) = \bar{B}_r((f, x)) (\neq \phi)$. Therefore the triple sequence spaces of Bernstein polynomials of $(B_{mnk}(f, x))$ is $\mathfrak{I}$-analytic. Let us suppose that (f, x) has got another $\mathfrak{I}$-cluster point (f', x) which is differing from (f, x). The point

$$\left(\bar{f}, x\right) = (f, x) + \frac{r}{|(f,x)-(f',x)|}((f, x) - (f', x))$$
$$\left(\bar{f}, x\right) - (f', x) = (f, x) - (f', x) + \frac{r}{|(f,x)-(f',x)|}\left[((f,x)-(f',x)) - ((f',x)-(f',x))\right]$$
$$\left|\left(\bar{f}, x\right) - (f', x)\right| = \left|(f, x) - (f', x)\right| + \frac{r}{|(f,x)-(f',x)|}\left|(f, x) - (f', x)\right|$$
$$\left|\left(\bar{f}, x\right) - (f', x)\right| = \left|(f, x) - (f', x)\right| + r > r.$$

Since $(f', x) \in \mathfrak{I}(\Gamma_x)$, by Theorem 3.2, $\left(\bar{f}, x\right) \notin \mathfrak{I} - LIM^r B_{mnk}(f, x)$. It is not possible as $\left|\left(\bar{f}, x\right) - (f, x)\right| = r$ and $\mathfrak{I} - LIM^r B_{mnk}(f, x) = \bar{B}_r((f, x))$. Because (f, x) is the unique $\mathfrak{I}$-cluster point of (f, x). Hence $\Longrightarrow B_{mnk}(f, x) \rightarrow^{\mathfrak{I}} f(x)$. ■

Corollary 3.1 *If $(X, |.,.|)$ is a strictly convex spaces and let us assume that f is a continuous function defined on* $[0, 1]$. *A triple sequence of Bernstein polynomials $(B_{mnk}(f,x)) \in X$, there exists $y_1, y_2 \in \mathfrak{I} - LIM^r B_{mnk}(f,x)$ so that $|y_1 - y_2| = 2r$, then this triple sequence $(f, x) \rightarrow^{\mathfrak{I}} \frac{y_1+y_2}{2}$.*

Theorem 3.4 *If $\mathfrak{I} - LIM^r \neq \phi$, then $\mathfrak{I} - \limsup B_{mnk}(f,x)$ and $\mathfrak{I} - \liminf B_{mnk}(f,x)$ belong to the set $\mathfrak{I} - LIM^{2r} B_{mnk}(f,x)$.*

Proof. We know that $\mathfrak{I} - LIM^r B_{mnk}(f,x) \neq \phi$, a triple sequence of Bernstein polynomials of $(B_{mnk}(f,x))$ is $\mathfrak{I}$-analytic. The number $\mathfrak{I} - \liminf B_{mnk}(f,x)$ is an $\mathfrak{I}$-cluster point of (f,x) and consequently, we have $|(f,x) - \mathfrak{I} - \liminf B_{mnk}(f,x)| \leq r \ \forall \ (f,x) \in \mathfrak{I} - LIM^r(f,x)$. Let $A = \left\{(m,n,k) \in \mathbb{N}^3 : |(f,x) - B_{mnk}(f,x)| \geq r + \epsilon\right\}$. Now if $(m,n,k) \notin A$, then

$$|B_{mnk}(f,x) - (\mathfrak{I} - \liminf B_{mnk}(f,x))| \leq |B_{mnk}(f,x) - (f,x)| + |(f,x) - (\mathfrak{I} - \liminf B_{mnk}(f,x))| < 2r + \epsilon.$$

Thus $\mathfrak{I} - \liminf B_{mnk}(f,x) \in \mathfrak{I} - LIM^{2r} B_{mnk}(f,x)$. Likewise, it can be seen that $\mathfrak{I} - \limsup x_{mnk} \in \mathfrak{I} - LIM^{2r} x_{mnk}$. ■

Corollary 3.2 *Assume that f is a continuous function given on the prescribed interval* $[0,1]$. *A triple sequence of Bernstein polynomials $(B_{mnk}(f,x))$ of real numbers, if $\mathfrak{I} - LIM^r B_{mnk}(f,x) \neq \phi$, then $\mathfrak{I} - core\{f(x)\} \subseteq \mathfrak{I} - LIM^{2r} B_{mnk}(f,x)$.*

Proof. We now have $\mathfrak{I} - LIM^r B_{mnk}(f,x) = [\mathfrak{I} - \limsup B_{mnk}(f,x) - 2r, \mathfrak{I} - \liminf B_{mnk}(f,x) + 2r]$. Then the result is followed from Theorem 3.4. ■

Theorem 3.5 *Assume that f is a continuous function defined on* $[0,1]$. *A triple sequence of Bernstein polynomials $(B_{mnk}(f,x))$ of real numbers. Under these conditions, the* $\operatorname{diam}(\mathfrak{I} - core\{B_{mnk}(f,x)\})$ *of the set*

$$\mathfrak{I} - core\{B_{mnk}(f,x)\} = r \Longleftrightarrow \mathfrak{I} - core\{f(x)\} = \mathfrak{I} - LIM^r B_{mnk}(f,x).$$

Proof. $\operatorname{diam}(\mathfrak{I} - core\{B_{mnk}(f,x)\}) = r \Longleftrightarrow (\mathfrak{I} - \limsup B_{mnk}(f,x)) - (\mathfrak{I} - \liminf x_{mnk}) = r$

$$\Longleftrightarrow \mathfrak{I} - core\{x_{mnk}\} = [\mathfrak{I} - \liminf x_{mnk}, \mathfrak{I} - \limsup B_{mnk}(f,x)]$$
$$= [\mathfrak{I} - \limsup B_{mnk}(f,x) - r, \mathfrak{I} - \liminf B_{mnk}(f,x) + r] = \mathfrak{I} - LIM^r B_{mnk}(f,x).$$

And it can easily be seen that
(i) $r > \operatorname{diam}(\mathfrak{I} - core\{B_{mnk}(f,x)\}) \Longleftrightarrow \mathfrak{I} - core\{B_{mnk}(f,x)\} \subset \mathfrak{I} - LIM^r B_{mnk}(f,x)$,
(ii) $r < \operatorname{diam}(\mathfrak{I} - core\{B_{mnk}(f,x)\}) \Longleftrightarrow \mathfrak{I} - LIM^r B_{mnk}(f,x) \subset \mathfrak{I} - core\{B_{mnk}(f,x)\}$. ■

Theorem 3.6 *Assume that f is a continuous function defined on* $[0,1]$. *A triple sequence of Bernstein polynomials $(B_{mnk}(f,x))$ of real numbers, if $\bar{r} = \inf\{r \geq 0 : \mathfrak{I} - LIM^r B_{mnk}(f,x) \neq \phi\}$, then $\bar{r} = \operatorname{radius}(\mathfrak{I} - core\{B_{mnk}(f,x)\})$.*

Proof. When the given set $\mathfrak{I} - core\{B_{mnk}(f,x)\}$ is singleton, then $\operatorname{radius}(\mathfrak{I} - core\{B_{mnk}(f,x)\}) = 0$ and the triple sequence of Bernstein polynomials is $\mathfrak{I}$-convergent, i.e., $\mathfrak{I} - LIM^0 B_{mnk}(f,x) \neq \phi$. Therefore we obtain $\bar{r} = \operatorname{radius}(\mathfrak{I} - core\{B_{mnk}(f,x)\}) = 0$.

Let us take the assumption that the set $\mathfrak{I} - core\{B_{mnk}(f,x)\}$ is now not a singleton. It can now be written as $\mathfrak{I} - core\{B_{mnk}(f,x)\} = [a,b]$ in which $a = \mathfrak{I} - \liminf B_{mnk}(f,x)$ and $b = \mathfrak{I} - \limsup B_{mnk}(f,x)$. Again, let us suppose that $\bar{r} \neq \operatorname{radius}(\mathfrak{I} - core\{B_{mnk}(f,x)\})$. When $\bar{r} < \operatorname{radius}(\mathfrak{I} - core\{x_{mnk}\})$, we can define $\bar{\epsilon} = \frac{\frac{b-a}{2} - \bar{r}}{3}$. Now, be definition of $\bar{r}$ implies that $\mathfrak{I} - LIM^{\bar{r}+\bar{\epsilon}} B_{mnk}(f,x) \neq \phi$, given $\epsilon > 0 \ \exists \ l \in \mathbb{R} : A = \{(m,n,k) \in \mathbb{N} : |B_{mnk}(f,x) - f(x)| \geq (\bar{r} + \bar{\epsilon}) + \epsilon\} \in \mathfrak{I}$. Since $\bar{r} + \bar{\epsilon} < \frac{b-a}{2}$ which is a contradiction of the definition of a and b.

When $\bar{r} > \operatorname{radius}(\mathfrak{I} - core\{B_{mnk}(f,x)\})$, then we can define $\bar{\epsilon} = \frac{\bar{r} - \frac{b-a}{2}}{3}$ and $r' = \bar{r} - 2\bar{\epsilon}$. It is obvious that $0 \leq r' \leq \bar{r}$ and by definitions of a and b, the number $\frac{b-a}{2} \in \mathfrak{I} - LIM^{r'} B_{mnk}(f,x)$. Under these conditions, we have $\bar{r} \in \{r \geq 0 : \mathfrak{I} - LIM^r B_{mnk}(f,x) \neq \phi\}$, which is a contradiction to the equality $\bar{r} = \inf\{r \geq 0 : \mathfrak{I} - LIM^r B_{mnk}(f,x) \neq \phi\}$ as $r' < r$. ■

Corollary 3.3 *Assume that f is a continuous function defined on* $[0, 1]$. *A triple sequence of Bernstein polynomials* $(B_{mnk}(f, x))$ *of real numbers, then* $\mathfrak{I} - core\{B_{mnk}(f, x)\} = \mathfrak{I} - LIM^{2\bar{r}} B_{mnk}(f, x)$.

Example 3.1 *Using the symbolic programming software Matlab, we have shown the comparisons and some easy-to-understand graphics for the convergence of operators given by (1.2) to the function* $f(x) = 1 + x^3 \sin(14x)$ *for various parameters. Looking at figure 1(a), it can be seen that as the q and p approaches towards 1 provided $0 < q < p \leq 1$, (p, q)-Bernstein-Stancu operators presented by (1.2) converges towards the function* $f(x) = 1 + x^3 \sin(14x)$. *From figure 1(a) and (b), it can be observed that for $\eta = \mu = 0$, as the value the (r, s, t) increases, (p, q)-Bernstein-Stancu operators (1.2) converges towards the function. Similarly from figure 2(a), it can be noticed that for $\eta = \mu = 5$, as the value the q and p approaches towards 1 or some thing else provided $0 < q < p \leq 1$, (p,q)-Bernstein-Stancu operators by (1.2) converges towards the function. From figure 2(a) and (b), it is seen that when the value the $[r,s,t]$ increases, (p, q)-Bernstein-Stancu operators which are presented by $f(x) = 1+x^3 \sin(14x)$ converges towards the function.*

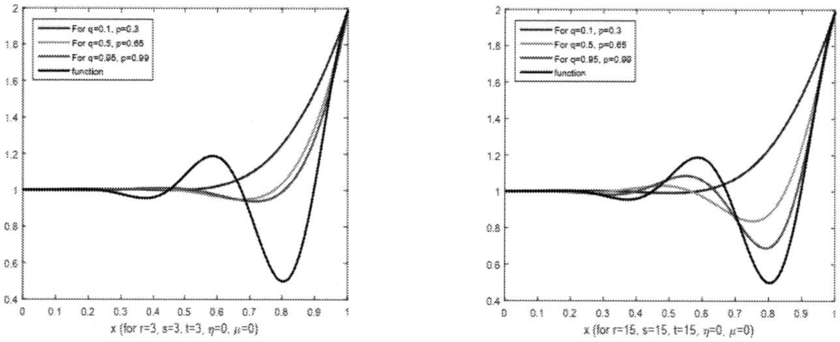

(a) (b)

Figure 1: (p, q)-Bernstein-Stancu operators

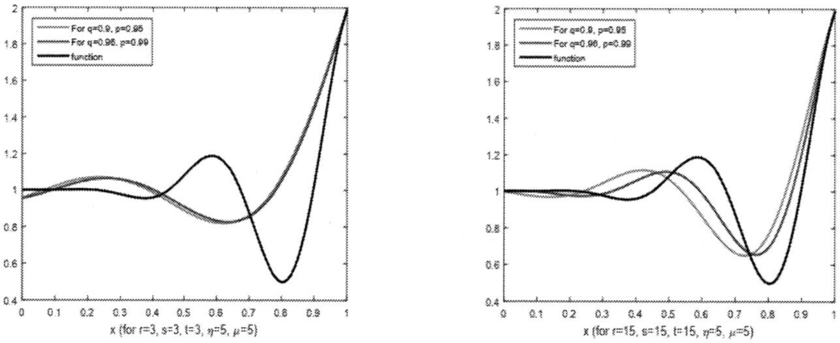

(a) (b)

Figure 2: (p, q)-Bernstein-Stancu operators

Competing Interests: We, the authors, hereby declare that there is not any conflict of interests in terms of the publication of this manuscript.

References

[1] Aytar S., Rough statistical convergence, Numer. Funct. Anal. Optim. **29(3-4)**, 291-303 (2008).

[2] Aytar S., The rough limit set and the core of a real sequence, Numer. Funct. Anal. Optim. **29(3-4)**, 283-290 (2008).

[3] Esi, A., On some triple almost lacunary sequence spaces defined by Orlicz functions, Research and Reviews: Discrete Mathematical Structures **1(2)**, 16-25 (2014).

[4] Esi, A. and Catalbas, M. N., Almost convergence of triple sequences, Global Journal of Mathematical Analysis **2(1)**, 6-10 (2014).

[5] Esi, A. and Savas, E., On lacunary statistically convergent triple sequences in probabilistic normed space, Appl. Math. Inf. Sci. **9(5)**, 2529-2534 (2015).

[6] Esi, A.; Araci, S. and Acikgoz, M., Statistical convergence of Bernstein operators, Appl. Math. Inf. Sci. **10(6)**, 2083-2086 (2016).

[7] Datta, A. J.; Esi, A. and Tripathy, B. C., Statistically convergent triple sequence spaces defined by Orlicz function, J. Math. Anal. **4(2)**, 16-22 (2013).

[8] Debnath, S.; Sarma, B. and Das, B. C., Some generalized triple sequence spaces of real numbers, J. Nonlinear Anal. Optim. **6(1)**, 71-79 (2015).

[9] Dundar, E.; Cakan, C., Rough $\mathcal{I}$-convergence, Demonstr. Math. **47(3)**, 638-651 (2014).

[10] Phu, H. X., Rough convergence in normed linear spaces, Numer. Funct. Anal. Optim. **22(1-2)**, 199-222 (2001).

[11] Phu, H. X., Rough continuity of linear operators, Numer. Funct. Anal. Optim. **23(1-2)**, 139 - 146 (2002).

[12] Phu, H. X., Rough convergence in infinite dimensional normed spaces, Numer. Funct. Anal. Optimiz. **24(3-4)**, 285-301 (2003).

[13] Sahiner, A.; Gurdal, M. and Duden, F. K., Triple sequences and their statistical convergence, Selcuk J. Appl. Math. **8(2)**, 49-55 (2007).

[14] Sahiner, A.; Tripathy, B. C., Some $\mathcal{I}$-related properties of triple sequences, Selcuk J. Appl. Math. **9(2)**, 9-18 (2008).

[15] Subramanian, N. and Esi, A., The generalized tripled difference of χ^3 sequence spaces, Global Journal of Mathematical Analysis **3(2)**, 54-60 (2015).

ITM Web of Conferences **22**, 01061 (2018) https://doi.org/10.1051/itmconf/20182201061
CMES-2018

On the exact and numerical solutions to a nonlinear model arising in mathematical biology

Asif Yokus[1], *Tukur Abdulkadir* Sulaiman, *Haci Mehmet* Baskonus[4], and *Sibel Pasali* Atmaca[5,*]

[1]Department of Actuary, Firat University, Elazig, Turkey
[2]Department of Mathematics, Firat University, Elazig, Turkey
[3]Department of Mathematics, Federal University, Dutse, Jigawa, Nigeria
[4]Department of Computer Engineering, Munzur University, Tunceli, Turkey
[5]Mugla Sitki Kocman University, Mugla, Turkey

Abstract. This study acquires the exact and numerical approximations of a reaction-convection-diffusion equation arising in mathematical biology namely; Murry equation through its analytical solutions obtained by using a mathematical approach; the modified $\exp(-\Psi(\eta))$-expansion function method. We successfully obtained the kink-type and singular soliton solutions with the hyperbolic function structure to this equation. We performed the numerical simulations (3D and 2D) of the obtained analytical solutions under suitable values of parameters. We obtained the approximate numerical and exact solutions to this equation by utilizing the finite forward difference scheme by taking one of the obtained analytical solutions into consideration. We investigate the stability of the finite forward difference method with the equation through the Fourier-Von Neumann analysis. We present the L_2 and L_∞ error norms of the approximations. The numerical and exact approximations are compared and the comparison is supported by a graphic plot. All the computations and the graphics plots in this study are carried out with help of the Matlab and Wolfram Mathematica softwares. Finally, we submit a comprehensive conclusion to this study.

1 Introduction

Partial differential equations (PDEs) serve as the tremendous tools that bond the realistic problems and mathematical theories. PDEs describe various complex real life aspects arising in the various fields of science, such as; mathematical biology, chemistry, mathematical physics and son. It is sometimes difficult to investigate the solutions of these type of models, more especially in their nonlinear forms. Various mathematical approaches have been developed and invested to solve such kind of models [1–18].

The aim of this research is to acquire the exact and numerical approximations of a nonlinear model that arises in mathematical biology through its analytical solutions

*Corresponding Author: sibela@mu.edu.tr

obtained by using the modified exp $(-\Omega(\eta))$-expansion function method (MEFM) [19, 20]. The finite forward difference method [21] is going to be used for the exact and numerical solutions approximation. Mathematical biology is one of the interesting and most exciting modern application of mathematics.

This study considers a particular case of reaction-convection-diffusion equations with a convection term given by [22]:

$$u_t = (A(u)u_x)_x + B(u)u_x + C(u), \tag{1}$$

where $A(u)$, $B(u)$ and $C(u)$ are arbitrary smooth functions.

Eq. (1) stands for the generalized form of various second order nonlinear evolution equations that describe various nonlinear processes in biology [23, 24].

The governing equation: The Murry equation which is the particular case of Eq. (1) and considered as a generalization of the Fisher and Burgers equations is given by [22, 25]:

$$u_t - u_{xx} - auu_x - bu + cu^2 = 0, \tag{2}$$

where a, b and c are nonzero constants.

2 The Analysis of MEFM

In this section, we present the description of the MEFM . The MEFM is designed by improving the popularly known exp $(-\Psi(\eta))$-expansion function method. To explore the search for the new solutions of any given nonlinear partial differential equation (Eq. (3)), we follow the following steps:

$$P(u, u_x, u_x u^2, u_{xx}, u_{xxt}, \ldots), \tag{3}$$

where $u = u(x,t)$ is unknown function, P is a polynomial in $u(x,t)$ and its derivatives and the subscripts stand for the partial derivatives.

Step 1: Consider the wave transformation given as follows

$$u(x,t) = U(\eta), \quad \eta = k(x - ct), \tag{4}$$

using Eq. (4) on Eq. (3), gives the following nonlinear ordinary differential equation (NODE):

$$D(U, U^2, U', U'', \ldots), \tag{5}$$

where D is a polynomial of U and its derivatives and the superscripts stand for the ordinary derivatives of U with respect to η.

Step 2: Assuming that the wave solutions of Eq. (5) can be written in the following form:

$$U(\eta) = \frac{\sum_{i=0}^{N} A_i \left[e^{-\Psi(\eta)}\right]^i}{\sum_{j=0}^{M} B_j \left[e^{-\Psi(\eta)}\right]^j} = \frac{A_0 + A_1 e^{-\Psi} + \ldots + A_N e^{-N\Psi}}{B_0 + B_1 e^{-\Psi} + \ldots + B_M e^{-M\Psi}}, \tag{6}$$

where $A_i, B_j, (0 \leq i \leq N, 0 \leq j \leq M)$ are constants to be found later, such that $A_N \neq o$, $B_M \neq o$, and $\Psi = \Psi(\eta)$ simplifies the following ODE:

$$\Psi'(\eta) = e^{-\Psi(\eta)} + \mu e^{\Psi(\eta)} + \lambda. \tag{7}$$

Eq.(7) has the following families of solutions [26–28]:

Family 1: When $\mu \neq 0$, $\lambda^2 - 4\mu > 0$,

$$\Psi(\eta) = ln\bigg(\frac{-\sqrt{\lambda^2 - 4\mu}}{2\mu} tanh\Big(\frac{\sqrt{\lambda^2 - 4\mu}}{2}(\eta + E)\Big) - \frac{\lambda}{2\mu}\bigg). \tag{8}$$

Family 2: When $\mu \neq 0$, $\lambda^2 - 4\mu < 0$,

$$\Psi(\eta) = ln\bigg(\frac{\sqrt{-\lambda^2 + 4\mu}}{2\mu} tan\Big(\frac{\sqrt{-\lambda^2 + 4\mu}}{2}(\eta + E)\Big) - \frac{\lambda}{2\mu}\bigg). \tag{9}$$

Family 3: When $\mu = 0$, $\lambda \neq 0$ and $\lambda^2 - 4\mu > 0$,

$$\Psi(\eta) = -ln\bigg(\frac{\lambda}{e^{\lambda(\eta+E)} - 1}\bigg). \tag{10}$$

Family 4: When $\mu \neq 0$, $\lambda \neq 0$ and $\lambda^2 - 4\mu = 0$,

$$\Psi(\eta) = ln\bigg(-\frac{2\lambda(\eta + E) + 4}{\lambda^2(\eta + E)}\bigg). \tag{11}$$

Family 5: When $\mu = 0$, $\lambda = 0$ and $\lambda^2 - 4\mu = 0$,

$$\Psi(\eta) = ln(\eta + E), \tag{12}$$

where $A_i, B_j, (0 \leq i \leq N, 0 \leq j \leq M)$, E, λ, μ are coefficients to be found later, and M, N are positive integers that can be obtained by using the balancing technique.

Step 3: Putting Eq. (6) and its derivatives along with the Eq. (7) into Eq. (5) and solving, we find an equation involving polynomial of $e^{-\Psi(\eta)}$. We extract system of equations from that polynomial of $e^{-\Psi(\eta)}$ by summing all the terms of the same power and equating each summation to zero. To find the new solutions of (3), we simplify the system of equations with aid of the Wolfram Mathematica 11 to find the values of the various coefficients $A_i, B_j, (0 \leq i \leq N, 0 \leq j \leq M)$, E, λ, μ. Putting the obtained values of the coefficients along with one of Eqs. (8-12) into Eq. (6), yields new solution to (3).

3 The Analysis of FDM

Presenting the finite forward difference method needs definitions of some notations given as

1. Δx stands for the spatial step
2. Δt stands for the time step

3. $x_i = a + i\Delta x$, $i = 0, 1, 2, ..., N$ are coordinates points of the mesh and $N = \frac{b-a}{\Delta x}$, $t_j = j\Delta x$, $j = 0, 1, 2, ..., M$ and $M = \frac{T}{\Delta t}$.
4. The functions $u(x,t)$ represents the values of the solutions at the grid points which are $u(x_i, t_j) \approx u_{i,j}$. the $u_{i,j}$ and $v_{i,j}$ stands for the numerical solutions of the exact values of $u(x,t)$ at the point (x_i, t_j)

We therefore present the difference operators as follows:

$$H_t u_{i,j} = u_{i,j+1} - u_{i,j}, \tag{13}$$

$$H_x u_{i,j} = u_{i+1,j} - u_{i,j}, \tag{14}$$

$$H_{xx} u_{i,j} = u_{i+1,j} - 2u_{i,j} + u_{i-1,j}. \tag{15}$$

Hence, the partial derivatives can be approximated through the finite difference operators as:

$$\left.\frac{\partial u}{\partial t}\right|_{i,j} = \frac{H_t u_{i,j}}{\Delta t} + O((\Delta t)), \tag{16}$$

$$\left.\frac{\partial u}{\partial x}\right|_{i,j} = \frac{H_x u_{i,j}}{\Delta x} + O((\Delta x)), \tag{17}$$

$$\left.\frac{\partial^2 u}{\partial x^2}\right|_{i,j} = \frac{H_{xx} u_{i,j}}{(\Delta x)^2} + O((\Delta x)^2), \tag{18}$$

Eq. (2) is given in operators form as:

$$\frac{H_t u_{i,j}}{\Delta t} - \frac{H_{xx} u_{i,j}}{(\Delta x)^2} - a u_{i,j}\frac{H_x u_{i,j}}{\Delta x} - b u_{i,j} + c u_{i,j}^2 = 0. \tag{19}$$

The following indexed form is produced by inserting Eq. (13-15) into Eq. (19)

$$u_{i+1,j} = \frac{1}{(\Delta t)(\Delta x) u_{i,j}}\Big(2(\Delta t)u_{i,j} - (\Delta t)u_{i-1,j} - (\Delta x)^2 u_{i,j} + 2\sqrt{5}(\Delta t)(\Delta x)^2 u_{i,j} + (\Delta t)(\Delta x)u_{i,j}^2 + (\Delta t)(\Delta x)^2 u_{i,j}^2 - (\Delta t)u_{i,j+1} + (\Delta x)^2 u_{i,j+1}\Big), \tag{20}$$

4 Von-Neumann Stability Analysis

In this section, we examine the stability of the numerical scheme with the use the Eq. (2) by using the Von-Neumann stability analysis. We consider η^n to be the amplication factor, and the growth factor of a typical Fourier mode to be

$$u_m^n = \eta^n e^{im\Phi}, \tag{21}$$

where $i = \sqrt{-1}$.

To investigate the stability of the numerical scheme with Eq. (2), we linearize the nonlinear terms uu_x, u^2 by making the quantity $\widehat{u} = u$ a local constant. Thus, the nonlinear terms in the equation becomes $\widehat{u}u_x$ and $\widehat{u}^2$ respectively. In this case Eq. (2) becomes

$$u_t - u_{xx} - a\widehat{u}u_x - bu + c\widehat{u}^2 = 0. \tag{22}$$

Inserting Eq. (16), (17), (18) and the Fourier mode (21) into the recurrence relation (22), yields

$$\eta = p + iq, \tag{23}$$

where

$$p = \frac{1}{(\Delta t) - (\Delta x)^2}\Big(2(\Delta t) + a\widehat{u}(\Delta x)(\Delta t) + (\Delta x)^2(c\widehat{u}(\Delta t) \\ -b(\Delta t) - 1) - (\Delta t)(a\widehat{u}(\Delta x))cos(\Phi)\Big) \tag{24}$$

and

$$q = \frac{1}{(\Delta t) - (\Delta x)^2}\Big(1 - a\widehat{u}(\Delta x)\Big)(\Delta t)sin(\Phi). \tag{25}$$

The numerical scheme is said to be unconditionally stable as $|\eta| \leq 1$ [29].

5 L_2 and L_∞ Error Norms

For the test problem used in the present study, exact and numerical approximations of Eq. (2) have been investigated. To see how the exact and numerical approximations are close to each other we use L_2 and L_∞ error norms. The error L_2 norms defined as [?]

$$L_2 = \left\|u^{exact} - u^{numeric}\right\|_2 = \sqrt{h\sum_{j=0}^{N}\left|u_j^{exact} - u_j^{numeric}\right|^2},$$

and L_∞ error norm defined as

$$L_\infty = ||u^{exact} - u^{numeric}||_\infty = \underset{j}{Max}|u_j^{exact} - u_j^{numeric}|.$$

6 Implementation of MEFM

Consider the Murry equation given in Eq. (2).

With the following wave transformation; $u(x,t) = U(\eta)$, $\eta = x - kt$, Eq. (2) is transformed to the following NODE:

$$U'' + kU' + aUU' + bU - cU^2 = 0. \tag{26}$$

We obtain $N = M + 2$ as relation between M and N under the terms U'' and U^2 by balancing technique.

Choosing $M = 1$, gives $N = 3$. Using $M = 1$, $N = 3$ and Eq. (6), yields the following equation:

$$U(\eta) = \frac{A_0 + A_1e^{-\Psi(\eta)} + A_2e^{-2\Psi(\eta)} + A_3e^{-3\Psi(\eta)}}{B_0 + B_1e^{-\Psi(\eta)}}. \tag{27}$$

We substitute Eq. (7), Eq. (27) with its first and second derivatives into Eq. (26) and obtain an equation in $e^{-\Psi(\eta)}$. We collect a set from algebraic equations from the obtained equation by setting each summation of the coefficients of $e^{-\Psi(\eta)}$ with the same power to zero. We simplify the set of equations with the aid of Wolfram Mathematica package and obtained the values of the coefficients involved in the equations in various cases. In order to obtain the wave solutions of Eq. (2), we insert the values of the coefficients into Eq. (27) along with suitable condition from family 1-5.

Case-1:

$$A_0=\frac{B_0}{a}\Big(\lambda-\sqrt{\lambda^2-4\mu}\Big), A_1=\frac{1}{a}\Big(2B_0+\lambda B_1-B_1\sqrt{\lambda^2-4\mu}\Big), A_2=\frac{2B_1}{a}, A_3=0,$$

$b=-\frac{2c}{a}\sqrt{\lambda^2-4\mu}, k=\sqrt{\lambda^2-4\mu}-\frac{2c}{a},$

When $\mu\neq 0$, $\lambda^2-4\mu>0$, we get:

$$u_{1.1}(x,t)=\frac{1}{a^2}\left(-a\sqrt{\lambda^2-4\mu}+a\left(\lambda-\frac{4\mu}{\Big(\lambda+\sqrt{\lambda^2-4\mu}\, tanh[\Psi_{1.1}(x,t)]\Big)}\right)\right), \quad (28)$$

where $\Psi_{1.1}(x,t)=\frac{1}{2}\sqrt{\lambda^2-4\mu}\left(E+x+\Big(\frac{2c}{a}-\sqrt{\lambda^2-4\mu}\Big)t\right)$.

When $\mu=0$, $\lambda\neq 0$ **and** $\lambda^2-4\mu>0$, we get:

$$u_{1.2}(x,t)=\frac{\lambda}{a}\left(coth\Big[\frac{1}{2}\lambda\Big(E+x+\Big(\frac{2c}{a}-\lambda\Big)t\Big)\Big]\right). \quad (29)$$

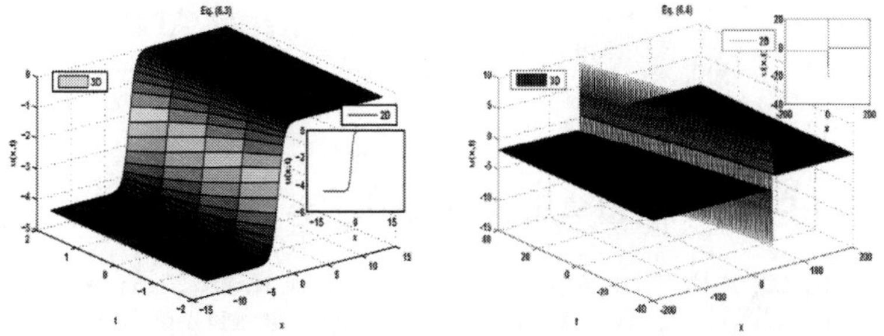

Figure 1. The kink-type and singular soliton surfaces of Eqs. (28) and (29) under the values $E=a=c=\mu=1$, $\lambda=3$, ($-15<x<15$, $-2<t<2$ for Eq. (28)), ($-200<x<200$, $-40<t<40$ for Eq. (29)) and $t=0.5$ for the 2D graphics.

Case-2:

$$A_0=\frac{b}{2c\sqrt{\lambda^2-4\mu}}\Big(B_0\Big(\sqrt{\lambda^2-4\mu}-\lambda\Big)+2\mu B_1\Big), A_2=0, A_3=0, a=-\frac{2c}{b}\sqrt{\lambda^2-4\mu}$$

$A_1 = \frac{b}{2c\sqrt{\lambda^2-4\mu}}\left(B_1\left(\sqrt{\lambda^2-4\mu}+\lambda\right)+2B_0\right), k = \frac{b+\lambda^2-4\mu}{\sqrt{\lambda^2-4\mu}},$

When $\mu \neq 0$**,** $\lambda^2 - 4\mu > 0$, we get:

$$u_{2.1}(x,t) = -\left(2c\sqrt{\lambda^2-4\mu}\left(-2\mu B_1 + B_0\left(\lambda + \sqrt{\lambda^2-4\mu}\, tanh[\Psi_{2.1}(x,t)]\right)\right)\right)^{-1} \times\left(b\left(\left(\lambda\left(\sqrt{\lambda^2-4\mu}-\lambda\right)+4\mu\right)B_0 - 2\mu B_1\sqrt{\lambda^2-4\mu}\right)\right)(tanh[\Psi_{2.1}(x,t)]-1), \tag{30}$$

where $\Psi_{2.1}(x,t) = \frac{1}{2}\left((E+x)\sqrt{\lambda^2-4\mu} - (b+\lambda^2-4\mu)t\right)$.

When $\mu = 0$**,** $\lambda \neq 0$ **and** $\lambda^2 - 4\mu > 0$**,** we get:

$$u_{2.2}(x,t) = \frac{b(\lambda B_1 - B_0)}{c\lambda}\left(\frac{B_0}{\lambda}\left(-1+e^{\lambda\left(E+x-\frac{(b+\lambda^2)}{\lambda}t\right)}\right)+B_1\right)^{-1}. \tag{31}$$

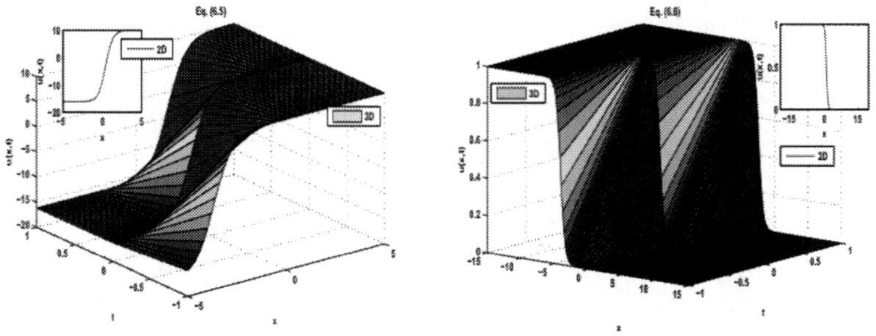

Figure 2. The kink-type surfaces of Eqs. (30) and (31) under the values $B_0 = B_1 = E = b = c = 1$, $\lambda = 3$, ($-5 < x < 5$, $-1 < t < 1$ for Eq. (30)), ($-15 < x < 15$, $-1 < t < 1$ for Eq. (31)) and $t = 0.5$ for the 2D graphics.

7 Exact and Numerical Approximations

In this section, we utilize the numerical scheme in obtaining the exact and numerical approximations to Eq. (2) by considering Eq. (28). Inserting $E = B_1 = a = c = \mu = 1$ and $\lambda = 3$ into Eq. (28), yields the following special exact solution:

$$u(x,t) = 3 - \sqrt{5} - \frac{4}{3+\sqrt{5}tanh\left[\frac{1}{2}\sqrt{5}\left(1+\left(2+\sqrt{5}\right)t+x\right)\right]}, \tag{32}$$

at $t = 0$, Eq. (32) becomes

$$u_0(x) = 3 - \sqrt{5} - \frac{4}{3+\sqrt{5}tanh\left[\frac{1}{2}\sqrt{5}(1+x)\right]}. \tag{33}$$

Inserting $(\Delta x) = (\Delta t) = 0.0001$ into Eq. (20), yields the following indexed form:

$$u_{i+1,j} = \frac{1.000 \times 10^8}{u_{i,j}}\Big(- 0.0001u_{i-1,j} + 0.00019999u_{i,j} + 1.0001 \times 10^{-8}u_{i,j}^2 - 0.00009999u_{i,j+1}\Big). \quad (34)$$

L_2 and L_∞ errors norm table

Table 1.

(Δx)	(Δt)	Numerical solution	Exact Solution	Eror
0	0.01	-0.0686494	-0.0686681	1.86873×10^{-5}
0.01	0.01	-0.0686343	-0.0686530	1.86833×10^{-5}
0.02	0.01	-0.0686192	-0.0686379	1.86792×10^{-5}
0.03	0.01	-0.0686041	-0.0686228	1.86752×10^{-5}
0.04	0.01	-0.0685890	-0.0686077	1.86712×10^{-5}
0.05	0.01	-0.0685739	-0.0685926	1.86671×10^{-5}
0.06	0.01	-0.0685588	-0.0685775	1.86631×10^{-5}

Table 1: Numerical, exact approximations and absolute errors of Eq. (2) by considering Eq. (28), under the value $\Delta(x) = 0.0001$ and $0 \leq x \leq 1$.

Table 2.

$\Delta(x) = \Delta(t)$	L_2	L_∞
0.1	1.0073×10^{-2}	1.9158×10^{-2}
0.01	1.0167×10^{-2}	1.9158×10^{-2}
0.001	5.3000×10^{-5}	1.8690×10^{-4}
0.0001	8.8763×10^{-6}	1.8687×10^{-5}

Figure 3. Numerical and exact approximations of Eq. (2) by considering Eq. (28).

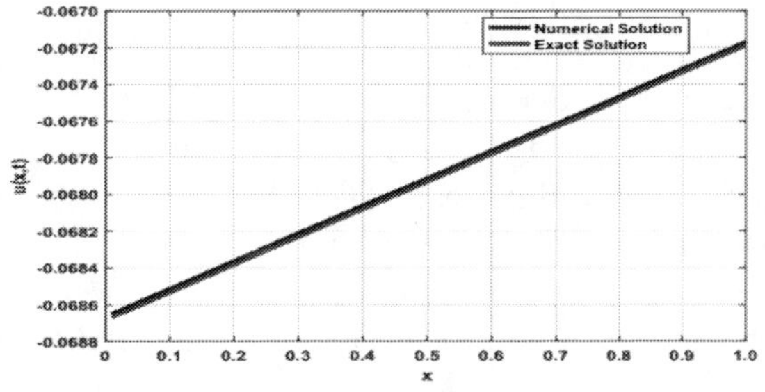

Fig. 3 illustrates the physical properties of the exact and numerical approximations, and it shows that the exact values are almost close to the numerically calculated values. We also observed that from Tab. 1 the truncation errors depend on the choice of Δx and Δt. As Δx and Δt approaching zero we can see that the truncation errors are getting smaller. Also from Tab. 2 as Δx and Δt approaching zero, we can see that the L_2 and L_∞ errors norm is getting smaller which is also indicating how close the exact and numerical approximations are. Fig. 4 illustrates the physical features of the absolute errors.

Figure 4. Absolute error graphic at $\Delta(x) = \Delta(t) = 0.0001$.

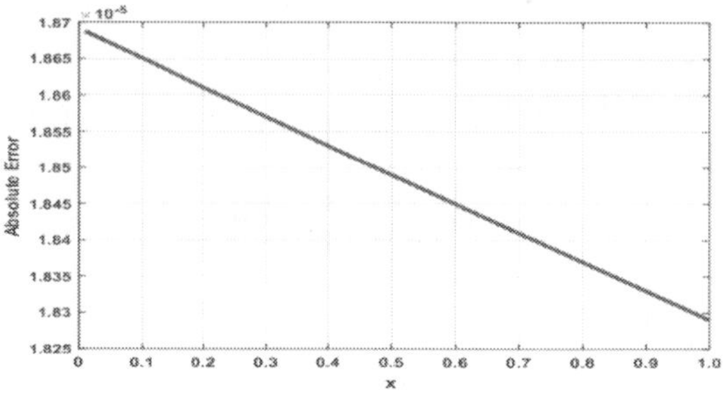

8 Conclusions

In this paper, with help of the Matlab and Wolfram Mathematica softwares, we investigate the solutions of a reaction-convection-diffusion equation, namely; Murry equation by using the modified $\exp(-\Psi(\eta))$-expansion function method. We successfully obtained some kink-type and singular soliton solutions with the hyperbolic function structure to this equation. We presented the interesting 2D and 3D graphics of all the obtained analytical solutions in this paper. Furthermore, through the obtained analytical solutions, we utilized the finite forward difference scheme in obtaining the exact and numerical approximations to this equation. We observed that the exact solutions is almost close to the calculated numerical solutions and we supported this comparison with a graphic plot (Fig. 3). We also observed that as Δx and Δt approaching zero, the truncation errors are getting smaller. The modified $\exp(-\Psi(\eta))$-expansion function method is efficient and simple mathematical tool that can be used to search for the analytical solutions of various nonlinear evolution equations. And, finite forward difference scheme is a powerful numerical method that can be used to find the approximate solutions to various nonlinear evolution equations through their analytical solutions.

References

[1] M. Matinfar, M. Eslami, S. Roshandel, Pramana-J Phys **85(4)**, 593-603 (2015)
[2] H. Bulut, T.A. Sulaiman and B. Demirdag, Nonlinear Dyn. **91** 1985–1991 (2018)
[3] H. Bulut, H.A. Isik and T.A. Sulaiman, ITM Web of Conferences **13**, 01019 (2017)
[4] H. Bulut, T.A. Sulaiman, H.M. Baskonus and T. Akturk, Opt Quant Electron **50**, 134 (2018)
[5] A.S. Alofi, International Mathematical Forum **7(53)**, 2639-2649 (2012)
[6] H. Zhang, Communications in Nonlinear Science and Numerical Simulation **12(5)**, 627-635 (2007)
[7] B. Bülbül and M. Sezer, Journal of Applied Mathematics **2013**, 691614 (2013)
[8] C. Cattani, T.A. Sulaiman, H.M. Baskonus and H. Bulut, **50**, Opt Quant Electron 138 (2018)
[9] L. Zhu and S.L. Wu, International Journal of Biomathematics **10(6)**, 1750086 (2017)
[10] C. Cattani, T.A. Sulaiman, H.M. Baskonus and H. Bulut, Eur. Phys. J. Plus **133**, 228 (2018)

[11] H.M. Baskonus, T.A. Sulaiman and H. Bulut, Opt Quant Electron **50**, 253 (2018)
[12] E. Fan and H. Zhang, Physics Letters A **246(5)**, 403-406 (1998)
[13] H. Bulut, T.A. Sulaiman, H.M. Baskonus, and A.A. Sandulyak, Optik **135**, 327-336 (2017)
[14] S. Guo, Y. Zhou, X Zhao, Applied Mathematics and Computation **215**, 3214-3221 (2010)
[15] L. Mei, Y. Gao, and Z. Chen, Numerical Methods for Partial Differential Equations **31(6)**, 1875-1889 (2015)
[16] A. Esen and S. Kutluay, International Journal of Computer Mathematics 83(3), 319-330 (2006)
[17] S. Kutluay, Applied mathematics and computation 168(2), 1159-1168 (2005)
[18] A. Esen, B. Karaagac and O. Tasbozan, Appl Math Inf Sci Lett 4(1), 1-4 (2016)
[19] S. Duran, M. Askin and T.A. Sulaiman, IJOCTA **7(3)**, 240–247 (2017)
[20] A. Yokus, H.M. Baskonus, T.A. Sulaiman and H. Bulut, Numer Methods Partial Differential Eq. **34**, 211–227 (2018)
[21] A. Yokus, T.A. Sulaiman and H. Bulut, Opt Quant Electron **50**, 31 (2017)
[22] E.H.M. Zahran and M.M.A. Khater, Journal of Research in Applied Sciences **2(4)** 101-107 (2015)
[23] J.D. Murry, *Nonlinear Differential Equation Models in Biology*, (Clarendon Press, Oxford, (1977)) 54–156
[24] J.D. Murry, *Mathematical Biology*, (Springer Berlin, (1989)) 45-123
[25] R.M. Cherniha, Symmetry in Nonlinear Mathematical Physics **1**, 138-146 (1997)
[26] H.O. Roshid and M.A. Rahman, Results Phys **4**, 150-155 (2014)
[27] A.E. Abdelrahman, E.H.M. Zahran and M.M.A. Khater, Int J Mod Nonlinear Theory Appl **4**, 37-47 (2015)
[28] M.G. Hafez, M.N. Alam, and M.A. Akbar, World Appl Sci J **32**, 2150-2155 (2014)
[29] A. Golbabai and E. Mohebianfar, Journal of Nonlinear Science and Applications, Comput Econ **49(2)**, 271-288 (2016)

ITM Web of Conferences **22**, 01062 (2018) https://doi.org/10.1051/itmconf/20182201062
CMES-2018

Construction of various soliton solutions via the simplified extended sinh-Gordon equation expansion method

Tukur Abdulkadir Sulaiman[1,2], *Hasan* Bulut[1,3,*], and *Haci Mehmet* Baskonus[4]

[1]Department of Mathematics, Firat University, Elazig, Turkey
[2]Department of Mathematics, Federal University, Dutse, Jigawa, Nigeria
[3]Department of Mathematics Education, Final International University, Kyrenia, Cyprus
[4]Department of Computer Engineering, Munzur University, Tunceli, Turkey

Abstract. In this paper, we present the simplified version of the extended sinh-Gordon equation expansion method. The newly proposed approach is based on the well-known sinh-Gordon equation and a travelling wave transformation. We successfully employed this approach to the (2+1)-dimensional nonlinear Chiral Schrodinger's and various solitary wave solutions to the studied nonlinear model are successfully constructed. The (2+1)-dimensional nonlinear Chiral Schrodinger's equation describes the edge states of the fractional quantum hall effect. The 2D and 3D surfaces of some of the obtained solutions are plotted.

1 Introduction

Nonlinear evolution equations (NEEs) and their solutions play a vita role in the field of nonlinear science. These equations describe various nonlinear physical aspects that arise in the field of nonlinear science such as fluid mechanics, plasma physics, optic fibers, solid state physics etc. Various studies have been conducted on such type of equations by different scientists from all over the world [1–9].

However, in this study, the extended sinh-Gordon equation expansion method (ShGEEM) is simplified, in the sense that all the ansatzs use for obtaining the values of the parameters involved are derived from one equation ($\theta' = \sqrt{\sigma}\ cosh(\theta)$), which leads to using single system of algebra equations from where the values of the parameters involved can be obtained. To test the simplicity and efficiency of the proposed approach, we apply it in investigating the (2+1)-dimensional Chiral nonlinear Schrödinger equation [10].

The (2+1)-dimensional Chiral nonlinear Schrödinger equation is given by [10]

$$iq_t + a(q_{xx} + q_{yy}) + i\Big(b_1(qq_x^* - q^*q_x) + b_2(qq_y^* - q^*q_y)\Big)q = 0, \tag{1}$$

where q is the complex function of x and t, a is the coefficient of the dispersion terms and b_1, b_2 are nonlinear coupling constants.

The (2+1)-dimensional Chiral nonlinear Schrödinger equation describes the edge states of the fractional quantum hall effect [10].

*Corresponding Author: hbulut@firat.edu.tr

2 The simplified extended ShGEEM

In this section, we present the general facts of the proposed scheme.

To search for the new travelling wave solutions to a given NEE by using the proposed scheme, the following analysis is carried out:

Consider the following equation:

$$F(q,\ qq_x,\ q^2q_t,\ qq_{tt},\ \ldots) = 0, \tag{2}$$

where q is a function of the independent variable x, t and the subscripts indicate the derivative of q with respect to the independent variables and F is the polynomial of the function q.

Consider the following sinh-Gordon equation [11]:

$$q_{xt} = \lambda sinh(q), \tag{3}$$

where q is the unknown function of x, t and $\lambda \in \mathbb{R} \setminus \{0\}$.

Using the following wave transformation

$$q = \Psi(\xi), \quad \xi = x + \vartheta t \tag{4}$$

on Eq. (3), we get the following nonlinear ordinary differential equation (NODE);

$$\Psi'' = \frac{\lambda}{\vartheta} sinh(\Psi), \tag{5}$$

where $\Psi = \Psi(\xi)$, ξ is the amplitude and ϑ the velocity of the travelling wave respectively. Integrating Eq. (5), yields

$$\left[\left(\frac{\Psi}{2}\right)'\right]^2 = \frac{2\lambda}{\vartheta} sinh^2\left(\frac{\Psi}{2}\right) + \frac{\lambda\delta}{\vartheta}, \tag{6}$$

where δ is the integration constant, resulting from integrating Eq. (5).

Setting $\theta = \frac{\Psi}{2}$ and $\sigma = \frac{2\lambda}{\vartheta} = \frac{\lambda\delta}{\vartheta}$, yields

$$\theta' = \sqrt{\sigma}\, cosh(\theta). \tag{7}$$

Eq. (7) is a variables separable equation, simplifying it, yields the following significant solutions:

$$cosh(\theta) = tan\Big(\sqrt{\sigma}(\xi + d)\Big) \text{ or } cosh(\theta) = cot\Big(\sqrt{\sigma}(\xi + d)\Big), \tag{8}$$

$$sinh(\theta) = sec\Big(\sqrt{\sigma}(\xi + d)\Big) \text{ or } sinh(\theta) = csc\Big(\sqrt{\sigma}(\xi + d)\Big), \tag{9}$$

where d is the integration constant, resulting from integrating Eq. (7).

When $\sigma = -1$ and $d = 0$, Eq. (8) and (9) become

$$cosh(\theta) = -i\ tanh(\xi) \text{ or } cosh(\theta) = i\ coth(\xi) \tag{10}$$

and

$$sinh(\theta) = sech(\xi)) \text{ or } sinh(\theta) = i\ csch(\xi) \tag{11}$$

respectively, where $i = \sqrt{-1}$.

To find the new solutions to Eq. (2), we consider the following equation:

$$\Psi(\theta) = \sum_{k=1}^{m} cosh^{k-1}(\theta)[B_k sinh(\theta) + A_k cosh(\theta)] + A_0, \tag{12}$$

and the set of series solutions

$$\Psi(\xi) = \sum_{k=1}^{m} tan^{k-1}\Big(\sqrt{\sigma}(\xi+d)\Big)\Big[B_k sec\Big(\sqrt{\sigma}(\xi+d)\Big) \pm A_k tan\Big(\sqrt{\sigma}(\xi+d)\Big)\Big] + A_0, \tag{13}$$

$$\Psi(\xi) = \sum_{k=1}^{m} cot^{k-1}\Big(\sqrt{\sigma}(\xi+d)\Big)\Big[B_k csc\Big(\sqrt{\sigma}(\xi+d)\Big) \pm A_k cot\Big(\sqrt{\sigma}(\xi+d)\Big)\Big] + A_0, \tag{14}$$

$$\Psi(\xi) = \sum_{k=1}^{m} tanh^{k-1}(\xi)\Big[B_k sech(\xi) \pm i\ A_k tanh(\xi)\Big] + A_0, \tag{15}$$

$$\Psi(\xi) = \sum_{k=1}^{m} coth^{k-1}(\xi)\Big[i\ B_k csch(\xi) \pm i\ A_k coth(\xi)\Big] + A_0. \tag{16}$$

The value of m is determined by balancing the highest power nonlinear term and the highest derivative in the reduced NODE. Setting each summation of the coefficients of $\theta'^s sinh^i(\theta) cosh^j(\theta)$, $(s = 0,\ 1,\ 0 \le k \le m,\ 0 \le j \le m)$ with the same power to zero, produces a set of algebraic equations. Solving this set of algebraic equations with the aid of the Wolfram Mathematical package, yields the values of the coefficients involved. Inserting the obtained values of coefficients into one of Eq. (13), (14), (15) or (16) along with m, gives the new travelling wave solutions to Eq. (2).

3 Application

In this section, we present the application of the simplified extended ShGEEM to the (2+1)-dimensional Chiral nonlinear Schrödinger equation [10] given in Eq. (1).

Consider the following complex wave transformation:

$$q = \Psi(\xi)e^{i\Omega},\ \xi = \alpha x + \beta y - \vartheta t,\ \Omega = rx + sy + \omega t + \varphi \tag{17}$$

into Eq. (1), we get the relation

$$\vartheta = 2a(\alpha r + \beta s) \tag{18}$$

from the imaginary part and

$$a(\alpha^2 + \beta^2)\Psi'' + 2(rb_1 + sb_2)\Psi^3 - (a(r^2 + s^2) + \omega)\Psi = 0 \tag{19}$$

from the real part.

Balancing the terms Ψ'' and Ψ^3 in Eq. (19), gives $m = 1$.

Thus, Eq. (12), (13), (14), (15) and (16) take the following forms

$$\Psi(\theta) = B_1 sinh(\theta) + A_1 cosh(\theta) + A_0, \tag{20}$$

$$\Psi(\xi) = B_1 sec\Big(\sqrt{\sigma}(\xi + d)\Big) \pm A_1 tan\Big(\sqrt{\sigma}(\xi + d)\Big) + A_0, \tag{21}$$

$$\Psi(\xi) = B_1 csc\Big(\sqrt{\sigma}(\xi + d)\Big) \pm A_1 cot\Big(\sqrt{\sigma}(\xi + d)\Big) + A_0, \tag{22}$$

$$\Psi(\xi) = B_1 sech(\xi) \pm i\, A_1 tanh(\xi) + A_0, \tag{23}$$

$$\Psi(\xi) = i\, B_1 csch(\xi) \pm i\, A_1 coth(\xi) + A_0. \tag{24}$$

Substituting Eq. (20) and its second derivative into Eq. (19), we get an equation in hyperbolic functions. After making some hyperbolic function identities substitutions, we collect a group of algebraic equations by equating each summation of the coefficients of the hyperbolic functions with the same power to zero. To get the solutions to Eq. (1), we insert the obtained values of the coefficients into one of Eq. (21), (22), (23) and (24).

Case-I: $A_0 = 0,\ A_1 = B_1 = -\frac{\sqrt{-\sigma\omega(\alpha^2+\beta^2)}}{\sqrt{2(-2s^2-2r^2+\sigma(\alpha^2+\beta^2))(rb_1+sb_2)}}$,
$a = \frac{2\omega}{\sigma(\alpha^2+\beta^2)-2s^2-2r^2}$.

Case-II: $A_0 = 0,\ A_1 = B_1 = -\frac{\sqrt{a^{3/2}\sigma(\alpha^2+\beta^2)\left(\sqrt{2(a(-2s^2+\sigma(\alpha^2+\beta^2))-2\omega)}\, b_1+2sb_2\sqrt{a}\right)}}{2\sqrt{(a(-2s^2+\sigma(\alpha^2+\beta^2))-2\omega)b_1^2-2as^2b_2^2}}$,
$r = -\frac{\sqrt{a(-2s^2+\sigma(\alpha^2+\beta^2))-2\omega}}{\sqrt{2a}}$.

Utilizing these coefficients with Eq. (21), (22), (23) and (24)), we pursue the following set of solutions:

Set-1: The singular periodic wave solutions

Substituting the coefficients in case-I into Eq. (21) and (22), we have

$$\begin{aligned} q_1(x,y,t) &= \frac{\sqrt{-\sigma\omega(\alpha^2+\beta^2)}}{\sqrt{2(-2s^2-2r^2+\sigma(\alpha^2+\beta^2))(rb_1+sb_2)}} \\ &\times\Bigg(\pm tan\Big[\sqrt{\sigma}\Big(d + \alpha x + \beta y - \frac{4\omega(\alpha r+\beta s)}{\sigma(\alpha^2+\beta^2)-2s^2-2r^2}t\Big)\Big] \\ &-sec\Big[\sqrt{\sigma}\Big(d + \alpha x + \beta y - \frac{4\omega(\alpha r+\beta s)}{\sigma(\alpha^2+\beta^2)-2s^2-2r^2}t\Big)\Big]\Bigg) e^{i(rx+sy+\omega t+\varphi)}, \end{aligned} \tag{25}$$

$$q_2(x,y,t) = -\frac{\sqrt{-\sigma\omega(\alpha^2+\beta^2)}}{\sqrt{2(-2s^2-2r^2+\sigma(\alpha^2+\beta^2))(rb_1+sb_2)}}$$
$$\times cot\Big[\frac{1}{2}\sqrt{\sigma}\Big(d+\alpha x+\beta y-\frac{4\omega(\alpha r+\beta s)}{\sigma(\alpha^2+\beta^2)-2s^2-2r^2}t\Big)\Big]e^{i(rx+sy+\omega t+\varphi)}, \qquad (26)$$

$$q_3(x,y,t) = -\frac{\sqrt{-\sigma\omega(\alpha^2+\beta^2)}}{\sqrt{2(-2s^2-2r^2+\sigma(\alpha^2+\beta^2))(rb_1+sb_2)}}$$
$$\times tan\Big[\frac{1}{2}\sqrt{\sigma}\Big(d+\alpha x+\beta y-\frac{4\omega(\alpha r+\beta s)}{\sigma(\alpha^2+\beta^2)-2s^2-2r^2}t\Big)\Big]e^{i(rx+sy+\omega t+\varphi)}, \qquad (27)$$

where $\sigma > 0$ for valid solution.

Substituting the coefficients in case-II into Eq. (21) and (22), we have

$$q_4(x,y,t) = \frac{\sqrt{a^{3/2}\sigma(\alpha^2+\beta^2)\Big(\sqrt{2(a(-2s^2+\sigma(\alpha^2+\beta^2))-2\omega)}\,b_1+2sb_2\sqrt{a}\Big)}}{2\sqrt{(a(-2s^2+\sigma(\alpha^2+\beta^2))-2\omega)b_1^2-2as^2b_2^2}}$$
$$\times\Bigg(\pm tan\Big[\sqrt{\sigma}\Big(d+\alpha x+\beta y-2a\Big(\beta s-\frac{\alpha\sqrt{a(\sigma(\alpha^2+\beta^2)-2s^2)-2\omega}}{\sqrt{2a}}\Big)t\Big)\Big]$$
$$-sec\Big[\sqrt{\sigma}\Big(d+\alpha x+\beta y-2a\Big(\beta s-\frac{\alpha\sqrt{a(\sigma(\alpha^2+\beta^2)-2s^2)-2\omega}}{\sqrt{2a}}\Big)t\Big)\Big]\Bigg)$$
$$\times e^{i\left(sy-\frac{\sqrt{a(-2s^2+\sigma(\alpha^2+\beta^2))-2\omega}}{\sqrt{2a}}x+\omega t+\varphi\right)}, \qquad (28)$$

$$q_5(x,y,t) = -\frac{\sqrt{a^{3/2}\sigma(\alpha^2+\beta^2)\Big(\sqrt{2(a(-2s^2+\sigma(\alpha^2+\beta^2))-2\omega)}\,b_1+2sb_2\sqrt{a}\Big)}}{2\sqrt{(a(-2s^2+\sigma(\alpha^2+\beta^2))-2\omega)b_1^2-2as^2b_2^2}}$$
$$\times cot\Big[\frac{1}{2}\sqrt{\sigma}\Big(d+\alpha x+\beta y-2a\Big(\beta s-\frac{\alpha\sqrt{a(\sigma(\alpha^2+\beta^2)-2s^2)-2\omega}}{\sqrt{2a}}\Big)t\Big)\Big]$$
$$\times e^{i\left(sy-\frac{\sqrt{a(-2s^2+\sigma(\alpha^2+\beta^2))-2\omega}}{\sqrt{2a}}x+\omega t+\varphi\right)}, \qquad (29)$$

$$q_6(x,y,t) = -\frac{\sqrt{a^{3/2}\sigma(\alpha^2+\beta^2)\Big(\sqrt{2(a(-2s^2+\sigma(\alpha^2+\beta^2))-2\omega)}\,b_1+2sb_2\sqrt{a}\Big)}}{2\sqrt{(a(-2s^2+\sigma(\alpha^2+\beta^2))-2\omega)b_1^2-2as^2b_2^2}}$$
$$\times tan\Big[\frac{1}{2}\sqrt{\sigma}\Big(d+\alpha x+\beta y-2a\Big(\beta s-\frac{\alpha\sqrt{a(\sigma(\alpha^2+\beta^2)-2s^2)-2\omega}}{\sqrt{2a}}\Big)t\Big)\Big]$$
$$\times e^{i\left(sy-\frac{\sqrt{a(-2s^2+\sigma(\alpha^2+\beta^2))-2\omega}}{\sqrt{2a}}x+\omega t+\varphi\right)}, \qquad (30)$$

where σ, $2a(a(\sigma(\alpha^2+\beta^2)-2s^2)-2\omega)>0$ for valid solution.

Set-2: Combine dark-bright soliton solutions

Substituting the coefficients in case-I into Eq. (23), we have

$$
\begin{aligned}
q_7(x,y,t) &= \frac{\sqrt{\omega(\alpha^2+\beta^2)}}{\sqrt{2(2s^2+2r^2+\alpha^2+\beta^2)(rb_1+sb_2)}} \\
&\times\Bigg(i\;sech\Big[\alpha x+\beta y+\frac{4(\alpha r+\beta s)}{2s^2+2r^2+\alpha^2+\beta^2}t\Big] \\
\pm tanh\Big[\alpha x+\beta y &+ \frac{4(\alpha r+\beta s)}{2s^2+2r^2+\alpha^2+\beta^2}t\Big]\Bigg)e^{i(rx+sy+\omega t+\varphi)},
\end{aligned} \tag{31}
$$

where $2\omega(\alpha^2+\beta^2)(2s^2+2r^2+\alpha^2+\beta^2)(rb_1+sb_2)>0$ for valid soliton.

Substituting the coefficients in case-II into Eq. (23), we have

$$
\begin{aligned}
q_8(x,y,t) &= \frac{\sqrt{-a^{3/2}(\alpha^2+\beta^2)\Big(\sqrt{-2a(2s^2+\alpha^2+\beta^2)-4\omega}\;b_1+2sb_2\sqrt{a}\Big)}}{2\sqrt{(a(2s^2+\alpha^2+\beta^2)+2\omega)b_1^2+2as^2b_2^2}} \\
&\times\Bigg(i\;sech\Big[\alpha x+\beta y-2a\Big(\beta s-\frac{\alpha\sqrt{-a(2s^2+\alpha^2+\beta^2)-2\omega}}{\sqrt{2a}}\Big)t\Big] \\
&\pm tanh\Big[\alpha x+\beta y-2a\Big(\beta s-\frac{\alpha\sqrt{-a(2s^2+\alpha^2+\beta^2)-2\omega}}{\sqrt{2a}}\Big)t\Big]\Bigg) \\
&\times e^{i\left(sy-\frac{\sqrt{-a(2s^2\alpha^2+\beta^2)}}{\sqrt{2a}}x+\omega t+\varphi\right)},
\end{aligned} \tag{32}
$$

where $a^{3/2}(\alpha^2+\beta^2)<0$, $(2a(2s^2+\alpha^2+\beta^2)+4\omega)<0$, $2a(a(2s^2+\alpha^2+\beta^2)+2\omega)<0$, $(a(2s^2+\alpha^2+\beta^2)+2\omega)b_1^2+2as^2b_2^2>0$ and $a>0$ for valid soliton.

Set-3: The dark soliton solutions

Substituting the coefficients in case-I into Eq. (24), we have

$$
\begin{aligned}
q_9(x,y,t) &= \frac{\sqrt{\omega(\alpha^2+\beta^2)}}{\sqrt{2(2s^2+2r^2+\alpha^2+\beta^2)(rb_1+sb_2)}} \\
\times tanh\Big[\frac{1}{2}\Big(\alpha x+\beta y &+ \frac{4\omega(\alpha r+\beta s)}{2s^2+2r^2+\alpha^2+\beta^2}t\Big)\Big]e^{i(rx+sy+\omega t+\varphi)},
\end{aligned} \tag{33}
$$

where $2\omega(\alpha^2+\beta^2)(2s^2+2r^2+\alpha^2+\beta^2)(rb_1+sb_2)>0$ for valid soliton.

Substituting the coefficients in case-II into Eq. (24), we have

$$q_{10}(x,y,t)=\frac{\sqrt{-a^{3/2}(\alpha^2+\beta^2)\Big(\sqrt{-2a(2s^2+\alpha^2+\beta^2)-4\omega}\;b_1+2sb_2\sqrt{a}\Big)}}{2\sqrt{(a(2s^2+\alpha^2+\beta^2)+2\omega)b_1^2+2as^2b_2^2}}$$
$$\times tanh\Big[\frac{1}{2}\Big(\alpha x+\beta y-2a\Big(\beta s-\frac{\alpha\sqrt{-a(2s^2+\alpha^2+\beta^2)-2\omega}}{\sqrt{2a}}\Big)t\Big)\Big]$$
$$\times e^{i\left(sy-\frac{\sqrt{-a(2s^2\alpha^2+\beta^2)}}{\sqrt{2a}}x+\omega t+\varphi\right)}, \tag{34}$$

where $a^{3/2}(\alpha^2+\beta^2)<0$, $(2a(2s^2+\alpha^2+\beta^2)+4\omega)<0$, $2a(a(2s^2+\alpha^2+\beta^2)+2\omega)<0$, $(a(2s^2+\alpha^2+\beta^2)+2\omega)b_1^2+2as^2b_2^2>0$ and $a>0$ for valid soliton.

Set-4: Singular soliton solutions

Substituting the coefficients in case-I into Eq. (24), we have

$$q_{11}(x,y,t)=-\frac{\sqrt{\omega(\alpha^2+\beta^2)}}{\sqrt{2(2s^2+2r^2+\alpha^2+\beta^2)}(rb_1+sb_2)}$$
$$\times coth\Big[\frac{1}{2}\Big(\alpha x+\beta y+\frac{4\omega(\alpha r+\beta s)}{2s^2+2r^2+\alpha^2+\beta^2}t\Big)\Big]e^{i(rx+sy+\omega t+\varphi)}. \tag{35}$$

Substituting the coefficients in case-I into Eq. (24), we have

$$q_{12}(x,y,t)=-\frac{\sqrt{-a^{3/2}(\alpha^2+\beta^2)\Big(\sqrt{-2a(2s^2+\alpha^2+\beta^2)-4\omega}\;b_1+2sb_2\sqrt{a}\Big)}}{2\sqrt{(a(2s^2+\alpha^2+\beta^2)+2\omega)b_1^2+2as^2b_2^2}}$$
$$\times coth\Big[\frac{1}{2}\Big(\alpha x+\beta y-2a\Big(\beta s-\frac{\alpha\sqrt{-a(2s^2+\alpha^2+\beta^2)-2\omega}}{\sqrt{2a}}\Big)t\Big)\Big]$$
$$\times e^{i\left(sy-\frac{\sqrt{-a(2s^2\alpha^2+\beta^2)}}{\sqrt{2a}}x+\omega t+\varphi\right)}, \tag{36}$$

where $2a(a(2s^2+\alpha^2+\beta^2)+2\omega)<0$ for valid soliton.

4 Results and Discussion

In this section, we discuss the previously presented analytical techniques which have been developed by using the concept of the sinh-Gordon equation and tell the differences with our newly presented simplified approach. Under the choice of suitable values of the parameters, we also present the two- and three-dimensional plots to some of the obtained solutions obtained in this study.

Yan [12] proposed a sinh-Gordon equation expansion method which obtains the Jacobi elliptic function solutions to some nonlinear wave equations. In Yan [12], Eq. (3) is transformed into nonlinear ordinary differential equation (Eq. (5)) by using Eq. (4). When simplifying Eq. (5) by Yan [12], the constant of integration is considered to be nonzero and the following Jacobi elliptic functions are obtained:

$$sinh(\theta) = cs(\xi; m) \text{ and } cosh(\theta) = ns(\xi; m). \tag{37}$$

It is known that when $m \to 1$, $cs(\xi; m) \to csch(\xi)$ and $ns(\xi; m) \to coth(\xi)$ and when $m \to 0$, $cs(\xi; m) \to cot(\xi)$ and $ns(\xi; m) \to csc(\xi)$. For details, see Yan [12].

Xian-Lin and Jia-Shi [13] extended sinh-Gordon equation expansion method developed by Yan [12]. In the extended sinh-Gordon equation expansion method trigonometric and hyperbolic functions solutions can directly be secured. When simplifying Eq. (5), Xian-Lin and Jia-Shi [13] also considered the constant of integration to be zero in the following way:

In Eq. (6), Xian-Lin and Jia-Shi [13] considered $\frac{2\lambda}{\vartheta} = a$ and $\frac{\lambda\delta}{\vartheta} = b$ so that the following equation is secured:

$$\theta' = \sqrt{a\ sinh^2(\theta) + b}. \tag{38}$$

Thus, Xian-Lin and Jia-Shi [13] solved Eq. (38) considering the following two cases:

Case-I: When $a = 1$, $b = 0$, Eq. (38) becomes

$$\theta' = sinh(\theta). \tag{39}$$

Case-II: When $b = 1$, $a = 1$, Eq. (38) becomes

$$\theta' = cosh(\theta). \tag{40}$$

Case-I is solved to give hyperbolic functions solutions and Case-II is solved to give trigonometric functions solutions. This leads to having two individual polynomial equations in the power of hyperbolic functions $(\theta'^s sinh^k(\theta) cosh^j(\theta))$, $(s = 0, 1$ and $k,\ j = 0,\ 1,\ 2,\ \ldots)$ as θ' varies for each case. For details, see Xian-Lin and Jia-Shi [13].

In this study, we transform the same sinh-Gordon equation Eq. (3) as in Yan [12], Xian-Lin and Jia-Shi [13] by using Eq. (4). We also consider a nonzero constant of integration when simplifying Eq. (5), but in different way with Yan [12], Xian-Lin and Jia-Shi [13], as it can be seen from Eq. (6) in section 2. Doing this, we obtain Eq. (7). Eq. (7) is a variable separable equation, we simplify it and obtain the following trigonometric function solutions:

$$cosh(\theta) = tan\Big(\sqrt{\sigma}(\xi + d)\Big) \ \text{ or } \ cosh(\theta) = cot\Big(\sqrt{\sigma}(\xi + d)\Big), \tag{41}$$

$$sinh(\theta) = sec\Big(\sqrt{\sigma}(\xi + d)\Big) \ \text{ or } \ sinh(\theta) = csc\Big(\sqrt{\sigma}(\xi + d)\Big). \tag{42}$$

When we consider $\sigma = -1$ and $d = 0$ in Eq. (41) and (42), we directly obtain the following hyperbolic function solutions:

$$cosh(\theta) = -i\ tanh(\xi) \ \text{ or } \ cosh(\theta) = i\ coth(\xi), \tag{43}$$

$$sinh(\theta) = sech(\xi)) \ \text{ or } \ sinh(\theta) = i\ csch(\xi). \tag{44}$$

Putting these solutions together, we have the forms of solutions to Eq. (2) presented in section 2 (that is Eq. (13), (14), (15) and (16)) which can be used to obtain different travelling wave solutions to various nonlinear evolution equations.

It can be seen that in this study, all the solutions can be generated from Eq. (7) ($\sqrt{\sigma}\ cosh(\theta)$) which overcome the use of two individual polynomial equations in the power of hyperbolic functions ($\theta'^{s} sinh^{k}(\theta) cosh^{j}(\theta)$), ($s = 0, 1$ and $k,\ j = 0,\ 1,\ 2,\ \ldots$) as we have only one θ'. Thus, one single polynomial equation can be used in collecting the system of algebraic equations and this is enough to obtained the solutions of the considered nonlinear model after obtaining the values of the parameters involved by solving the system of algebraic equations.

5 Conclusion

In this study, we present the simplified version of the extended sinh-Gordon equation expansion method. The efficiency of the simplified approach is tested on the (2+1)-dimensional nonlinear Chiral Schrödinger equation. We successfully construct singular periodic wave, dark soliton, combine dark-bright soliton, singular soliton and combine singular soliton solutions to these nonlinear complex models. We observed that the simplified sinh-Gordon equation expansion method is easy, direct and powerful mathematical approach that can be used to solved various nonlinear evolution equations.

References

[1] H. Bulut, H.A. Isik and T.A. Sulaiman, ITM Web of Conferences **13**, 01019 (2017)

[2] X.J. Yang, J.A.T. Machado and D. Baleanu, Fractals **25(4)**, 1740006 (2017)

[3] A. Neirameh, Fractals, **55**, 2839-2847 (2016)

[4] S. Bibi and S.T. Mohyud-Din, Journal of the Association of Arab Universities for Basic and Applied Sciences **15** 90-93 (2014)

[5] T.A. Sulaiman, T. Akturk, H. Bulut and H.M. Baskonus, Journal of Electromagnetic Waves and Applications **32(9)**, 1093–1105 (2017)

[6] S. Duran, M. Askin and T.A. Sulaiman, IJOCTA **7(3)**, 240–247 (2017)

[7] A. Yokus, H.M. Baskonus, T.A. Sulaiman and H. Bulut, Numer Methods Partial Differential Eq. **34**, 211–227 (2018)

[8] A.M. Wazwaz and S.A. El-Tantawy, Nonlinear Dynamics **83(3)** 1529-1534 (2016)

[9] R.I. Nuruddeena, L. Muhammad, A.M. Nass and T.A. Sulaiman, Palestine Journal of Mathematics **7(1)** 262–280 (2018)

[10] H. Bulut, T.A. Sulaiman and B. Demirdag, Nonlinear Dyn. **91** 1985–1991 (2018)

[11] M.A. Ablowitz and P.A. Clarkson, *Solitons, Nonlinear Evolution Equations and Inverse Scattering*, (University Press, Cambridge, (1991)) 115-235

[12] Z. Yan, MMRC, AMSS, Academia Sinica, Beijing **22** 363-375 (2003)

[13] X. Xian-Lin and T. Jia-Shi, Commun. Theor. Phys. **50**, 1047 (2008)

On the exact solitary wave solutions to the long-short wave interaction system

Haci Mehmet Baskonus[1,*], *Tukur Abdulkadir* Sulaiman[2,3], and *Hasan* Bulut[2,4]

[1]Department of Computer Engineering, Munzur University, Tunceli, Turkey
[2]Department of Mathematics, Firat University, Elazig, Turkey
[3]Department of Mathematics, Federal University, Dutse, Jigawa, Nigeria
[4]Department of Mathematics Education, Final International University, Kyrenia, Cyprus

Abstract. In this paper, the application of the simplified the extended sinh-Gordon equation expansion method to the long-short-wave interaction system. We successfully construct various solitary wave solutions to this nonlinear complex model. The long-short-wave interaction system describes the interaction between one long longitudinal wave and one short transverse wave propagating in a generalized elastic medium. The 2D and 3D surfaces to some of the obtained solutions are plotted.

1 Introduction

Nonlinear evolution equations (NEEs) and their solutions play a vita role in the field of nonlinear science. These equations describe various nonlinear physical aspects that arise in the field of nonlinear science such as fluid mechanics, plasma physics, optic fibers, solid state physics etc. Various studies have been conducted on such types of equation by different scientists from all over the world [1–17].

However, in this study, the extended sinh-Gordon equation expansion method (ShGEEM) is simplified, in the sense that all the ansatzs use for obtaining the values of the parameters involved are derived from one equation ($\theta' = \sqrt{\sigma}\ cosh(\theta)$), which leads to using single system of algebra equations from where the values of the parameters involved can be obtained. To test the simplicity and efficiency of the proposed approach, we apply it in investigating the long-short-wave interaction system (LSWIS) [18].

The long-short-wave interaction system is given by [18]

$$\begin{aligned} iq_t + q_{xx} - qv &= 0, \\ v_t + v_x + (|q|^2)_x &= 0, \end{aligned} \tag{1}$$

where q is a complex function, v is a real function, x is the spatial coordinate, and t is the time respectively.

Eq. (1) describes the interaction between one long longitudinal wave and one short transverse wave propagating in a generalized elastic medium. the function v

*Corresponding Author: hmbaskonus@gmail.com

portrays the long longitudinal wave and q is the slowly varying envelope of the short transverse wave [18].

2 The simplified extended ShGEEM

In this section, we present the general facts of the proposed scheme.

To search for the new travelling wave solutions to a given NEE by using the proposed scheme, the following analysis is carried out:

Consider the following equation:

$$F(q,\ qq_x,\ q^2q_t,\ qq_{tt},\ \ldots) = 0, \tag{2}$$

where q is a function of the independent variable x, t and the subscripts indicate the derivative of q with respect to the independent variables and F is the polynomial of the function q.

Consider the following sinh-Gordon equation [19]:

$$q_{xt} = \lambda sinh(q), \tag{3}$$

where q is the unknown function of x, t and $\lambda \in \mathbb{R} \setminus \{0\}$.

Using the following wave transformation

$$q = \Psi(\xi), \quad \xi = x + \vartheta t \tag{4}$$

on Eq. (3), we get the following nonlinear ordinary differential equation (NODE);

$$\Psi'' = \frac{\lambda}{\vartheta} sinh(\Psi), \tag{5}$$

where $\Psi = \Psi(\xi)$, ξ is the amplitude and ϑ the velocity of the travelling wave respectively. Integrating Eq. (5), yields

$$\left[\left(\frac{\Psi}{2}\right)'\right]^2 = \frac{2\lambda}{\vartheta} sinh^2\left(\frac{\Psi}{2}\right) + \frac{\lambda\delta}{\vartheta}, \tag{6}$$

where δ is the integration constant, resulting from integrating Eq. (5).

Setting $\theta = \frac{\Psi}{2}$ and $\sigma = \frac{2\lambda}{\vartheta} = \frac{\lambda\delta}{\vartheta}$, yields

$$\theta' = \sqrt{\sigma}\, cosh(\theta). \tag{7}$$

Eq. (7) is a variables separable equation, simplifying it, yields the following significant solutions:

$$cosh(\theta) = tan\Big(\sqrt{\sigma}(\xi + d)\Big) \ \text{ or } \ cosh(\theta) = cot\Big(\sqrt{\sigma}(\xi + d)\Big), \tag{8}$$

$$sinh(\theta) = sec\Big(\sqrt{\sigma}(\xi + d)\Big) \ \text{ or } \ sinh(\theta) = csc\Big(\sqrt{\sigma}(\xi + d)\Big), \tag{9}$$

where d is the integration constant, resulting from integrating Eq. (7).

When $\sigma = -1$ and $d = 0$, Eq. (8) and (9) become

$$cosh(\theta) = -i\ tanh(\xi) \text{ or } cosh(\theta) = i\ coth(\xi) \tag{10}$$

and

$$sinh(\theta) = sech(\xi)) \text{ or } sinh(\theta) = i\ csch(\xi) \tag{11}$$

respectively, where $i = \sqrt{-1}$.

To find the new solutions to Eq. (2), we consider the following equation:

$$\Psi(\theta) = \sum_{k=1}^{m} cosh^{k-1}(\theta)[B_k sinh(\theta) + A_k cosh(\theta)] + A_0, \tag{12}$$

and the set of series solutions

$$\Psi(\xi) = \sum_{k=1}^{m} tan^{k-1}\Big(\sqrt{\sigma}(\xi+d)\Big)\Big[B_k sec\Big(\sqrt{\sigma}(\xi+d)\Big) \pm A_k tan\Big(\sqrt{\sigma}(\xi+d)\Big)\Big] + A_0, \tag{13}$$

$$\Psi(\xi) = \sum_{k=1}^{m} cot^{k-1}\Big(\sqrt{\sigma}(\xi+d)\Big)\Big[B_k csc\Big(\sqrt{\sigma}(\xi+d)\Big) \pm A_k cot\Big(\sqrt{\sigma}(\xi+d)\Big)\Big] + A_0, \tag{14}$$

$$\Psi(\xi) = \sum_{k=1}^{m} tanh^{k-1}(\xi)\Big[B_k sech(\xi) \pm i\ A_k tanh(\xi)\Big] + A_0, \tag{15}$$

$$\Psi(\xi) = \sum_{k=1}^{m} coth^{k-1}(\xi)\Big[i\ B_k csch(\xi) \pm i\ A_k coth(\xi)\Big] + A_0. \tag{16}$$

The value of m is determined by balancing the highest power nonlinear term and the highest derivative in the reduced NODE. Setting each summation of the coefficients of $\theta'^s sinh^i(\theta) cosh^j(\theta)$, $(s = 0,\ 1,\ 0 \le k \le m,\ 0 \le j \le m)$ with the same power to zero, produces a set of algebraic equations. Solving this set of algebraic equations with the aid of the Wolfram Mathematical package, yields the values of the coefficients involved. Inserting the obtained values of coefficients into one of Eq. (13), (14), (15) or (16) along with m, gives the new travelling wave solutions to Eq. (2).

3 Application

In this section, we present the application of the simplified extended ShGEEM to the long-short-wave interaction system [18] given in Eq. (1).

Consider the following complex wave transformation:

$$q = \Psi(\xi)e^{i\Omega},\ v = V(\xi),\ \xi = x - \vartheta t,\ \Omega = px + rt. \tag{17}$$

Substituting Eq. (17) into Eq. (1), the relation $\vartheta = 2p$ is obtained from the imaginary part and the following single NODE is obtained from the real part:

$$(1-2p)\Psi'' + \Psi^3 - (1-2p)(p^2+r)\Psi = 0. \tag{18}$$

Balancing the terms Ψ'' and Ψ^3 in Eq. (18), we get $m = 1$.

Thus, Eq. (12), (13), (14), (15) and (16) take the following forms:

$$\Psi(\theta) = B_1 sinh(\theta) + A_1 cosh(\theta) + A_0, \tag{19}$$

$$\Psi(\xi) = B_1 sec\Big(\sqrt{\sigma}(\xi + d)\Big) \pm A_1 tan\Big(\sqrt{\sigma}(\xi + d)\Big) + A_0, \tag{20}$$

$$\Psi(\xi) = B_1 csc\Big(\sqrt{\sigma}(\xi + d)\Big) \pm A_1 cot\Big(\sqrt{\sigma}(\xi + d)\Big) + A_0, \tag{21}$$

$$\Psi(\xi) = B_1 sech(\xi) \pm i\ A_1 tanh(\xi) + A_0, \tag{22}$$

$$\Psi(\xi) = i\ B_1 csch(\xi) \pm i\ A_1 coth(\xi) + A_0. \tag{23}$$

Substituting Eq. (19) and its second derivative into Eq. (18), we get an equation in hyperbolic functions. After making some hyperbolic function identities substitutions, we collect a group of algebraic equations by equating each summation of the coefficients of the hyperbolic functions with the same power to zero. To get the solutions to Eq. (1), we insert the obtained values of the coefficients into one of Eq. (20), (21), (22) and (23).

Case-I:

$$A_0 = 0,\ A_1 = -\sqrt{\sigma(p - \frac{1}{2})},\ B_1 = \sqrt{\sigma(p - \frac{1}{2})},\ r = \frac{1}{2}(\sigma - 2p^2).$$

Case-II:

$$A_0 = 0,\ A_1 = -\sqrt{(p^2 + r)(2p - 1)},\ B_1 = \sqrt{(p^2 + r)(2p - 1)},\ \sigma = 2(p^2 + r).$$

The following set of solutions are obtained with respect to each case:

Set-1: Singular periodic waves solutions.

Substituting the coefficients in case-I into Eq. (20) and (21), we have

$$q_1(x,t) = \sqrt{\sigma\Big(p - \frac{1}{2}\Big)}\Big(sec\Big[\sqrt{\sigma}(x-2pt+d)\Big] \pm tan\Big[\sqrt{\sigma}(x-2pt+d)\Big]\Big)e^{i\left(px+\frac{1}{2}(\sigma-2p^2)t\right)}, \tag{24}$$

$$v_1(x,t) = -\frac{\sigma\Big(p-\frac{1}{2}\Big)}{1-2p}\Big(sec\Big[\sqrt{\sigma}(x-2pt+d)\Big] \pm tan\Big[\sqrt{\sigma}(x-2pt+d)\Big]\Big)^2. \tag{25}$$

$$q_2(x,t) = \sqrt{\sigma\Big(p-\frac{1}{2}\Big)}tan\Big[\frac{1}{2}\sqrt{\sigma}(x-2pt+d)\Big]e^{i\left(px+\frac{1}{2}(\sigma-2p^2)t\right)}, \tag{26}$$

$$v_2(x,t) = -\frac{\sigma\Big(p-\frac{1}{2}\Big)}{1-2p}tan^2\Big[\frac{1}{2}\sqrt{\sigma}(x-2pt+d)\Big]. \tag{27}$$

$$q_3(x,t)=\sqrt{\sigma\Big(p-\frac{1}{2}\Big)}cot\Big[\frac{1}{2}\sqrt{\sigma}(x-2pt+d)\Big]e^{i\left(px+\frac{1}{2}(\sigma-2p^2)t\right)}, \tag{28}$$

$$v_3(x,t)=-\frac{\sigma\left(p-\frac{1}{2}\right)}{1-2p}cot^2\Big[\frac{1}{2}\sqrt{\sigma}(x-2pt+d)\Big], \tag{29}$$

where $\sigma>0$ for valid solutions.

Substituting the coefficients in case-II into Eq. (20) and (21), we have

$$q_4(x,t)=\sqrt{(p^2+r)(2p-1)}\Big(sec\Big[\sqrt{2(p^2+r)}(x-2pt+d)\Big] \pm tan\Big[\sqrt{2(p^2+r)}(x-2pt+d)\Big]\Big)e^{i(px+rt)}, \tag{30}$$

$$v_4(x,t)=-\frac{(p^2+r)(2p-1)}{1-2p}\Big(sec\Big[\sqrt{2(p^2+r)}(x-2pt+d)\Big] \pm tan\Big[\sqrt{2(p^2+r)}(x-2pt+d)\Big]\Big)^2, \tag{31}$$

$$q_5(x,t)=\sqrt{(p^2+r)(2p-1)}tan\Big[\sqrt{\frac{p^2+r}{2}}(x-2pt+d)\Big]e^{i(px+rt)}, \tag{32}$$

$$v_5(x,t)=-\frac{(p^2+r)(2p-1)}{1-2p}tan^2\Big[\sqrt{\frac{p^2+r}{2}}(x-2pt+d)\Big], \tag{33}$$

$$q_6(x,t)=\sqrt{(p^2+r)(2p-1)}cot\Big[\sqrt{\frac{p^2+r}{2}}(x-2pt+d)\Big]e^{i(px+rt)}, \tag{34}$$

$$v_6(x,t)=-\frac{(p^2+r)(2p-1)}{1-2p}cot^2\Big[\sqrt{\frac{p^2+r}{2}}(x-2pt+d)\Big], \tag{35}$$

where $2(p^2+r)>0$ for valid solution.

Set-2: The soliton solutions

Substituting the coefficients in case-I into Eq. (22) and (23), we have

$$q_7(x,t)=\sqrt{p-\frac{1}{2}}\Big(i\ sech[2pt-x]\pm tanh[2pt-x]\Big)e^{i(px-\frac{1}{2}(1+2p^2)t)}, \tag{36}$$

$$v_7(x,t)=-\frac{\left(p-\frac{1}{2}\right)}{1-2p}\Big(i\ sech[2pt-x]\pm tanh[2pt-x]\Big)^2. \tag{37}$$

$$q_8(x,t) = \sqrt{p - \frac{1}{2}}tanh\Big[\frac{1}{2}(x - 2pt)\Big]e^{i(px - \frac{1}{2}(1+2p^2)t)}, \tag{38}$$

$$v_8(x,t) = -\frac{\left(p - \frac{1}{2}\right)}{1-2p}tanh^2\Big[\frac{1}{2}(x - 2pt)\Big]. \tag{39}$$

Substituting the coefficients in case-II into Eq. (22) and (23), we have

$$q_9(x,t) = \frac{1}{\sqrt{2}}\sqrt{-1+\sqrt{-2-4r}}\Big(i\ sech\Big[\sqrt{2(-1-2r)}\ t - x\Big] \pm tanh\Big[\sqrt{2(-1-2r)}\ t - x\Big]\Big)e^{i\left(\frac{\sqrt{-1-2r}}{\sqrt{2}}x + rt\right)}, \tag{40}$$

$$v_9(x,t) = -\frac{\Big(-1+\sqrt{-2-4r}\Big)}{2\Big(1-\sqrt{2(-1-2r)}\Big)}\Big(i\ sech\Big[\sqrt{2(-1-2r)}\ t - x\Big] \pm tanh\Big[\sqrt{2(-1-2r)}\ t - x\Big]\Big)^2, \tag{41}$$

$$q_{10}(x,t) = \frac{1}{\sqrt{2}}\sqrt{-1+\sqrt{-2-4r}}\ tanh\Big[\frac{1}{2}\Big(x - \sqrt{2(-1-2r)}\ t\Big)\Big]e^{i\left(\frac{\sqrt{-1-2r}}{\sqrt{2}}x + rt\right)}, \tag{42}$$

$$v_{10}(x,t) = -\frac{\Big(-1+\sqrt{-2-4r}\Big)}{2\Big(1-\sqrt{2(-1-2r)}\Big)}tanh^2\Big[\frac{1}{2}\Big(x - \sqrt{2(-1-2r)}\ t\Big)\Big], \tag{43}$$

where $r \leq -1$ for valid soliton solution.

Set-3: The combine singular soliton solutions

Substituting the coefficients in case-I into Eq. (23), we have

$$q_{11}(x,t) = \sqrt{p - \frac{1}{2}}\Big(coth[2pt - x] + csch[2pt - x]\Big)e^{i\left(px - \frac{1}{2}(1+2p^2)t\right)}, \tag{44}$$

$$v_{11}(x,t) = -\frac{\left(p - \frac{1}{2}\right)}{1-2p}\Big(coth[2pt - x] + csch[2pt - x]\Big)^2. \tag{45}$$

Substituting the coefficients in case-II into Eq. (23), we have

$$q_{12}(x,t) = \frac{1}{\sqrt{2}}\sqrt{-1+\sqrt{-2-4r}}\Big(coth\Big[\sqrt{2(-1-2r)}\ t - x\Big] + csch\Big[\sqrt{2(-1-2r)}\ t - x\Big]\Big)e^{i\left(\frac{\sqrt{-1-2r}}{\sqrt{2}}x + rt\right)}, \tag{46}$$

$$v_{12}(x,t) = -\frac{\left(-1+\sqrt{-2-4r}\right)}{2\left(1-\sqrt{2(-1-2r)}\right)}\left(coth\left[\sqrt{2(-1-2r)}\,t-x\right]\right.$$
$$\left.+csch\left[\sqrt{2(-1-2r)}\,t-x\right]\right)^2, \qquad (47)$$

where $r \leq -1$ for valid soliton solution.

4 Results and Discussion

In this section, we discuss the previously presented analytical techniques which have been developed by using the concept of the sinh-Gordon equation and tell the differences with our newly presented simplified approach. Under the choice of suitable values of the parameters, we also present the two- and three-dimensional plots to some of the obtained solutions obtained in this study.

Yan [20] proposed a sinh-Gordon equation expansion method which obtains the Jacobi elliptic function solutions to some nonlinear wave equations. In Yan [20], Eq. (3) is transformed into nonlinear ordinary differential equation (Eq. (5)) by using Eq. (4). When simplifying Eq. (5) by Yan [20], the constant of integration is considered to be nonzero and the following Jacobi elliptic functions are obtained:

$$sinh(\theta) = cs(\xi;m) \text{ and } cosh(\theta) = ns(\xi;m). \qquad (48)$$

It is known that when $m \to 1$, $cs(\xi;m) \to csch(\xi)$ and $ns(\xi;m) \to coth(\xi)$ and when $m \to 0$, $cs(\xi;m) \to cot(\xi)$ and $ns(\xi;m) \to csc(\xi)$. For details, see Yan [20].

Xian-Lin and Jia-Shi [21] extended sinh-Gordon equation expansion method developed by Yan [20]. In the extended sinh-Gordon equation expansion method trigonometric and hyperbolic functions solutions can directly be secured. When simplifying Eq. (5), Xian-Lin and Jia-Shi [21] also considered the constant of integration to be zero in the following way:

In Eq. (6), Xian-Lin and Jia-Shi [21] considered $\frac{2\lambda}{\vartheta} = a$ and $\frac{\lambda\delta}{\vartheta} = b$ so that the following equation is secured:

$$\theta' = \sqrt{a\ sinh^2(\theta) + b}. \qquad (49)$$

Thus, Xian-Lin and Jia-Shi [21] solved Eq. (49) considering the following two cases:

Case-I: When $a = 1$, $b = 0$, Eq. (49) becomes

$$\theta' = sinh(\theta). \qquad (50)$$

Case-II: When $b = 1$, $a = 1$, Eq. (49) becomes

$$\theta' = cosh(\theta). \qquad (51)$$

Case-I is solved to give hyperbolic functions solutions and Case-II is solved to give trigonometric functions solutions. This leads to having two individual polynomial equations in the power of hyperbolic functions $(\theta'^s sinh^k(\theta)cosh^j(\theta))$, $(s =$

$0, 1$ and $k,\ j = 0,\ 1,\ 2,\ \ldots$) as θ' varies for each case. For details, see Xian-Lin and Jia-Shi [21].

In this study, we transform the same sinh-Gordon equation Eq. (3) as in Yan [20], Xian-Lin and Jia-Shi [21] by using Eq. (4). We also consider a nonzero constant of integration when simplifying Eq. (5), but in different way with Yan [20], Xian-Lin and Jia-Shi [21], as it can be seen from Eq. (6) in section 2. Doing this, we obtain Eq. (7). Eq. (7) is a variable separable equation, we simplify it and obtain the following trigonometric function solutions:

$$cosh(\theta) = tan\Big(\sqrt{\sigma}(\xi + d)\Big) \text{ or } cosh(\theta) = cot\Big(\sqrt{\sigma}(\xi + d)\Big), \tag{52}$$

$$sinh(\theta) = sec\Big(\sqrt{\sigma}(\xi + d)\Big) \text{ or } sinh(\theta) = csc\Big(\sqrt{\sigma}(\xi + d)\Big). \tag{53}$$

When we consider $\sigma = -1$ and $d = 0$ in Eq. (52) and (53), we directly obtain the following hyperbolic function solutions:

$$cosh(\theta) = -i\ tanh(\xi) \text{ or } cosh(\theta) = i\ coth(\xi), \tag{54}$$

$$sinh(\theta) = sech(\xi)) \text{ or } sinh(\theta) = i\ csch(\xi). \tag{55}$$

Putting these solutions together, we have the forms of solutions to Eq. (2) presented in section 2 (that is Eq. (13), (14), (15) and (16)) which can be used to obtain different travelling wave solutions to various nonlinear evolution equations.

It can be seen that in this study, all the solutions can be generated from Eq. (7) ($\sqrt{\sigma}\ cosh(\theta)$) which overcome the use of two individual polynomial equations in the power of hyperbolic functions ($\theta'^{s} sinh^{k}(\theta) cosh^{j}(\theta)$), ($s = 0, 1$ and $k,\ j = 0,\ 1,\ 2,\ \ldots$) as we have only one θ'. Thus, one single polynomial equation can be used in collecting the system of algebraic equations and this is enough to obtained the solutions of the considered nonlinear model after obtaining the values of the parameters involved by solving the system of algebraic equations.

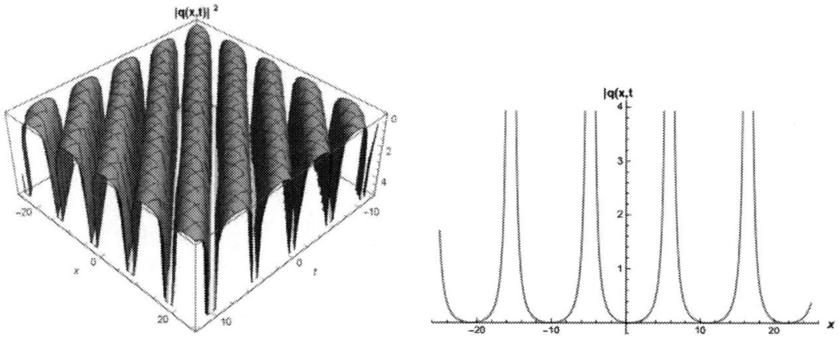

Figure 1. The 3D and 2D surfaces of Eq. (26) under the values $\sigma = 0.35$, $p = 1$, $d = 0.5$, $-25 < x < 25$, $-12 < t < 12$ and $t = 0.5$ for the 2D graphic.

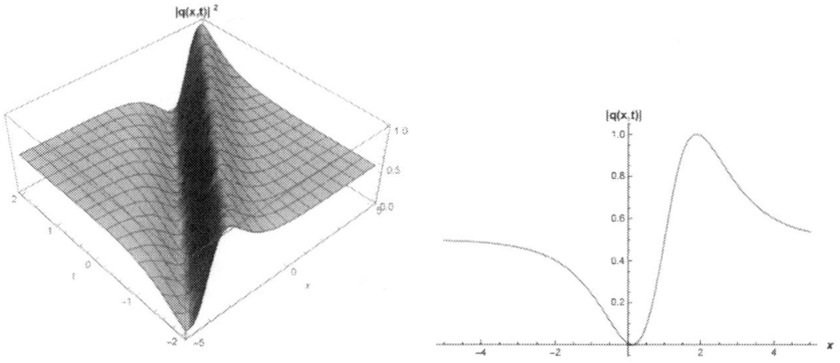

Figure 2. The 3D and 2D surfaces of Eq. (36) under the values $p = 1$, $-5 < x < 5$, $-2 < t < 2$ and $t = 0.5$ for the 2D graphic.

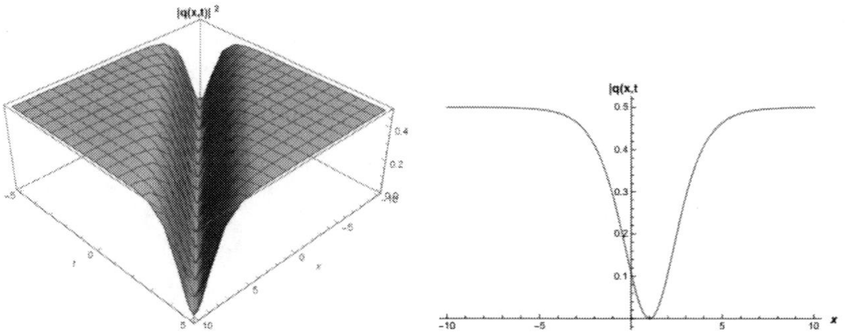

Figure 3. The 3D and 2D surfaces of Eq. (38) under the values $p = 1$, $-10 < x < 10$, $-5 < t < 5$ and $t = 0.5$ for the 2D graphic.

5 Conclusion

In this study, we present the simplified version of the extended sinh-Gordon equation expansion method. The efficiency of the simplified approach is tested with the long-short-wave interaction system. We successfully construct singular periodic waves, soliton and singular soliton solutions to this nonlinear model. We observed that the simplified sinh-Gordon equation expansion method is easy, direct and powerful mathematical approach that can be used to solved various nonlinear evolution equations.

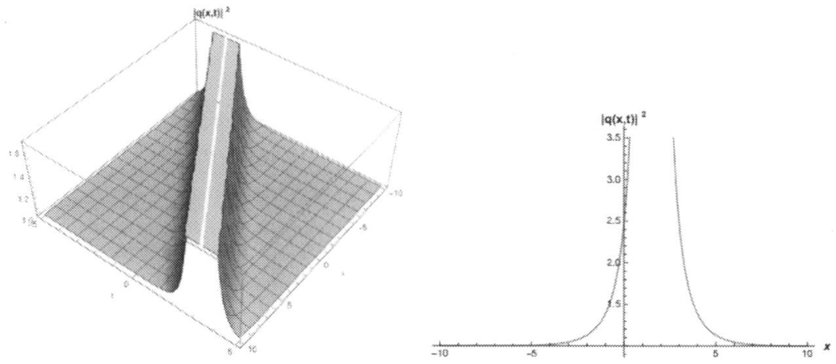

Figure 4. The 3D and 2D surfaces of Eq. (44) under the values $p = 1.5$, $-10 < x < 10$, $-5 < t < 5$ and $t = 0.5$ for the 2D graphic.

References

[1] H. Bulut, T.A. Sulaiman and H.M. Baskonus, Optical and Quantum Electronics **50**, 87 (2018)
[2] M.A. Abdou and A.A. Soliman, Journal of Computational and Applied Mathematics **181**, 245-251 (2005)
[3] J. Lu, Computers and Mathematics with Applications **54**, 1133-1138 (2007)
[4] H. Bulut, T.A. Sulaiman, H.M. Baskonus and A.A. Sandulyak, Optik **135**, 327-336 (2017)
[5] H.M. Baskonus, H. Bulut and T.A. Sulaiman, Indian J. Phys. **135**, 327-336 (2017)
[6] M.M.A. Khater, E.H.M. Zahran and M.S.M. Shehata, Journal of the Egyptian Mathematical Society **25(1)**, 8-12 (2017)
[7] H. Bulut, T.A. Sulaiman, H.M. Baskonus and F. Erdogan, Eur. Phys. J. Plus **132**, 350 (2017)
[8] M. Eslami, M.A. Mirzazadeh and A. Neirameh, Pramana **84(1)**, 3–8 (2015)
[9] S. Duangpithak, Int. Journal of Math. Analysis **6(22)**, 1071-1077 (2012)
[10] S. Duran, M. Askin and T.A. Sulaiman, IJOCTA **7(3)**, 240–247 (2017)
[11] H. Bulut, H.A. Isik and T.A. Sulaiman, ITM Web of Conferences **13**, 01019 (2017)
[12] A. Yokus, H.M. Baskonus, T.A. Sulaiman and H. Bulut, Numer Methods Partial Differential Eq. **34**, 211–227 (2018)
[13] M. Eslami, A. Neyrame and M. Ebrahimi, Journal of King Saud University-Science **24(1)**, 69–71 (2012)
[14] O.A. Ilhan, H. Bulut, T.A. Sulaiman and H.M. Baskonus, Indian Journal of Physics **92(8)**, 999-1007 (2018)
[15] H.M. Baskonus and H. Bulut, Physics Letters A **25(4)**, 720-728 (2015)
[16] T.A. Sulaiman, T. Akturk, H. Bulut and H.M. Baskonus, Journal of Electromagnetic Waves and Applications **32(9)**, 1093–1105 (2017)
[17] H. Bulut, T.A. Sulaiman and B. Demirdag, Nonlinear Dyn. **91** 1985–1991 (2018)
[18] A.H. Khater, M.M. Hassan and D.K. Callebaut, Reports on Mathematical Physics **66(1)**, 1-19 (2010)

[19] M.A. Ablowitz and P.A. Clarkson, *Solitons, Nonlinear Evolution Equations and Inverse Scattering*, (University Press, Cambridge, (1991)) 115-235
[20] Z. Yan, MMRC, AMSS, Academia Sinica, Beijing **22**, 363-375 (2003)
[21] X. Xian-Lin and T. Jia-Shi, Commun. Theor. Phys. **50** 1047 (2008)

Stability Analysis, Numerical and Exact Solutions of the (1+1)-Dimensional NDMBBM Equation

Asif Yokus[1], *Tukur Abdulkadir* Sulaiman[2,3], *Mehmet Tahir* Gulluoglu[4,*], and *Hasan* Bulut[2,5]

[1]Department of Actuary, Firat University, Elazig, Turkey
[2]Department of Mathematics, Firat University, Elazig, Turkey
[3]Department of Mathematics, Federal University, Dutse, Jigawa, Nigeria
[4]Department of Mathematics, University of Harran, Sanlinurfa, Turkey
[5]Department of Mathematics Education, Final International University, Kyrenia, Cyprus

Abstract. A newly propose mathematical approach is presented in this study. We utilize the new approach in investigating the solutions of the (1+1)-dimensional nonlinear dispersive modified Benjamin-Bona-Mahony equation. The new analytical technique is based on the popularly known sinh-Gordon equation and a wave transformation. In developing this new technique at each every steps involving integration, the integration constants are considered to not be zero which gives rise to new form of travelling wave solutions. The (1+1)-dimensional nonlinear dispersive modified Benjamin-Bona-Mahony is used in modelling an approximation for surface long waves in nonlinear dispersive media. We construct some new trigonometric function solution to this equation. Moreover, the finite forward difference method is utilized in investigating the numerical behavior of this equation by taking one of the obtained analytical solutions into consideration. We finally, give a comprehensive conclusions.

1 Introduction

Various complex aspects such as plasma physics, fluid dynamics, biological and chemical phenomena can be expressed in form of nonlinear partial differential equations (NPDEs). Because of the nonlinearity nature of these equations its sometimes difficult to find their exact and numerical solutions, but their solutions play a vital roles in our real life. Therefore, its important to seek their solutions. Various analytical and numerical technique have been formulated and applied to acquire the solutions of such type of models [1–16].

This study investigates the analytical solutions of the (1+1)-dimensional nonlinear dispersive modified Benjamin-Bona-Mahony (NDMBBM) equation arising in nonlinear dispersive media [18] by using a newly proposed technique called the New sinh-Gordon function method (ShGFM). The (1+1)-dimensional NDMBBM equation is used in modelling an approximation for surface long waves in nonlinear dispersive

*Corresponding Author: thrgll@gmail.com

media [18]. The new ShGFM is based on the sinh-Gordon equation and a wave transformation. In this new technique, when simplifying the sinh-Gordon equation, we consider the integration constant to not be zero. On the other hand, we utilize the finite difference method (FDM) in examining the numerical behavior of (1+1)-dimensional NDMBBM.

2 The ShGFM

In this section, we present the analysis of the proposed approach on how it can be applied in exploring search for the new solutions of various nonlinear models. To obtain the new travelling wave solutions of any given nonlinear model by using the proposed approach, we do the following;

Consider the following equation;

$$F(u,\ uu_x,\ u^2u_t,\ uu_{tt},\ \ldots) = 0, \tag{1}$$

where u is a function of the independent variable x, t and the subscripts indicate the derivative of u with respect to the independent variables and F is the polynomial of the function u.

Consider the following sinh-Gordon equation [17]:

$$u_{xt} = \lambda sinh(u), \tag{2}$$

where u is the unknown function of x, t and $\lambda \in \mathbb{R} \setminus \{0\}$.

Using the following wave transformation

$$u = U(\zeta), \quad \zeta = \mu(x + ct) \tag{3}$$

on Eq. (2), we get the following nonlinear ordinary differential equation (NODE);

$$U'' = \frac{\lambda}{c\mu^2} sinh(U), \tag{4}$$

where $U = U(\zeta)$, ζ is the width, μ is the height and c the velocity of the travelling wave. Integrating Eq. (4), we get the following;

$$\left[\left(\frac{U}{2}\right)'\right]^2 = \frac{2\lambda}{c\mu^2} sinh^2\left(\frac{U}{2}\right) + \frac{\lambda q}{c\mu^2}, \tag{5}$$

where q is the integration constant, resulting from integrating Eq. (4).

Setting $\varphi = \frac{U}{2}$ and $p = \frac{2\lambda}{c\mu^2} = \frac{\lambda q}{c\mu^2}$, we get the following;

$$\varphi' = \sqrt{p}\, cosh(\varphi). \tag{6}$$

Eq. (6) is a variables separable equation, simplifying it, produces the following two significant equations;

$$cosh(\varphi) = tan\Big(\sqrt{p}(\zeta + d)\Big), \tag{7}$$

$$sinh(\varphi) = sec\Big(\sqrt{p}(\zeta + d)\Big), \tag{8}$$

where d is the integration constant, resulting from integrating Eq. (6).

To find the new solutions to Eq. (1), we consider the following two equations;

$$U(\varphi) = \sum_{i=1}^{m} cosh^{i-1}(\varphi)[B_i sinh(\varphi) + A_i cosh(\varphi)] + A_0, \tag{9}$$

$$U(\zeta) = \sum_{i=1}^{m} tan^{i-1}\Big(\sqrt{p}(\zeta + d)\Big)\Big[B_i sec\Big(\sqrt{p}(\zeta + d)\Big) + A_i tan\Big(\sqrt{p}(\zeta + d)\Big)\Big] + A_0. \tag{10}$$

The value of m in Eq. (9) and (10) is determined by using the balancing technique thereby considering the highest power nonlinear term and the highest derivative in the reduced NODE. Setting each summation of the coefficients of $sinh^i(\varphi)cosh^j(\varphi)$, $(0 \leq i \leq n,\ 0 \leq j \leq n)$ with the same power to zero, yields a group of algebraic equations. Simplifying this group of algebraic equations with the aid of the Wolfram Mathematical package, gives the values of the coefficients A_i, B_i, μ, c and p. Substituting the values of A_i, B_i, μ, c and p into Eq. (10) along with the value of m, yields the new travelling wave solutions to Eq. (1).

3 The FDM

To use the finite forward difference method their is needs to give the following notations;

- Δx, which is the spatial step.
- Δt, which is the time step.
- $x_i = a + i\Delta x$, $i = 0, 1, 2, ..., N$ points are the coordinates of mesh and $N = \frac{b-a}{\Delta x}$, $t_j = j\Delta x$, $j = 0, 1, 2, ..., M$ and $M = \frac{T}{\Delta t}$.
- The functions $u(x,t)$ stands for the value of this solution at the grid point which is $u(x_i, t_j) \approx u_{i,j}$.
- $u_{i,j}$ stands for the numerical solution of the exact value of $u(x,t)$ at the point (x_i, t_j).

We have the following finite difference operators that are associated to the considered equation:

$$H_t u_{i,j} = u_{i,j+1} - u_{i,j}, \tag{11}$$

$$H_x u_{i,j} = u_{i+1,j} - u_{i,j}, \tag{12}$$

$$H_{xxx} u_{i,j} = u_{i+2,j} - 2u_{i+1,j} + 2u_{i-1,j} - u_{i-2,j}. \tag{13}$$

Thus, the approximate form of the partial derivatives through the finite difference operators is given by:

$$\left.\frac{\partial u}{\partial t}\right|_{i,j} = \frac{H_t u_{i,j}}{\Delta t} + O((\Delta t)), \tag{14}$$

$$\left.\frac{\partial u}{\partial x}\right|_{i,j} = \frac{H_x u_{i,j}}{\Delta x} + O((\Delta x)), \tag{15}$$

$$\left.\frac{\partial^3 u}{\partial x^3}\right|_{i,j} = \frac{H_{xxx} u_{i,j}}{(\Delta x)^3} + O((\Delta x)^2), \tag{16}$$

Therefore, we can write Eq. (24) in the finite difference operator as:

$$\frac{H_t u_{i,j}}{\Delta t} + \frac{H_x u_{i,j}}{\Delta x} - \alpha u_{i,j}^2 \frac{H_x u_{i,j}}{\Delta x} + \frac{H_{xxx} u_{i,j}}{(\Delta x)^3} = 0. \tag{17}$$

Utilizing Eqs. (11), (12) and (13), transforms Eq. (17) to the following indexed form:

$$u_{i+1,j} = \frac{1}{2(\Delta t)\Big(1 - (\Delta t)^2 + \alpha(\Delta x)^2 u_{i,j}^2\Big)} \Big(- (\Delta t) u_{i-2,j} + 2(\Delta t) u_{i-1,j}$$
$$-2(\Delta x)^3 u_{i,j} - 2(\Delta x)^2(\Delta t) u_{i,j} + 2\alpha(\Delta x)^2(\Delta t) u_{i,j}^3 + 2(\Delta x)^3 u_{i,j+1} + (\Delta t) u_{i+2,j}\Big), \tag{18}$$

where the initial value $u_{i,0}$ is $u_0(x_i)$.

4 Von-Neumann Stability Analysis

Here, we utilize the Von-Neumann stability analysis technique in investigating the stability of Eq. (24). When the Fourier method of analyzing stability is used, and ξ^n is considered as the amplication factor, the growth factor of a typical Fourier mode is given as:

$$u_m^n = \xi^n e^{im\Phi}, \tag{19}$$

where $i = \sqrt{-1}$, ξ^n is taken as the amplication factor.

To examine the stability of the numerical scheme, the nonlinear term $u^2 u_x$ in the (1+1)-dimensional nonlinear dispersive modified Benjamin-Bona-Mahony equation has been linearized by making the quantity $\widehat{u} = u^2$ a local constant. Thus the nonlinear term in the equation changes into $\widehat{u} u_x$ in this case Eq. (24) becomes

$$u_t + u_x - \alpha \widehat{u} u_x + u_{xxx} = 0. \tag{20}$$

Inserting Eq. (14), (15), (16) and the Fourier mode (19) into the recurrence relation (20), yields

$$\xi = X + iY, \tag{21}$$

where

$$X = \frac{1}{(\Delta x)}\Big((\Delta t) - \alpha(\Delta t)\widehat{u} + (\Delta x) - (\Delta t) cos(\Phi)(1 - \alpha \widehat{u})\Big) \tag{22}$$

and

$$Y = \frac{(\Delta t)}{(\Delta x)^3}\Big(2 - (\Delta x)^2 + \alpha(\Delta x)^2 \widehat{u} - sin(2\Phi)\Big). \tag{23}$$

According to the Fourier stability, for the given scheme to be stable, the condition $|\xi| \leq 1$ must be satisfied and $|\xi|^2 = X^2 + Y^2$.

5 L_2 and L_∞ Error Norms

For the test problem used in this study, numerical solutions of the (1+1)-dimensional NDMBBM equtaion have been investigated and all the computations are carried out in the Wolfram Mathematica 11. To show that how the analytical results and numerical results are close to each other we use L_2 and L_∞ error norms. The error L_2 norms defined as [20]

$$L_2 = \left\|u^{exact} - u^{numeric}\right\|_2 = \sqrt{h\sum_{j=0}^{N}\left|u_j^{exact} - u_j^{numeric}\right|^2},$$

and L_∞ error norm defined as

$$L_\infty = ||u^{exact} - u^{numeric}||_\infty = \underset{j}{Max}|u_j^{exact} - u_j^{numeric}|.$$

6 Theoretical Calculations

In this section, we present the applications of the ShGFM to the (1+1)-dimensional NDMBBM equation [18] given by:

$$u_t + u_x - \alpha u^2 u_x + u_{xxx} = 0, \tag{24}$$

carrying the following wave transformation on Eq. (24)

$$u(x,t) = U(\zeta), \quad \zeta = \mu(x - ct), \tag{25}$$

yields the following NODE:

$$3(1-c)U - \alpha U^3 + 3\mu^2 U'' + K = 0, \tag{26}$$

where K is an integration constant.

Balancing U^3 and U'', yields $m = 1$. With Eq. (35) and $m = 1$, we obtain the following:

$$U(\varphi) = B_1 sinh(\varphi) + A_1 cosh(\varphi) + A_0, \tag{27}$$

differentiating Eq. (27) twice, we obtain the following:

$$U''(\varphi) = 2pB_1 cosh^2(\varphi)sinh(\varphi) + pA_1 sinh^2(\varphi)cosh(\varphi) + pA_1 cosh^3(\varphi). \tag{28}$$

Inserting Eq. (27) and (28) into Eq. (26), gives an equation in hyperbolic functions form. From this equation, we collect a set of algebraic equations by equating each sum of the coefficients of the hyperbolic functions with the same power to zero. We solve the set of algebraic equations and obtain the values of the coefficients involved. We obtain the following two cases of the values for the coefficients;

Case 1.

$$A_0 = 0, A_1 = -\frac{\sqrt{\frac{3}{2}p}}{\sqrt{\alpha}}\mu, B_1 = \frac{\sqrt{\frac{3}{2}p}}{\sqrt{\alpha}}\mu, K = 0, c = 1 + \frac{p\mu^2}{2},$$

Case 2.

$$A_0 = 0, A_1 = -\frac{\sqrt{3(c-1)}}{\sqrt{\alpha}}, B_1 = -\frac{\sqrt{3(c-1)}}{\sqrt{\alpha}}, K = 0, \mu = \frac{\sqrt{2(c-1)}}{\sqrt{p}}.$$

With the values of the coefficients in case 1, we obtain the following solution

$$u_1(x,t) = \frac{\sqrt{\frac{3}{2}p}}{\sqrt{\alpha}}\mu\Bigg(sec\Big[\sqrt{p}\Big(d+\mu\Big(x-\Big(1+\frac{p\mu^2}{2}\Big)t\Big)\Big)\Big] \\ -tan\Big[\sqrt{p}\Big(d+\mu\Big(x-\Big(1+\frac{p\mu^2}{2}\Big)t\Big)\Big)\Big]\Bigg). \quad (29)$$

With the values of the coefficients in case 2, we obtain the following solution

$$u_2(x,t) = -\frac{\sqrt{3(c-1)}}{\sqrt{\alpha}}\Bigg(sec\Big[\sqrt{p}\Big(d+\frac{\sqrt{2(c-1)}}{\sqrt{p}}(x-ct)\Big)\Big] \\ +tan\Big[\sqrt{p}\Big(d+\frac{\sqrt{2(c-1)}}{\sqrt{p}}(x-ct)\Big)\Big]\Bigg). \quad (30)$$

Remark 1. We observed that the values of the integration constant "K" in both the two cases is zero, but we still attain to the solutions of the (1+1)-dimensional NDMBBM which we claim to be new solutions.

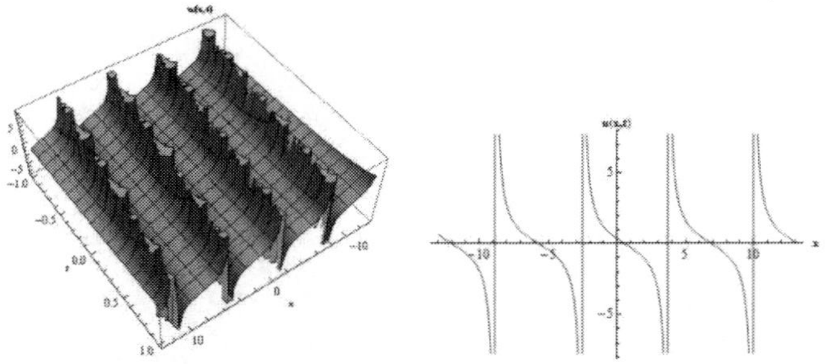

Figure 1. The singular periodic waves surface of Eq. (29) under the values $p = d = \alpha = \mu = 1$, $-13 < x < 13$, $-1 < t < 1$ and $t = 0$ for the 2D graphic.

7 Numerical Example

In this section, we investigate the numerical solutions of Eq. (24) by considering Eq. (29). Substituting these datum $p = 1, \mu = 1, d = 1, \alpha = 1$ $0 < x < 1$ and $0 < t < 1$ into Eq. (29), we obtain the following special exact solution:

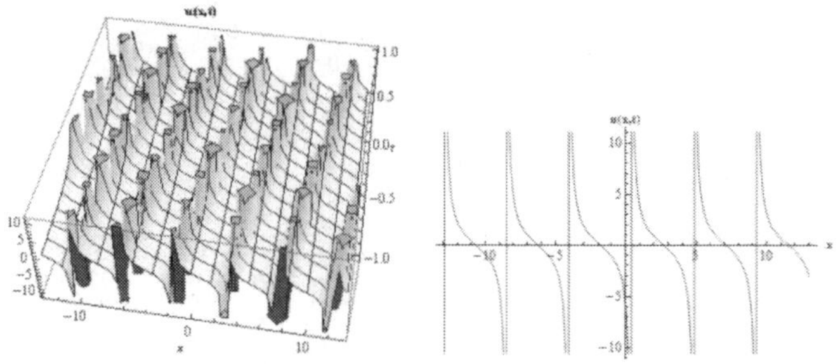

Figure 2. The singular periodic waves surface of Eq. (30) under the values $p = d = \alpha = 1$, $c = 2$, $-13 < x < 13$, $-1 < t < 1$ and $t = 0$ for the 2D graphic.

$$u(x,t) = \sqrt{\frac{3}{2}}\Big(sec\Big[1 - \frac{3}{2}t + x\Big] - tan\Big[1 - \frac{3}{2}t + x\Big]\Big). \tag{31}$$

At $t = 0$, Eq. (31) becomes:

$$u_0(x) = \sqrt{\frac{3}{2}}\Big(sec\Big[1 + x\Big] - tan\Big[1 + x\Big]\Big). \tag{32}$$

Inserting $(\Delta x) = (\Delta t) = 0.02$ into Eq. (18), yields the following indexed form:

$$u_{i+1,j} = \frac{1}{-124950. - 50.u_{i,j}^2}\Big(62500. - u_{i-2,j} - 125000.u_{i-1,j} + 50.u_{i,j} \\ -50.u_{i,j}^3 - 50.(-u_{i,j} + u_{i,j+1}) - 62500.u_{i+2,j}\Big). \tag{33}$$

Table 1.

(Δx)	(Δt)	Numerical solution	Exact Solution	Eror
0	0.01	0.379271	0.379392	1.20954×10^{-4}
0.01	0.01	0.365894	0.366011	1.16185×10^{-4}
0.02	0.01	0.352597	0.352709	1.11472×10^{-4}
0.03	0.01	0.339376	0.339483	1.06814×10^{-4}
0.04	0.01	0.326229	0.326331	1.02208×10^{-4}
0.05	0.01	0.313151	0.313248	9.76527×10^{-5}
0.06	0.01	0.300139	0.300232	9.31453×10^{-5}

Table 1: Numerical, exact solutions and absolute errors of Eq. (24) by considering Eq. (29), under the value $\Delta(x) = 0.02$ and $0 \leq x \leq 1$.

L_2 and L_∞ error norm table

Table 2.

$x_i = t_j$	L_2	L_∞
0.2	6.22622×10^{-3}	1.14559×10^{-2}
0.02	2.49150×10^{-4}	3.95029×10^{-4}
0.05	3.74149×10^{-4}	7.48387×10^{-4}
0.01	1.52415×10^{-5}	3.03436×10^{-5}
0.001	1.63392×10^{-7}	1.18747×10^{-6}

Table 2: L_2 and L_∞ error norm when $0 \leq h \leq 1$ and $0 \leq x \leq 1$

Figure 3. Numerical and exact solution of Eq. (24) by considering Eq. (29).

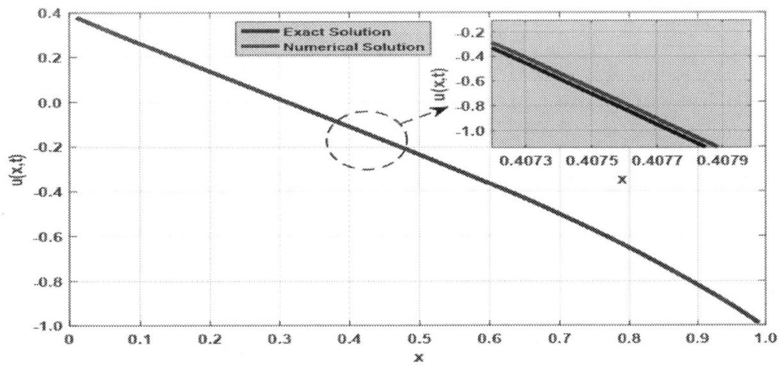

8 Results and Discussions

Various analytical techniques have been invested to solve the (1+1)-dimensional nonlinear dispersive modified Benjamin-Bona-Mahony equation and solutions of different structures were obtained. Khan *et al.* [18] used the modified simple equation method in solving this equation and some hyperbolic and trigonometric solutions were obtained. Baskonus *et al.* [19] used the generalized Kudryashov method in investigating some new solutions to this equation and some hyperbolic, trigonometric and exponential function solutions were obtained. In this study, we used the ShGFM and construct some new analytical solutions with trigonometric structure. We present below the trigonometric function solutions obtained in the three above mentioned references.

Results obtained by Khan *et al.* [18]:

$$u_1(x,t) = \pm\sqrt{\frac{3(c-1)}{\alpha}}tan[\varpi_1(x,t)], \tag{34}$$

$$u_2(x,t) = \pm\sqrt{\frac{3(c-1)}{\alpha}}cot[\varpi_1(x,t)], \tag{35}$$

where $\varpi_1(x,t) = \sqrt{\frac{(c-1)}{2}}(x - ct)$.

Results obtained by Baskonus *et al.* [19]:

$$u_1(x,t) = \pm i\sqrt{\frac{6}{\alpha}} cosec[i(-x + 2t)], \tag{36}$$

$$u_2(x,t) = \pm 3\sqrt{\frac{3}{2\alpha}}\Big(- i\ tan\Big[\frac{3}{4}i(2x + 7t)\Big]\Big), \tag{37}$$

$$u_3(x,t) = \pm\sqrt{\frac{6}{\alpha}}(i\ cot[i(x + t)]). \tag{38}$$

We observed that the results obtained by using ShGFM are newly constructed solutions when compared the available results in the literature. Although, in Eq. (30) when $sec\Big[\sqrt{p}\Big(d + \frac{\sqrt{2(c-1)}}{\sqrt{p}}(x - ct)\Big)\Big] \to 0$ and we choose $d = 0$, we get the following:

$$u(x,t) = -\sqrt{\frac{3(c-1)}{\alpha}} tan\Big[\sqrt{2(c-1)}(x - ct)\Big] \tag{39}$$

which is closely the same with the result obtained in Eq. (34) by Khan *et al.* [18].

In Table 1, we present the values of the numerical and exact solutions obtained by using finite forward difference method.

In table 2, we present the L_2 and L_∞ error norms for the numerical solutions of Eq. (24).

9 Conclusions

In this study, we proposed a new analytical technique called a new sinh-Gordon function method. With the aid of the Wolfram Mathematica package, we utilized this new technique in investigating some new solutions to the (1+1)-dimensional nonlinear dispersive modified Benjamin-Bona-Mahony equation. We obtained some new analytical solutions with trigonometric function structure. We observed that our results are newly constructed when compared with the existing results in the literature. We performed the numerical simulations of the obtained solutions by choosing suitable parameters. We performed the numerical simulation of all the obtained analytical solutions in this study. On the other hand we compute the exact and numerical approximations of the studied equation using the finite forward difference method. We checked the stability of the numerical scheme by using the Fourier-Von Neumann analysis. In table 1, we present the results for both the numerical and exact solutions obtained by considering Eq. (29). We give the L_2 and L_∞ error norms analysis of this equation in table 2. The comparison between the numerical and analytical solutions can be observed from table 1, which is also supported by fig. 3. To the best of our knowledge, the application of this new method which is giving new results has not been applied and submitted to the literature.

References

[1] Z. Fu, S. Liu, Q. Zhao, Physics Letters A **290**, 72-76 (2001)
[2] A. Yokus, H.M. Baskonus, T.A. Sulaiman and H. Bulut, Numer Methods Partial Differential Eq. **34**, 211–227 (2018)
[3] S. Duran, M. Askin and T.A. Sulaiman, IJOCTA **7(3)**, 240–247 (2017)
[4] M.A. Akbar and N.H.M. Ali, Cogent Mathematics **4**, 1282577 (2017)
[5] Y. Zhao, Journal of Applied Mathematics **2013**, 895760 (2013)
[6] H. Bulut, T.A. Sulaiman, H.M. Baskonus and T. Akturk, Opt Quant Electron **50**, 134 (2018)
[7] H.M. Baskonus, T.A. Sulaiman and H. Bulut, Opt Quant Electron **50**, 253 (2018)
[8] H.M. Baskonus, T.A. Sulaiman and H. Bulut, Optik **131**, 1036-1043 (2017)
[9] H. Bulut, T.A. Sulaiman and H.M. Baskonus, Opt. Quant. Electron **48**, 564 (2016)
[10] H. Bulut, T.A. Sulaiman, H.M. Baskonus and A.A. Sandulyak, Optik **135**, 327-336 (2017)
[11] C. Cattani, T.A. Sulaiman, H.M. Baskonus and H. Bulut, **50**, Opt Quant Electron 138 (2018)
[12] H.N. Hassan, Z. Naturforsch **65a**, 305-314 (2009)
[13] H. Borluk and G.M. Muslu, Numerical Methods for Partial Differential Equations **31**, 995-1008 (2014)
[14] Y. Dereli, Int. J. Nonlinear Sci **13(1)**, 28-38 (2012)
[15] C. Cattani, T.A. Sulaiman, H.M. Baskonus and H. Bulut, Eur. Phys. J. Plus **133**, 228 (2018)
[16] B. Saka and I. Dag, Commun. Numer. Methods Eng. Biomed. Appl. **24(11)**, 1339-1361 (2008)
[17] M.A. Ablowitz and P.A. Clarkson, *Solitons, Nonlinear Evolution Equations and Inverse Scattering*, (Cambridge University Press, Cambridge, (1991)) 234-345
[18] K. Khan, M.A. Akbar and S.M.R. Islam, SpringerPlus **3**, 724 (2014)
[19] H.M. Baskonus and H. Bulut, Entropy **17**, 4255-4270 (2015)
[20] A. Yokus, T.A. Sulaiman and H. Bulut, Opt Quant Electron **50**, 31 (2017)

ITM Web of Conferences **22**, 01065 (2018) https://doi.org/10.1051/itmconf/20182201065
CMES-2018

A New Method for (4+1) Dimensional Fokas Equation

Seyma Tuluce Demiray[1,*] and *Hasan Bulut*[2]

[1] Department of Mathematics, Osmaniye Korkut Ata University, Osmaniye, Turkey

[2] Department of Mathematics, Fırat University, Elazığ, Turkey

Abstract. In this paper, modified $\exp(-\Omega(\xi))$-expansion function method (MEFM) has been tackled for procuring exact solutions of (4+1) dimensional Fokas equation. Hyperbolic function solutions and dark soliton solutions of (4+1) dimensional Fokas equation have been found by means of this method. Moreover, by the help of Mathematica 9, some graphical simulations were given to clarify the behavior of these solutions.

1 Introduction

Nonlinear evolution equations (NLEEs) are considerably used to identify a variety of physical circumstances in the areas such as quantum field theory, hydrodynamics, chemical kinematics, geochemistry, electricity, elastic media and plasma physics.

Recently, most of researchers have submitted to acquire exact solutions of NLEEs many methods such as G'/G-expansion method [1], modified extended tanh-function method [2], sine-cosine method [3], exp-function method [4], modified simple equation method [4], extended trial equation method [5], generalized Kudryashov method [6] . In this study, MEFM [7] will be implemented to seek exact solutions of (4+1) dimensional Fokas equation.

We consider (4+1) dimensional Fokas equation [8-14],

$$4u_{tx} - u_{xxxy} + u_{xyyy} + 12u_x u_y + 12uu_{xy} - 6u_{zw} = 0. \quad (1)$$

This equation has been obtained by Fokas by expanding the Lax pairs of the integrable Kadomtsev–Petviashvili (KP) and Davey–Stewartson (DS) equations to some higher-dimensional nonlinear wave equations [8]. The importance of Eq. (1) suggests that the idea of complexifying time can be considered in the context of modern field theories via the existence of integrable nonlinear equations in four spatial dimensions involving complex time [8].

* Corresponding author: seymatuluce@gmail.com

Our target in this work is to gain exact solutions of (4+1) dimensional Fokas equation. In Sec. 2, we clarify general structure of MEFM. In Sec. 3, we get exact solutions of this equation by means of MEFM.

2 General Structure of Method

Step 1. We handle PDE as follows:

$$P\left(u,u_x,u_y,u_z,u_w,u_t,u_{xx},u_{yy},u_{zz},u_{ww},u_{tt},\cdots\right)=0, \tag{2}$$

where $u=u(x,y,z,w,t)$ is an unknown function, P is a polynomial in $u(x,y,z,w,t)$ and its derivatives, in which the highest order derivatives and nonlinear terms are included and the subscripts demonstrate the partial derivatives. Then, we get traveling transformation

$$u(x,y,z,w,t)=U(\xi),\quad \xi=\alpha x+\beta y+\gamma z+\delta w+\varepsilon t. \tag{3}$$

Using Eq. (3), we can turn Eq. (2) into a nonlinear ordinary differential equation (NODE) described by:

$$NODE\left(U,U',U'',U''',\cdots\right)=0, \tag{4}$$

where $NODE$ is a polynomial of U and its derivatives and the superscripts demonstrate the ordinary derivatives according to ξ.

Step 2. Assume the traveling wave solution of Eq. (4) can be shown as follows:

$$U(\xi)=\frac{\sum_{i=0}^{N}A_i\left[\exp(-\Omega(\xi))\right]^i}{\sum_{j=0}^{M}B_j\left[\exp(-\Omega(\xi))\right]^j}=\frac{A_0+A_1\exp(-\Omega)+\cdots+A_N\exp(\mathrm{N}(-\Omega))}{B_0+B_1\exp(-\Omega)+\cdots+B_M\exp(\mathrm{M}(-\Omega))}, \tag{5}$$

where $A_i, B_j, (0\le i\le N, 0\le j\le M)$ are constants to be described later, such that $A_N\neq 0, B_M\neq 0,$ and $\Omega=\Omega(\xi)$ is solution of the following ordinary differential equation:

$$\Omega'(\xi)=\exp(-\Omega(\xi))+\mu\exp(\Omega(\xi))+\lambda. \tag{6}$$

There are the following solution families of Eq. (6):

Family1: When $\mu\neq 0,\ \lambda^2-4\mu>0,$

$$\Omega(\xi)=\ln\left(\frac{-\sqrt{\lambda^2-4\mu}}{2\mu}\tanh\left(\frac{\sqrt{\lambda^2-4\mu}}{2}(\xi+E)\right)-\frac{\lambda}{2\mu}\right). \tag{7}$$

Family2: When $\mu\neq 0,\ \lambda^2-4\mu<0,$

$$\Omega(\xi)=\ln\left(\frac{\sqrt{-\lambda^2+4\mu}}{2\mu}\tan\left(\frac{\sqrt{-\lambda^2+4\mu}}{2}(\xi+E)\right)-\frac{\lambda}{2\mu}\right). \tag{8}$$

Family3: When $\mu=0,\ \lambda\neq 0,$ and $\lambda^2-4\mu>0,$

$$\Omega(\xi)=-\ln\left(\frac{\lambda}{\exp(\lambda(\xi+E))-1}\right). \tag{9}$$

Family4: When $\mu\neq 0,\ \lambda\neq 0,$ and $\lambda^2-4\mu=0,$

$$\Omega(\xi)=\ln\left(-\frac{2\lambda(\xi+E)+4}{\lambda^2(\xi+E)}\right). \tag{10}$$

Family5: When $\mu=0,\ \lambda=0,$ and $\lambda^2-4\mu=0,$

$$\Omega(\xi)=\ln(\xi+E). \tag{11}$$

such that $A_0, A_1, A_2, \cdots A_N, B_0, B_1, B_2, \cdots B_M, E, \lambda, \mu$ are constants to be described later. The positive integers N and M can be identified by taking into consideration the homogeneous balance between the highest order derivatives and the nonlinear terms arising in Eq. (5).

Step 3: Embedding Eqs. (6) and (7–11) into Eq. (5), we attain a polynomial of $\exp(-\Omega(\xi))$. We compensate all the coefficients of same power of $\exp(-\Omega(\xi))$ to zero. This process provides a system of equations which can be solved to obtain $A_0, A_1, A_2, \cdots A_N, B_0, B_1, B_2, \cdots B_M, E, \lambda, \mu$ by using Wolfram Mathematica 9. Putting the values of $A_0, A_1, A_2, \cdots A_N, B_0, B_1, B_2, \cdots B_M, E, \lambda, \mu$ into Eq. (5), the general solutions of Eq. (5) fulfil the determination of the solution of Eq. (1).

3 MEFM for (4+1) Dimensional Fokas Equation

In order to solve Eq. (1), we use the travelling wave transform

$$u(x,y,z,w,t)=U(\xi),\quad \xi=\alpha x+\beta y+\gamma z+\delta w+\varepsilon t. \tag{12}$$

where $\alpha, \beta, \gamma, \delta$ and ε are arbitrary constants. Then, we reduce Eq. (1) to following equation

$$(\alpha\beta^3-\alpha^3\beta)u''+(4\alpha\varepsilon-6\gamma\delta)u+6\alpha\beta u^2=0. \tag{13}$$

Using balance principle in Eq. (13), we obtain

$$N=M+2. \tag{14}$$

If we take $M=1$ so $N=3$, we can acquire

$$U=\frac{A_0+A_1\exp(-\Omega)+A_2\exp(2(-\Omega))+A_3\exp(3(-\Omega))}{B_0+B_1\exp(-\Omega)}, \tag{15}$$

$$\begin{aligned} U'=&\frac{\left[A_1\exp(-\Omega)(-\Omega')+A_2\exp(2(-\Omega))(-2\Omega')+A_3\exp(3(-\Omega))(-3\Omega')\right]\left[B_0+B_1\exp(-\Omega)\right]}{\left[B_0+B_1\exp(-\Omega)\right]^2}\\ &-\frac{\left[A_0+A_1\exp(-\Omega)+A_2\exp(2(-\Omega))+A_3\exp(3(-\Omega))\right]\left[B_1\exp(-\Omega)(-\Omega')\right]}{\left[B_0+B_1\exp(-\Omega)\right]^2}=\frac{\Upsilon}{\Psi},\\ U''=&\frac{\Upsilon'\Psi-\Upsilon\Psi'}{\Psi^2},\\ &\vdots \end{aligned} \tag{16}$$

where $A_3\neq 0$ and $B_1\neq 0$. Substituting Eqs. (15) and (16) in Eq. (13), we attain a system of algebraic equations from the coefficients of polynomial of $\exp(-\Omega(\xi))$. By solving this system of algebraic equations by using Wolfram Mathematica 9, it yields us the following coefficients:

Case 1.

$$\begin{aligned} &A_0=\frac{\mu A_3 B_0}{B_1}, A_1=A_3\left(\mu+\frac{\lambda B_0}{B_1}\right), A_2=A_3\left(\lambda+\frac{B_0}{B_1}\right), A_3=A_3, B_0=B_0, B_1=B_1,\\ &\alpha=-\frac{\sqrt{A_3+\beta^2 B_1}}{\sqrt{B_1}}, \varepsilon=\frac{\beta(\lambda^2-4\mu)A_3-\dfrac{6\gamma\delta B_1^{3/2}}{\sqrt{A_3+\beta^2 B_1}}}{4B_1}, \lambda=\lambda, \mu=\mu. \end{aligned} \tag{17}$$

Embedding Eq. (17) together with Eqs. (3) and (7) in Eq. (15), we find new hyperbolic function solutions for Eq. (1) as follows:

$$u_1(x,y,z,w,t)=\frac{\mu\left(-\lambda^2+4\mu\right)A_3}{B_1\left(\lambda\cosh\left[f(x,y,z,w,t)\right]+\sqrt{\lambda^2-4\mu}\,\sinh\left[f(x,y,z,w,t)\right]\right)^2},\qquad(18)$$

where

$$f(x,y,z,w,t)=\frac{1}{2}\sqrt{\lambda^2-4\mu}\left[E+y\beta+z\gamma+w\delta+\frac{t\beta\left(\lambda^2-4\mu\right)A_3}{4B_1}-\frac{2xA_3+B_1\left(2x\beta^2+3t\gamma\delta\right)}{2\sqrt{B_1}\left(\sqrt{A_3+\beta^2B_1}\right)}\right],$$

and $\mu\neq 0,\lambda^2-4\mu>0$.

Case 2.

Substituting Eq. (19) together with Eqs. (3) and (7) in Eq. (15), we obtain new dark soliton solutions for Eq. (1) as follows:

$$u_2(x,y,z,w,t)=\frac{\left(\lambda^2-4\mu\right)A_3\left(\lambda^2-6\mu+2\lambda\sqrt{\lambda^2-4\mu}\tanh\left[g(x,y,z,w,t)\right]+\left(\lambda^2+2\mu\right)\tanh^2\left[g(x,y,z,w,t)\right]\right)}{6B_1\left(\lambda+\sqrt{\lambda^2-4\mu}\,\tanh\left[g(x,y,z,w,t)\right]\right)^2},\qquad(20)$$

where

$$g(x,y,z,w,t)=\frac{1}{2}\sqrt{\lambda^2-4\mu}\left[E+y\beta+z\gamma+w\delta-\frac{t\beta\left(\lambda^2-4\mu\right)A_3}{4B_1}-\frac{2xA_3+B_1\left(2x\beta^2+3t\gamma\delta\right)}{2\sqrt{B_1}\left(\sqrt{A_3+\beta^2B_1}\right)}\right],$$

and $\mu\neq 0,\lambda^2-4\mu>0$.

Case 3.

$$A_0=\left(\alpha^2-\beta^2\right)\mu B_0,A_1=\frac{A_2B_0-\left(\alpha^2-\beta^2\right)\left(B_0^2-\mu B_1^2\right)}{B_1},A_3=\left(\alpha^2-\beta^2\right)B_1,$$

$$\lambda=\frac{A_2-\left(\alpha^2-\beta^2\right)B_0}{\left(\alpha^2-\beta^2\right)B_1},\varepsilon=\frac{1}{4}\left(\frac{6\gamma\delta}{\alpha}-4\alpha^2\beta\mu+4\beta^3\mu+\frac{\beta\left(A_2-B_0\left(\alpha^2-\beta^2\right)\right)^2}{\left(\alpha^2-\beta^2\right)B_1^2}\right),\qquad(21)$$

$$A_2=A_2,B_0=B_0,B_1=B_1,\lambda=\lambda,\mu=\mu.$$

Putting Eq. (21) together with Eqs. (3) and (7) in Eq. (15), we find new hyperbolic function solutions for Eq. (1) as follows:

$$u_3(x,y,z,w,t)=-\frac{\left(\alpha^2-\beta^2\right)\mu\left(\left(A_2-\left(\alpha^2-\beta^2\right)B_0\right)^2-4\left(\alpha^2-\beta^2\right)^2\mu B_1^2\right)\operatorname{sec}h^2\left[h(x,y,z,w,t)\right]}{\left[A_2-\left(\alpha^2-\beta^2\right)\left(B_0-B_1\sqrt{-4\mu+\frac{\left(A_2-\left(\alpha^2-\beta^2\right)B_0\right)^2}{\left(\alpha^2-\beta^2\right)^2B_1^2}}\tanh\left[h(x,y,z,w,t)\right]\right)\right]^2},\qquad(22)$$

where

$$h(x,y,z,w,t)=\frac{1}{2}\sqrt{-4\mu+\frac{\left(A_2-\left(\alpha^2-\beta^2\right)B_0\right)^2}{\left(\alpha^2-\beta^2\right)^2B_1^2}}\left[E+x\alpha+y\beta+z\gamma+w\delta+\frac{1}{4}t\left(\frac{6\gamma\delta}{\alpha}-4\alpha^2\beta\mu+4\beta^3\mu+\frac{\beta\left(A_2-B_0\left(\alpha^2-\beta^2\right)\right)^2}{\left(\alpha^2-\beta^2\right)B_1^2}\right)\right],$$

and $\mu\neq 0,\lambda^2-4\mu>0$.

Remark. The exact solutions of Eq. (1) were obtained via MEFM and were controlled by use of Mathematica Release 9. As far as we know, the solutions of Eq. (1) that we found in this study, are new and are not indicated before.

Fig.1. The 3D and 2D surfaces of Eq. (18) for
$\lambda = 0.3,\ \mu = -0.5, \beta = 1, \gamma = 2, \delta = 3, A_3 = 2, B_1 = 1$
$E = 0.5, y = 0.3, z = 0.2, w = 0.1,\ -15 < x < 15,\ -15 < t < 15$ and $-25 < x < 25, t = 0.01$ for 2D surface.

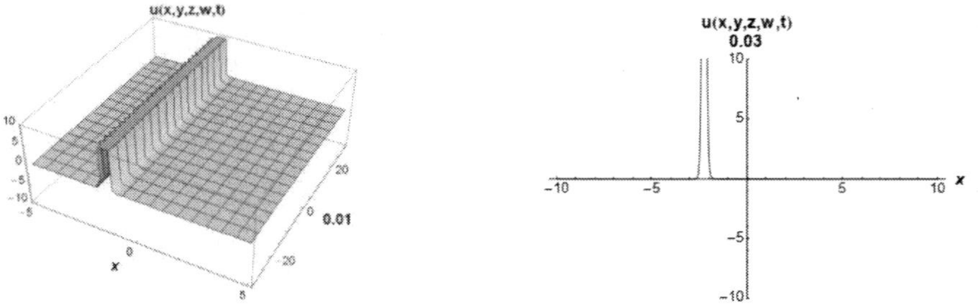

Fig.2. The 3D and 2D surfaces of Eq. (22) for
$\mu = -0.4, \alpha = 2, \beta = 1, \gamma = 3, \delta = 2, A_2 = 0.4, B_0 = 0.03, B_1 = 0.01$
$E = 0.25, y = 0.3, z = 0.2, w = 0.1,\ -5 < x < 5,\ -30 < t < 30$ and $-10 < x < 10, t = 0.03$ for 2D surface.

4 Conclusion

In this paper, we apply MEFM to obtain exact solutions of (4+1) dimensional Fokas equation. Then, for proper parameters, we draw 2D and 3D surfaces of some exact solutions of this equation by using Mathematica Release 9. This method yields us to make complicated and tedious algebraic calculations. According to these informations, this method has been influential for the analytical solutions of (4+1) dimensional Fokas equation. Also, MEFM is a strong mathematical device in the way of finding new exact solutions. The graphical demonstrations clearly show validness of proposed method.

References

1. M. Younis, S. T. R. Rizvi, Optik, **126**, 5812-5814 (2015).
2. E. H. M. Zahran, M. M. A. Khater, Appl. Math. Model., **40**, 1769-1775 (2016).
3. S. Bibi, S. T. Mohyud-Din, J. Assoc. Arab Univ. Basic Appl., **15**, 90-93 (2014).
4. K. Khan, M. A. Akbar, Ain Shams Engineering Journal, **5**, 247-256 (2014).
5. S. Tuluce Demiray, H. Bulut, Wave Random Complex, **25(1)**, 75-90 (2015)
6. S. Tuluce Demiray, H. Bulut,. Opt. Quant. Electron., **49(117)**, 1-8 (2017).
7. H. M. Baskonus, H Bulut, A. Atangana, Smart Mater. Struct., **25(3)**, 1-8 (2016).
8. A. S. Fokas, Phys. Rev. Lett., **96(19),** Article ID:190201 (2006).
9. Z. Z. Yang, Z. Y. Yan, Commun. Theor. Phys. (Beijing,) **51(5),** 876-880 (2009).
10. J. Lee, R. Sakthivel, L. Wazzan, Modern Phys. Lett. B, **24(10),** 1011-1021 (2010).
11. H. Kim, R. Sakthivel, Rep. Math. Phys., **70(1),** 39-50 (2012).
12. Y. He, Math. Probl. Eng., **2014,** Article ID: 972519, 11 pages (2014).
13. S. Zhang, C. Tian, W. Y. Qian, Pramana J. Phys., **86(6),** 1259-1267 (2016).
14. M. O. Al-Amr, S. El-Ganaini, Comput. Math. Appl., **74**, 1274-1287 (2017).

AUTHOR INDEX

AUTHOR INDEX